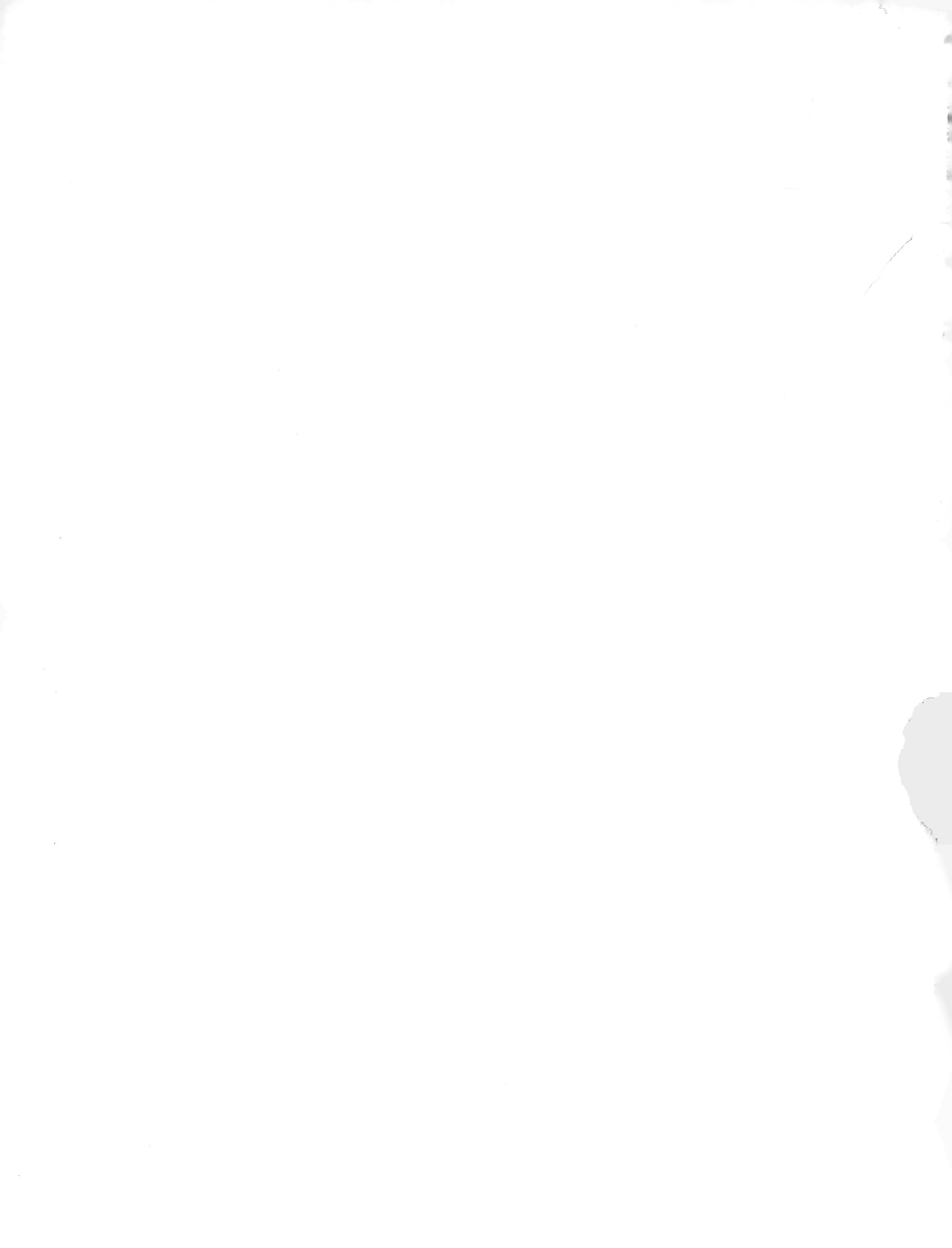

Botany: Growth and Developments of Plants

Botany: Growth and Developments of Plants

Editor: Geoffrey Watkins

R CALLISTO REFERENCE

www.callistoreference.com

Callisto Reference,
118-35 Queens Blvd., Suite 400,
Forest Hills, NY 11375, USA

Visit us on the World Wide Web at:
www.callistoreference.com

ISBN: 978-1-64116-161-9 (Hardback)

Trademark Notice: Registered trademark of products or corporate names are used only for explanation and identification without intent to infringe.

Cataloging-in-Publication Data

Botany : growth and developments of plants / edited by Geoffrey Watkins.
    p. cm.
Includes bibliographical references and index.
ISBN 978-1-64116-161-9
1. Botany. 2. Growth (Plants). 3. Plants--Development. I. Watkins, Geoffrey.
QK47 .B68 2019
580--dc23

Table of Contents

Preface

Botany is the branch of biology that studies plant life. Various agricultural products as well as medicinal and recreational drugs are directly derived from plants. Alcoholic beverages and stimulants such as caffeine, tea, chocolate and nicotine are also derived from plants. Besides such applications, plants are utilized for the production of dyes and pigments, sugar, cotton, rope, oil, wax, rubber, etc. The technological tools of optical microscopy, electron microscopy, live cell imaging, molecular genetic analysis are significant to the development of botany. Due to the rise of molecular scale biological approaches of genomics, metabolomics and proteomics, various aspects of plant genome, physiology, morphology and behavior of plants are now explored in a detailed manner. This book brings forth some of the most innovative concepts and elucidates the unexplored aspects of botany. It strives to provide a fair idea about this discipline and to help develop a better understanding of the latest advances within this field. It is an essential guide for both academicians and those who wish to pursue this discipline further.

This book is the end result of constructive efforts and intensive research done by experts in this field. The aim of this book is to enlighten the readers with recent information in this area of research. The information provided in this profound book would serve as a valuable reference to students and researchers in this field.

At the end, I would like to thank all the authors for devoting their precious time and providing their valuable contribution to this book. I would also like to express my gratitude to my fellow colleagues who encouraged me throughout the process.

Editor

Effects of water availability and pest pressures on tea (*Camellia sinensis*) growth and functional quality

Selena Ahmed[1,2]*, Colin M. Orians[2], Timothy S. Griffin[3], Sarabeth Buckley[4], Uchenna Unachukwu[5], Anne Elise Stratton[2], John Richard Stepp[6], Albert Robbat Jr.[7], Sean Cash[3] and Edward J. Kennelly[5,8]

[1] Sustainable Food and Bioenergy Systems Program, Department of Health and Human Development, Montana State University, Bozeman, MT 59715, USA
[2] Department of Biology, Tufts University, Medford, MA 02155, USA
[3] Friedman School of Nutrition Science and Policy, Tufts University, Boston, MA 02111, USA
[4] Department of Earth Sciences, Boston University, Boston, MA 02215, USA
[5] Department of Biochemistry, The Graduate Center of the City University of New York, New York, NY 10016, USA
[6] Department of Anthropology, University of Gainesville, Gainesville, FL 32611, USA
[7] Department of Chemistry, Tufts University, Medford, MA 02155, USA
[8] Department of Biological Sciences, Lehman College, Bronx, NY 10468, USA

Abstract. Extreme shifts in water availability linked to global climate change are impacting crops worldwide. The present study examines the direct and interactive effects of water availability and pest pressures on tea (*Camellia sinensis*; Theaceae) growth and functional quality. Manipulative greenhouse experiments were used to measure the effects of variable water availability and pest pressures simulated by jasmonic acid (JA) on tea leaf growth and secondary metabolites that determine tea quality. Water treatments were simulated to replicate ideal tea growing conditions and extreme precipitation events in tropical southwestern China, a major centre of tea production. Results show that higher water availability and JA significantly increased the growth of new leaves while their interactive effect was not significant. The effect of water availability and JA on tea quality varied with individual secondary metabolites. Higher water availability significantly increased total methylxanthine concentrations of tea leaves but there was no significant effect of JA treatments or the interaction of water and JA. Water availability, JA treatments or their interactive effects had no effect on the concentrations of epigallocatechin 3-gallate. In contrast, increased water availability resulted in significantly lower concentrations of epicatechin 3-gallate but the effect of JA and the interactive effects of water and JA were not significant. Lastly, higher water availability resulted in significantly higher total phenolic concentrations but there was no significant impact of JA and their interaction. These findings point to the fascinating dynamics of climate change effects on tea plants with offsetting interactions between precipitation and pest pressures within agro-ecosystems, and the need for future climate studies to examine interactive biotic and abiotic effects.

Keywords: *Camellia sinensis*; catechins; climate change; herbivory; methylxanthines; precipitation; tea; total phenolic concentrations.

* Corresponding author's e-mail address: selena.ahmed@montana.edu

Introduction

Crops around the world are being impacted by extreme shifts in water availability linked to global climate change. For example, droughts and floods are reducing the yields of many crops (Porter and Semenov 2005; Lobell *et al.* 2011) as well as altering their quality (Coley 1998; Jamieson *et al.* 2012). In fact, precipitation is the most important climatic determinant, along with temperature, for plant growth and survival (Boisvenue and Running 2006). Future climatic projections show strong precipitation heterogeneity depending on geographic location, including an increase in the number of heavy precipitation events as well as longer and more intense droughts (Orlowsky and Seneviratne 2012; Seneviratne *et al.* 2012). Crop performance is further impacted by indirect climatic influences via alterations in ecological interactions such as pest pressures (Berggren *et al.* 2009; Brenes-Arguedas *et al.* 2009; Schepp 2009). Although the magnitude and direction of future climatic-induced alterations to water availability remain uncertain, it is recognized that these changes will be notable and often exceed plant adaptive capacity (IPCC 2007).

Given present and future water availability scenarios, research is needed to understand crop responses to both direct and indirect effects of climate change for future food security. While previous research has documented the impact of extreme precipitation events on crop yields (Ewert *et al.* 2005; Porter and Semenov 2005; Nelson *et al.* 2009; Schlenker and Lobell 2010; Lobell *et al.* 2011), less is known about the direct and interactive effects of water availability and pest pressures on crop quality. Crop quality is largely determined by nutrient and secondary metabolite profiles via their effects on functional and sensory characteristics for human consumers. Secondary metabolites serve as defence compounds in plants that vary in concentration with a range of environmental, genetic and management conditions, including water availability and pest pressures (Herms and Mattson 1992; Glynn *et al.* 2007; Gutbrodt *et al.* 2011, 2012; Tharayil *et al.* 2011; Atkinson and Urwin 2012; Kruidhof *et al.* 2012; Ahmed *et al.* 2013). Changes induced by both water availability and pest pressures are mediated via signalling pathways (Atkinson and Urwin 2012) that can cause an increase or decrease in the concentrations of secondary metabolites (Gutbrodt *et al.* 2011; Kruidhof *et al.* 2012).

The present study examines the direct and interactive effects of water availability and pest pressures on the functional quality of tea (*Camellia sinensis*; Theaceae). Tea plants, the source of the world's most widely consumed beverage after water, are geographically located in high-risk regions for climate change. Our preliminary work has suggested that tea functional quality drops significantly with extreme precipitation events that accompany the annual onset of the East Asian monsoon and that monsoon patterns are shifting. Tea functional quality is largely determined by polyphenolic catechin and methylxanthine secondary metabolites that are responsible for its antioxidant, anti-inflammatory, cardioprotective and stimulant properties for human consumers (Lin *et al.* 2003). Catechins and methylxanthines are found in the highest concentrations in young expanding leaves, those harvested for commercial tea, and human consumers are able to perceive changes in the concentrations of these metabolites by their bitterness, astringency and sweet aftertaste (Ahmed *et al.* 2010). Since the concentrations of these compounds are predicted to increase following herbivory, increasing pest pressures during the rainy season (Coley 1998) could offset the effects of heavy rainfall.

In this study, manipulative greenhouse experiments were used to measure the effects of variable water availability and pest pressures on secondary metabolites that determine tea quality. Water treatments were simulated to replicate ideal tea growing conditions and extreme precipitation events in tropical southwestern China, a major centre of tea production located in a high-risk region for climate change (Maplecroft 2011). Pest pressures were experimentally simulated here through the application of the plant hormone jasmonic acid (JA) to young tea leaves (McDowell and Dangl 2000; Kruidhof *et al.* 2012). It is well known that an increase in water availability can cause an increase in growth and a decline in secondary metabolites (Brenes-Arguedas *et al.* 2006); whether simulated pest pressures would counter this response is unknown. We hypothesized that increased water availability would indeed lead to lower concentrations of tea secondary metabolites, but that simulated pest pressures would offset these direct effects of water availability.

Methods

Plant material

Tea plants (*C. sinensis*; Theaceae) of ~2 years of age were purchased from Logee's Greenhouse (Danielson, CT, USA). Plants were transplanted into 6-inch plastic pots (total volume 1800 mL) with four drainage holes at the base of each pot. A total of 1300 mL of soil mix that comprised 50 % pearlite and 50 % peat moss was added to each pot. The soil mixture was selected to facilitate quick drainage. Plants were fertilized (Osmocote© Plus 15-9-12, Marysville, OH, USA) 1 week prior to the experimental period. A total of 120 tea plants were included in the experiment.

Greenhouse set-up

Tea plants were maintained and treated at the greenhouse facility of the Weld Hill Research Building at the Arnold Arboretum, Harvard University (Jamaica Plain, MA, USA). One greenhouse room was used for the present experiment. Temperature, humidity and shade conditions were selected to reflect ideal tea growing conditions. The temperature was maintained at a range of 20–22 °C with a humidity range of 60–70 % and steady air circulation. Shade was set at 50 % over-storey density. Plants were randomly assigned to each water availability and pest pressure treatment and were labelled with treatment identifiers. Tea plants were moved on a weekly basis to eliminate any possible location effects within the greenhouse.

Water availability treatments

Water availability treatments involved altering the soil moisture content of tea plants to simulate conditions that exist during the spring harvest in tropical southwestern China and extreme precipitation events of drought and heavy monsoon rains (Dou et al. 2007), hereafter termed moderate water, low water and high water, respectively. A total of 120 tea plants were treated under each of the three water availability treatments (40 tea plants per treatment) on the basis of field capacity of the experimental soil mixture (32 %) as well as soil moisture of field conditions at the reference location in southwestern China during mean and extreme precipitation levels. The moderate-water treatment was maintained at 12–16 % soil moisture content with drainage, the low-water treatment was maintained at 4–8 % soil moisture content with drainage and the high-water treatment was maintained at 28–32 % soil moisture content with no drainage. Water treatments were applied for 6 weeks before experimental harvest to quantify leaf secondary metabolites.

Simulated pest pressure treatments with JA

The application of JA to tea leaves was used to simulate pest pressure on the basis of previous studies that have shown JA application to produce induced resistance, marked by an upregulation of secondary metabolic activity that simulates plant response by actual herbivory leaves (McDowell and Dangl 2000; Kruidhof et al. 2012). Using standard methods (Babst et al. 2005), half of the plants randomly assigned to each of the three water availability treatments were designated as having the presence of pest pressure and treated with a solution of 0.125 % JA and 0.0625 % Triton X-100 surfactant (both purchased from Sigma-Aldrich Co. LLC, St Louis, MO, USA) in distilled water prior to the experimental period and then 2 days prior to the harvest period. Triton was added to the solution to improve the penetration of the JA through the waxy cuticles of tea leaves. Jasmonic acid was applied to the upper and lower surface of the newest leaf on each branch of tea plants designated with the presence of pest pressure. The plants designated with the absence of pest pressure were treated with a solution of 0.0625 % Triton surfactant in distilled water.

Plant growth

Growth was measured by quantifying the number of new leaves and the height of tea plants during the experimental period.

Sample collection

A sub-sample of 40 tea plants equally representing each of the water availability and pest pressure treatments was harvested by clipping three new leaves at their base using sharp shearing scissors. Samples were stored on ice and transferred to a lyophilizer (VirTis, SP Scientific) for a drying period of 48 h. Dry weights were recorded upon removal from the lyophilizer.

Sample extraction

Leaf material was finely ground using a ball mill (Kleco pulverizer). Twenty milligrams of pulverized leaf material from each sample were extracted in 1.5 mL of 80 % aqueous HPLC-grade methanol (Fisher Scientific). The resulting mixture was vortexed for 30 s (Genie 2) and sonicated for 30 min at 20 °C (Quantrex 280, L&R Ultrasonics). Samples were centrifuged following sonication for 15 min at 15 000 rpm (Marathin Micro A, Fisher Scientific) and the supernatant was transferred to high-performance liquid chromatography (HPLC) vials for analyses of tea quality.

Chemical analyses of tea functional quality

Tea quality was measured using HPLC to determine the concentration of eight antioxidant polyphenol compounds and three methylxanthine compounds linked to tea functional quality, including its health claims and stimulant properties. Individual methylxanthine compounds were aggregated into a measure of total methylxanthine concentrations (TMCs). In addition, total phenolic concentrations (TPCs) of tea leaves were measured. High-performance liquid chromatography was performed as previously described to measure antioxidant polyphenol and methylxanthine secondary metabolites (Unachukwu et al. 2010). The polyphenols measured include catechin (C), catechin gallate (CG), epicatechin 3-gallate (ECG), epigallocatechin (EGC), epigallocatechin 3-gallate (EGCG), gallic acid (GA) and gallocatechin 3-gallate (GCG; ChromaDex). The methylxanthines measured include caffeine, theobromine and theophylline (ChromaDex). A Waters 2695 (Milford, MA, USA) module equipped with a 996 photodiode array detector and a 4 μm, 250 × 4.6 mm ID, C-18 Synergi Fusion, reversed-phase column (Phenomenex, Torrance, CA, USA) was used for the HPLC analysis. Prior to the experimental run, the HPLC method was validated with respect to accuracy, precision, sensitivity and selectivity. For each sample, 5 μL were injected using a mobile phase of 0.05 % (v/v) trifluoroacetic acid in distilled water (Solvent A) and 0.05 % (v/v) trifluoroacetic acid in acetonitrile (Solvent B). The solvent gradient was set at a flow rate of 1 mL min^{-1} as follows: 12–21 % Solvent B from 0 to 25 min; 21–25 % Solvent B from 25 to 30 min. The column and autosampler temperatures were maintained at 38 and 4 °C, respectively. At the end of each run, the column was flushed with 100 % Solvent B for 10 min and was re-equilibrated for 5 min to starting conditions. Spectra were recorded from 254 to 400 nm and relevant peaks were detected at 280 nm on the basis of characteristic absorbance spectra and retention time. Analyte concentrations were determined using peak areas and the linearity determined by plotting signal versus concentration standard curve equations with the limit of detection and the limit of quantification in the ranges of 0.05–1 and 0.1–5 g mL^{-1}, respectively.

Total phenolic concentration was determined spectrophotometrically using Folin–Ciocalteau reagent as previously described (Unachukwu et al. 2010). Samples were analysed in triplicate. Absorbance values were measured at 765 nm using a Benchmark Plus microplate spectrometer (Bio-Rad) and results expressed as gallic acid equivalents (GAE) in mg g^{-1} dry plant material. The concentration of polyphenols in tea samples was derived from a standard curve of GA concentration versus absorbance between 31.25 and 500 g mL^{-1}.

Statistical analysis

A fit model using a standard least squares means personality function and analysis of variance was performed using JMP 10.0 (SAS Institute Inc.) to determine how leaf growth and secondary metabolite concentrations vary among the precipitation and JA treatments. Data were analysed for the overall effect of water availability, JA treatment and their interactive effects. In addition, a multiple comparison using the least squares means Tukey's HSD method was applied to look at the difference between the three water availability treatments.

Results

Plant growth

Both higher water availability ($P < 0.001$) and JA ($P < 0.001$) significantly increased the growth of new leaves while their interactive effect was not significant ($P = 0.94$; Fig. 1). Overall, high-water plants had significantly more

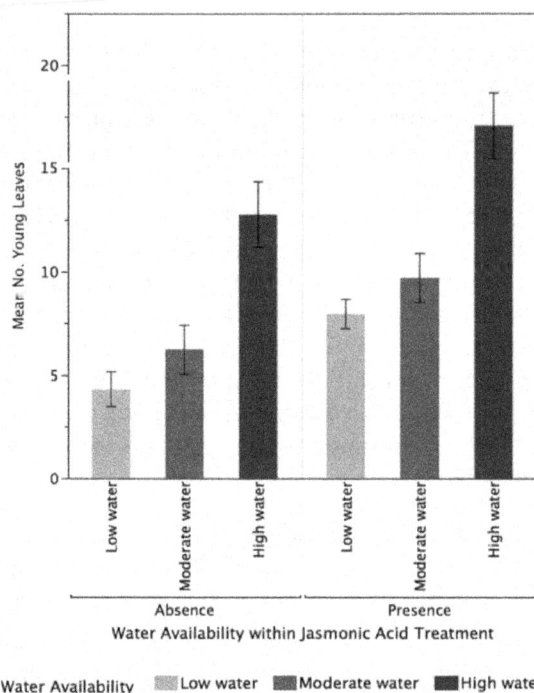

Figure 1. Effects of water availability and JA on leaf growth. Higher water availability ($P < 0.001$) and JA ($P < 0.001$) significantly increased the growth of new leaves while their interactive effect was not significant ($P = 0.94$). Values are means ± 1 standard error.

leaves than moderate-water plants ($P < 0.0001$) and low-water plants ($P < 0.0001$). The moderate-water plants and low-water plants did not differ significantly in the growth of new leaves ($P = 0.24$). Tea plants under the JA treatments had a significantly greater number of new leaves compared with plants that were not treated with JA ($P = 0.0003$). Higher water availability ($P = 0.001$) but not JA ($P = 0.54$) or their interaction ($P = 0.90$) resulted in significantly increased plant height (Fig. 2). The high-water-availability plants had significantly greater leaf growth than the low-water plants ($P = 0.001$) but did not differ significantly from the moderate-water-availability plants (0.059). While the low-water plants differed significantly in leaf growth from the high-water plants, they did not differ from the moderate-water plants ($P = 0.22$).

Chemical analyses of tea functional quality

Higher water availability ($P < 0.001$) significantly increased TMCs of tea plants but there was no significant effect of JA treatments ($P = 0.53$) or the interaction between water and JA ($P = 0.06$; Fig. 3). High-water plants ($P < 0.0217$) and moderate-water plants ($P < 0.0009$) had significantly higher concentrations of TMC compared with low-water plants but did not differ significantly from each other ($P = 0.41$). For the concentrations of EGCG, there was no significant effect for water availability

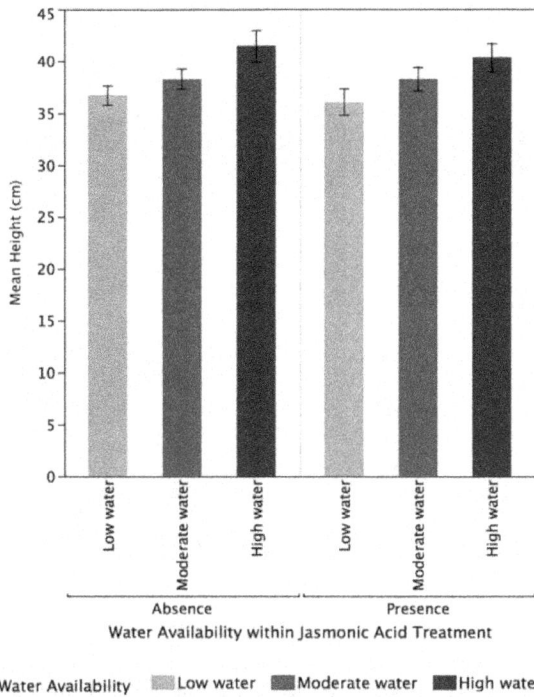

Figure 2. Effects of water availability and JA on plant height. Higher water availability ($P = 0.001$) but not JA ($P = 0.54$) or their interaction ($P = 0.90$) resulted in significantly increased plant height. Values are means $\pm$ 1 standard error.

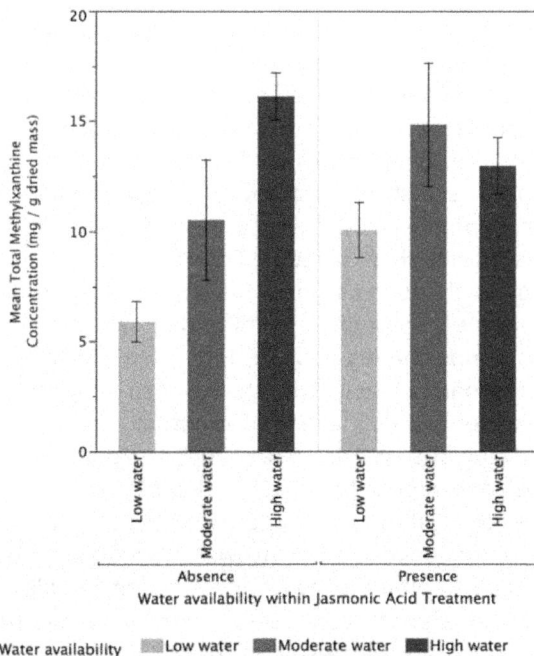

Figure 3. Effects of water availability and JA on TMC. Higher water availability ($P < 0.001$) significantly increased the TMCs of tea plants but there was no significant effect of JA treatments ($P = 0.53$) or the interaction between water and JA ($P = 0.06$). Values are means $\pm$ 1 standard error.

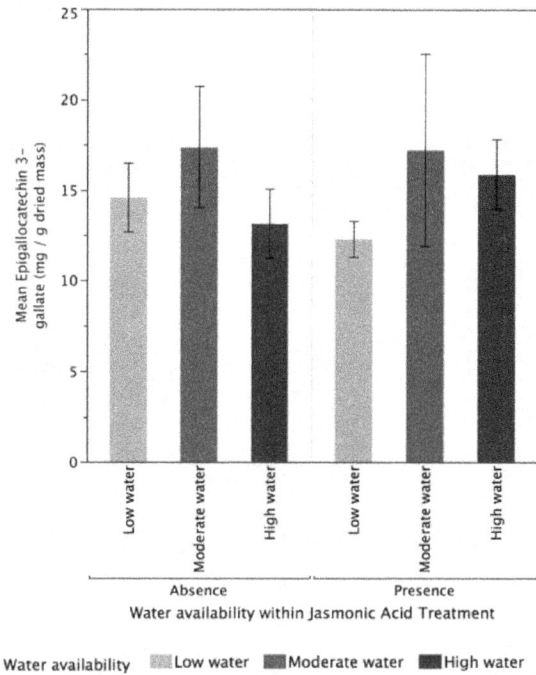

Figure 4. Effects of water availability and JA on the concentration of EGCG. Higher water availability ($P = 0.37$), JA treatments ($P = 0.95$) and their interactive effects had no significant effect on concentrations of EGCG. Values are means $\pm$ 1 standard error.

($P = 0.37$), JA treatments ($P = 0.95$) or their interactive effects ($P = 0.68$; Fig. 4). Neither the low-water ($P = 0.28$) nor the high-water ($P = 0.49$) treatments were significantly different from the moderate-water plants for EGCG concentrations. Additionally, there was no significant difference in EGCG concentrations between high- and low-water plants ($P = 0.8891$). In contrast, for ECG concentrations (Fig. 5), increased water availability ($P = 0.02$) resulted in significantly lower ECG but the effect of JA ($P = 0.982$) and the interactive effects of water and JA were not significant ($P = 0.138$). The high-water-availability treatments had significantly greater ECG concentrations compared with the low-water treatments ($P = 0.0117$) but did not differ significantly from the moderate-water treatments ($P = 0.29$). While the high- and low-water treatments differed significantly in their ECG concentrations, the moderate-water treatment did not differ significantly from either ($P = 0.29$). For TPC, higher water availability resulted in significantly higher TPC ($P < 0.0001$) but there was no significant impact of JA ($P = 0.89$) and their interaction (0.09; Fig. 6). High-water treatments had significantly greater TPC compared with moderate-water treatments ($P = 0010$) and low-water treatments ($P < 0.0001$). Moderate-water treatments had significantly higher TPC compared with low-water treatments ($P = 0.0107$).

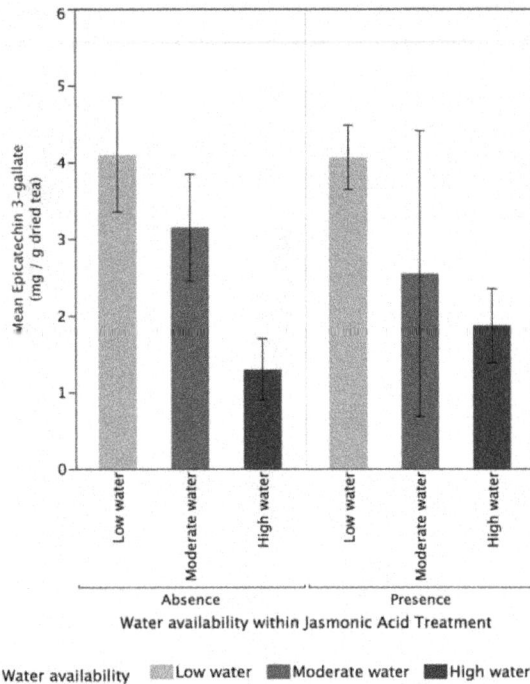

Water availability   Low water   Moderate water   High water

Figure 5. Effects of water availability and JA on the concentration of ECG. Higher water availability ($P = 0.02$) resulted in significantly lower ECG but the effect of JA ($P = 0.982$) and their interactive effects were not significant ($P = 0.138$). Values are means $\pm$ 1 standard error.

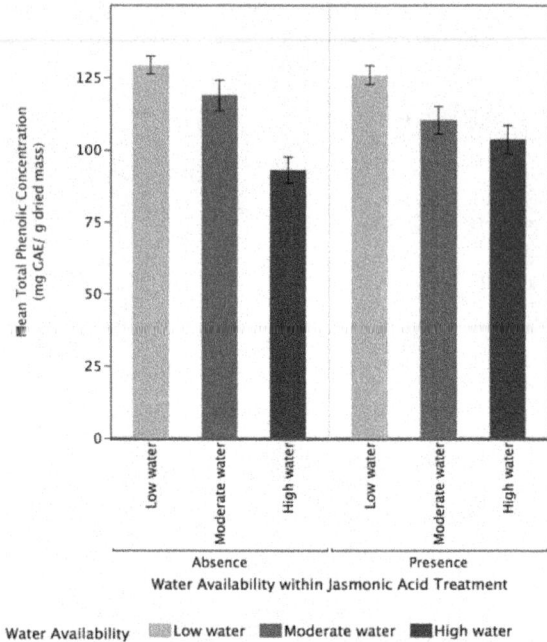

Water Availability   Low water   Moderate water   High water

Figure 6. Effects of water availability and JA on TPC. Higher water availability ($P < 0.0001$) resulted in significantly higher TPC but there was no significant impact of JA ($P = 0.89$) and their interaction (0.09). Values are means $\pm$ 1 standard error.

Discussion

This study supports the view that an increase in water availability results in a significant increase in growth of potted tea plants on the basis of both plant height and new leaves while the effects on secondary metabolites vary depending on chemical class. Higher water availability increased TMCs, decreased ECG levels and decreased TPCs of tea leaves. Epigallocatechin 3-gallate was the only tea functional quality parameter measured that was not significantly impacted by water availability treatments. Surprisingly, pest pressures as simulated by JA increased plant growth on the basis of new leaves, indicating that potted tea plants in a greenhouse setting may respond to pest pressures by prioritizing new leaf growth. Unexpectedly, JA had no significant effect on secondary metabolite chemistry. However, the interactive effects of water availability and simulated pest pressures show a trend to offset the direct effects of water availability on TMC and TPC. These findings point to the fascinating dynamics of climate change effects on tea plants with offsetting interactions within agro-ecosystems and the need for future climate studies to examine climate variables and pest pressures as well as their interactive effects.

In general, our findings concur with previous studies which found that altered water availability is a key driver of plant performance (Gulati and Ravindranath 1996;

Dou et al. 2007) and significantly impacts both growth and secondary metabolite concentrations of tea plants (Gulati and Ravindranath 1996; Yao et al. 2005; Schepp 2009; Honow et al. 2010; CIAT 2011). Given the slow-growing nature of woody tea plants, the less notable effect of the treatments on plant height compared with leaves is expected. The reduced growth of plants under drought treatment in this study concurs with the widely accepted recognition that lower soil moisture content reduces photosynthesis, growth and survivability of plants (Kozlowski et al. 1991; Condit 1998). Shrubs with shallow roots, such as clonal tea shrubs, are particularly susceptible to drought effects and show severe water stress during the dry season (Tobin et al. 1997). Plants may respond to drought by closing their stomata to reduce water loss at the cost of eventually facing carbon starvation, or may keep their stomates open and face the risk of hydraulic failure (Zeppel et al. 2011).

The variability of the response of specific secondary metabolite concentrations to water variability emphasizes the complex changes in tea functional quality with forecasted climate change and concurs with studies showing idiosyncratic responses of individual compounds to environmental stress (Glynn et al. 2007). Caffeine is the primary secondary metabolite responsible for tea's stimulant properties and contributes to its bitter taste. Epicatechin 3-gallate and EGCG are prominent polyphenolic catechins in tea that contribute to tea's bitter taste as well as its sweet aftertaste, which is highly desirable. In addition, these compounds contribute to its antioxidant and anti-inflammatory properties and other medicinal attributes. Total phenolic concentration and antioxidant activity further contribute to the overall functional properties of tea. Consumers can discern changes in these compounds that influence their purchasing decisions (Ahmed et al. 2010). The methylxanthine caffeine is a nitrogen-based compound, while individual polyphenolic catechins along with the cumulative TPC measure represent carbon-based compounds.

We expected JA treatments to result in a large increase in these key secondary metabolites (Karban and Baldwin 1997; Kruidhof et al. 2012). Kruidhof et al. showed that proteinase inhibitors are highly induced by a second jasmonate, methyl jasmonate (124 % increase). Interestingly, they found that proteinase inhibitors were not expressed in glasshouse-grown plants. They suggest that the UV filtering properties prevent expression. Although they did not induce these glasshouse-grown plants with methyl jasmonate, it is possible that induction of many compounds is dependent on light quality. We suggest that future experiments should test the effects of jasmonates and pest pressures on tea plants grown in the field. Furthermore, this study used JA to simulate pest pressures that may provide an indication of what might happen when leaf-chewing caterpillars attack the plant. Tea is also attacked by leaf-sucking herbivores such as leaf hoppers, which induce different signalling pathways and thus may have very different effects on tea secondary chemistry and ultimately tea quality.

The significant impact of water availability on tea functional quality found in this study represents a conservative estimate of what would happen under field conditions, as manipulative studies are likely to underestimate plant responses to climate change for at least two reasons (Wolkovich et al. 2012). Field plants are exposed to many abiotic stressors (e.g. wind) that change plant chemistry and being older plants they typically have higher concentrations of many secondary metabolites. In addition, the interaction with additional climate variables, including temperature and carbon dioxide levels, would further exacerbate complexity with opposing or enhancing effects. In summary, future studies are needed that examine the interactive effects of multiple climatic factors with specialist tea pests in both controlled and field conditions.

Conclusions

This study provides some of the first evidence on the multi-directionality of shifts in water availability, pest pressures and their interactive effects on tea quality. While numerous studies have documented the impact of climate change on crop yield, this study contributes to the knowledge gap on climate effects on crop quality that are crucial to examine for food security. Results indicate that while extreme drought and precipitation conditions might decrease or increase plant growth and functional quality, pest pressures may offset these effects. For example, drought conditions may result in a decline of both tea growth and stimulant properties of tea but pest pressures may offset these effects. If the changes in tea functional quality with water availability and herbivore pressures are indicative of broader climate change, tea production areas face increased heterogeneity with forecasted prolonged and more frequent droughts along with increased heavy precipitation events (Orlowsky and Seneviratne 2012; Seneviratne et al. 2012). Future research in both controlled and natural settings across spatial and temporal scales is needed to better understand the interplay between a range of climatic conditions, tea plants, herbivore pressures and other multi-trophic interactions.

Sources of Funding

Our work was funded by a Tufts Collaborates Seed Grant, TEACRS Program at Tufts University (NIH National Institute of General Medical Sciences IRACDA-K12GM074869), National Science Foundation Research Experiences for Undergraduates Program (NSF DBI 1005082) and Tufts Institute for the Environment.

Contributions by the Authors

All authors contributed to the overall study design. S.A. and T.G. designed water availability manipulations. S.A. and C.O. designed simulated pest pressure treatments. S.A. conducted greenhouse manipulations, coordinated study logistics and harvested samples. S.A., S.B. and U.U. were involved in the chemical and statistical analysis. A.S. helped with the experimental set-up of the pilot study. S.A. and C.O. primarily wrote the manuscript with contributions from all authors.

Acknowledgements

We thank the Arnold Arboretum at Harvard University for greenhouse facility use and support of this project; Angus Schaefer for assisting with plant growth measurements; and Amanda Kowalsick, Feven Asefaha and Matt Reynolds for their work on the pilot experiments that informed this study.

Literature Cited

Ahmed S, Unachukwu U, Stepp JR, Peters CM, Long C, Kennelly E. 2010. Pu-erh tea tasting in Yunnan, China: correlation of drinkers' perceptions to phytochemistry. *Journal of Ethnopharmacology* **132**:176–185.

Ahmed S, Peters CM, Long C, Meyer R, Unachukwu U, Litt A, Kennelly E, Stepp JR. 2013. Biodiversity and phytochemical quality in indigenous and state-supported tea management systems of Yunnan, China. *Conservation Letters* **5**:28–36.

Atkinson NJ, Urwin PE. 2012. The interaction of plant biotic and abiotic stresses: from genes to the field. *Journal of Experimental Botany* **63**:3523–3544.

Babst BA, Ferrieri RA, Gray DW, Lerdau M, Schlyer DJ, Schueller M, Thorpe MR, Orians CM. 2005. Jasmonic acid induces rapid changes in carbon transport and partitioning in *Populus*. *New Phytologist* **167**:63–72.

Berggren A, Bjorkman C, Bylund H, Ayres MP. 2009. The distribution and abundance of animal populations in a climate of uncertainty. *Oikos* **118**:1121–1126.

Boisvenue C, Running S. 2006. Impacts of climate change on natural forest productivity—evidence since the middle of the 20th century. *Global Change Biology* **12**:862–882.

Brenes-Arguedas T, Horton MW, Coley PD, Lokvam J, Waddell RA, Meizoso-O'Meara BE, Kursar TA. 2006. Contrasting mechanisms of secondary metabolite accumulation during leaf development in two tropical tree species with different leaf expansion strategies. *Oecologia* **149**:91–100.

Brenes-Arguedas T, Coley PD, Kursar TA. 2009. Pests vs. drought as determinants of plant distribution along a tropical rainfall gradient. *Ecology* **90**:1751–1761.

CIAT. 2011. *Future climate scenarios for Kenya's tea growing areas.* International Center for Tropical Agriculture Report. Cali, Colombia.

Coley P. 1998. Possible effects of climate change on plant/herbivore interactions in moist tropical forests. *Climatic Change* **39**: 445–472.

Condit R. 1998. Ecological implications of changes in drought patterns: shifts in forest composition in Panama. *Climatic Change* **39**:413–427.

Dou J, Zhang Y, Yu G, Zhao S, Song Q. 2007. Interannual and seasonal variations of energy and water vapor fluxes above a tropical seasonal rain forest in Xishuangbanna, SW China. *Acta Ecologica Sinica* **27**:3099–3109.

Ewert F, Rounsevell MDA, Reginster I, Metzger MJ, Leemans R. 2005. Future scenarios of European agricultural land use I. Estimating changes in crop productivity. *Agriculture, Ecosystems and Environment* **107**:101–116.

Glynn C, Herms DA, Orians CM, Hansen RC, Larsson S. 2007. Testing the growth-differentiation balance hypothesis: Dynamic responses of willows to nutrient availability. *New Phytologist* **176**: 623–634.

Gulati A, Ravindranath SD. 1996. Seasonal variations in quality of Kangra tea (*Camellia sinensis* (L) O Kuntze) in Himachal Pradesh. *Journal of the Science of Food and Agriculture* **71**: 231–236.

Gutbrodt B, Mody K, Dorn S. 2011. Drought changes plant chemistry and causes contrasting responses in lepidopteran herbivores. *Oikos* **120**:1732–1740.

Gutbrodt B, Dorn S, Mody K. 2012. Drought stress affects constitutive but not induced herbivore resistance in apple plants. *Arthropod-Plant Interactions* **6**:171–179.

Herms DA, Mattson WJ. 1992. The dilemma of plants—to grow or defend. *The Quarterly Review of Biology* **67**:283–335.

Honow R, Gu KR, Hesse A, Siener R. 2010. Oxalate content of green tea of different origin, quality, preparation and time of harvest. *Urological Research* **38**:377–381.

IPCC. 2007. Climate change 2007: impacts, adaptation and vulnerability. In: Parry M, Canziani O, Palutikof J, Van Der Linden P, Hanson C, eds. *Contribution of Working Group II to the fourth assessment report of the Intergovernmental Panel on Climate Change.* Cambridge: Cambridge University Press.

Jamieson MA, Trowbridge AM, Raffa KF, Lindroth RL. 2012. Consequences of climate warming and altered precipitation patterns for plant–insect and multitrophic interactions. *Plant Physiology* **160**:1719–1727.

Karban R, Baldwin IT. 1997. *Induced responses to herbivory.* Chicago: University of Chicago Press.

Kozlowski TT, Kramer PJ, Pallardy SG. 1991. *The physiological ecology of woody plants.* New York: Academic Press.

Kruidhof HM, Allison JD, Hare D. 2012. Abiotic induction affects the costs and benefits of inducible herbivore defenses in *Datura wrightii*. *Journal of Chemical Ecology* **38**:1215–1224.

Lin Y, Tsai Y, Tsay J, Lin J. 2003. Factors affecting the levels of tea polyphenols and caffeine in tea leaves. *Journal of Agriculture and Food Chemistry* **51**:1864–1873.

Lobell DB, Schlenker W, Costa-Roberts J. 2011. Climate trends and global crop production since 1980. *Science* **333**:616–620.

Maplecroft. 2011. Climate Change and Environmental Risk. www.maplecroft.com

McDowell JM, Dangl JL. 2000. Signal transduction in the plant immune response. *Trends in Biochemical Sciences* **25**:79–82.

Nelson GC, Rosegrant MW, Koo J, Robertson R, Sulser T, Zhu T, Ringler C, Msangi S, Palazzo A, Batka M, Magalhaes M, Valmonte-Santos R, Ewing M, Lee D. 2009. *Climate change: impact on agriculture and costs of adaptation.* Food Policy Report, International Food Policy Research Institute (IFPRI), Washington, DC.

Orlowsky B, Seneviratne S. 2012. Global changes in extreme events: regional and seasonal dimension. *Climatic Change* **110**: 669–696.

Porter JR, Semenov MA. 2005. Crop responses to climatic variation. *Philosophical Transactions of the Royal Society B* **360**: 2021–2035.

Schepp K. 2009. *Strategy to adapt to climate change for Michimikuru tea farmers in Kenya.* AdapCC Report. Deutsche Gesellschaft für Technische Zusammenarbeit (GTZ) GmbH, Eschborn, Germany.

Schlenker W, Lobell DB. 2010. Robust negative impacts of climate change on African agriculture. *Environmental Research Letters* **5**:014010. doi:10.1088/1748-9326/5/1/014010.

Seneviratne S, Nicholls N, Easterling D, Goodess C, Kanae S, Kossin J, Luo Y, Marengo J, McInnes K, Rahimi M, Reichstein M, Sorteberg A, Vera C, Zhang X. 2012. Changes in climate extremes and their impacts on the natural physical environment. In: Field CB, Barros V, Stocker TF, Qin D, Dokken DJ, Ebi KL, Mastrandrea MD, Mach KJ, Plattner G-K, Allen SK, Tignor M, Midgley PM, eds. *Managing the risks of extreme events and disasters to advance climate change adaptation. A special report of Working Groups I and II of the Intergovernmental Panel on Climate Change (IPCC).* Cambridge, UK: Cambridge University Press, 109–230.

Tharayil N, Suseela V, Triebwasser DJ, Preston CM, Gerartd PD, Dukes JS. 2011. Changes in the structural composition and reactivity of *Acer rubrum* leaf litter tannins exposed to warming and altered precipitations: climatic stress-induced tannins are more reactive. *New Phytologist* **191**:132–145.

Tobin MF, Lopez OR, Kursar TA. 1997. Drought response of tropical understory species with long and short leaf lifespans. *Biotropica* **31**:570–578.

Unachukwu U, Ahmed S, Kavalier A, Lyles J, Kennelly E. 2010. Variation of phenolic and methylxanthine composition and anti-oxidant activity among white and green teas (*Camellia sinensis* var. *sinensis* (L.) Kuntze Theaceae). *Journal of Food Science* **75**:C541–C548.

Wolkovich EM, Cook BI, Allen JM, Crimmins TM, Betancourt JL, Travers SE, Pau S, Regetz J, Davies TJ, Kraft NJB, Ault TR, Bolmgren K, Mazer SJ, McCabe GJ, McGill BJ, Parmesan C, Salamin N, Schwartz MD, Cleland EE. 2012. Warming experiments underpredict plant phenological responses to climate change. *Nature* **485**:494–497.

Yao L, Caffin N, D'Arcy B, Jiang Y, Shi J, Singanusong R, Liu X, Datta N, Kakuda Y, Xu Y. 2005. Seasonal variations of phenolic compounds in Australia-grown tea (*Camellia sinensis*). *Journal of Agriculture and Food Chemistry* **53**:6477–6483.

Zeppel MJB, Adams HD, Anderegg WRL. 2011. Mechanistic causes of tree drought mortality: recent results, unresolved questions and future research needs. *New Phytologist* **192**:800–803.

Bark and leaf chlorophyll fluorescence are linked to wood structural changes in *Eucalyptus saligna*

Denise Johnstone[1]*, Michael Tausz[2], Gregory Moore[1] and Marc Nicolas[3]

[1] Department of Resource Management and Geography, University of Melbourne, Burnley Campus, Richmond 3012, Australia
[2] Department of Forest and Ecosystem Science, University of Melbourne, Creswick Campus, Creswick 3363, Australia
[3] Department of Agriculture and Food Systems, University of Melbourne, Parkville Campus, Parkville 3010, Australia

Abstract. Wood structure and wood anatomy are usually considered to be largely independent of the physiological processes that govern tree growth. This paper reports a statistical relationship between leaf and bark chlorophyll fluorescence and wood density. A relationship between leaf and bark chlorophyll fluorescence and the quantity of wood decay in a tree is also described. There was a statistically significant relationship between the leaf chlorophyll fluorescence parameter F_v/F_m and wood density and the quantity of wood decay in summer, but not in spring or autumn. Leaf chlorophyll fluorescence at 0.05 ms (the O step) could predict the quantity of wood decay in trees in spring. Bark chlorophyll fluorescence could predict wood density in spring using the F_v/F_m parameter, but not in summer or autumn. There was a consistent statistical relationship in spring, summer and autumn between the bark chlorophyll fluorescence parameter F_v/F_m and wood decay. This study indicates a relationship between chlorophyll fluorescence and wood structural changes, particularly with bark chlorenchyma.

Keywords: Bark; chlorophyll fluorescence; photosynthesis; stress physiology; wood decay; wood structure.

Introduction

Tree physiology and wood structure and anatomy are often considered to be independent, as wood occurs primarily in what is sometimes described as the non-functioning heartwood of the tree (Zweifel *et al.* 2006). On the other hand, wood as a tissue (i.e. the secondary xylem of trees) determines long-distance water transport in trees. During water transport, if xylem vessels are under water stress, air bubbles in the xylem can expand due to tension, a process known as cavitation (Hacke *et al.* 2001; Taiz and Zeiger 2010). Once a xylem vessel cavitates it fills with water vapour and then forms an embolism in quick succession, slowing xylem hydraulic conductivity (Tyree and Sperry 1989). Therefore, wood density is increasingly

being measured in conjunction with water-use properties, as low stem wood density can make angiosperms more vulnerable to cavitation, especially during drought (Hacke *et al.* 2001; Holste *et al.* 2006; Bobich *et al.* 2010). However, conifers do not necessarily follow this pattern as their xylem conduits are shorter and narrower. In a study of *Picea abies* (Norway spruce), wood density was unrelated to xylem cavitation (Rosner *et al.* 2007). The relationship between wood decay and physiological measurements not directly related to water use has rarely been assessed. Wood structural changes are frequently caused by wood decay organisms (Rayner and Boddy 1988). Decayed wood shows decreased density as a result of degradation by fungi or bacteria (Harris *et al.* 2004).

* Corresponding author's e-mail address: denisej@unimelb.edu.au

Weight loss or dry weight is a common means by which to evaluate wood decay, particularly in the early stages of decay (Wilcox 1978; Pandey and Pitman 2003; Wei et al. 2010). Despite ongoing methodological difficulties, wood decay can be quantified by a variety of methods, such as with devices using electrical conductivity, drilling resistance, core sampling or acoustic methods (Johnstone et al. 2010a). It appears logical that wood decay, leading to decreased wood density, can affect tree water transport and, consequently, canopy physiology, mainly under periods of increased demand on water transport. Because wood decay involves invading organisms such as fungi or bacteria, it may also be speculated that biochemical changes (e.g. defence reactions) can affect the physiological function of other tissues.

Trees have chlorenchyma, i.e. photosynthetically active tissue, in their bark below the rhytidomal or outer peridermal layers (Strain and Johnson 1963; Pfanz et al. 2002). Such cortical or peridermal chlorenchyma is able to utilize CO_2 from gaseous xylem efflux and from mitochondrial respiration to photosynthesize (Wittmann et al. 2006; Pfanz 2008). Bark photosynthesis can be strongly shade adapted, particularly in deciduous trees (Pfanz et al. 2002; Damesin 2003; Manetas 2004). Eucalyptus globulus bark behaved as a shade leaf in a study by Eyles et al. (2009); however, Tausz et al. (2005) found that parts of sun-exposed Eucalyptus nitens bark had photosynthetic pigments of similar quantity and composition to that of sun leaves. Bark photosynthetic activity in stems is generally lower than in the leaves of broadleaf trees such as Betula pendula, Quercus robur and Fagus sylvatica, but it could be a way of improving the carbon balance of stems, particularly where water is limiting (Wittmann and Pfanz 2008b).

Chlorophyll fluorescence (CF) is an excellent tool to assess the physiological state of photosynthetic tissues (Govindjee 2004). F_v/F_m is the most commonly cited CF parameter, where F_v is the difference between maximum (F_m) and minimum (F_o) fluorescence (Maxwell and Johnson 2000). F_v/F_m is the theoretical measure of the quantum efficiency of photosystem II (PSII) if all the PSII reaction centres are open (Maxwell and Johnson 2000). The average F_v/F_m value for healthy tissues is believed to be around 0.83 (Bjorkman and Demmig 1987; Johnson et al. 1993). Decreased values indicating reduced maximum quantum efficiency commonly occur upon impact of environmental stress. F_v/F_m is therefore commonly used to assess stress impacts on plants (Maxwell and Johnson 2000).

The analysis of the intermediate data points of the fast fluorescence rise (i.e. the determination of F_m in the calculation of F_v/F_m) is called the O–J–I–P polyphasic fast fluorescence rise analysis or the O–K–J–I–P polyphasic

fast fluorescence rise analysis (Susplugas et al. 2000; Strasser and Stirbet 2001; Govindjee 2004; Strasser et al. 2004; Percival 2005). The phases are O at the origin (0.05 ms), K at ~0.2 ms, J at ~2 ms, I at ~20 ms and P at ~200 ms, depending on the curve (Strasser and Stirbet 2001). O or F_o fluorescence is measured when all the plastoquinone Q_A electron carrier molecules are in their oxidized state (Krause and Weis 1984; Percival 2005). The K step, not apparent in all cases, may be the result of an imbalance in electron flow coming to the reaction centre from PSII in some species of plants (Strasser et al. 2004). The O–J phase is believed to represent the reduction of the Q_A molecule from Q_A to Q_A^- (Hsu and Leu 2003; Strasser et al. 2004; Percival 2005). J–I may be fluorescence from the abaxial layer of the sample in some plants (Hsu and Leu 2003), or both the J–I and I–P phases could reflect the existence of fast and slow reducing plastoquinone centres (Percival 2005). P or F_m occurs when all the plastoquinone Q_A electron carrier molecules are in their reduced state (Krause and Weis 1984; Percival 2005). The characteristics of the fast fluorescence rise also change upon stress impact, and are therefore used to assess stress impacts on plants.

There is evidence that leaf photosynthetic capacity and the hydraulic properties of tree stems are related (Brodribb and Feild 2000; Brodribb et al. 2007), yet any direct relationship between wood properties or wood decay and photosynthetic properties has rarely been examined. The symptoms of 'esca' disease in Vitis vinifera (grapevines) and CF parameters have been linked (Christen et al. 2007). Esca disease infects the xylem and causes the white rot decay and/or necrosis of woody tissues and, subsequently, wilting of the leaves. However, no investigations using tree species prior to the current study have attempted to relate photosynthetic properties to wood decay.

In a previous study, the authors investigated a relationship between crown condition and leaf and bark CF (Johnstone et al. 2012). There was little evidence to support a relationship between leaf CF and crown condition. On the other hand, there was a strong relationship between bark CF and crown condition. The current study uses the leaf and bark CF data from the above-mentioned study, but compares it with wood density and wood decay. In this study, the relationship between CF and wood structural properties is examined, rather than CF and growth parameters.

The current study investigated plantation-grown Eucalyptus saligna trees exhibiting a range of wood decay from virtually none to moderately decayed. We chose trees already decayed as inducing decay in trees can be a slow process, dependent on tree species and the causal agent of decay (Schwarze 2008). Trees were chosen to represent the best possible range of decay under otherwise uniform conditions. We examined the relationships between wood decay and density and CF in leaf and bark tissues to test the following hypotheses: (i) increasing wood decay is related to stress symptoms in leaves, particularly in summer when demand on xylem water transport is greatest, and (ii) increasing wood decay is related to stress symptoms in bark chlorenchyma.

Methods

The trees used in this study were E. saligna (Bateman's Bay). They were ~20 years old in 2008, between 17 and 27 m high, and with diameters at 1.3 m of between 142 and 318 mm. The 36 selected trees were part of a larger species/provenance study covering a total area of ~10 ha in a eucalypt plantation at Tostaree in rural Victoria, Australia (latitude 37°47′; longitude 148°11′). Sample trees were chosen to represent a range of wood decay and excluded any break or edge trees. In this investigation, CF measurements in both leaves and bark were compared with wood density and the percentage of decay over three seasons (spring, summer and autumn).

Chlorophyll fluorescence measurements

Chlorophyll fluorescence data were collected and analysed according to the method described in Johnstone et al. (2012). Branches ~10 mm in diameter were harvested from the upper canopy with a 12-gauge shotgun in the morning, between 0600 and 0800 h depending on the season. Leaf fluorescence measurements were taken between 13 September and 21 September 2007 (spring), 28 January and 1 February 2008 (summer) and 5 April and 13 April 2008 (autumn). Most eucalypts can have two or three different leaf ages present in the crown at any one season, with leaves lasting up to 18 months. Eucalypts have opportunistic crown phenology dependent on their environmental conditions (Jacobs 1955).

Leaf CF measurements were taken on mature sun leaves from upper canopy branches using a Hansatech-handy plant efficiency analyser (Hansatech Instruments, King's Lynn, Norfolk, UK). Ten leaves from each tree were dark adapted for 30 min with leaf clips. A saturating flash of red light onto the leaf after the period of darkness induced a time-dependent fluorescence kinetic known as the Kautsky effect (Govindjee 2004; Percival 2005). All trees were tested within 2–3 h of being harvested as recommended by Epron and Dreyer (1992).

Bark CF testing was performed in a 350-mm strip in a cross-section of the trunk on the north half of the trees, 35 mm apart. The test area on the bark was circular and 4.5 mm in diameter. Eight to 10 tests were performed on each tree after material had been dark adapted for 30 min. The bark was not damaged or removed in any way. Test results were excluded if the bark was damaged, decorticating or had only recently been exposed to sunlight. The height at which trees were measured was variable as it was necessary to measure above the sock of rough bark at the base. Bark fluorescence measurements were taken between 24 September and 28 September 2007 (spring), 22 January and 26 January 2008 (summer) and 31 March and 4 April 2008 (autumn).

The CF data were averaged from 8–10 measurements from each tree in each tissue (bark and leaf) and in each season. The ratio F_v/F_m was calculated from the raw CF data. F_v/F_m is a derived measure $F_v = F_m - F_o$, where F_v is the difference between maximum (F_m) and minimum (F_o) CF (Maxwell and Johnson 2000). In addition to calculating the F_v/F_m ratio, time data taken over a 1-s period were logarithmically transformed and the O–J–I–P CF phases were allocated following the method devised by Strasser and Stirbet (2001).

Each polyphasic increase in fluorescence was characterized by examining logarithmic graphs for each season and in both leaf and bark tissues. After an exponential rise in graphed data, each phase was deemed complete, with the next phase being deemed to start at the critical point (O, J, I or P). Every step is followed by a characteristic temporary decrease or dip (Strasser et al. 2004). There was no 'K' step observed on the graphs. 'O' was at the origin, taken at 0.05 ms, as in many other studies (Krause and Weis 1984; Susplugas et al. 2000; Strasser and Stirbet 2001; Govindjee 2004; Strasser et al. 2004; Percival 2005). The O–J phase was characterized as ending at 4 ms (J step). The 'I' step in leaf fluorescence data was observed at 60 ms and in bark at 90 ms. The 'P' step was observed at ~700 ms on leaf fluorescence graphs, previously observed at 200–300 ms in other studies. The 'P' step was not observed in bark fluorescence as the last recording point taken by the instrument was at 1000 ms, and fluorescence was still increasing at this time. The JIP test was not applied to the data; comparisons were made using the raw fluorescence values for O (0.05 ms all data), J (4 ms all data), I (60 ms leaf data, 90 ms bark data), P (700 ms leaf data) and the 1000 ms data point on bark.

Wood density measurement and wood decay estimation

The 36 E. saligna were tested for basic wood density from a small sample collected from the trunk at 1.5 m in height from the trees when they were felled in 2008. Basic wood density was estimated as oven dry mass of wood/volume of wood when 'green' (Walker et al. 1993). Wood decay in the trees was quantified using the Resi system utilizing the IML-Resi constant feed drill described in Johnstone et al. (2007, 2010b). The method begins with cross-sectional drilling measurements of the trunk at 0.3 m. The method combines the IML-Resi raw data and Shigo's (1979) compartmentalization of decay in trees (CODIT) model to predict the quantity of wood decay beyond the linear drill locations of the IML-Resi. The method relied on the experienced use of the IML-Resi, knowledge of models of decay in trees and image analysis software (Johnstone et al. 2007, 2010b).

Statistical analysis of data

A comparison was made between spring, summer and autumn CF data and wood density and wood decay data using simple linear regression analysis. Simple linear regression analyses were performed using the software package SAS (Statistical Analysis System) version 9.2 (SAS Institute Inc., Cary, NC, USA). Although multiple comparisons were made, Bonferroni corrections were not applied in order to maximize statistical power and minimize Type II errors in the analysis (Moran 2003).

One tree had no leaves and could not be included in leaf CF analysis, and the bark of this tree had died by the autumn sampling date. Data more than two standard deviations away from the next nearest result were considered outliers and eliminated from analysis, resulting in 34–35 individual replicate trees for regression analysis.

Results

Comparing leaf and bark fluorescence and basic wood density

There was a statistically significant and positive relationship between summer leaf F_v/F_m and basic wood density (Table 1 and Fig. 1A). There was a statistically significant

Table 1. Summarized results from simple linear regression analyses comparing spring, summer and autumn leaf or bark F_v/F_m with basic wood density data. n, the number of samples; P, the probability for the t-test that the coefficient of the independent variable is equal to zero; r^2, the variation in the dependent variable that can be explained by the fluorescence data. [a]The dependent variable is the spring basic wood density data in all cases. [b]The statistical relationship is significant and positive. Bold values indicate statistical significance.

Independent variable[a]	n	P	r^2
Spring leaf fluorescence—F_v/F_m	34	0.531	0.012
Summer leaf fluorescence—F_v/F_m	34	**0.001[b]**	**0.291**
Autumn leaf fluorescence—F_v/F_m	35	0.387	0.023
Spring bark fluorescence—F_v/F_m	35	**0.035[b]**	**0.128**
Summer bark fluorescence—F_v/F_m	35	0.512	0.013
Autumn bark fluorescence—F_v/F_m	35	0.249	0.040

and positive relationship between spring bark F_v/F_m and basic wood density (Table 1 and Fig. 1B). There was no statistical relationship between spring and autumn leaf F_v/F_m or summer and autumn bark F_v/F_m and basic wood density (Table 1). There was no statistical relationship between spring, summer and autumn leaf CF at the O, J, I or P step and basic wood density (Table 2). There was also no statistically significant relationship between spring, summer and autumn bark CF at the O, J, I or 1000 ms step and basic wood density (Table 2).

Comparing leaf and bark fluorescence and wood decay

There was a statistically significant and positive relationship between spring leaf CF at the O step and wood decay (Table 3 and Fig. 2A). There was a statistically significant and negative relationship between the summer leaf F_v/F_m ratio and wood decay (Table 3). There was a statistically significant and negative relationship between spring, summer and autumn bark F_v/F_m and wood decay (Table 3 and Fig. 2B).

There was no statistically significant relationship between the spring leaf F_v/F_m ratio or CF at the J, I and P step and wood decay (Table 4). There was no statistically significant relationship between summer leaf CF at the O, J, I, and P step and wood decay (Table 4). There was no statistically significant relationship between autumn leaf CF and wood decay (Table 4). There was no statistically significant relationship between bark CF at the O, J, I or 1000 ms step and wood decay, in spring, summer or autumn (Table 4).

Discussion

Weight loss or its corollary wood density has been used to assess wood decay for many years (Kennedy 1958; Wilcox 1978; Wei et al. 2010). Wood decay organisms can be responsible for weight losses as small as 5 % or less (Noguchi et al. 1986). In instances of very early decay, even a light microscope may not be able to detect wood decay visually (Wilcox 1978). Hence there is clearly a strong relationship between measured wood density and wood decay, even in assumed sound or intact wood.

There was a statistically significant and positive relationship between summer leaf F_v/F_m and basic wood density, but not in spring or autumn. In this study, the summer period of investigation coincided with maximum seasonal tree stress in southern Australia, when the mean average maximum temperature at the test site in January 2008 was 27 °C (minimum average 16 °C, Bureau of Metrology Australia 2008). However, summer predawn leaf and stem water potentials were not significantly different from spring values, although values of around

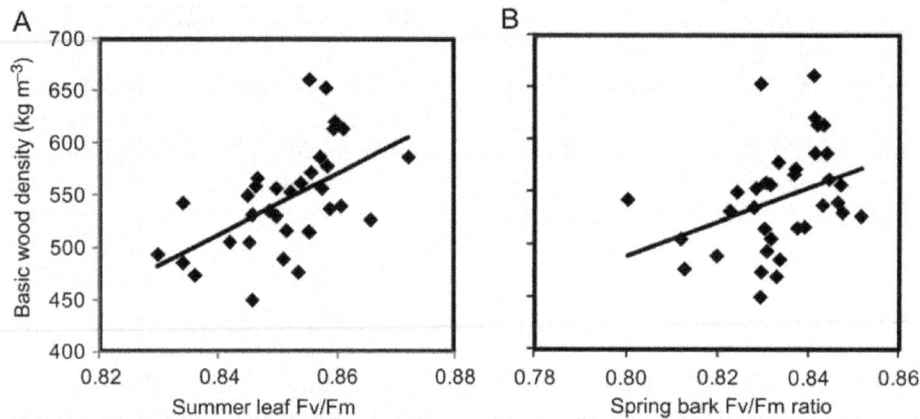

Figure 1. (A) Basic wood density in kg m^{-3} versus summer leaf F_v/F_m. Trend line = linear regression, $P = 0.001$, $r^2 = 0.291$. F_v/F_m ratio data begin at 0.820, and basic density data begin at 400 kg m^{-3}. (B) Basic wood density in kg m^{-3} versus spring bark F_v/F_m. F_v/F_m ratio data begin at 0.7900, and basic density data begin at 400 kg m^{-3}. Trend line = linear regression, $P = 0.035$, $r^2 = 0.128$.

1.2 MPa for all trees in summer indicated mild to moderate drought stress at this site (White *et al.* 2000; Johnstone *et al.* 2012). Predawn water potentials did not show a relationship with wood density or decay, but water potentials were not measured during the day when water deficit becomes more noticeable. Summer leaf CF at the O step also correlated with a visual vitality measurement in summer (Johnstone *et al.* 2012), which suggests that the trees were suffering some type of stress during the seasonal summer drought and that the leaf O step was sensitive to the stress. Wood density is sometimes measured in conjunction with other parameters for assessing the water status of trees (O'Grady *et al.* 2009; Gotsch *et al.* 2010). Low stem wood density in angiosperms is sometimes thought to be indicative of increased vulnerability to xylem cavitation during drought stress (Holste *et al.* 2006; Bobich *et al.* 2010). If cavitation is occurring in the xylem of the *E. saligna* in the current study, it could establish a favourable environment for fungal pathogens (Rayner and Boddy 1988), or the pathogens may assist with the cavitation process (Tyree and Sperry 1989; Tyree and Zimmermann 2002). This may explain why there is a relationship between leaf CF and wood decay in two seasons (spring and summer) rather than just one as was the case with wood density.

Changes in the availability of water for the *E. saligna* may have contributed directly to the relationship between photosynthetic efficiency and wood decay discovered in this study. The water saturation of wood has long been known to prevent the development of wood decay, and air is necessary for the development of decay in wood (Rayner and Boddy 1988). The barrier zones in Shigo's CODIT model (Shigo 1979) are said to be a response to xylem embolism by Rayner and Boddy (1988), rather than the incursion of decay organisms *per se*. Cavitation

during moisture stress is one way a xylem vessel can develop an embolism (Tyree and Sperry 1989). The direct relationship between xylem cavitation and wood decay has not been evaluated, but the introduction of a gaseous phase during the compartmentalization process, according to Rayner and Boddy (1988), is a primary component in the development of wood decay in trees. It is when tree wood dries out that compartmentalization barriers are breached (Rayner and Boddy 1988).

Trees with lower wood density have also been associated with an increased risk of cavitation (Holste *et al.* 2006; Bobich *et al.* 2010). Therefore, it is not surprising that in this study, in the hot Australian summer, *E. saligna* showed an inverse relationship between leaf CF and wood density, and an even stronger relationship between leaf CF and wood decay. Unlike many other studies, the link between moisture stress, cavitation, embolism and wood density/decay described here is a within-species effect, rather than the ecological inter-species effect of low wood density and water relations/growth discussed in other studies (Bucci *et al.* 2004; O'Grady *et al.* 2009). The lower density wood is produced due to stressful environmental conditions or is a result of very early wood decay in the *E. saligna*; there are no genetic differences or predispositions at play. The link between wood density, leaf CF and wood decay within species established in this study has not been previously reported.

There was a statistically significant and positive relationship between leaf CF values at the 'O' step and wood decay in spring. The O–J step is believed to represent the reduction of the plastoquinone Q_A molecule from Q_A to Q_A^- between PSII and photosystem I (PSI) (Hsu and Leu 2003; Strasser *et al.* 2004; Percival 2005); therefore, it appears that the reduction of Q_A between PSII and PSI during leaf photosynthesis is associated with wood decay in *E. saligna*. F_v/F_m is the theoretical

Table 2. Summarized results from simple linear regression analyses comparing spring, summer and autumn leaf or bark net chlorophyll fluorescence with basic wood density data. Results from these analyses were not significant. n, the number of samples; P, the probability for the t-test that the coefficient of the independent variable is equal to zero; r^2, the variation in the dependent variable that can be explained by the fluorescence data. [a]The dependent variable is the spring basic wood density data in all cases.

Independent variable[a]	n	P	r^2
Spring leaf fluorescence—'O' step	34	0.741	0.004
Spring leaf fluorescence—'J' step	34	0.620	0.008
Spring leaf fluorescence—'I' step	34	0.462	0.017
Spring leaf fluorescence—'P' step	34	0.891	0.001
Spring bark fluorescence—'O' step	35	0.702	0.005
Spring bark fluorescence—'J' step	35	0.691	0.005
Spring bark fluorescence—'I' step	35	0.298	0.033
Spring bark fluorescence—1000 ms	35	0.173	0.056
Summer leaf fluorescence—'O' step	34	0.072	0.097
Summer leaf fluorescence—'J' step	34	0.085	0.090
Summer leaf fluorescence—'I' step	34	0.134	0.069
Summer leaf fluorescence—'P' step	34	0.913	0.000
Summer bark fluorescence—'O' step	35	0.309	0.031
Summer bark fluorescence—'J' step	35	0.832	0.001
Summer bark fluorescence—'I' step	35	0.256	0.039
Summer bark fluorescence—1000 ms	35	0.191	0.051
Autumn leaf fluorescence—'O' step	34	0.810	0.002
Autumn leaf fluorescence—'J' step	34	0.558	0.011
Autumn leaf fluorescence—'I' step	34	0.905	0.001
Autumn leaf fluorescence—'P' step	34	0.747	0.003
Autumn bark fluorescence—'O' step	35	0.427	0.020
Autumn bark fluorescence—'J' step	35	0.461	0.017
Autumn bark fluorescence—'I' step	35	0.734	0.004
Autumn bark fluorescence—1000 ms	35	0.828	0.002

Table 3. Summarized results from simple linear regression analyses comparing spring, summer and autumn leaf or bark F_v/F_m and 'O' step fluorescence values with wood decay data. n, the number of samples; P, the probability for the t-test that the coefficient of the independent variable is equal to zero; r^2, the variation in the dependent variable that can be explained by the fluorescence data. [a]The dependent variable is wood decay in all cases. [b]The statistical relationship is significant and positive. [c]The statistical relationship is significant and negative. Bold values indicate statistical significance.

Independent variable[a]	n	P	r^2
Spring leaf fluorescence—F_v/F_m	34	0.505	0.014
Spring leaf fluorescence—'O' step	34	**0.004[b]**	**0.230**
Spring bark fluorescence—F_v/F_m	35	**0.036[b]**	**0.127**
Spring bark fluorescence—'O' step	35	0.363	0.025
Summer leaf fluorescence—F_v/F_m	34	**0.025[c]**	**0.148**
Summer leaf fluorescence—'O' step	34	0.080	0.093
Summer bark fluorescence—F_v/F_m	35	**0.037[b]**	**0.125**
Summer bark fluorescence—'O' step	35	0.101	0.079
Autumn leaf fluorescence—F_v/F_m	35	0.853	0.001
Autumn leaf fluorescence—'O' step	34	0.870	0.001
Autumn bark fluorescence—F_v/F_m	35	**0.034[b]**	**0.129**
Autumn bark fluorescence—'O' step	35	0.363	0.025

measure of the quantum efficiency of PSII if all the PSII reaction centres are open (Maxwell and Johnson 2000). There was a significant and negative relationship between the leaf CF F_v/F_m ratio and wood decay in summer, suggesting that wood decay may also be associated with the quantum efficiency of PSII in leaves.

This study further emphasizes the link between the operation of photosynthesis in leaves and environmental stress. The O step in the OJIP fluorescence transient in leaves, which relates to the part of the photosynthetic light reaction where plastoquinone Q_A electron carrier molecules are in their oxidized state between PSII and PSI, is particularly affected by moisture stress in other

studies of trees (Epron et al. 1992; Percival and Sheriffs 2002). This study establishes a new link between the quantum efficiency of PSII (F_v/F_m) in leaves, wood density and wood decay. The study also establishes a new and consistent pattern of correlation between the quantum efficiency of PSII (F_v/F_m) in bark and environmental stress, wood decay and to a lesser extent wood density. Further research could examine the link between the quantum efficiency of PSII in bark in relation to other tree species, and other environmental stressors.

Christen et al. (2007) investigated the esca disease in V. vinifera (grapevines) and the relationship between the white rot decay and/or necrosis of woody tissues, the wilting of leaves and CF parameters. They used four categories of white rot decay and eight categories of necrosis, rather than percentages of decay. Necrosis and white rot were more widespread in Cabernet Sauvignon than in Merlot plants. The more decayed Cabernet Sauvignon plants showed decreased efficiency in PSII and the PI_{ABS} (performance index) value according to the CF results, compared with the Merlot population. However, the statistical relationship between CF and wood decay was only significant at a cultivar level in V. vinifera, rather than at an individual plant level, as was the case in the E. saligna from the current study.

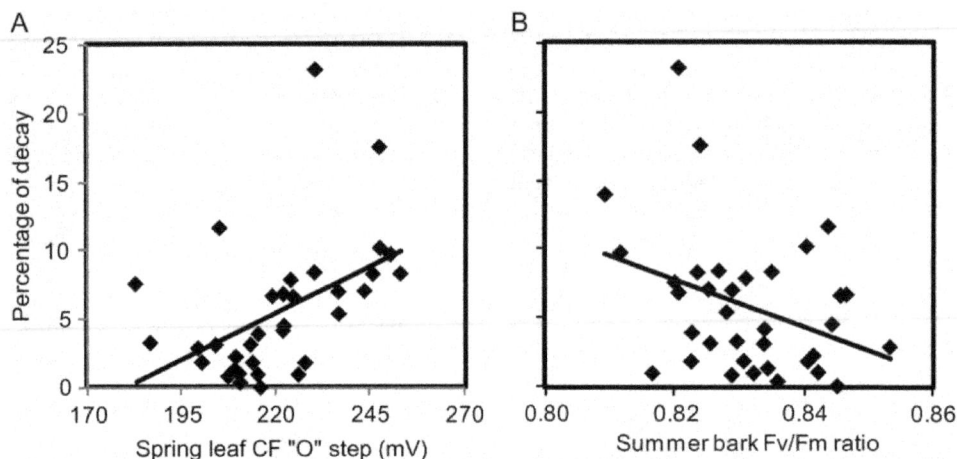

Figure 2. (A) Percentage of decay using the Resi system versus spring leaf chlorophyll fluorescence at the 'O' step in millivolts. Chlorophyll fluorescence data begin at 100 mV. Trend line = linear regression, $P = 0.004$, $r^2 = 0.230$. (B) Percentage of decay using the Resi system versus summer bark F_v/F_m. F_v/F_m ratio data begin at 0.800. Trend line = linear regression, $P = 0.021$, $r^2 = 0.148$.

Table 4. Summarized results from simple linear regression analyses comparing spring, summer and autumn leaf or bark 'J', 'I', 'P' or 1000 ms fluorescence values with wood decay data. Results from these analyses were not significant. n, the number of samples; P, the probability for the t-test that the coefficient of the independent variable is equal to zero; r^2, the variation in the dependent variable that can be explained by the fluorescence data. [a]The dependent variable is the wood decay data in all cases.

Independent variable[a]	n	P	r^2
Spring leaf fluorescence—'J' step	34	0.076	0.095
Spring leaf fluorescence—'I' step	34	0.456	0.018
Spring leaf fluorescence—'P' step	34	0.158	0.062
Spring bark fluorescence—'J' step	35	0.207	0.048
Spring bark fluorescence—'I' step	35	0.617	0.008
Spring bark fluorescence—1000 ms	35	0.901	0.001
Summer leaf fluorescence—'J' step	34	0.104	0.081
Summer leaf fluorescence—'I' step	34	0.452	0.018
Summer leaf fluorescence—'P' step	34	0.660	0.006
Summer bark fluorescence—'J' step	35	0.095	0.082
Summer bark fluorescence—'I' step	35	0.295	0.033
Summer bark fluorescence—1000 ms	35	0.430	0.019
Autumn leaf fluorescence—'J' step	34	0.969	0.000
Autumn leaf fluorescence—'I' step	34	0.350	0.027
Autumn leaf fluorescence—'P' step	34	0.319	0.031
Autumn bark fluorescence—'J' step	35	0.478	0.015
Autumn bark fluorescence—'I' step	35	0.691	0.005
Autumn bark fluorescence—1000 ms	35	0.987	0.000

The question as to why trees with decreased photosynthetic efficiency are more decayed is not easy to answer. Lorio (1986) suggested that the production of oleoresin, a

protective agent against *Dendroctonus frontalis* (southern pine beetle) in *Pinus taeda* (loblolly pine), is lower in suppressed trees when the production of wood is depressed. Hence, one reason why *E. saligna* trees with decreased photosynthetic efficiency may be more decayed may be because when the growth of wood is depressed, the synthesis of protective chemical compounds produced in the wood is also decreased. In addition, research on the progression of wood decay in trees suggests that the origin of wood decay can be in the sapwood, rather than from saprotrophic growth in non-functioning heartwood (Boddy and Rayner 1983; Parfitt et al. 2010). It is possible that during the seasonal challenge inherent in hot summers, there is pressure on photosynthesis as a result of stomatal closure and the resultant high light stress on photosystems (Faria et al. 1998). Therefore, under such conditions trees with partly decayed primary xylem suffer more because they may have to close their stomata more or more often because their water transport system is less efficient.

Induction curves did not reach their maximum in the bark fluorescence measurements; however, high F_v/F_m ratios indicate that values were close to maximum and were not significantly biased by low light intensities (Johnstone et al. 2012). Bark F_v/F_m ratios were negatively correlated with wood density in *E. saligna* in spring only, and in this instance the bark CF statistical relationships were weaker than those for leaf CF (Tables 1 and 2). The statistical relationships between bark F_v/F_m ratios and wood decay were weak but more consistent over three seasons than correlations with leaf parameters (Tables 3 and 4). Therefore, PSII in leaves may be more sensitive to the immediate effects of water flow disruption than bark photosynthesis, but the longer-term sustained effects of moisture stress, such as cavitation and the subsequent entry of wood decay pathogens, affect PSII in bark in a more consistent pattern than in the leaves. Stem photosynthesis is believed to use gaseous xylem efflux as a source of CO_2 (Pfanz 2008); therefore, if the xylem is not fully functioning it may affect the health of bark chlorenchyma, and thus PSII. Some tree species have been found to have elevated CO_2 in decayed wood tissues, while CO_2 was depressed in winter in other species (Rayner and Boddy 1988); thus the complex interactions of the metabolism of xylem, bark and wood decay organisms warrant further investigation.

The transpirational xylem stream supplies inorganic nutrients (and water) to bark chlorenchyma (Pfanz 2008), so if the xylem stream is disrupted it may affect stem photosynthesis. *Eucalyptus* sp. may be sensitive to factors that affect stem photosynthesis as stem photosynthesis may be a more important source of photosynthates for them than for other broadleaved trees, because they have a low leaf area index and are prone to defoliation by insects, diseases or drought (Tausz *et al.* 2005; Eyles *et al.* 2009). Interestingly, unlike the leaf CF measurements, only the quantum efficiency (F_v/F_m) of PSII within bark chlorenchyma was associated with wood decay; the reduction of the plastoquinone Q_A molecule between PSII and PSI (O–J step) was not affected.

The PI_{ABS} CF value has been used to successfully quantify drought stress in trees (Percival and AlBalushi 2007; Swoczyna *et al.* 2010). The PI_{ABS} value was not calculated in the current study, as it is not as widely used as the F_v/F_m value. Future studies could examine the effect of wood decay and wood density in trees on the PI_{ABS} value in relation to bark photosynthesis and other derived measures that form part of the 'JIP test', such as the apparent rates of photosynthetic electron transport and non-photochemical quenching (Lüttge *et al.* 2003).

Conclusions

The CF measurements in this study clearly support the hypothesis that there is a relationship between CF and wood structural changes. The results suggest that when photosynthesis is impaired, trees are more prone to wood decay and low wood density. Although chlorenchymes are present in bark and indeed in many woody tree tissues (Pfanz *et al.* 2002), it may not be possible to test trunk tissue in many tree species due to peridermal thickening (Aschan *et al.* 2001). However in some studies photosynthesis has been measured successfully in stems (Damesin 2003; Wittmann and Pfanz 2008a). This raises the possibility of testing larger branches on many temperate species where the bark is not as thick as on the trunk of the tree. This study showed that the reduced functioning of

PSII in bark chlorenchyma in particular is an indication that a tree may have a larger quantity of decay in the xylem tissues.

Sources of Funding

Our work was funded by the University of Melbourne, Australia.

Contributions by the Authors

D.J. performed all of the collection and analysis of data, was responsible for the experimental design and wrote most of the introduction, methods, results, discussion and conclusion. M.T. assisted in the design of the experiments and wrote sections of the introduction and discussion. G.M. offered advice on experimental design and assistance with the introduction and discussion sections. M.N. contributed to the experimental design.

Acknowledgements

We thank the arboricultural consultancy company 'Tree Logic' for the loan of the Handy PEA chlorophyll fluorescence meter.

Literature Cited

Aschan G, Wittmann C, Pfanz H. 2001. Age-dependent bark photosynthesis of aspen twigs. *Trees—Structure and Function* **15**: 431–437.

Bjorkman O, Demmig B. 1987. Photon yield of O_2 evolution and chlorophyll fluorescence characteristics at 77-K among vascular plants of diverse origins. *Planta* **170**:489–504.

Bobich EG, Barron-Gafford GA, Rascher KG, Murthy R. 2010. Effects of drought and changes in vapour pressure deficit on water relations of *Populus deltoides* growing in ambient and elevated CO_2. *Tree Physiology* **30**:866.

Boddy L, Rayner A. 1983. Origins of decay in living deciduous trees: the role of moisture content and a reappraisal of the expanded concept of tree decay. *New Phytologist* **94**:623–641.

Brodribb T, Feild T. 2000. Stem hydraulic supply is linked to leaf photosynthetic capacity: evidence from New Caledonian and Tasmanian rainforests. *Plant, Cell and Environment* **23**:1381–1388.

Brodribb TJ, Feild TS, Jordan GJ. 2007. Leaf maximum photosynthetic rate and venation are linked by hydraulics. *Plant Physiology* **144**: 1890.

Bucci S, Goldstein G, Meinzer F, Scholz F, Franco A, Bustamante M. 2004. Functional convergence in hydraulic architecture and water relations of tropical savanna trees: from leaf to whole plant. *Tree Physiology* **24**:891–899.

Christen D, Schonmann S, Jermini M, Strasser RJ, Defago G. 2007. Characterization and early detection of grapevine (*Vitis vinifera*) stress responses to esca disease by *in situ* chlorophyll fluorescence and comparison with drought stress. *Environmental and Experimental Botany* **60**:504–514.

Damesin C. 2003. Respiration and photosynthesis characteristics of current-year stems of *Fagus sylvatica*: from the seasonal pattern to an annual balance. *New Phytologist* **158**:465–475.

Epron D, Dreyer E. 1992. Effects of severe dehydration on leaf photosynthesis in *Quercus petraea* (Matt) Liebl.: photosystem II efficiency, photochemical and nonphotochemical fluorescence quenching and electrolyte leakage. *Tree Physiology* **10**:273–284.

Epron D, Dreyer E, Breda N. 1992. Photosynthesis of oak trees [*Quercus petraea* (Matt) Liebl] during drought under field conditions—diurnal course of net CO_2 assimilation and photochemical efficiency of photosystem-II. *Plant, Cell and Environment* **15**: 809–820.

Eyles A, Pinkard EA, O'Grady AP, Worledge D, Warren CR. 2009. Role of corticular photosynthesis following defoliation in *Eucalyptus globulus*. *Plant, Cell and Environment* **32**:1004–1014.

Faria T, Silvério D, Breia E, Cabral R, Abadia A, Abadia J, Pereira JS, Chaves MM. 1998. Differences in the response of carbon assimilation to summer stress (water deficits, high light and temperature) in four Mediterranean tree species. *Physiologia Plantarum* **102**:419–428.

Gotsch SG, Geiger EL, Franco AC, Goldstein G, Meinzer FC, Hoffmann WA. 2010. Allocation to leaf area and sapwood area affects water relations of co-occurring savanna and forest trees. *Oecologia* **163**:291–301.

Govindjee. 2004. Chlorophyll *a* fluorescence: a bit of basics and history. In: Papageorgiou GC, Govindjee, eds. *Chlorophyll a fluorescence a signature of photosynthesis*. Dordrecht: Springer, 1–41.

Hacke UG, Sperry JS, Pockman WT, Davis SD, McCulloh KA. 2001. Trends in wood density and structure are linked to prevention of xylem implosion by negative pressure. *Oecologia* **126**: 457–461.

Harris R, Clark J, Matheny N. 2004. *Arboriculture—integrated management of landscape trees, shrubs and vines*. New Jersey: Prentice Hall.

Holste EK, Jerke MJ, Matzner SL. 2006. Long term acclimatization of hydraulic properties, xylem conduit size, wall strength and cavitation resistance in *Phaseolus vulgaris* in response to different environmental effects. *Plant, Cell and Environment* **29**: 836–843.

Hsu BD, Leu KL. 2003. A possible origin of the middle phase of polyphasic chlorophyll fluorescence transient. *Functional Plant Biology* **30**:571–576.

Jacobs MR. 1955. *Growth Habits of the Eucalypts*. Canberra: Government Press.

Johnson G, Young A, Scholes J, Horton P. 1993. The dissipation of excess excitation energy in British plant species. *Plant, Cell and Environment* **16**:673–679.

Johnstone D, Moore G, Tausz M, Nicolas M. 2010a. The measurement of wood decay in landscape trees. *Arboriculture and Urban Forestry* **36**:121–127.

Johnstone D, Tausz M, Moore G, Nicolas M. 2010b. Quantifying wood decay in Sydney bluegum (*Eucalyptus saligna*) trees. *Arboriculture and Urban Forestry* **36**:243–252.

Johnstone D, Tausz M, Moore G, Nicolas M. 2012. Chlorophyll fluorescence of the trunk rather than leaves indicates visual vitality in *Eucalyptus saligna*. *Trees—Structure and Function* **26**:1565–1576.

Johnstone DM, Ades PK, Moore GM, Smith IW. 2007. Predicting wood decay in eucalypts using an expert system and the IML-Resistograph drill. *Arboriculture and Urban Forestry* **33**:76–82.

Kennedy R. 1958. Strength retention in wood decayed to small weight losses. *Forest Products Journal* **8**:308–314.

Krause GH, Weis E. 1984. Chlorophyll fluorescence as a tool in plant physiology. *Photosynthesis Research* **5**:139–157.

Lorio PL Jr. 1986. Growth-differentiation balance: a basis for understanding southern pine beetle-tree interactions. *Forest Ecology and Management* **14**:259–273.

Lüttge U, Berg A, Fetene M, Nauke P, Peter D, Beck E. 2003. Comparative characterization of photosynthetic performance and water relations of native trees and exotic plantation trees in an Ethiopian forest. *Trees—Structure and Function* **17**:40–50.

Manetas Y. 2004. Probing corticular photosynthesis through in vivo chlorophyll fluorescence measurements: evidence that high internal CO_2 levels suppress electron flow and increase the risk of photoinhibition. *Physiologia Plantarum* **120**:509–517.

Maxwell K, Johnson GN. 2000. Chlorophyll fluorescence—a practical guide. *Journal of Experimental Botany* **51**:659.

Metrology. 2008. Bureau of Metrology Commonwealth of Australia. http://www.bom.gov.au/ (accesssed 13 February 2008).

Moran MD. 2003. Arguments for rejecting the sequential Bonferroni in ecological studies. *Oikos* **100**:403–405.

Noguchi M, Nishimoto K, Imamura Y, Fijii Y, Okumura S, Miyauchi T. 1986. Detection of very early stages of decay in western hemlock wood using acoustic emissions. *Forest Products Journal* **36**:35–36.

O'Grady A, Cook P, Eamus D, Duguid A, Wischusen JDH, Fass T, Worldege D. 2009. Convergence of tree water use within an arid-zone woodland. *Oecologia* **160**:643–655.

Pandey K, Pitman A. 2003. FTIR studies of the changes in wood chemistry following decay by brown-rot and white-rot fungi. *International Biodeterioration & Biodegradation* **52**:151–160.

Parfitt D, Hunt J, Dockrell D, Rogers HJ, Boddy L. 2010. Do all trees carry the seeds of their own destruction? PCR reveals numerous wood decay fungi latently present in sapwood of a wide range of angiosperm trees. *Fungal Ecology* **3**:338–346.

Percival GC. 2005. *Identifying tree decline using 'fluorescence fingerprinting'*. Hobart: International Society of Arboriculture Australia Chapter.

Percival GC, AlBalushi AMS. 2007. Paclobutrazol-induced drought tolerance in containerized English and evergreen oak. *Arboriculture and Urban Forestry* **33**:397–409.

Percival GC, Sheriffs CN. 2002. Identification of drought-tolerant woody perennials using chlorophyll fluorescence. *Journal of Arboriculture* **28**:215–223.

Pfanz H. 2008. Bark photosynthesis. *Trees—Structure and Function* **22**:137–138.

Pfanz H, Aschan G, Langenfeld-Heyser R, Wittmann C, Loose M. 2002. Ecology and ecophysiology of tree stems: corticular and wood photosynthesis. *Naturwissenschaften* **89**:147–162.

Rayner ADM, Boddy L. 1988. *Fungal decomposition of wood. Its biology and ecology*. Chichester: John Wiley & Sons Ltd.

Rosner S, Klein A, Müller U, Karlsson B. 2007. Hydraulic and mechanical properties of young Norway spruce clones related to growth and wood structure. *Tree Physiology* **27**:1165.

Schwarze FWMR. 2008. *Diagnosis and prognosis of the development of wood decay in urban trees*. Rowville: Enspec.

Shigo AL. 1979. Tree decay an expanded concept. *USDA Forest Service Information Bulletin* **419**:72.

Strain BR, Johnson PL. 1963. Corticular photosynthesis and growth in *Populus tremuloides*. *Ecology* **44**:581–584.

Strasser RJ, Stirbet AD. 2001. Estimation of the energetic connectivity of PS II centres in plants using the fluorescence rise OJIP: fitting of experimental data to three different PS II models. *Mathematics and Computers in Simulation* **56**:451–462.

Strasser RJ, Tsimilli-Michael M, Srivastava A. 2004. Analysis of the chlorophyll *a* fluorescence transient. In: Papageorgiou GC, Govindjee, eds. *Chlorophyll a fluorescence a signature of photosynthesis*. Dordrecht: Springer, 321–362.

Susplugas S, Srivastava A, Strasser RJ. 2000. Changes in the photosynthetic activities during several stages of vegetative growth of *Spirodela polyrhiza*: effect of chromate. *Journal of Plant Physiology* **157**:503–512.

Swoczyna T, Kalaji HM, Pietkiewicz S, Borowski J, Zaras-Januszkiewicz E. 2010. Photosynthetic apparatus efficiency of eight tree taxa as an indicator of their tolerance to urban environments. *Dendrobiology* **63**:65–75.

Taiz L, Zeiger E. 2010. *Plant physiology*. Massachusetts: Sinauer Associates.

Tausz M, Warren CR, Adams MA. 2005. Is the bark of shining gum (*Eucalyptus nitens*) a sun or a shade leaf? *Trees—Structure and Function* **19**:415–421.

Tyree MT, Sperry JS. 1989. Vulnerability of xylem to cavitation and embolism. *Annual Review of Plant Biology* **40**:19–36.

Tyree MT, Zimmermann MH. 2002. *Xylem structure and the ascent of sap*. Berlin: Springer.

Walker JCF, Butterfield B, Langrish T, Harris J, Uprichard J. 1993. *Primary wood processing*. London: Chapman and Hall.

Wei D, Houtman CJ, Kapich AN, Hunt CG, Cullen D, Hammel KE. 2010. Laccase and its role in production of extracellular reactive oxygen species during wood decay by the brown rot basidiomycete *Postia placenta*. *Applied and Environmental Microbiology* **76**: 2091–2097.

White DA, Turner NC, Galbraith JH. 2000. Leaf water relations and stomatal behavior of four allopatric *Eucalyptus* species planted in Mediterranean southwestern Australia. *Tree Physiology* **20**: 1157–1165.

Wilcox WW. 1978. Review of literature on the effects of early stages of decay on wood strength. *Wood and Fiber Science* **9**: 252–257.

Wittmann C, Pfanz H. 2008a. Antitranspirant functions of stem periderms and their influence on corticular photosynthesis under drought stress. *Trees—Structure and Function* **22**:187–196.

Wittmann C, Pfanz H. 2008b. General trait relationships in stems: a study on the performance and interrelationships of several functional and structural parameters involved in corticular photosynthesis. *Physiologia Plantarum* **134**:636–648.

Wittmann C, Pfanz H, Loreto F, Centritto M, Pietrini F, Alessio G. 2006. Stem CO_2 release under illumination: corticular photosynthesis, photorespiration or inhibition of mitochondrial respiration? *Plant, Cell and Environment* **29**:1149–1158.

Zweifel R, Zimmermann L, Zeugin F, Newbery DM. 2006. Intra-annual radial growth and water relations of trees: implications towards a growth mechanism. *Journal of Experimental Botany* **57**: 1454–1459.

3

Influence of root-bed size on the response of tobacco to elevated CO$_2$ as mediated by cytokinins

Ulrike Schaz[1,2], Barbara Düll[1], Christiane Reinbothe[1] and Erwin Beck[1*]

[1] Department of Plant Physiology, University of Bayreuth, Universitätsstrasse 30, 95440 Bayreuth, Germany
[2] Present address: Department of Anatomy and Cell Biology, University of Ulm, Albert-Einstein-Allee 11, D-89081 Ulm, Germany

Abstract. The extent of growth stimulation of C$_3$ plants by elevated CO$_2$ is modulated by environmental factors. Under optimized environmental conditions (high light, continuous water and nutrient supply, and others), we analysed the effect of an elevated CO$_2$ atmosphere (700 ppm, EC) and the importance of root-bed size on the growth of tobacco. Biomass production was consistently higher under EC. However, the stimulation was overridden by root-bed volumes that restricted root growth. Maximum growth and biomass production were obtained at a root bed of 15 L at ambient and elevated CO$_2$ concentrations. Starting with seed germination, the plants were strictly maintained under ambient or elevated CO$_2$ until flowering. Thus, the well-known acclimation effect of growth to enhanced CO$_2$ did not occur. The relative growth rates of EC plants exceeded those of ambient-CO$_2$ plants only during the initial phases of germination and seedling establishment. This was sufficient for a persistently higher absolute biomass production by EC plants in non-limiting root-bed volumes. Both the size of the root bed and the CO$_2$ concentration influenced the quantitative cytokinin patterns, particularly in the meristematic tissues of shoots, but to a smaller extent in stems, leaves and roots. In spite of the generally low cytokinin concentrations in roots, the amounts of cytokinins moving from the root to the shoot were substantially higher in high-CO$_2$ plants. Because the cytokinin patterns of the (xylem) fluid in the stems did not match those of the shoot meristems, it is assumed that cytokinins as long-distance signals from the roots stimulate meristematic activity in the shoot apex and the sink leaves. Subsequently, the meristems are able to synthesize those phytohormones that are required for the cell cycle. Root-borne cytokinins entering the shoot appear to be one of the major control points for the integration of various environmental cues into one signal for optimized growth.

Keywords: Biomass portioning; C/N ratio; cytokinins; elevated CO$_2$; growth; root-bed volume; tobacco.

Introduction

Current CO$_2$ scenarios (IPCC 2013) predict CO$_2$ concentrations of 700 ppm at the end of the 21st century (Prentice *et al.* 2001). Such an increase will affect natural and agricultural ecosystems as 'green CO$_2$ sinks' (Drake *et al.* 1997; Krupa 2003; Lindroth 2010; Hikosaka *et al.* 2011) and thus may have far-reaching economic and political consequences. Knowledge of the reaction of plants to elevated CO$_2$ concentrations is essential for assessing the potential capacity of green CO$_2$ sinks.

In numerous laboratory and field studies, different responses of plants to elevated CO$_2$ concentrations have been observed (Makino and Mae 1999; Poorter and Navas 2003; Long *et al.* 2006; for a review see Körner

* Corresponding author's e-mail address: erwin.beck@uni-bayreuth.de

2006). Apart from the CO_2 concentration itself, these effects were modulated by a variety of factors, e.g. light intensity, temperature, and water and nutrient (especially nitrogen) supply (McConnaughay et al. 1993; Stitt and Krapp 1999; Ainsworth and Long 2005; Johnsen 2006; Reich et al. 2006; for a review see Kant et al. 2012). Differences in the reactions of plants to elevated CO_2 depend also on the experimental approach: chamber studies with plants grown in pots versus field studies using free-air concentration enrichment (Long et al. 2006; Ainsworth et al. 2008).

For C_3 plants, an initial and transitory increase of the rates of photosynthesis and plant growth was generally observed upon exposure to elevated CO_2 concentrations, a phenomenon that has been termed 'acclimation' (Long et al. 2004). Acclimation to elevated CO_2 results from a decrease of Rubisco activity, for which several explanations have been presented, in particular an internal feedback mechanism termed 'sink limitation'. If production of photosynthates in the leaves exceeds utilization by the various sink activities of a plant, e.g. root and shoot growth, export via the phloem decreases and carbohydrates accumulate as starch in the chloroplasts (Thomas and Strain 1991). This in turn inhibits chloroplastic metabolism, in particular photosynthesis (Stitt 1991; Paul and Foyer 2001). Thus, the response of a plant to elevated CO_2 should be governed by the activities of its sinks (Reekie et al. 1998). High sink activities can only be expected under optimal growth conditions, which hardly occur in nature. Under optimal light, high CO_2 concentration and sufficient water supply, macronutrients may readily become limiting. Deterioration of the plant's nutrient status (in particular nitrogen) due to an attenuation of nutrient uptake might affect growth and thus sink activities, and in turn cause 'acclimation' to elevated CO_2 (Makino et al. 1997; Nakano et al. 1997; Sicher and Bunce 1997; Curtis et al. 2000). Decreasing nutrient uptake could result from a declining nutrient concentration in the root bed (Geiger et al. 1999; Stitt and Krapp 1999) or by a dwindling uptake capacity of the roots (Thomas and Strain 1991). Both reasons often coincide in pot experiments with too small root-bed volumes that restrict root growth (Arp 1991; Berntson et al. 1993; McConnaughay et al. 1993; Rabha and Uprety 1998; Zhu et al. 2006; Yang et al. 2007, 2010). This is important as elevated CO_2 concentrations stimulate root growth, in particular fine root production (Norby et al. 2004; Stiling et al. 2013), and thus a small root-bed size should restrict root growth much more at elevated than at ambient CO_2 concentrations.

Both increased root growth in response to elevated CO_2 on the one hand and limited root growth by root-bed size on the other hand must be balanced by the plant. It must be able to recognize the 'nutrient status' of the root and of the shoot and 'translate' this information into a growth response (Berntson et al. 1993; McConnaughay et al. 1993). To adapt growth to the resource status, plants rely on specific long-distance growth signals that mediate the communication between the root and the shoot. At least with respect to nitrogen as the most important macronutrient, translation of the 'resource status' of the root into a long-distance signal—cytokinins—has been shown (Beck and Wagner 1994; Beck 1999; Yong et al. 2000; Miyawaki et al. 2004).

Cytokinins are mitogenic signals that control the cell cycle (Francis and Sorrell 2001; Hartig and Beck 2005; Harashima and Schnittger 2010; Dudits et al. 2011) and thus activity of meristems. Reduced cytokinin contents in transgenic tobacco and Arabidopsis plants resulted in slow-growing, stunted shoots with small leaves but an enhanced root system (Werner et al. 2001, 2003, 2008; Yang et al. 2003).

Of course, cytokinins are not the only long-distance and cellular signals that adapt plant growth to its resource status, but they may be the most prominent ones. Evidence is increasing that it is not the cytokinin load exported from the root to the shoot which directly controls the activity of the shoot meristems, but that the meristematic cells convert such exogenous into intracellular cytokinin signals, which however are also under the control of other signals (Bürkle et al. 2003; Motyka et al. 2003). Bishopp et al. (2011) showed that cytokinins moving from the shoot to the root via the phloem play a role in vascular patterning of the root apex.

In the present study, the growth response of tobacco plants to an elevated CO_2 concentration was analysed under controlled conditions (optimal water and nutrient supply as well as light conditions). In contrast to similar studies reported so far, an optimized nutrient solution was continuously flushed, day and night, through the root beds of pure quartz sand, keeping the nutrient concentration around the roots constant and thus avoiding effects of changing nutrient concentrations.

With that approach, especially the improved nutrient supply, the following hypotheses were examined: (i) under our conditions, the size of the root bed is the crucial factor that controls growth of the plant under ambient as well as elevated CO_2 concentrations, and therefore (ii) acclimation of plants to elevated CO_2 can be prevented by a sufficiently large root bed. (iii) We further hypothesize that cytokinins are involved in growth stimulation under conditions where acclimation is avoided. To this end, cytokinins in different plant parts and in the xylem sap were quantified comparing tobacco plants grown under ambient and elevated CO_2 and cultivated in growth limiting and non-limiting root beds.

Methods

Plant growth and experimental setup

Plant growth. *Nicotiana tabacum* cv. *Samsun* was grown in two accurately controlled walk-in climate chambers (York International, volume 11.5 m³). The light period of 14 h at 26 °C and 70 % relative humidity was followed by a 10-h dark period at 19 °C and 60 % relative humidity. The light intensity at the top of the pots was 700 μmol photons $m^{-2} s^{-1}$ (MT 400 DL/BH E-40 lamps; Iwasaki Electric Co., Tokyo, Japan). The dark phase included 30 min 'dawn' and 'dusk' with increasing or decreasing light intensities. The CO_2 concentration in the control chamber was 360 ppm (atmospheric ambient concentration, 'AC') whereas in the other chamber a permanent CO_2 concentration of 700 ppm CO_2 (elevated, 'EC') was provided (the CO_2 concentration was automatically controlled with a gas exchange analyser, BINOS; Leybold-Heraeus).

Tobacco seeds were germinated in transparent boxes (Phytotray II; Sigma) on agar in the respective climate chambers at 360 or 700 ppm CO_2. The medium for germination contained 0.5 % (w/v) inorganic Murashige–Skoog medium, pH 5.7 (Murashige and Skoog 1962), and 0.5 % (w/v) agar to which a small quantity of active carbon powder was added to avoid infestation by fungi. After sowing, the boxes were closed and covered with a net (mesh size 8 mm) to mitigate illumination strength. After 1 week, the boxes were opened once per day to allow exchange with the atmosphere of the climate chamber.

Twelve to 15 days after sowing, when the cotyledons had completely unfolded, the seedlings were transferred to sand culture. The root bed consisted of pure quartz sand (grain size 0.7–1.2 mm) that—after washing with one pot volume of deionized water—was rinsed with half of the pot volume of diluted nutrient solution (nutrient solution : water = 1 : 3). The nutrient solution contained 3 mM K_2HPO_4, 4 mM KNO_3, 4 mM $Mg(NO_3)_2$, 2 mM $MgSO_4$, 4 mM $CaSO_4$, 0.02 mM Fe-EDTA, 50 μM KCl, 20 μM H_3BO_4,

2 μM $MnSO_4$, 2 μM $ZnSO_4$, 0.5 μM $CuSO_4$ and 0.5 μM MoO_3 (pH 6.0, adjusted with H_2SO_4; Dertinger *et al.* 2003). After transfer to the sand culture, the pots were initially covered with cellophane for maintaining high air humidity.

Two days later, the cellophane was replaced by a shading net, under which the plantlets were kept for 12 days. The root bed was continuously percolated with nutrient solution with a pump (IPC-N; Ismatec Laboratoriumstechnik GmbH, Wertheim, Germany). To maintain the nutrient concentration around the roots constant, the root bed was percolated with 300 mL of nutrient solution per 1 L of root-bed volume and day. During the first 2 weeks, the plantlets were supplied with a half concentrated nutrient solution. Thereafter, undiluted nutrient solution was supplied. To provide equal growth conditions for all plants, the pots were turned around and their places in the climate chambers were exchanged twice a week.

Experimental setup. For close comparison, tobacco plants were cultivated at the same time under the two CO_2 concentrations in pots of 1, 5, 10, 15 and 20 L volume until an age of 61 days. Five replicates were grown per pot volume (triplicates in the 20-L pot experiment). A more detailed time kinetics of growth and biomass partitioning under AC and EC was made with the plants in the 15-L pots.

Biomass (dry weights) and relative growth rates (RGRs) of the plantlets were investigated at the ages of 1, 2, 3, 4, 5, 6, 7, 8, 9, 12 and 15 days after sowing in 5-fold replicates samples of 10 plants each. The older plants were examined in triplicate at 21, 28, 35, 42, 48 and 61 days after sowing. The leaves were numbered from the bottom to the top, beginning with the first leaf following the cotyledons. For comparing parameters of older plants, the parameter 'physiological age' was used, which was defined by the number of leaves and the area of the biggest leaf (see Table 1).

Xylem sap was collected at two sites of 35-day-old plants: at the base of the stem just above leaf 4 and at

Table 1. Number of leaves, leaf length, leaf width and plant age of *N. tabacum* cv. *Samsun* grown at 360 and 700 ppm CO_2, respectively, in 15-L sand culture. The leaves were counted from bottom to top. Leaf length and leaf width of the biggest leaf per plant were determined.

Number of leaves		Biggest leaf (leaf number)	Leaf length × leaf width (mm × mm)		Plant age (days)
360 ppm CO_2	700 ppm CO_2		360 ppm CO_2	700 ppm CO_2	
11	12	7	105 × 80	130 × 100	28
15	16	8	170 × 125	195 × 145	35
22	23	9	210 × 180	250 × 200	42
25	25	10	240 × 200	280 × 220	49

the petiole of a source leaf (leaf 8). The effect of transpiration (measured with a porometer) was compensated by pressurizing the root with nitrogen in a root pressure chamber (Passioura 1980). Before cutting, the pressure in the root chamber was slightly higher than necessary for compensating the transpiration to avoid embolism. Immediately after cutting, the flow rate was adjusted by reducing the pressure on the root. The first 100 μL of the sap samples were discarded. Xylem sap was subsequently collected for 1 h and stored at $-20\,°C$ until cytokinin analysis.

Measurements

Leaf area was calculated by the following equation:

$$\text{leaf area} = 0.74 \times \text{length} \times \text{width}$$

[see Supporting Information].

Plants were consistently harvested in the middle of the light period. Dry weights were determined after drying the material at $80\,°C$ to constant weight. Relative growth rate was calculated by the equation

$$R = \ln(dw_1/dw_0)/(t_1 - t_0)$$

where dw_1 is the dry weight of the plant at the time of measurement t_1 and dw_0 the dry weight at the starting time t_0.

Carbon and nitrogen contents were determined with an element analyser (CHN-O-Rapid; Elementar Analysesysteme GmbH, Hanau, Germany). The dried material was homogenized in a ball mill (Schwingmühle MM2000; Retsch GmbH & Co. KG, Haan, Germany) and aliquots of the powder were analysed. Acetanilide (71.09 % carbon and 10.36 % nitrogen) was used as the standard.

For cytokinin analysis, the procedure of Wagner and Beck (1993) modified by Hartig and Beck (2005) was followed. This method is more complex than the currently used detection by liquid chromatography-mass spectrometry. It has carefully been elaborated for several plant tissues—among others for tobacco—and for analysis of xylem sap using cross-reactivity of the diverse antibodies as internal control **[see Supporting Information]**. It separates the cytokinins (also the aromatic ones), its nucleotides and glucosides and thus provides a comprehensive view of the entire cytokinin pattern. Plant material, as shown in **Supporting Information**, was extracted with 80 % aqueous methanol in the cold and the extracts were purified by reverse-phase (RP) column chromatography (Bakerbond spe™ octadecyl (C_{18}) disposable extraction column; J. T. Baker, Deventer, Holland) with 80 % methanol. The cytokinins were fractionated by anion exchange chromatography on a DEAE Sephadex™ A-25 column (Amersham Pharmacia Biotech AB, Uppsala, Sweden) into two fractions, using 40 mM NH_4OAc buffer (pH 6.5) as the eluent: fraction (i) containing the bases (t-Z, DHZ, IP) + ribosides (t-ZR, DHZR, IPA) + glucosides (Z9G, ZOG, ZROG, DHZ9G, DHZOG, DHZROG) and fraction (ii) with the nucleotides (ZN, DHZN, IPN). To collect fraction (i), a 1.5-mL RP cartridge (Sep-Pak®; Waters Corporation, Milford, MA, USA) was coupled to the anion exchange column. The nucleotides were subsequently eluted from the anion exchanger with 6 % formic acid (v/v) into another RP cartridge from which they were collected with 80 % aqueous methanol. The individual cytokinins were separated by preparative RP-HPLC (column: 250 mm long, 4.6 mm i.d., 5 μm Hypersil® ODS; Muderer & Wochele, Berlin, Germany) and collected in 36 fractions. Elution was carried out with a gradient of acetonitrile as the non-polar phase and 0.1 % (v/v) aqueous triethylammonium acetate in bi-distilled water, pH 6, as the polar phase (for the protocol, see Wagner and Beck 1993). The different cytokinins were quantified by competitive enzyme-linked immunosorbent assay (ELISA) using three phosphatase-coupled antibodies (anti-t-ZR, anti-DHZR and anti-IPA; Wagner and Beck 1993). The bases and 9-glucosides could be determined due to their cross-reactivity with the three antibodies (Weiler and Zenk 1976; Wagner and Beck 1993). The O-glucosides, which did not show cross-reactivity, were quantified as free bases or ribosides after removal of the O-glucosyl residues with β-glucosidase (Wang et al. 1977). The nucleotides were analysed as ribosides after dephosphorylation with alkaline phosphatase followed by ELISA.

The yield of the whole procedure determined with internal standards was more than 90 % and the reliability of the ELISA was examined according to Pengelly (1986) and Crozier et al. (1986). The sources of the chemicals and the standard substances as well as the preparation of the standard solutions have been described in detail by Wagner and Beck (1993).

Statistics

The approach encompassed two variables: two CO_2 concentrations and five different volumes of the root beds. Assessment of the effects of root-bed size on plant growth required comparison under identical environmental conditions, i.e. five root-bed variants (with replications) each under ambient and elevated CO_2, respectively. Owing to spatial limitations, the entire setup encompassing the subprojects 'Age-dependence of the effect of the CO_2 concentration on the growth of tobacco plants' and 'Cytokinin patterns in tobacco plants grown under limiting/non-limiting root-bed conditions and at ambient or elevated CO_2' could be performed only once. Therefore, the question of pseudoreplication (Hurlbert

1984; Waller *et al.* 2013) arises. Since we had only two walk-in climate chambers with sophisticated control of environmental parameters, we cannot principally rule out pseudoreplication of the CO_2 effect with our sample and replication sets. However, due to the fact that for a given root-bed volume all environmental factors including the composition of the root bed were carefully controlled and maintained while only the CO_2 concentration was different, we think that the presented results are trustworthy. Furthermore, the CO_2 concentrations were addressed as 'ambient' and 'elevated' and correlations with the absolute values of these concentrations were not established, and thus the danger of erroneous results due to pseudoreplication should be low.

Statistical analyses were carried out with SigmaStat (Systat Software GmbH, Germany). The number of replicates varied between 3 and 30 (and between 2 and 5 for cytokinin determination), as shown in the figures. As the data were not Gaussian distributed and the variances were not homogeneous (as tested by Kolmogorov–Smirnov's test), statistical differences between data from AC and EC plants (weights, shoot–root and shoot–leaf ratios, the leaves' carbon and nitrogen contents and their ratios, and the concentration of cytokinins) were examined using the Wilcoxon–Mann–Whitney test. Statistically significant results were indicated in the figures by asterisks: $0 < P \leq 0.05^*$, $0.05 < P \leq 0.01^{**}$ and $0.01 < P \leq 0.001^{***}$.

Results

Influence of root-bed volume and CO_2 concentration on plant growth rates, morphology, biomass production and carbon/nitrogen status

Tobacco plants were grown under ambient or elevated CO_2 concentration in root-bed volumes ranging from 1 to 20 L. Figure 1A illustrates that the size of the plants and the leaf areas of the 61-day-old plants increased with the pot size of up to 15 L. The dry weights of plants increased linearly with the pot size at both CO_2 concentrations (Fig. 1D). At the same pot size, plants grown at elevated CO_2 (Fig. 1C) were larger and had higher dry weights, but had only one leaf more than plants grown at ambient CO_2 (Fig. 1B). The pot size of 15 L was optimal for growth at both CO_2 concentrations and a further increase to 20 L had no further effect on the size and biomass production.

Depending on their position on the stem and their physiological role as source, expanding or sink leaves, the net CO_2 uptake rates of the individual leaves differed. While the CO_2 concentration had no influence on net CO_2 uptake by older, fully expanded leaves, younger, still growing leaves showed higher assimilation rates under elevated CO_2 **[see Supporting Information]**.

In order to investigate the influence of root-bed size at ambient and elevated CO_2 on biomass partitioning, the shoot and root dry weights were determined (Fig. 1E). All plants allocated more biomass to the shoot than to the root.

Alleviating root growth limitation by increasing the volume of the root bed and promoting root growth by elevated CO_2 should result in enhanced nutrient uptake from the continuously supplied nutrients. This was examined by the carbon/nitrogen (C/N) ratio of the plants, of which three leaves (leaves 6, mature; 10, still expanding; and 20, young) were selected for determination of their carbon and nitrogen contents (Fig. 2A–C). Mainly due to the enhanced nitrogen content, the C/N ratios declined with increasing root-bed volumes, demonstrating the expected effect (Fig. 2C). Leaves of high-CO_2 plants had consistently higher C/N ratios than those of ambient-CO_2 plants. At both CO_2 conditions, the C/N ratios increased with leaf age.

Cytokinin patterns in tobacco plants grown under limiting/non-limiting root-bed conditions and at ambient or elevated CO_2

Cytokinins are known as signals from the root to the shoot, which respond to the supply of nutrients to and their concentrations in the root. However, also the meristems of the shoot are capable of cytokinin production and the fully expanded leaves, which upon transpiration receive cytokinins via the xylem sap, must be able to metabolize these phytohormones or reload it to the phloem (Beck 1999; Yong *et al.* 2000). For a better understanding of the effect of root growth restriction and the response to the CO_2 concentration on the cytokinin signal, we investigated (i) the cytokinin patterns of tobacco plants grown in a restricted or optimal root-bed volume under ambient or elevated CO_2 and (ii) examined the effect of the CO_2 concentration separately in an experiment designed for an assessment of the significance of cytokinins as a root signal.

Cytokinin patterns in organs of tobacco plants. Cytokinins of plants from 1- and 15-L pots that had been grown in the described climate chambers under 360 and 700 ppm CO_2, respectively, were analysed (Fig. 3).

The age of 42 days was chosen, as it still represents the late phase of vegetative growth. For the sake of clarity, the concentrations of the so-called active cytokinins, namely the free bases and the ribosides of each of the three examined cytokinin families, were combined (Mok and Mok 2001). Nucleotides which by dephosphorylation can readily give rise to active cytokinin species as well as the inactive glucosylated species are shown separately.

Figure 1. (A) Sixty-one-day-old tobacco plants grown in different pot volumes at 700 ppm CO_2, (B) 35-day-old tobacco plants grown at 350 ppm or (C) at 700 ppm CO_2. (D) Dry weights of entire 61-day-old tobacco plants grown at 360 ppm (open circles) or 700 ppm CO_2 (closed circles), respectively (means of $n = 5$ and $n = 3$ for the 20-L pot and standard deviation), and (E) shoot (open upright triangle, black upright triangle) and root (open inverted triangle, black inverted triangle) dry weights of the 61-day-old tobacco plants grown in the indicated pot volumes under ambient (360 ppm; open upright triangle, open inverted triangle) and elevated (700 ppm; black upright triangle, black inverted triangle) CO_2 concentrations. Mean values $\pm$ standard deviation are presented ($n = 5$ for 1-, 5-, 10- and 15-L pots, $n = 3$ for 20-L pots). Asterisks show statistical significances between plants grown under the two CO_2 concentrations.

A clear positive effect of root-bed size was observed in the concentrations of members of all three cytokinin families in the meristematic tissues of the shoot, namely the apices and the unfolding small leaves around the shoot apex ('sink leaves'). On average, concentrations were up to 6-fold higher in meristems of plants from the 15-L pots, with extremes of 28- and 35-fold higher concentrations for individual cytokinin species (compare Fig. 3A and B and Fig. 3C and D). trans-Z and its derivatives

were the most prominent cytokinin species in apices and sink leaves with concentrations up to 37 pmol g^{-1} fresh weight (t-zeatinriboside) in plants from 1-L pots (Fig. 3A) and up to 140 pmol g^{-1} fresh weight (t-zeatin) in those from 15-L pots (Fig. 3B). Concentrations of dihydrozeatin and isopentenyladenine and their derivatives were on average 4- to 5-fold lower. Cytokinin concentrations were much lower in mature leaves, stems and roots, and with one exception (t-zeatinriboside in

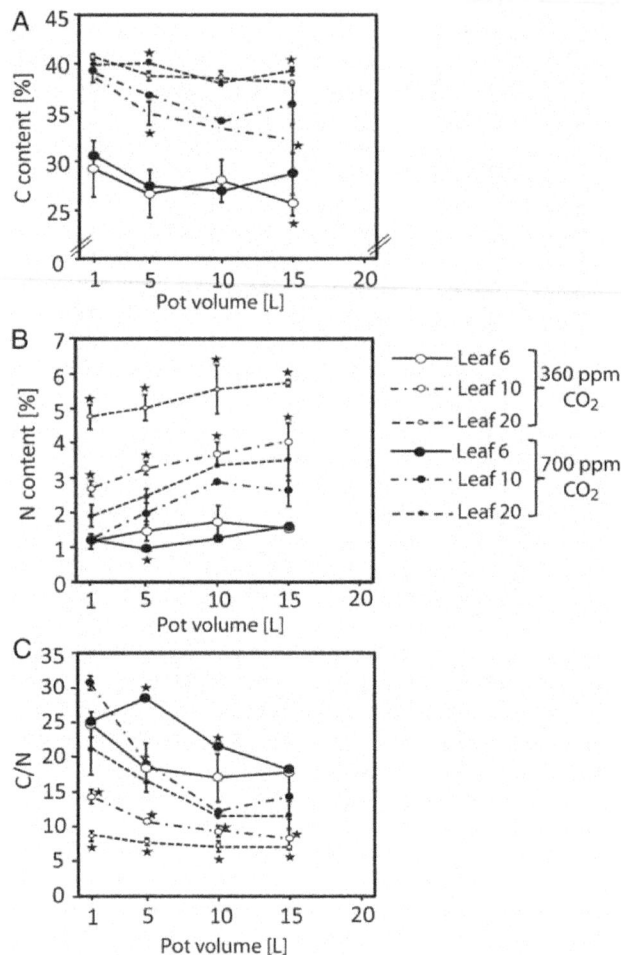

Figure 2. (A and B) Carbon and nitrogen content (%), and (C) C/N ratio of leaves 6, 10 and 20 (numbered from the bottom) of 61-day-old tobacco plants grown in different pot volumes under ambient (360 ppm) and elevated (700 ppm) CO_2 concentration (means of $n = 5$ and standard deviation). Asterisks as in Fig. 1.

the stems of plants from 15-L pots) out of more than 100 values lower than 2.5 pmol g^{-1} fresh weight. Differences between the cytokinin concentrations in mature leaves, stems and roots of plants from 1- and 15-L pots were small with a slightly higher concentration in plants from the larger pots. The concentrations in roots as the major cytokinin-producing organs were surprisingly low, whereby representatives of the isopentenyladenine group equalled those of the t-Z family. It should be mentioned, however, that the concentrations do not reflect the total amounts, which—due to the different biomasses—are less conclusive. Substantial concentrations of the inactive O- and N-glycosides of the t-zeatin and dihydrozeatin families were not found in any of the organs, irrespective of pot size or CO_2 concentration. Also, a consistent effect of CO_2 concentration on the cytokinin patterns could not be detected in spite of several significant differences in the concentrations of individual cytokinin species.

Effect of CO_2 concentration on cytokinin signal

While an increase of root-bed size had only a small positive effect on cytokinin concentrations in the roots, a strong effect was observed in the shoot apex and the sink leaves. Also, the difference between the cytokinin concentrations in ambient- and high-CO_2 plants from the same root-bed volume was negligible in the roots but high in the meristematic tissues of the shoot. Thus, the question arises as to how cytokinin translocation from the roots by the xylem stream can control growth and development of the shoot. Using the root pressure chamber (Passioura 1980), we collected xylem fluid from 35-day-old tobacco plants grown in root beds of 10 L in the described climate chambers. The technical limitation in this experiment was the size and number of root pressure chambers. Xylem fluid from the entire stem was collected for 1 h at the natural flow rates above the fourth node and from comparable plants from the petiole of a mature leaf (leaf 8). The cytokinin

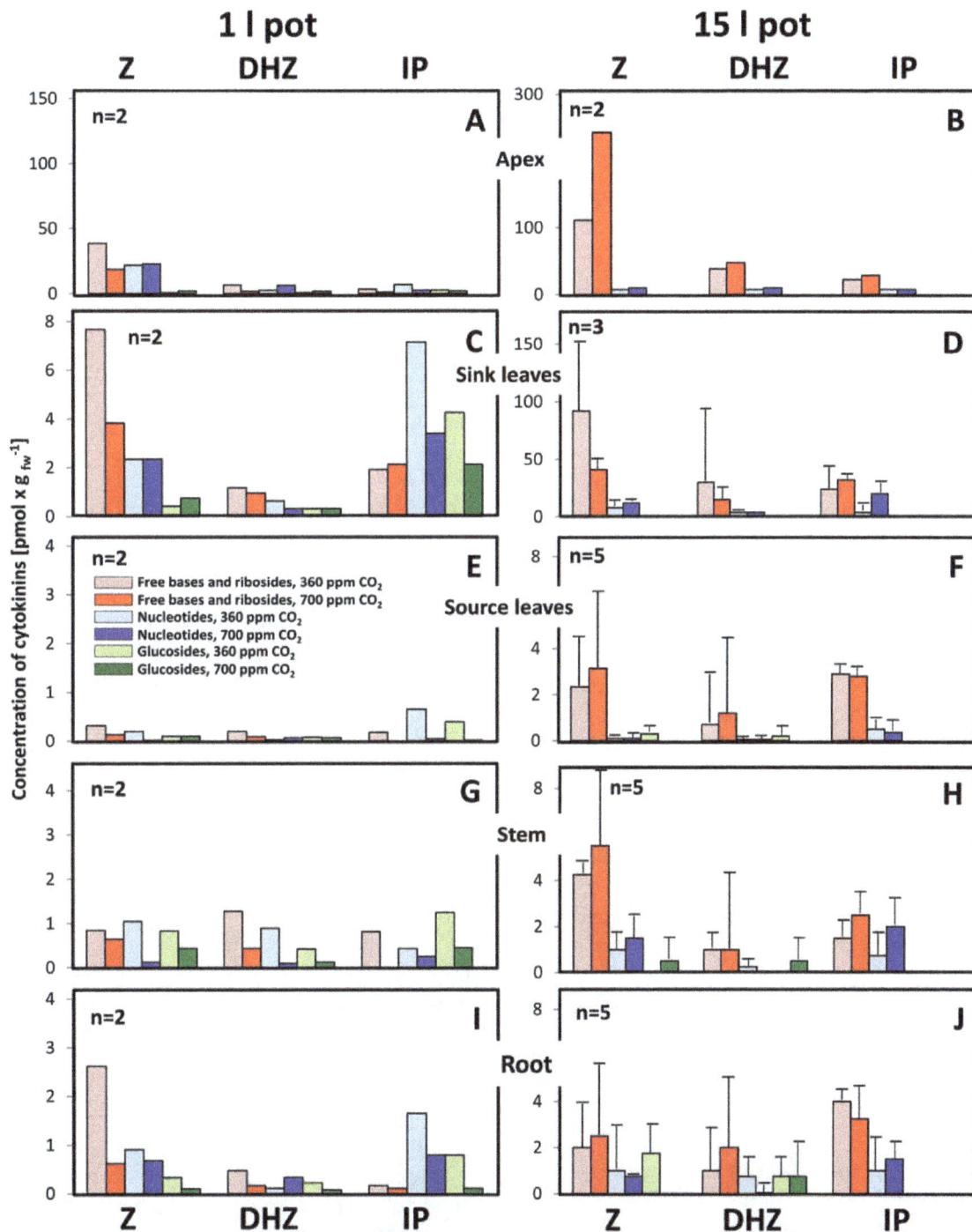

Figure 3. Concentrations of cytokinins in the apices (A, B), sink leaves (C, D), source leaves (E, F), stems (G, H) and roots (I, J) of 42-day-old tobacco plants grown in 1-L sand (A, C, E, G, I) or 15-L sand culture (B, D, F, H, J) under 360 and 700 ppm CO_2, respectively (means of at least two independent experiments $\pm$ standard deviation). The concentrations of the free bases and ribosides as well as of the glucosides of each of the measured families were combined. Z, members of the *t*-zeatin family; DHZ, members of the dihydrozeatin family; IP, members of the isopentenyladenine family.

patterns of these xylem fluids **[Supporting Information]** were compared with those of the roots and the corresponding mature leaves, respectively (Table 2). Multiplication of the cytokinin concentrations with the fresh weight

of the investigated plant organs yielded the total amounts. Likewise, the flow rates of cytokinins were calculated from the concentrations in the xylem fluid and the respective measured transpiration rates. The cytokinin concentrations

Table 2. Effect of CO_2 concentration on the concentration of cytokinins (Cks; Z, members of the t-zeatin family; DHZ, members of the dihydrozeatin family; IP, members of the isopentenyladenine family) in mature leaves and roots as well as the xylem sap of petioles from mature leaves and stems (see explanations in the text for further details on the collection of xylem sap) from 35-day-old tobacco plants grown in root beds of 10 L. Units corresponding to italic values are indicated in italics.

	Cks in mature leaf ($pmol\ g\ FW^{-1}$) (pmol leaf^{-1})			Transpiration ($nL\ cm^{-2}\ s^{-1}$) (mL h^{-1} leaf^{-1})	Cks in xylem sap (nM) (pmol leaf^{-1} h^{-1})		
	Z	DHZ	IP		Z	DHZ	IP
360	*1.11*	*1.76*	*2.31*	*10.6 ± 1.6*	*0.89*	*0.44*	*0.42*
	4.91	7.80	10.23	5.51 ± 0.92	4.73	2.42	2.31
700	*0.78*	*1.17*	*1.51*	*8.04 ± 1.61*	*0.74*	*0.44*	*0.28*
	7.58	11.37	14.68	5.84 ± 1.11	4.32	2.57	1.63
	Cks in the roots ($pmol\ g\ FW^{-1}$) (pmol per root system)			Transpiration (mL h^{-1} shoot^{-1})	Cks in xylem sap (nM) (pmol shoot^{-1} h^{-1})		
	Z	DHZ	IP		Z	DHZ	IP
360	*11.1*	*4.3*	*4.3*	*15.2 ± 0.68*	*4.5*	*3.3*	*4.3*
	160	62.1	62.5		68.2	49.9	64.6
700	*7.7*	*1.9*	*2.1*	*14.5 ± 2.37*	*6.2*	*3.8*	*4.7*
	193	47.3	53.3		90.0	55.2	67.4

in the transpiration stream and the total amounts of transported cytokinins were higher in the high-CO_2 plants than in the ambient-CO_2 plants. The cytokinin patterns in the fluid from the base of the stem did not exactly match the patterns in the root where the concentrations and amounts of the t-zeatin group were significantly higher. Comparison of the cytokinin patterns of the xylem fluids from the petioles of mature leaf no. 8 with those of the fluids collected at the bottom of the stems showed two major differences: the concentrations of the cytokinins were considerably lower and the xylem fluid in the petiole appeared to be substantially depleted of the dihydrozeatin and the isopentenyladenine cytokinins [**Supporting Information**]. But the quantitative cytokinin patterns in the xylem fluid from the petioles from ambient- and high-CO_2 plants were almost identical.

Age dependence of the effect of CO_2 concentration on the growth of tobacco plants

The results described so far revealed a growth-stimulating effect of elevated CO_2 on the growth of tobacco plants. However, the question remained about the onset and age dependence of that effect. Germination, seedling development and growth of tobacco plants were therefore followed under ambient or elevated CO_2 concentrations at a sufficiently large root bed (Fig. 4). Germination (on agar) started 3 days after sowing irrespective of the CO_2 concentration (Fig. 4A). Under both CO_2 concentrations, apparent RGRs increased from day 1 to day 4 mainly due to water uptake. Thereafter, RGR decreased to zero prior to

unfolding of the cotyledons (Fig. 4C). After development of the cotyledons (days 6 and 7 under elevated CO_2 and days 7 and 8 under ambient CO_2), the seed coat was shed decreasing RGR to values below zero, which also indicated biomass loss by respiration. After the start of photosynthesis, dry weights and RGRs increased substantially, whereby the increase at elevated CO_2 showed a head start of 1 day over that at ambient CO_2. Already 8 days after sowing, seedlings grown under 700 ppm CO_2 were significantly heavier than those grown under 360 ppm CO_2. The higher initial RGR of high-CO_2 seedlings, however, was caught up within 24 h by the ambient-CO_2 plantlets, leading to equal RGRs from day 9 to day 13 after sowing. Nevertheless, the initially higher RGR of the high-CO_2 seedlings was sufficient to produce higher biomasses than the ambient-CO_2 plants during the entire preflowering development.

After transfer of the seedlings from agar to the 15-L sand culture, a third wave of increasing and decreasing RGRs was observed for both sets of plants between day 15 and day 21 after sowing (Fig. 4C). The maximum RGR of the high-CO_2 plantlets exceeded that of the plantlets grown under ambient CO_2 until day 22 while it was slightly lower during the following 2 weeks. From day 35 to day 61, the RGRs of plants of both sets were identical but decreased slightly.

Taken together, absolute biomass production was consistently higher under elevated than under ambient CO_2. However, the major effects of elevated CO_2 on RGR were observed during unfolding of the cotyledons and

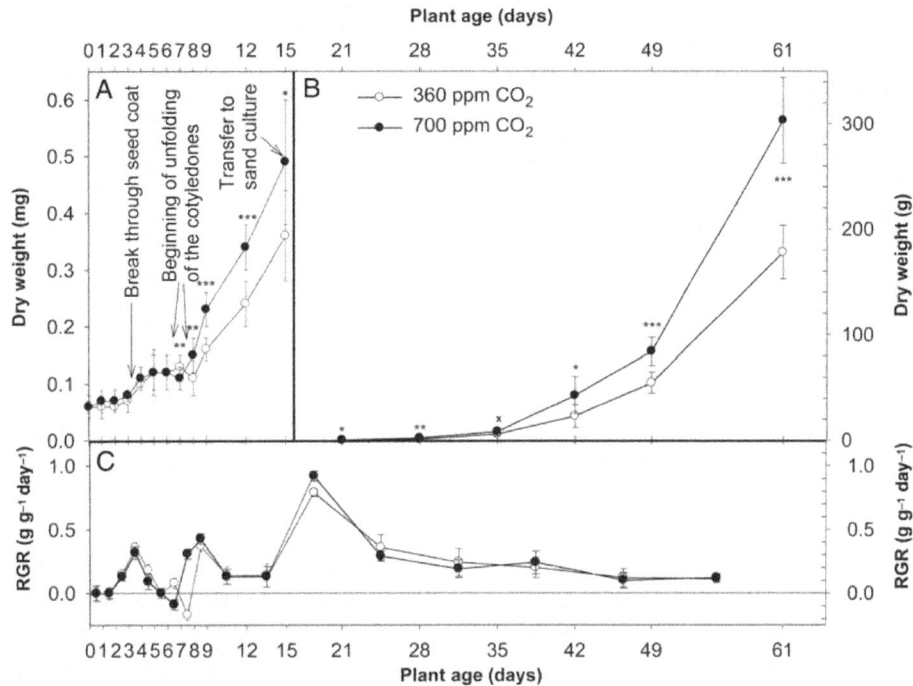

Figure 4. Increase of dry weights and the RGRs of tobacco plants after germination and growth under ambient (360 ppm) and elevated (700 ppm) CO_2 concentration. (A) Dry weights (mg) of the seedlings after sowing (mean of $n = 30$ and standard deviation), (B) dry weights (g) of the tobacco plants after transfer to 15-L sand culture (mean of $n = 5$ and standard deviation) and (C) RGR (g g^{-1} day^{-1}) calculated from the dry weights. Asterisks as in Fig. 2.

after transfer of the seedlings from the agar to the sand culture.

Biomass allocation within the shoot and morphological characterization of tobacco plants grown at ambient or elevated CO_2 concentrations

As shown before, the effect of elevated CO_2 on growth and biomass production differed with the developmental stages of the plants. In order to examine the morphological differences of plants grown at ambient or elevated CO_2 in more detail, the leaf number and leaf size of the plants (grown in 15-L pots) were examined at the ages of 28, 35, 42 and 49 days (Table 1). Plants grown under elevated CO_2 concentration developed faster until the age of 42 days since they had one leaf more than those under ambient CO_2. Elevated CO_2 also stimulated the expansion of individual leaves (Table 1). However, the final leaf number at the emergence of flower buds was not increased by elevated CO_2: at the age of 49 days, plants under both growth conditions had an equal leaf number.

Biomass allocation to the stem and leaves, expressed as the stem/:leaves ratio (St/L in Fig. 5), was initially identical in high-CO_2 and ambient-CO_2 plants (Fig. 5). From 42 days on, the high-CO_2 plants allocated more biomass to the stem than the ambient-CO_2 plants. These differences were significant and increased slightly with plant age.

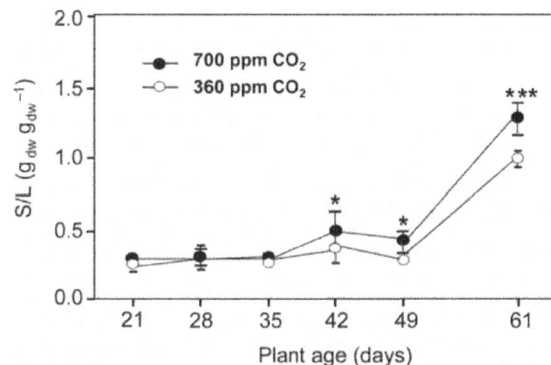

Figure 5. Stem:leaves ratios (St/L) of tobacco plants grown in 15-L sand culture under ambient (360 ppm) and elevated (700 ppm) CO_2 concentration at different plant ages (means of $n = 5$ and standard deviation). Asterisks as in Fig. 2.

Influence of CO_2 concentration on the nutrient status of tobacco leaves

The C/N ratios of two leaves each (leaves 6 and 10) were determined at the ages of 35, 42 and 61 days (Fig. 6). The contents of both carbon and nitrogen of all leaves decreased with age (Fig. 6A and B), but the carbon content of the EC leaves decreased less than that of the AC plants. An effect of the CO_2 concentration on the nitrogen content was only observed in young leaves, the nitrogen content of which was lower in the high-CO_2 plants. Since

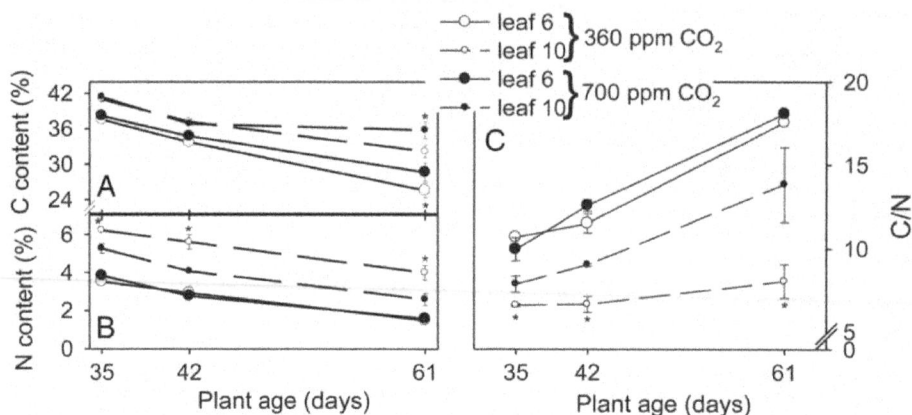

Figure 6. Age-dependent change of (A) carbon and (B) nitrogen content (%), and (C) C/N ratio of leaf 6 and leaf 10 of tobacco plants grown in 15-L sand culture under ambient (360 ppm) and elevated (700 ppm) CO_2 concentrations (means of $n = 3$ and standard deviation). Asterisks as in Fig. 1.

the age-dependent decrease of the portion of carbon was considerably larger than that of nitrogen in high-CO_2 plants, the C/N ratio decreased substantially with increasing plant age. This decrease was less pronounced in ambient-CO_2 plants (Fig. 6C).

Discussion

In the present comprehensive study, the question of pseudoreplication must be addressed (see also the subsection Statistics): rather than duplicating the entire experimental setup, replicates were confined to the (great number of) samples. This appears permissible, as the results are interpreted to reflect the plants' response to an elevated but not to a particular CO_2 concentration. In addition, growth conditions were carefully controlled during the entire experiment to minimize errors resulting from an unexpected environmental factor. Although the problem of pseudoreplication (*sensu* Hurlbert 1984) thus cannot be principally ruled out, the results are not at odds with the general view of the effects of an increasing CO_2 atmosphere on the growth of plants. They provide, however, further insight into the role of nutrient acquisition in the response of the plant to an improved carbon source and in the signals involved in the adaptation of plants to the expected future environment.

Elevated CO_2 concentration enhanced growth at all root-bed sizes by a factor of between 2.1 and 1.6, with a mean of 1.8. Root-bed volumes smaller than 15 L greatly inhibited biomass production irrespective of the CO_2 concentration. When reviewing the effects of elevated CO_2 on photosynthesis and biomass production of a variety of plants that were cultivated under controlled conditions in different root-bed volumes or in the field, Arp (1991) found a significant negative effect of a small root bed on both parameters, especially under high CO_2. Small

pots inhibit root growth and thus reduce its sink strength for biomass allocation. This effect was taken as evidence of a feedback regulation of a plant's photosynthetic rate by the demand of its sinks (Thomas and Strain 1991). On that background, root-bed restriction and elevated CO_2 should result in an additive negative effect, as was observed by Arp (1991) as well as in our study by a stronger inhibition of biomass production at elevated than under ambient CO_2 (steeper slope at elevated CO_2 in Fig. 1D). Because our tobacco plants were grown from the very beginning, i.e. seed germination, under the two CO_2 concentrations, the so-called acclimation at the transition from ambient to high CO_2 could neither be expected nor was it observed during plant development (Fig. 4).

Our experiments showed that irrespective of the CO_2 concentration, biomass allocation to the root increased with decreasing restriction of root growth by the pot (Fig. 1E). While this observation could be expected, another finding was at first glance surprising: the allocation of biomass to the root was higher under elevated than under ambient CO_2. With one exception: the smallest pot, where the high-CO_2 plants had a higher shoot-to-root biomass ratio. A lower ratio of shoot-to-root biomass in high-CO_2 plants than in ambient-CO_2 plants has been observed with a variety of plant species (for reviews, see Pritchard *et al.* 1999; Stitt and Krapp 1999; Yong *et al.* 2000; Yang *et al.* 2010; Wang *et al.* 2013) and appears to be a general phenomenon irrespective of root-bed size. Data from forested ecosystems revealed that elevated CO_2 leads to an increased fine root production (Norby *et al.* 2004; Stiling *et al.* 2013) and to deeper rooting (for a review, see Iversen 2010). Consequently, the negative impact of a limiting root-bed size on root growth should be more pronounced at elevated CO_2, which was indeed observed. But the increase of limitation under high CO_2 was not as strong as supposed because the rates of net CO_2 uptake of older, fully expanded leaves did not respond to the CO_2 concentration **[see Supporting Information]**. A reason for that might be seen in a partial closure of the stomates of the high-CO_2 leaves as a response to the elevated CO_2 concentration. This explanation was corroborated by the transpiration rates of those leaves that were significantly higher under ambient than under high CO_2 (Table 2). The stronger promotion of shoot than root growth of tobacco plants in 1-L pots at elevated CO_2 was linked to a changed allocation pattern of the assimilates. One-litre pots were completely packed with roots, especially under elevated CO_2. Root growth was more or less completely inhibited and the shoot received an excessive share of assimilates as indicated by the high C/N ratios of these plants (Fig. 2C).

A sequence of processes that take part in the regulation of growth have their origin in the uptake of macronutrients, in particular of nitrogen (McConnaughay et al. 1993; Stitt and Krapp 1999; Yang et al. 2007). Nutrient uptake is dependent on nutrient supply as well as on nutrient uptake associated with growth of the roots. Restriction of root growth results not only in an enhanced allocation of biomass to the root (decreasing S/R ratio) but also in a drop of the plant's nitrogen status (Ronchi et al. 2006; Yang et al. 2007). Both phenomena have been interpreted to reflect a decline of specific root functions in small pots (Yang et al. 2010). They were also observed in our experiments, in which the nitrogen content of the leaves increased with pot size (Fig. 2B). As expected, the nitrogen content of the youngest leaves (no. 20, as counted from the base) was much higher than that of leaves at the end of the expansion process (no. 10) or the oldest leaves at the bottom of the stem (no. 20). An age-dependent decline of the nitrogen content of the leaves is quite normal, and in tobacco the carbon content also decreases, due to the deposition of calcium oxalate as sand especially in the older leaves. The nitrogen contents of the leaves of plants grown at ambient CO_2 were significantly higher than those of the leaves of high-CO_2 plants irrespective of the position of the leaves. This finding reflects the increased demand for nutrients of the plants when grown under elevated CO_2. Under that condition, the observed C/N ratios are understandable by assuming carbon limitation of growth under ambient CO_2 and nitrogen limitation under enhanced CO_2. Similar effects have been reported for a variety of other plant species (Yin 2002; for a review, see Taub and Wang 2008), and in more detail recently for wheat (Gutiérrez et al. 2013; Wang et al. 2013). Leaves of tobacco plants grown under high CO_2 were consistently bigger than

corresponding leaves of plants growing at ambient CO_2 (Table 1), and if nutrient uptake by the roots does not match plant growth a reduced nutrient content must result. Another consequence of a reduced nutrient—especially nitrogen—availability under elevated CO_2 is the enhanced formation of axial tissue, whose structural elements contain less nitrogen compared with leaf tissue (Fig. 5). In that context, it must be underlined that in contrast to all variants of nutrient supply described in the literature, our plants grew in purified quartz sand, which was continuously (day and night) flushed with a nutrient solution that had been optimized for the growth of tobacco. Therefore, restricted uptake rather than availability was the reason for nutrient limitation by the size of the root bed. Assuming the 15-L root-bed volume as sufficient for optimal root growth and maximal nutrient acquisition, the C/N ratios of plants grown in 15-L pots would reflect a kind of standard for optimal plant growth under the various CO_2 concentrations. For the high-CO_2 variant, this could even be the maximal achievable biomass production and growth of *N. tabacum* cv. *Samsun*.

Cytokinins as potential signals in the realization of the effects of a restricted root bed and of growth under elevated CO_2

Cytokinins as one group of phytohormones are associated with regulation of the size and activity of the shoot and root apical meristem (for a review, see Skylar and Wu 2011). The effective concentrations to promote root growth are low whereas shoot growth is favoured by relatively high concentrations (Skoog and Miller 1957; Cary et al. 1995; Werner et al. 2001, 2003, 2010).

Our experiments showed a clear positive correlation between root-bed size, cytokinin concentrations in the apices and developing leaves and growth of the shoot. They also showed, irrespective of root-bed volume, much lower concentrations in the roots than in the growth regions of the shoot. While cytokinin concentrations in the roots of plants from the 15-L pots were only slightly higher than those in the roots of corresponding plants from the 1-L pots, the concentration gradients between the roots and the shoots were much lower in the small plants from the 1-L pots. With *Urtica*, a close correspondence of the daily cytokinin export from the root to the shoot and the nutrient (nitrogen) status of the roots has been reported (Beck 1999), and molecular models for translation of the nitrogen status of the root into the cytokinin signal were presented (Takei et al. 2004; Sakakibara et al. 2006). However, a recent study with transgenic tobacco and *Arabidopsis* plants with a reduced cytokinin concentration in the roots showed stimulation of root growth by the lowered cytokinin concentration, but no change of the shoot phenotype. This was interpreted as an indication that the shoot growth is not directly controlled by the cytokinin supply from the roots. Instead, shoot meristems themselves seem to produce cytokinins in sufficient amounts to maintain growth (Werner et al. 2010).

At first glance, the identity of the cytokinin patterns and flow rates of the xylem fluids of ambient- and high-CO_2 plants as measured at the base of the stems seems to corroborate the conclusions from studies with transgenic tobacco. In addition, the identity of the cytokinins in the xylem fluids of the petioles of mature leaves of plants grown under 360 and 700 ppm CO_2, respectively, underlines this notion. In that case, however, the distinct differences in the qualitative and quantitative patterns of cytokinins between the xylem fluids of the stems and petioles require further explanation because both sampling positions were at most 10 cm apart. The fact that tobacco as a member of the Solanaceae family has a bicollateral vascular system with an interior and an exterior phloem might provide an explanation. Studies with tomato (Houngbossa and Bonnemain 1985) and Nicotiana benthamiana (Cheng et al. 2000) have suggested that the meristems of the shoots are supplied by the internal phloem, while export from source leaves to the roots is mainly by the external phloem. Thus the 'xylem fluid' collected at the bottom of the stem from an adequately pressurized root system is composed of true xylem fluid and the likewise upwards flowing content of the internal phloem. In contrast, xylem fluid collected from a petiole may only contain negligible amounts of (internal) phloem sap due to the comparably few internal phloem elements. The bulk of the content of the internal phloem of the shoot obviously bypasses the petioles of mature leaves and hence the exudates from the petiole rather reflect true xylem fluid. Transport from the external to the internal phloem via ray parenchyma cells is possible but slow as observed with transport of viruses (Cheng et al. 2000). It is also known that assimilates can move from the external phloem via stem ray cells to the xylem. The cytokinin content of that sap from high-CO_2 plants was higher than from ambient-CO_2 plants, irrespective of the shares at which the xylem and the internal phloem contributed to the exudates from pressurized roots. This finding could reflect the stronger root signal hypothesized for plants grown under elevated CO_2 (Hypothesis iii; see also Yong et al. 2000). Because the patterns of the cytokinin species in the root exudates did not match those in the apical meristems or sink leaves, a direct contribution of the cytokinin root signal to the cytokinin content of the meristems is unlikely. Rather, this signal could trigger cell division associated with endogenous cytokinin production in the meristems.

Elevated CO_2 concentration already accelerated growth during germination and seedling development

Contrary to some reports in the literature (e.g. Miller et al. 1997; Ludewig and Sonnewald 2000), an accelerated ontogeny under elevated CO_2 could only be observed at the very early stage when the cotyledons opened. They unfolded 1 day earlier under high than under ambient CO_2. This advance turned out to be the basis for the persistently higher biomass production of plants grown at 700 ppm CO_2 (Fig. 4B). Also, leaf formation was faster under high CO_2, but the total leaf number per plant was equal under both CO_2 concentrations until onset of flower bud formation (Table 1).

Conclusions

Keeping several environmental factors constant, in particular nutrient concentration in the root bed, the effects of two variables, size of the root bed and atmospheric CO_2 concentration, on the growth of tobacco plants could be compared. Elevated CO_2 consistently stimulated growth but the effect of root-bed volume still overrode that effect (Hypothesis i). Limiting the production of new fine roots, spatial restriction of root growth in turn curtails nutrient uptake, as indicated by a higher C/N ratio of high-CO_2 plants. Our experiments did not comprise a transfer of plants from an atmosphere of ambient CO_2 into one with elevated CO_2; thus the classical acclimation effect was not in the scope of this work. Nevertheless, higher RGRs could be observed during germination and seedling development, which after 3 weeks declined and subsequently equalled those of ambient-CO_2 plants. In spite of this decrease in RGR, biomass production was consistently higher under elevated CO_2 and thus acclimation did not take place (Hypothesis ii). The effect of root-bed volume was strongly mirrored by the cytokinin concentrations of the meristems of the shoot, but less so of the stem, mature leaves and roots. A similar but less pronounced effect on the cytokinin concentrations was seen from the CO_2 concentration. In spite of the overall low cytokinin concentration in roots, the amounts of cytokinins moving from the root to the shoot were substantially higher in high-CO_2 plants (Hypothesis iii). Part of this root signal most probably migrates via the internal phloem of the bicollateral vascular system of the tobacco plant. The composition of the cytokinin patterns appears to be one of the major control points in which various environmental cues are integrated into one signal for optimized growth of the (tobacco) plants.

Sources of Funding

Our work was funded by the German Research Foundation with grant BE 473/23-4 to E.B.

Contributions by the Authors

U.S. and B.D. conducted research, C.R. analysed data and wrote the manuscript, and E.B. designed the experiments, is the senior author and finalized the manuscript.

Acknowledgements

The authors thank Mrs Tina Leistner and Mr Jörg Kastner for skilful technical assistance.

Supporting Information

The following **Supporting Information** is available in the online version of this article –

Figure SI 1. Regression lines resulting by plotting the products of length and width of the leaves of 42-day-old plants grown at 360 or 700 ppm CO_2 against their areas determined with an area meter (means of $n = 3$ and standard deviations).

Figure SI 2. Concentrations of cytokinins in the xylem sap taken from the shoot base (representing the location of loading from the root into the shoot) or the petioles of source leaves (representing the location of unloading into the source leaf) of 35-day-old tobacco plants.

Table SI 1. Reactivities ('cross-reactivities') of the antibodies against DHZR, ZR and 2iPA with various cytokinin standards. The intensity of the reaction in ELISA with the immediate antigen was set at 100 %.

Table SI 2. Minimum amounts of fresh material used for cytokinin determination.

Table SI 3. CO_2 net assimilation rates of a typical source (leaf no. 10) and a still growing leaf (leaf no. 15) of 42-day-old tobacco plants grown at ambient and 700 ppm CO_2, respectively, in 15-L pots. Carbon dioxide gas exchange of the leaves was measured *in situ*. Measurements were performed with a portable porometer (HCM 1000; Heinz Walz GmbH, Effeltrich, Germany), which was placed in the climate cabinets. Since leaf no. 15 was ∼25 cm above leaf no. 10, it received a higher quantum flux density. The rates were means of five plants each with SE.

Literature Cited

Ainsworth EA, Long SP. 2005. What have we learned from 15 years of free-air CO_2 enrichment (FACE)? A meta-analytic review of the responses of photosynthesis, canopy properties and plant production to rising CO_2. *New Phytologist* **165**:351–372.

Ainsworth EA, Leakey ADB, Ort DR, Long SP. 2008. FACE-ing the facts: inconsistencies and interdependence among field, chamber and modeling studies of elevated [CO_2] impacts on crop yield and food supply. *New Phytologist* **179**:5–9.

Arp WJ. 1991. Effects of source–sink relations on photosynthetic acclimation to elevated CO_2. *Plant, Cell and Environment* **14**:869–875.

Beck E. 1999. Towards an understanding of plant growth regulation: cytokinins as major signals for biomass distribution. In: Strnad M, Pec P, Beck E, eds. *Advances in regulation of plant growth and development*. Prague: Peres Publishers, 97–110.

Beck E, Wagner BM. 1994. Quantification of the daily cytokinin transport from the root to the shoot of *Urtica dioica*. *Botanica Acta* **107**:342–348.

Berntson GM, McConnaughay KDM, Bazzaz FA. 1993. Elevated CO_2 alters deployment of roots in 'small' growth containers. *Oecologia* **94**:558–564.

Bishopp A, Lehesranta S, Vatén A, Help H, El-Showk S, Scheres B, Helariutta K, Mähönen AP, Sakakibara H, Helariutta Y. 2011. Phloem-transported cytokinin regulates polar auxin transport and maintains vascular pattern in the root meristem. *Current Biology* **21**:927–932.

Bürkle L, Cedzich A, Döpke C, Stransky H, Okumoto S, Gillissen B, Kühn C, Frommer WB. 2003. Transport of cytokinins mediated by purine transporters of the PUP family expressed in phloem, hydathodes, and pollen of *Arabidopsis*. *The Plant Journal* **34**:13–26.

Cary AJ, Liu W, Howell SH. 1995. Cytokinin action is coupled to ethylene in its effects on the inhibition of root and hypocotyls elongation in *Arabidopsis thaliana* seedlings. *Plant Physiology* **107**:1075–1082.

Cheng N-H, Su C-L, Carter SA, Nelson RS. 2000. Vascular invasion routes and systemic accumulation patterns of tobacco mosaic virus in *Nicotiana benthamiana*. *The Plant Journal* **23**:349–362.

Crozier A, Sandberg G, Monteiro AM, Sundberg B. 1986. The use of immunological techniques in plant hormone analysis. In: Bopp M, ed. *Plant growth substances 1985*. Berlin, Heidelberg, New York, Tokyo: Springer, 13–21.

Curtis PS, Vogel CS, Wang XZ, Pregitzer KS, Zak DR, Lussenhop J, Kubiske M, Teeri JA. 2000. Gas exchange, leaf nitrogen, and growth efficiency of *Populus tremuloides* in a CO_2 enriched atmosphere. *Ecological Applications* **10**:3–17.

Dertinger U, Schaz U, Schulze E-D. 2003. Age-dependence of the antioxidative system in tobacco with enhanced glutathione reductase activity or senescence-induced production of cytokinin. *Physiologia Plantarum* **119**:19–29.

Drake BG, Gonzàlez-Mehler MA, Long SP. 1997. More efficient plants: a consequence of rising atmospheric CO_2? *Annual Review of Plant Physiology and Plant Molecular Biology* **48**:609–639.

Dudits D, Ábraham E, Miskolci P, Ayaydin F, Bilgin M, Horváth G. 2011. Cell-cycle control as a target for calcium, hormonal and developmental signals: the role of phosphorylation in the retinoblastoma-centred pathway. *Annals of Botany* **107**:1193–1202.

Francis D, Sorrell DA. 2001. The interface between the cell cycle and plant growth regulators: a mini review. *Plant Growth Regulation* **33**:1–12.

Geiger M, Haake V, Ludewig F, Sonnewald U, Stitt M. 1999. The nitrate and ammonium nitrate supply have a major influence on the response of photosynthesis, carbon metabolism, nitrogen metabolism and growth to elevated carbon dioxide in tobacco. *Plant, Cell and Environment* **22**:1177–1199.

Gutiérrez D, Morcuende R, Del Pozo A, Martínez-Carrasco R, Pérez P. 2013. Involvement of nitrogen and cytokinins in photosynthetic acclimation to elevated CO_2 of spring wheat. *Journal of Plant Physiology* **170**:1337–1343.

Harashima H, Schnittger A. 2010. The integration of cell division, growth and differentiation. *Current Opinion in Plant Biology* **13**: 66–74.

Hartig K, Beck E. 2005. Endogenous cytokinin oscillations control cell cycle progression of tobacco BY-2 cells. *Plant Biology* **7**:1–8.

Hikosaka K, Kinugasa T, Oikawa S, Onada Y, Hirose T. 2011. Effects of elevated CO_2 concentration on seed production in C_3 annual plants. *Journal of Experimental Botany* **62**:1523–1530.

Houngbossa S, Bonnemain J-L. 1985. Réexportation d'úne fraction du carbone importé lors de la phase de transition importation-exportation chez la feuille de tomate (*Lycopersicon esculentum*). *Comptes Rendus de l'Académie des Sciences- Paris Séries* **3300**:131–136.

Hurlbert SH. 1984. Pseudoreplication and the design of ecological field experiments. *Ecological Monographs (Ecological Society of America)* **54**:187–211.

IPCC. 2013. *Fifth Assessment Report of the Intergovernmental Panel on Climate Change*. http://www.ipcc.ch/report/ar5/wg2/.

Iversen CM. 2010. Digging deeper: fine-root responses to rising atmosphere CO_2 concentration in forested ecosystems. *New Phytologist* **186**:346–357.

Johnsen DW. 2006. Progressive N limitation in forests: review and implications for long-term responses to elevated CO_2. *Ecology* **87**:64–75.

Kant S, Seneweera S, Rodin J, Materne M, Burch D, Rothstein SJ, Spangenberg G. 2012. Improving yield potential in crops under elevated CO_2: integrating the photosynthetic and nitrogen utilization efficiencies. *Frontiers in Plant Science* **3**:162.

Körner C. 2006. Plant CO_2 responses; an issue of definition, time and resource supply. *New Phytologist* **172**:393–411.

Krupa S. 2003. Atmosphere and agriculture in the new millennium. *Environmental Pollution* **126**:293–300.

Lindroth RL. 2010. Impacts of elevated CO_2 and O_3 on forests: phytochemistry, trophic interactions, and ecosystem dynamics. *Journal of Chemical Ecology* **36**:2–21.

Long SP, Ainsworth EA, Rogers A, Ort DR. 2004. Rising atmospheric carbon dioxide: plants face the future. *Annual Review of Plant Physiology and Plant Molecular Biology* **55**:591–628.

Long SP, Ainsworth EA, Leakey ADB, Nösberger J, Ort DR. 2006. Food for thought: lower-than-expected crop yield stimulation with rising CO_2 concentrations. *Science* **312**:1918–1921.

Ludewig F, Sonnewald U. 2000. High CO_2-mediated down-regulation of photosynthesis gene transcripts is caused by accelerated leaf senescence rather than sugar accumulation. *FEBS Letters* **479**: 19–24.

Makino A, Mae T. 1999. Photosynthesis and plant growth at elevated CO_2. *Plant and Cell Physiology* **40**:999–1006.

Makino A, Harada M, Sato T, Nakano H, Mae T. 1997. Growth and N-allocation in rice plants under CO_2 enrichment. *Plant Physiology* **115**:199–203.

McConnaughay KDM, Berntson GM, Bazzaz FA. 1993. Limitations to CO_2-induced growth enhancement in pot studies. *Oecologia* **94**: 550–557.

Miller A, Tsai C-H, Hemphill D, Endress M, Rodermel S, Spalding M. 1997. Elevated CO_2 effects during leaf ontogeny. *Plant Physiology* **115**:1195–1200.

Miyawaki K, Matsumoto-Kitano M, Kakimoto T. 2004. Expression of cytokinin biosynthetic isopentenyltransferase genes in *Arabidopsis*: tissue specificity and regulation by auxin, cytokinin, and nitrate. *The Plant Journal* **37**:128–138.

Mok DWW, Mok MC. 2001. Cytokinin metabolism and action. *Annual Review of Plant Physiology* **52**:89–118.

Motyka V, Vankova R, Capkova V, Petrasek J, Kaminek M, Schmülling T. 2003. Cytokinin-induced upregulation of cytokinin oxidase activity in tobacco includes changes in enzyme glycosylation and secretion. *Physiologia Plantarum* **117**:11–21.

Murashige T, Skoog F. 1962. A revised medium for rapid growth and bioassays with tobacco tissue culture. *Journal of Plant Physiology* **15**:473–479.

Nakano H, Makino A, Mae T. 1997. The effect of elevated partial pressures of CO_2 on the relationship between photosynthetic capacity and N content in rice leaves. *Plant Physiology* **115**: 191–198.

Norby RJ, Ledford J, Reilly CD, Miller NE, O'Neill G. 2004. Fine-root production response of a decidious forest to atmospheric CO_2 enrichment. *Proceedings of the National Academy of Sciences of the USA* **101**:9689–9693.

Passioura JB. 1980. The transport of water from soil to shoot in wheat seedlings. *Journal of Experimental Botany* **31**:333–345.

Paul M, Foyer C. 2001. Sink regulation of photosynthesis. *Journal of Experimental Botany* **52**:1383–1400.

Pengelly WL. 1986. Validation of immunoassays. In: Bopp M, ed. *Plant growth substances 1985*. Berlin: Springer, 35–43.

Poorter H, Navas M-L. 2003. Plant growth and competition at elevated CO_2: on winners, losers and functional groups. *New Phytologist* **157**:175–198.

Prentice IC, Farquhar GD, Fasham MJR, Goulden ML, Heimann M, Jaramillo VJ, Kheshgi HS, LeQuere C, Scholes RJ, Wallace DWR. 2001. The carbon cycle and atmospheric carbon dioxide. In: Houghton JT, Ding Y, Griggs DJ, Noguer M, Van der Linder PJ, Dai X, Maskell K, Johnson CA, eds. *Climate change 2001: the scientific basis. Contributions of Working Group I to the Third Assessment Report of the Intergovernmental Panel on Climate Change*. Cambridge, UK: Cambridge University Press, 183–238.

Pritchard SG, Rogers HH, Prior SA, Peterson CM. 1999. Elevated CO_2 and plant structure: a review. *Global Change Biology* **5**:807–837.

Rabha BK, Uprety DC. 1998. Effects of elevated CO_2 and moisture stress on *Brassica juncea*. *Photosynthetica* **34**:597–602.

Reekie ED, MacDougall G, Wong I, Hicklenton PR. 1998. Effect of sink size on growth response to elevated atmospheric CO_2 within the genus *Brassica*. *Canadian Journal of Botany* **76**:826–835.

Reich PB, Hobbie SE, Lee T, Ellsworth DS, West JB, Tilman D, Knops JMH, Naeem S, Trost J. 2006. Nitrogen limitation constrains sustainability of ecosystem responses to CO_2. *Nature* **440**:922–925.

Ronchi CP, DaMatta FM, Batista KD, Moraes G, Loureiro ME, Ducatti C. 2006. Growth and photosynthetic down-regulation in *Coffea arabica* in response to restricted root volume. *Functional Plant Biology* **33**:1013–1023.

Sakakibara H, Takei K, Hirose N. 2006. Interactions between nitrogen and cytokinin in the regulation of metabolism and development. *Trends in Plant Science* **11**:440–448.

Sicher RC, Bunce JA. 1997. Relationship of photosynthetic acclimation to changes of Rubisco activity in field-grown winter wheat and barley during growth in elevated carbon dioxide. *Photosynthesis Research* **52**:27–38.

Skoog F, Miller CO. 1957. Chemical regulation of growth and organ formation in plant tissue cultures *in vitro*. *Symposia of the Society for Experimental Biology* **11**:118–131.

Skylar A, Wu X. 2011. Regulation of the meristem size by cytokinin signaling. *Journal of Integrative Plant Biology* **53**:446–454.

Stiling P, Moon D, Rossi A, Forkner R, Hungate BA, Day FP, Schroeder RE, Drake B. 2013. Direct and legacy effects of long-term elevated CO_2 on fine root growth and plant insect interactions. *New Phytologist* **200**:788–795.

Stitt M. 1991. Rising CO_2 levels and their potential significance for carbon flow in photosynthetic cells. *Plant, Cell and Environment* **14**:741–762.

Stitt M, Krapp A. 1999. The interaction between elevated carbon dioxide and nitrogen nutrition: the physiological and molecular background. *Plant, Cell and Environment* **22**:583–621.

Takei K, Ueda N, Aoki K, Kuromori T, Hirayama T, Shinozaki K, Yamaya T, Sakakibara H. 2004. AtIPT3 is a key determinant of nitrate-dependent cytokinin biosynthesis in *Arabidopsis*. *Plant and Cell Physiology* **45**:1053–1062.

Taub DR, Wang F. 2008. Why are nitrogen concentrations in plant tissues lower under elevated CO_2? A critical examination of the hypotheses. *Journal of Integrative Plant Biology* **50**:1365–1374.

Thomas RB, Strain BR. 1991. Root restriction as a factor in photosynthetic acclimation of cotton seedlings grown in elevated carbon dioxide. *Plant Physiology* **96**:627–634.

Wagner BM, Beck E. 1993. Cytokinins in the perennial herb *Urtica dioica* as influenced by its nitrogen status. *Planta* **190**:511–518.

Waller BM, Warmelink L, Liebal K, Micheletta J, Slocombe KE. 2013. Pseudoreplication: a widespread problem in primate communication research. *Animal Behaviour* **86**:483–486.

Wang L, Feng Z, Schjoerring JK. 2013. Effects of elevated atmospheric CO_2 on physiology and yield of wheat (*Triticum aestivum*): a meta-analytic test of current hypotheses. *Agriculture, Ecosystems and Environment* **178**:57–63.

Wang TL, Thompson AG, Horgan R. 1977. A cytokinin glucoside from the leaves of *Phaseolus vulgaris*. *Planta* **135**:285–288.

Weiler EW, Zenk MH. 1976. Radioimmunoassay for the determination of digoxin and related compounds in *Digitalis lanata*. *Phytochemistry* **15**:1537–1545.

Werner T, Motyka V, Strnad M, Schmülling T. 2001. Regulation of plant growth by cytokinin. *Proceedings of the National Academy of Sciences of the USA* **28**:10487–10492.

Werner T, Motyka V, Laucou V, Smets R, Van Onckelen H, Schmülling T. 2003. Cytokinin-deficient transgenic *Arabidopsis* plants show multiple developmental alterations indicating opposite functions of cytokinins in the regulation of shoot and root meristem activity. *The Plant Cell* **15**:2532–2550.

Werner T, Holst K, Pörs Y, Giuvarc'h A, Mustroph A, Chriqui D, Grimm B, Schmülling T. 2008. Cytokinin deficiency causes distinct changes of sink and root source parameters in tobacco shoots and roots. *Journal of Experimental Botany* **59**:2659–2672.

Werner T, Nehnevajova E, Köllmer I, Novák O, Strnad M, Krämer U, Schmülling T. 2010. Root-specific reduction of cytokinin causes enhanced root growth, drought tolerance, and leaf mineral enrichment in *Arabidopsis* and tobacco. *The Plant Cell* **22**:3905–3920.

Yang S, Yu H, Xu Y, Goh CJ. 2003. Investigation of cytokinin-deficient phenotypes in *Arabidopsis* by ectopic expression of orchid *DSCKX1*. *FEBS Letters* **555**:291–296.

Yang TZ, Zhu LN, Wang SP, Gu WJ, Huang DF, Xu WP, Jiang AL, Li SC. 2007. Nitrate uptake kinetics of grapevine under root restriction. *Scientia Horticulturae* **111**:358–364.

Yang Z, Hammer G, van Oosterom E, Rochais D, Deifel K. 2010. Effects of the pot size on growth of maize and sorghum plants. *Proceedings of the First Australian Summer Grains Conference*, Gold Coast, Australia, 21–24 June 2010.

Yin X. 2002. Responses of leaf nitrogen and specific leaf area to atmospheric CO_2 enrichment: a retrospective synthesis across 62 species. *Global Change Biology* **8**:631–642.

Yong JWH, Wong SC, Letham DS, Hocart CH, Farquhar GD. 2000. Effects of elevated [CO_2] and nitrogen nutrition on cytokinins in the xylem sap and leaves of cotton. *Plant Physiology* **124**:767–779.

Zhu LN, Wang SP, Yang TY, Zhang CX, Xu WP. 2006. Vine growth and nitrogen metabolism of 'Fujiminori' grapevines in response to root restriction. *Scientia Horticulturae* **107**:143–149.

4

Growing up or growing out? How soil pH and light affect seedling growth of a relictual rainforest tree

Catherine A. Offord[1]*, Patricia F. Meagher[1] and Heidi C. Zimmer[2]

[1] The Royal Botanic Gardens and Domain Trust, The Australian Botanic Garden, Mount Annan, NSW 2567, Australia
[2] Department of Forest and Ecosystem Science, University of Melbourne, Richmond, VIC 3121, Australia

Associate Editor: James F. Cahill

Abstract. Seedling growth rates can have important long-term effects on forest dynamics. Environmental variables such as light availability and edaphic factors can exert a strong influence on seedling growth. In the wild, seedlings of Wollemi pine (*Wollemia nobilis*) grow on very acid soils (pH ~4.3) in deeply shaded sites (~3 % full sunlight). To examine the relative influences of these two factors on the growth of young *W. nobilis* seedlings, we conducted a glasshouse experiment growing seedlings at two soil pH levels (4.5 and 6.5) under three light levels: low (5 % full sun), medium (15 %) and high (50 %). Stem length and stem diameter were measured, stem number and branch number were counted, and chlorophyll and carotenoid content were analysed. In general, increased plant growth was associated with increased light, and with low pH irrespective of light treatment, and pigment content was higher at low pH. Maximum stem growth occurred in plants grown in the low pH/high light treatment combination. However, stem number was highest in low pH/medium light. We hypothesize that these differences in stem development of *W. nobilis* among light treatments were due to this species' different recruitment strategies in response to light: greater stem growth at high light and greater investment in multiple stem production at low light. The low light levels in the *W. nobilis* habitat may be a key limitation on stem growth and hence *W. nobilis* recruitment from seedling to adult. Light and soil pH are two key factors in the growth of this threatened relictual rainforest species.

Keywords: Araucariaceae; conifer; conservation; light; rainforest; relictual species; soil pH; threatened species; Wollemi pine.

Introduction

Light is a fundamental factor limiting the growth and survival of seedlings in closed forests (Chazdon *et al.* 1996; Whitmore 1996; Nicotra *et al.* 1999). Recent work has suggested that soil pH may be of equal or greater importance than light (Holste *et al.* 2011). Species-specific growth and survival responses to resources, particularly soil and light, play a key role in determining forest composition (Grubb 1977; Davies *et al.* 2005; Holste *et al.* 2011). Central to

understanding tree species persistence is an understanding of their strategy for recruitment from understorey to canopy. Tree species range from those that are slow growing and shade tolerant, to fast-growing pioneers that typically dominate the post-disturbance environment (Denslow 1980; Whitmore 1989). It is competition for resources, governed by differences in survival and growth rate, that results in forest changes through time (i.e. stand development).

The influence of soil in defining species and plant community distributions is well known (Beadle 1954; Russo

* Corresponding author's e-mail address: cathy.offord@rbgsyd.nsw.gov.au

et al. 2008). Soil pH governs many plant–soil chemical relations, particularly the availability of micronutrients and toxic ions, due to its influence on solubility. At low pH, the availability of essential micronutrients Fe, Mn, Cu and Zn is increased, as is the availability of potentially toxic Al and Mn (Atwell *et al.* 2003). Alternatively, the availability of P and Mo decreases. High-pH soils, however, are high in Cr, Co, Ni, Fe and Mg, and deficient in N, P, K and Ca (Atwell *et al.* 2003). Plants with optimal growth and survival below and above pH 5–7 are known as acidophiles and calciphiles, respectively (Ehrenfeld *et al.* 2005); these plants employ strategies to avoid or tolerate otherwise suboptimal conditions. For example, acidophiles avoid the stresses of nutrient deficiency by conservation of minerals via slow growth, high storage in seeds, high root surface area and relationships with rhizosphere microorganisms (Marschner 1991). Soil pH can also influence plant community dynamics: low pH can prevent invasion of exotic species into native acid-tolerant plant communities (Thompson *et al.* 2001), while some species are apparently restricted to high-pH environments (e.g. ultramafics in New Caledonia; Jaffré 1992).

Wollemi pine (*Wollemia nobilis*, Araucariaceae) is a rare conifer with a highly restricted distribution. Araucariaceae is a family with origins in the early Triassic (Kershaw and Wagstaff 2001), and the earliest fossil record of *Wollemia* is from 91 million years ago (*Dilwynites* pollen; Macphail *et al.* 1995). *Wollemia nobilis* has unique architecture with only first-order plagiotropic branches and is capable of producing multiple stems without injury (Hill 1997). The plagiotropic branches can grow up to 150 cm long and are shed whole, forming a dense litter layer. Branches are short-lived (<15 years) and have either adult or juvenile leaf types above and below the rainforest canopy, respectively. Wollemi pine grows in several small stands in the Wollemi National Park, part of the Greater Blue Mountains World Heritage area, New South Wales, Australia (Jones *et al.* 1995; NSW Department of Environment and Conservation 2006). Wollemi National Park has predominantly sandstone geology (Jones *et al.* 1995), which is typically associated with acid soils (Binkley and Fisher 2000). *Wollemia nobilis* exists at the base, and on low terraces, of deep narrow canyons within a warm temperate rainforest community with co-dominant *Ceratopetalum apetalum* (Benson and Allen 2007). Recruitment from seed to adult is rare, even though seed production is relatively high (NSW Department of Environment and Conservation 2006). Fewer than 100 adult trees have been discovered, and some 300 seedlings have been observed, the majority under 500-mm stem length. Height growth of *W. nobilis* seedlings is very slow in the wild (5–20 mm per year; Zimmer *et al.* 2014). The presence of seedlings indicates that *W. nobilis* is capable of

producing viable seed (Offord *et al.* 1999). Yet lack of intermediate-sized trees (i.e. 5–20 m) and slow growth indicates that there are other factors limiting their establishment (Whitmore and Page 1980).

The overarching aim of this study was to explore the relative importance of light and soil pH in determining *W. nobilis* seedling success. To do this we investigated *W. nobilis* growth in response to the natural light and soil pH conditions where *W. nobilis* grows in the wild, and then to a wider range of light and pH conditions, in a glasshouse experiment.

Methods

Field observations

A preliminary study of the soil characteristics associated with the areas in which *W. nobilis* grows indicated that the pH is very acidic (~pH 4) (NSW Department of Environment and Conservation 2006). Soil samples (10 × 500 g) were collected adjacent to *W. nobilis* seedlings growing in the wild. Soil pH was measured in water 1:2.

Photosynthetic photon flux density (PPFD) was measured around 42 *W. nobilis* seedlings with a hand-held Licor quantum light meter around midday on two typical sunny days in summer. These were compared with full-sun light meter readings in nearby areas.

Glasshouse treatments

Wollemia nobilis seeds collected from multiple trees in the wild were germinated in Petri dishes in growth cabinets set at 24 °C (Offord and Meagher 2001). Freshly germinated seedlings were grown in 75-mm (0.44-L) pots containing steam-pasteurized peat and sand (1:2 v/v) at pH 4.5 or adjusted to pH 6.5 with lime and dolomite (1:1 w/w). When seedlings were 4–5 months of age they were potted in 140-mm (1.5-L) pots containing the same potting mix and pH treatments with the addition of the fertilizer Nutricote Total N_{13} (13:5.7:10.8 N:P:K) 270 day type added to the mix at a rate of 3 g L^{-1} prior to steam pasteurization (time zero). Pasteurization was undertaken such that it did not affect fertilizer release rates (temperatures did not exceed 60 °C for >30 min). Plants were fertilized and re-potted after the 12-month assessment. Plants were watered daily or as needed.

In a glasshouse at the Australian Botanic Garden, Mount Annan (ABGMA, 34°05′S, 150°47′E), plants were randomly assigned to areas with 5, 15 or 50 % full sunlight (low, medium or high relative light). This was achieved by using different grades of shade cloth with the same wavelength transmission properties, in addition to light attenuated by the glasshouse. The light at plant level relative to ambient was determined using a pyranometer (Environdata P/L). The temperatures within the glasshouse were controlled to a mean of 24 °C (standard error [SE] = 3 °C) during the day and 16 °C (SE = 3 °C) at night. There were 20 replicate potted plants of each light and pH treatment combination (in a full factorial design). Plants were randomly positioned, and randomly repositioned after each measurement, within each light treatment.

Plant growth characteristics were recorded at 6, 12 and 24 months after time zero. This included number and length of orthotropic (vertical) stems and plagiotropic (horizontal) branches, diameter at the base of the plant and general health characteristics. Destructive methods could not be used on the seedlings because of their rarity. Stem length was calculated as total length of all stems. Measurements were taken over 24 months because in a similar study of *Araucaria angustifolia*, measurements were made after only 4 months of growth, at which time there were no differences in the stem length or chlorophyll variables measured (Duarte and Dillenburg 2000). Average daily accumulation of incident photosynthetically active radiation (PAR) by month was calculated using 5 years of solar radiation data for ABGMA collected using a pyranometer (Fig. 1). Solar radiation (energy) was converted to PAR (quanta) using the correction factor $c = 2.3$ (Monteith and Unsworth 1990).

Pigment extraction and concentration

At 24 months four leaf samples taken from the new leaves of three plants per treatment combination were analysed for chlorophyll (chlorophyll *a* and chlorophyll *b*, protochlorophyll) and carotenoids (Chen *et al.* 1998).

Statistical analysis

Two-way analysis of variance (ANOVA) was conducted on all variables (SYSTAT; SPSS Inc.). Where there were no significant interactions, *post hoc* tests, specifically least significant difference (LSD), were undertaken. Where significant interactions were found, one-way ANOVAs and/or *t*-tests were conducted on the variables, and these results, and associated LSDs, were reported. Data for stem and branch number were log-transformed to normalize the data for analysis. Untransformed data are presented in the tables.

Results

Field observations

Mean soil pH in the field was 4.32 (SE = 0.12, $n = 10$) in water. Light levels at *W. nobilis* seedlings in the field were highly variable and were as low as 1 % of full sunlight even at the brightest time of the day. Light penetration into the canyon was restricted by its depth, the angle of the sun and the dense canopy of other tree species growing within it. On the canyon floor, PPFD at approximately midday on a sunny day in February averaged 60 μmol m^{-2} s^{-1} (SE = 9, $n = 43$), which represents around 3 % of full sunlight measured in adjacent open areas (mean = 2000 μmol m^{-2} s^{-1}, SE = 163, $n = 4$).

Growth measurements

For clarity, only the 24-month data are presented for growth and pigment variables. Significant interactions between pH and light were found for stem length, stem diameter and number of stems ($P < 0.05$; Table 1);

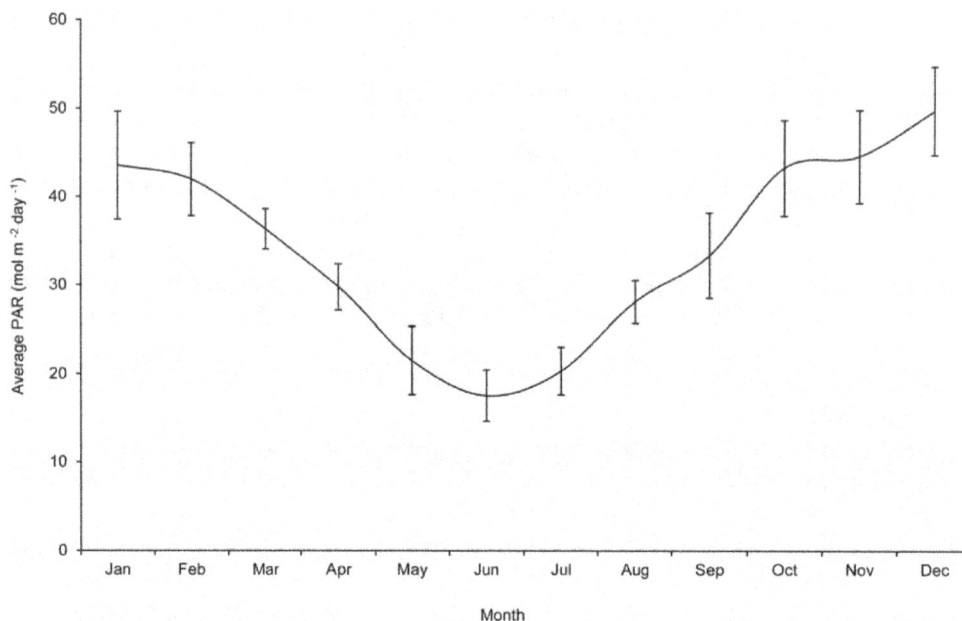

Figure 1. Average daily PAR (mol m^{-2}) received at the Australian Botanic Garden Mount Annan (5-year average 1998–2002) (error bars = SE).

Table 1. Analysis of variance of measured *W. nobilis* seedling characters (growth and leaf pigment) according to treatment variables (light and pH). Where there were significant interactions between the treatments, simple main effects are presented (lower table). **$P < 0.01$; *$P < 0.05$; NS, not significant; #$P = 0.050$.

F-test significance	Stem length (mm)	Diameter at base (mm)	Number of stems	Number of branches	Chlorophyll a	Chlorophyll b	Chlorophyll a + b	Chlorophyll a/b	Carotenoids/ chlorophyll	Protochlorophyll
Light	**	**	*	**	**	NS	*	**	NS	NS
pH	**	**	NS	**	**	**	**	**	*	**
Light × pH	*	**	*	NS	NS	NS	NS	NS	NS	NS
Low light	**	**	NS							
Med light	**	**	NS#							
High light	**	**	NS							
pH 4.5	**	**	**							
pH 6.5	**	**	NS							

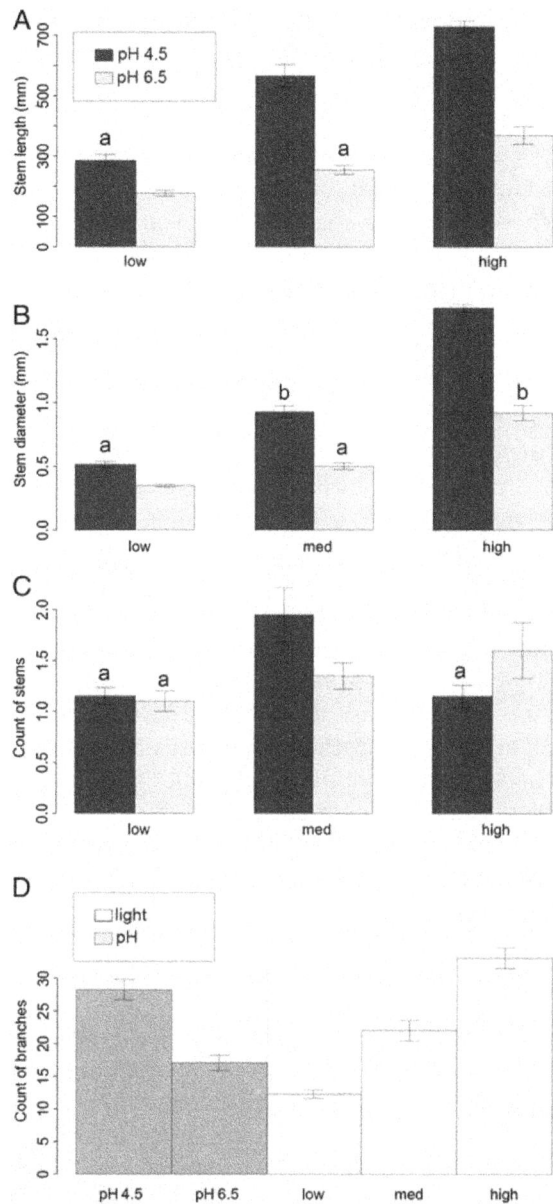

Figure 2. Growth characteristics of *W. nobilis* seedlings grown at low, medium and high light in potting mix at pH 4.5 or 6.5 for 24 months. (A) Stem length, (B) stem diameter and (C) stem count means taken across combined light and pH treatments ($\pm$ SE). (D) Mean branch count is presented separately for (i) light treatments and (ii) pH treatments ($\pm$ SE). Within each response variable, means sharing the same letter are not significantly different by LSD$_{5\%}$. Note: *y*-axis varies.

therefore, simple main effects were investigated for these variables (Table 1, Fig. 2). At the higher pH, the light level made little difference to the stem length of the plant compared with the large difference made by increased light in low-pH treatments, particularly at medium light. The significant interaction of light and pH for stem number can be accounted for by the higher number of stems found in the medium-light/low-pH treatment. Branch number was significantly higher at low pH. The increases in branch number with increasing light were approximately proportional to the increases in stem length.

Leaf pigment content

No significant interactions between light and pH were detected for the pigment concentrations ($P > 0.05$; Table 1). The concentrations of chlorophyll a, chlorophyll a + b and chlorophyll a/b were significantly associated with light ($P < 0.05$; Table 1). Chlorophyll a and chlorophyll b were higher in low light. The highest chlorophyll a + b concentration was in plants under the higher light treatments. The chlorophyll a/b ratio was also significantly higher at low and medium light.

The concentrations of chlorophyll a and chlorophyll a + b in the leaves of W. nobilis were also highly significantly associated with pH ($P < 0.01$): there were higher concentrations at the lower pH (Table 1, Fig. 3). The chlorophyll a/b ratio was significantly higher in the high-pH treatments, compared with low-pH treatments. The protochlorophyll levels were significantly higher in the low-pH treatments, compared with high-pH treatments. In line with this, chlorophyll a and chlorophyll b were significantly higher at low pH. Protochlorophyll tended to be higher at low- and medium-light treatments, but this variation was not significant. The carotenoid-to-chlorophyll ratio was higher in the high-pH and low-light treatments, which was reflected in the generally chlorotic appearance of these plants.

Discussion

Wollemia nobilis is an acidophile. Growth of seedlings was maximal at a soil pH considered suboptimal for many species (pH 4.5; Handreck and Black 2002). *Wollemia*

nobilis growth also increased with increased light, but this response was moderated by pH—higher pH resulted in growth suppression. Moreover, the chlorophyll content (chlorophyll a + b) of W. nobilis leaves was higher in the low soil pH treatments, indicating increased growth and, perhaps, potential for growth.

Acid soils are a feature of natural stands of Araucariaceae, particularly *Agathis australis* (Mirams 1957; Bieleski 1959; Ecroyd 1982; Weaver 1988; Wyse 2012), and Araucariaceae plantations (Curlevski et al. 2010). Changes in soil acidity under different plants can be due to differences among species in nutrient accumulation (Alban 1982; Finzi et al. 1998), nitrogen fixation (van Miegroet and Cole 1984), litter chemical composition (Ovington 1953; Alban 1982; Finzi et al. 1998) and the stimulation of mineral weathering (Tice et al. 1996). The source of acid soils associated with Araucariaceae is hypothesized to be litterfall (Bieleski 1959; Silvester and Orchard 1999). Acid soils (pH < 7) are also common in eastern New South Wales, but soils of pH ≤ 4.5 are much less common (Helyar et al. 1990). Sandstone-derived soils are also typically acidic (Binkley and Fisher 2000). Because we did not test the soils beyond the W. nobilis stand, we cannot say what effect (if any) W. nobilis has on the soil. Instead, our results suggest that low pH, and associated changes in soil nutrients, is likely to be a major factor in enhancing the growth of established W. nobilis seedlings. Indeed, low pH may also indirectly benefit W. nobilis by reducing competition from other species not adapted to acid soils.

Wollemia nobilis seedling growth increased with increasing light availability. Seedling growth was suppressed by low light (5 %), resulting in stem lengths half that attained by the high-light (50 %) treatment at 24 months. Previous studies have demonstrated that

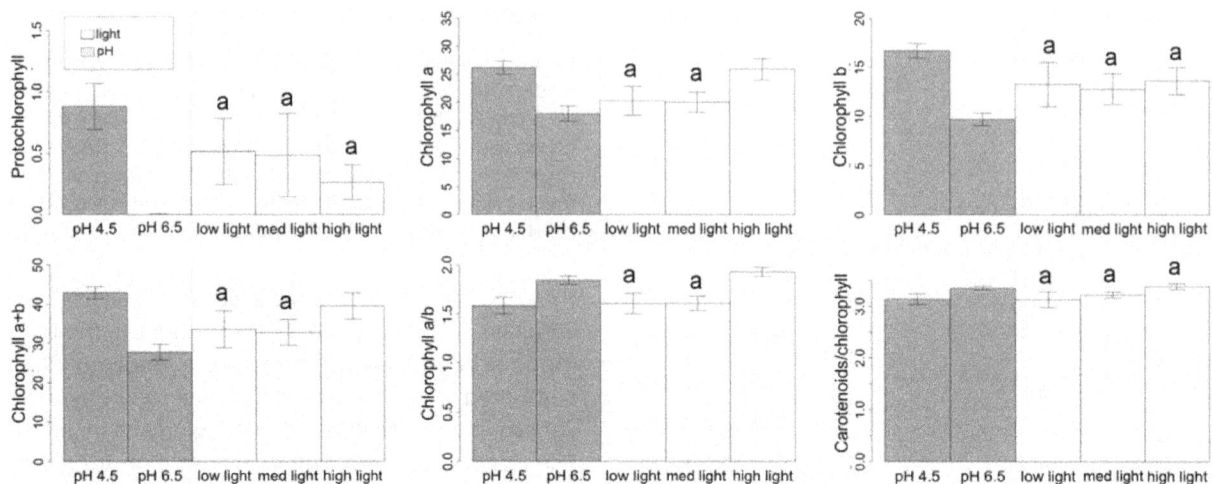

Figure 3. Pigment concentration ($\mu g\ cm^{-2}$) or ratio in leaves of W. nobilis seedlings grown at low, medium and high light in potting mix at pH 4.5 or 6.5 for 24 months (mean $\pm$ SE). (A–F) Means for pH and light treatments (means sharing the same letter are not significantly different by LSD$_{5\%}$). Note: y-axis varies.

rainforest species, including *C. apetalum*, have lower growth at high light, but higher mean biomass accumulation at low light, when compared with eucalypt or ecotone species (Barrett and Ash 1992). This study revealed that while stem length increased with increased light, the number of stems was highest in medium light. We hypothesize that this species has a low-light strategy of producing more stems and a high-light (i.e. large gap) strategy to invest more in growth of primary stems. However, it appears that multiple stem production in low light increases substantially above a threshold light level, between 5 and 15 %; this threshold is yet to be defined.

Wollemia nobilis seedlings respond to increased light by increasing growth, but they can also grow slowly and survive in low light. In low light, *W. nobilis* can maintain its photosynthetic apparatus in a state in which it can take advantage of changes in light levels, indicated by the chlorophyll pigment concentrations and ratio. However very low light (and associated low growth) in the wild may contribute to seedling mortality by fungal pathogens, herbivory, litterfall and drought, in line with models of increased juvenile risk in slow-growing species (Bond 1989).

Young *W. nobilis* leaves are adapted to shaded understorey conditions, maximizing light interception by arranging leaves in a single plane, while adult foliage is arranged in two ranks to capture light from all around (Hill and Brodribb 2003). Moreover, *W. nobilis* juveniles had a high proportion of branches to stems; this is typical of many understorey species, and demonstrates structural flexibility (Givnish 1988). This study indicates that *W. nobilis* has the ability to produce multiple leaders at all light levels, but particularly at medium-light and low-pH soil growth conditions. The difference among light treatments, in the production of multiple stems, was only apparent at the 24-month measurement (compared with the 12-month measurement; data not presented). This is comparable to the growth lag phase, which has been observed previously in *W. nobilis* (Offord *et al.* 1999). Hence, *W. nobilis* can take advantage of light gaps by extending leaders, in effect seeking light and then producing the branches at a later time to maximize growth in the gap (Givnish 1988).

The maximum light treatment used in our study was 50 %, rather than 100 %, full sun. Previous research has shown that young *Agathis* trees can be damaged by full sunlight (Whitmore 1977), and along with other temperate rainforest species, *Agathis* have been shown to have lower chlorophyll concentrations in full sunlight compared with trees grown in medium and heavy shade (Langenheim *et al.* 1984; Read 1985). Moreover, photoinhibition may occur before obvious leaf damage. *Wollemia nobilis* is highly susceptible to heat stress

(compared with other Araucariaceae species; Offord 2011), and anecdotal observations of young *W. nobilis* (<5 years) suggest that the leaves can become chlorotic under full sun. The quality of light in the canopy, especially the ratio of red to far red, may also influence the growth of shade-adapted *Agathis* species (Warrington *et al.* 1988). Although this study has found a positive correlation between light and growth in *W. nobilis* seedlings, this effect is likely to be curtailed at higher light availabilities due to light saturation.

The positive effects of light on growth were moderated by pH. Growth of stems (height and diameter), and stem number, was highest in the low-pH treatments, regardless of light. In contrast, in the high-pH treatments, increased light was associated with increased growth. While both pH and light affected pigments, there was no modulating effect detected. Over the range of light availabilities assessed, the chlorophyll content and the *a* to *b* ratio varied little, compared with the variation associated with pH. Protochlorophyll is a precursor to chlorophyll and is readily converted with light exposure (Lancer *et al.* 1976; Huq *et al.* 2004). Protochlorophyll was significantly higher in the low-pH treatment. The direct effects of pH on protochlorophyll are unknown, but low levels (in the high-pH treatment) suggest that the plants would be less able to take advantage of increases in light. Higher carotenoid-to-chlorophyll ratios in the high-pH treatment may also reflect growth suppression. Soil fertility (influenced by pH) can also affect plants' capacity to capture light (Baltzer and Thomas 2005). However, the combined effects of light and pH were not significant in defining pigment concentrations in this study.

Implications for *W. nobilis* stand dynamics

The need for open canopy conditions, with >10 % light (i.e. large canopy gaps), for recruitment of juveniles to canopy trees is evident in a number of Araucariaceae species (Bergin and Kimberley 1987; Enright *et al.* 1993; Fincke and Paulsch 1995; Rigg *et al.* 1998). Furthermore, infrequent large-scale landscape disturbance is required for substantial recruitment of other Araucariaceae species (Ogden *et al.* 1992; Burns 1993; Enright *et al.* 1999). Likewise, very low rates of seedling recruitment are a feature of mature undisturbed Araucariaceae stands (in association with increasing dominance of angiosperm tree species; Enright *et al.* 1999). The demonstrated increase in *W. nobilis* growth with light availability is consistent with growth responses observed in other Araucariaceae species.

Our results imply that canopy gaps may be required for significant increases in stem length and hence recruitment from *W. nobilis* juveniles to canopy trees. How much light required is unknown, but positive responses to 50 % light availability (i.e. large gaps) were recorded in this study. If *W. nobilis* growth and recruitment to larger size classes is dependent on light, this may explain the lack of intermediate-sized *W. nobilis* juveniles in the wild, where *W. nobilis* is competing with rainforest angiosperms (*sensu* Enright *et al.* 1999). *Wollemia nobilis* response to increased light availability after long-term suppression in the shade is also unknown. The self-coppicing habit of *W. nobilis* allows it to survive in low light (Givnish 1988) and may aid recovery from disturbance (Dietze and Clark 2008). Its architecture may also mean that it can respond quickly (change morphologically) to intercept light (Canham 1989; Whitmore 1989), similar to *A. angustifolia*, which can quickly colonize gaps from both seedling banks and resprouts of damaged trees via plastic growth patterns (Duarte *et al.* 2002).

The ability of *W. nobilis* to respond to increased light was strongly moderated by soil pH. Although there is evidence to suggest that Araucariaceae can modify soil pH, we suggest that pH is unlikely to be limiting *W. nobilis* success in the wild as acid soils are widespread in eastern New South Wales; hence the question remains, what factors limit the distribution of *W. nobilis* in the wild? Despite the strong limiting effect of pH on growth demonstrated in this study, light availability (i.e. the presence of large canopy gaps) may be more limiting in the wild.

Implications for conservation of *W. nobilis*

Understanding the climatic and edaphic factors governing plant growth is important for the conservation management of species in the wild, and especially species like *W. nobilis* that have restricted distributions and are under threats such as disease, fire and climate change. Where management of wild stands cannot provide sufficient protection, translocation of threatened species is one complementary conservation measure (Vallee *et al.* 2004). Selection of suitable soil types and light regimes provided by topography, aspect and vegetation assemblages is essential to translocation success. Additionally, optimal growth of plants *ex situ*, in gardens or germplasm collections, necessitates knowledge of fundamental growth requirements. This and several other studies partially provide this knowledge for *W. nobilis* (vide Offord 2011). Importantly, our results indicate that *W. nobilis* recruitment may be light limited in the wild. Our results also suggest that *W. nobilis* translocation efforts should be focused on sites with low-pH soils and outside the deeply shaded conditions of a closed rainforest.

Conclusions

Seedling growth responses to varying light regimes suggest that *W. nobilis* is a shade-tolerant, gap-responding species: tolerating low light and increasing growth at higher light. However, this response is strongly moderated by soil pH; *W. nobilis* growth is significantly enhanced on low-pH soils. While these factors clearly influence seedling growth of this species, at least under glasshouse conditions, other factors, such as drought, herbivory and microbial interactions, may also strongly influence the recruitment of this species in the wild.

Sources of Funding

Our work was funded by the Royal Botanic Gardens and Domain Trust, Sydney.

Contributions by the Authors

C.O. and P.M. were responsible for the experimental design. P.M. conducted the experimental components including data collection. C.O. and H.Z. analysed data and wrote the manuscript.

Acknowledgements

The authors wish to acknowledge the input of Faye Cairncross, Felicity Keiper, Graeme Errington, Glen Fensom, the Sydney Environment and Soil Laboratory and the staff of the NSW National Parks and Wildlife Service.

Literature Cited

Alban DH. 1982. Effects of nutrient accumulation by aspen, spruce, and pine on soil properties. *Soil Science Society of America Journal* **46**:853–861.

Atwell B, Kriedemann P, Turnbull C. 2003. *Plants in action: adaptations in nature, performance in cultivation*. South Yarra, Australia: MacMillian Publishers Pty Ltd.

Baltzer JL, Thomas SC. 2005. Leaf optical responses to light and soil nutrient availability in temperate deciduous trees. *American Journal of Botany* **92**:214–223.

Barrett D, Ash J. 1992. Growth and carbon partitioning in rainforest and eucalypt forest species of south coastal New South Wales, Australia. *Australian Journal of Botany* **40**:13–25.

Beadle N. 1954. Soil phosphate and the delimitation of plant communities in eastern Australia. *Ecology* **35**:370–375.

Benson J, Allen C. 2007. Vegetation associated with *Wollemia nobilis* (Araucariaceae). *Cunninghamia* **10**:255–262.

Bergin DO, Kimberley MO. 1987. Establishing kauri in a pine stand and in scrub. *New Zealand Journal of Forestry Science* **17**:3–11.

Bieleski RL. 1959. Factors affecting growth and distribution of kauri (*Agathis australis* Salisb.). III. Effect of temperature and soil conditions. *Australian Journal of Botany* **7**:279–294.

Binkley D, Fisher R. 2000. *Ecology and management of forest soils*. New York: John Wiley and Sons.

Bond W. 1989. The tortoise and the hare: ecology of angiosperm dominance and gymnosperm persistence. *Biological Journal of the Linnean Society* **36**:227–249.

Burns BR. 1993. Fire-induced dynamics of *Araucaria araucana—Nothofagus antartica* forest in the southern Andes. *Journal of Biogeography* **20**:669–685.

Canham CD. 1989. Different responses to gaps among shade-tolerant tree species. *Ecology* **70**:548–550.

Chazdon RL, Pearcy RW, Lee DW, Fetcher N. 1996. Photosynthetic responses of tropical forest plants to contrasting light environments. In: Mulkey SS, Chazdon RL, Smith AP, eds. *Tropical forest plant ecophysiology.* New York: Chapman and Hall, 5–55.

Chen DM, Keiper FJ, De Filippis LF. 1998. Physiological changes accompanying the induction of salt tolerance in *Eucalyptus microcorys* stems in tissue culture. *Journal of Plant Physiology* **152**: 555–563.

Curlevski NJA, Xu Z, Anderson IC, Cairney JWG. 2010. Converting Australian tropical rainforest to native Araucariaceae plantations alters soil fungal communities. *Soil Biology and Biochemistry* **42**: 14–20.

Davies S, Tan S, LaFrankie JV, Potts MD. 2005. Soil-related floristic variation in a hyperdiverse dipterocarp forest. In: *Pollination ecology and the rain forest.* Ecological Studies Vol. 174, 22–34.

Denslow JS. 1980. Patterns of plant species diversity during succession under different disturbance regimes. *Oecologia* **46**:18–21.

Dietze MC, Clark JS. 2008. Changing the gap dynamics paradigm: vegetative regeneration control on forest response to disturbance. *Ecological Monographs* **78**:331–347.

Duarte LS, Dillenburg LR. 2000. Ecophysiological responses of *Araucaria angustifolia* (Araucariaceae) seedlings to different irradiance levels. *Australian Journal of Botany* **48**:531–537.

Duarte LS, Dillenburg LR, Rosa LMG. 2002. Assessing the role of light availability in the regeneration of *Araucaria angustifolia* (Araucariaceae). *Australian Journal of Botany* **50**:741–751.

Ecroyd CE. 1982. Biological flora of New Zealand. 8. *Agathis australis.* D.Don. Lindl. (Araucariaceae) Kauri. *New Zealand Journal of Botany* **20**:17–36.

Ehrenfeld JG, Ravit B, Elgersma K. 2005. Feedback in the plant–soil system. *Annual Review of Environment and Resources* **30**:75–115.

Enright NJ, Bartlett RM, De Freitas CR. 1993. Patterns of species composition, recruitment, and growth within canopy gaps in two New Zealand kauri (*Agathis australis*) forests. *New Zealand Journal of Botany* **31**:361–373.

Enright NJ, Ogden J, Rigg LS. 1999. Dynamics of forests and Araucariaceae in the western Pacific. *Journal of Vegetation Science* **10**: 793–804.

Fincke M, Paulsch A. 1995. The ecological strategy of *Araucaria araucana. Flora* **190**:365–382.

Finzi AC, Canham CD, van Breemen N. 1998. Canopy tree–soil interactions within temperate forests: species effects on pH and cations. *Ecological Applications* **8**:447–454.

Givnish TJ. 1988. Adaptation to sun and shade: a whole-plant perspective. *Journal of Plant Physiology* **15**:63–92.

Grubb PJ. 1977. The maintenance of species-richness in plant communities: the importance of the regeneration niche. *Biological Reviews* **52**:107–145.

Handreck K, Black N. 2002. *Growing media for ornamental plants and turf.* 3rd edn. Sydney, Australia: University of New South Wales Press Ltd.

Helyar K, Cregan P, Godyn D. 1990. Soil acidity in New-South-Wales—current pH values and estimates of acidification rates. *Soil Research* **28**:523–537.

Hill KD. 1997. Architecture of the Wollemi pine (*Wollemia nobilis,* Araucariaceae), a unique combination of model and reiteration. *Australian Journal of Botany* **45**:817–826.

Hill RS, Brodribb TJ. 2003. Evolution of conifer foliage in the southern hemisphere. *Acta Horticulturae* **615**:53–58.

Holste EK, Kobe RK, Vriesendorp CF. 2011. Seedling growth responses to soil resources in the understory of a wet tropical forest. *Ecology* **92**:1828–1838.

Huq E, Al Sady B, Hudson M, Kim C, Apel K, Quail PH. 2004. Phytochrome-interacting factor 1 is a critical bHLH regulator of chlorophyll biosynthesis. *Science* **305**:1937–1941.

Jaffré T. 1992. Floristic and ecological diversity of the vegetation on ultramafic rocks in New Caledonia. *The Vegetation of Ultramafic (Serpentine) Soils: Proceedings of the First International Conference on Serpentine Ecology,* 101–107.

Jones WG, Hill KD, Allen JM. 1995. *Wollemia nobilis,* a new living Australian genus and species in the Araucariaceae. *Telopea* **6**: 173–176.

Kershaw P, Wagstaff B. 2001. The southern conifer family Araucariaceae: history, status, and value for paleoenvironmental reconstruction. *Annual Review of Ecology and Systematics* **32**:397–414.

Lancer HA, Cohen CE, Schiff JA. 1976. Changing ratios of phototransformable protochlorophyll and photochlorophyllide of bean seedlings developing in the dark. *Plant Physiology* **57**: 369–374.

Langenheim JH, Osmond CB, Borrks A, Ferrar PJ. 1984. Photosynthetic responses to light in seedlings of selected Amazonian and Australian rainforest tree species. *Geologia* **63**:215–224.

Macphail M, Hill K, Partridge A, Truswell E, Foster C. 1995. Wollemi Pine 'old pollen records for a newly discovered genus of gymnosperm'. *Geology Today* **11**:48–50.

Marschner H. 1991. Mechanisms of adaptation of plants to acid soils. *Plant and Soil* **134**:1–20.

Mirams RV. 1957. Aspects of the natural regeneration of the kauri (*Agathis australis* Salisb.). *Transactions of the Royal Society of New Zealand* **84**:661–680.

Monteith JL, Unsworth M. 1990. *Principles of environmental physics.* 2nd edn. London: Edward Arnold.

Nicotra AB, Chazdon RL, Iriarte SVB. 1999. Spatial heterogeneity of light and woody seedling regeneration in tropical wet forests. *Ecology* **80**:1908–1926.

NSW Department of Environment and Conservation. 2006. Wollemia nobilis *(Wollemi pine) recovery plan.* Hurstville, Australia: NSW Department of Environment and Conservation.

Offord CA. 2011. Pushed to the limit: consequences of climate change for the Araucariaceae: a relictual rain forest family. *Annals of Botany* **108**:347–357.

Offord CA, Meagher PF. 2001. The effects of temperature, light and stratification on seed germination of Wollemi pine (*Wollemia nobilis,* Araucariaceae). *Australian Journal of Botany* **49**:699–704.

Offord CA, Porter CL, Meagher PF, Errington G. 1999. Sexual reproduction and early plant growth of the Wollemi pine (*Wollemia nobilis*), a rare and threatened Australian conifer. *Annals of Botany* **84**:1–9.

Ogden J, Wilson A, Hendy C, Newnham RM. 1992. The late Quaternary history of kauri (*Agathis australis*) in New Zealand and its climatic significance. *Journal of Biogeography* **19**:611–622.

Ovington JD. 1953. Studies of the development of wood-land conditions under different trees. I. Soil pH. *Journal of Ecology* **41**: 13–52.

Read J. 1985. Photosynthetic and growth responses to different light regimes of the major canopy species of Tasmanian cool temperate rainforest. *Australian Journal of Ecology* **10**: 327–334.

Rigg LS, Enright NJ, Jafffré T. 1998. Stand structure of the emergent conifer *Araucaria laubenfelsii*, in maquis and rainforest, Mont Do, New Caledonia. *Australian Journal of Ecology* **23**: 528–538.

Russo SE, Brown P, Tan S, Davies S. 2008. Interspecific demographic trade-offs and soil-related habitat associations of tree species along resource gradients. *Journal of Ecology* **96**:192–203.

Silvester WB, Orchard TA. 1999. The biology of kauri (*Agathis australis*) in New Zealand. Production, biomass, carbon storage, and litter fall in four forest remnants. *New Zealand Journal of Botany* **37**: 553–571.

Thompson K, Hodgson JG, Grime JP, Burke MJW. 2001. Plant traits and temporal scale: evidence from a 5-year invasion experiment using native species. *Journal of Ecology* **89**:1054–1060.

Tice KR, Graham RC, Wood HB. 1996. Transformations of 2:1 phyllosilicates in 41-year-old soils under oak and pine. *Geoderma* **70**: 49–62.

Vallee L, Hogbin T, Monks L, Makinson B, Matthes M, Rossetto M. 2004. *Guidelines for the translocation of threatened plants in Australia*. 2nd edn. Canberra: Australian Network for Plant Conservation.

Van Miegroet H, Cole DW. 1984. The impact of nitrification on soil acidification and cation leaching in a red alder ecosystem. *Journal of Environmental Quality* **13**:586–590.

Warrington IJ, Rook DA, Morgan DC, Turnbull HL. 1988. The influence of simulated shadelight and daylight on growth, development and photosynthesis of *Pinus radiata, Agathis australis* and *Dacrydium cuperssinum*. *Plant, Cell and Environment* **11**:343–356.

Weaver SA. 1988. Soil differences between secondary and old growth *Agathis macrophylla* forest at Nadarivatu, Fiji. *Tuatara* **30**:55–61.

Whitmore TC. 1977. *A first look at Agathis*. Tropical Forestry Papers, No. 11. Commonwealth Forestry Institute. Oxford: University of Oxford.

Whitmore TC. 1989. Canopy gaps and the two major groups of forest trees. *Ecology* **70**:536–538.

Whitmore TC. 1996. A review of some aspects of tropical rain forest seedling ecology with suggestions for further enquiry. In: Swaine MD, ed. *The ecology of tropical forest tree seedlings*. Paris, France: Parthenon Publishing Group, 3–39.

Whitmore TC, Page CN. 1980. Evolutionary implications of the distribution and ecology of the tropical conifer *Agathis*. *New Phytologist* **84**:407–416.

Wyse SV. 2012. Growth responses of five forest plant species to the soils formed beneath New Zealand kauri (*Agathis australis*). *New Zealand Journal of Botany* **50**:411–421.

Zimmer HC, Auld TD, Benson J, Baker PJ. 2014. Recruitment bottlenecks in the rare Australian conifer. *Wollemia nobilis. Biodiversity and Conservation* **23**:203–215.

5

Effects of pollination limitation and seed predation on female reproductive success of a deceptive orchid

Ryan P. Walsh*, Paige M. Arnold and Helen J. Michaels

Department of Biological Sciences, Bowling Green State University, Bowling Green, OH 43402, USA

Associate Editor: Dennis F. Whigham

Abstract. For many species of conservation significance, multiple factors limit reproduction. This research examines the contributions of plant height, number of flowers, number of stems, pollen limitation and seed predation to female reproductive success in the deceit-pollinated orchid, *Cypripedium candidum*. The deceptive pollination strategy employed by many orchids often results in high levels of pollen limitation. While increased floral display size may attract pollinators, *C. candidum*'s multiple, synchronously flowering stems could promote selfing and also increase attack by weevil seed predators. To understand the joint impacts of mutualists and antagonists, we examined pollen limitation, seed predation and the effects of pollen source over two flowering seasons (2009 and 2011) in Ohio. In 2009, 36 pairs of plants size-matched by flower number, receiving either supplemental hand or open pollination, were scored for fruit maturation, mass of seeds and seed predation. Pollen supplementation increased proportion of flowers maturing into fruit, with 87 % fruit set when hand pollinated compared with 46 % for naturally pollinated flowers. Inflorescence height had a strong effect, as taller inflorescences had higher initial fruit set, while shorter stems had higher predation. Seed predation was seen in 73 % of all fruits. A parallel 2011 experiment that included a self-pollination treatment and excluded seed predators found initial and final fruit set were higher in the self and outcross pollination treatments than in the open-pollinated treatment. However, seed mass was higher in both open pollinated and outcross pollination treatments compared with hand self-pollinated. We found greater female reproductive success for taller flowering stems that simultaneously benefited from increased pollination and reduced seed predation. These studies suggest that this species is under strong reinforcing selection to increase allocation to flowering stem height. Our results may help explain the factors limiting seed production in other *Cypripedium* and further emphasize the importance of management in orchid conservation.

Keywords: Conservation; orchid; plant reproduction; plant–insect interactions; pollen limitation; pollination ecology; reproductive trade-offs; seed predation; supplemental pollination.

Introduction

The complex dynamics between plants, pollinators and seed predators, and how these interactions affect plant reproduction, are important in understanding the evolution of plant floral displays (Irwin and Brody 2011). While many have examined the roles of both mutualists and antagonists in plant reproduction (Gómez 2003; Strauss and Irwin 2004; Abdala-Roberts *et al.* 2009;

* Corresponding author's e-mail address: rpwalsh@bgsu.edu

Burkhardt *et al.* 2009; Carlson and Holsinger 2010; Kolb and Ehrlén 2010; Ågren *et al.* 2013), relatively few have examined the effect of these interactions on the selection of floral traits (Cariveau *et al.* 2004; Parachnowitsch and Caruso 2008). Both mutualists and antagonists may forage on individuals based on floral display size, where larger floral displays provide concentrated resources for pollinators (Peakall 1989; Brody and Mitchell 1997; Mitchell *et al.* 2004) as well as seed predators (Stephens and Myers 2012). The goal of this study was to assess the relationships among mutualists, antagonists and floral display size, and how these biotic interactions influence reproductive success in a deceptive pollination system of conservation concern.

Maximizing pollen transfer efficiency has greatly shaped the evolution of a multitude of floral forms and functions in angiosperms (Barrett 2003). Increased floral display size is expected to increase pollinator attraction and visitation (Peakall 1989; Burd 1995; Aragón and Ackerman 2004; Grindeland *et al.* 2005; Li *et al.* 2011; Sletvold and Ågren 2011). In an experimental manipulation of floral display size in *Mimulus ringens*, Mitchell *et al.* (2004) found that bumblebee pollinators strongly respond to floral display size, probing more flowers in sequence on plants with large numbers of flowers. Larger floral displays also increase plant visibility, thereby increasing the attraction of pollinators from greater distances (Kindlmann and Jersáková 2006). Flowering stem height may similarly increase visibility and pollination efficiency. Synchronous flowering within a plant may also effectively increase floral display size, potentially increasing individual visibility, while synchronous flowering on the population level may increase conspecific competition for the limited pollinator resources (Crone and Lesica 2004; Crone *et al.* 2009). Measures of plant size such as numbers of stems and leaves, which may or may not correlate with floral display size, while not directly affecting the recruitment of pollinators, represent increased photosynthetic resources available for subsequent fruit maturation (Bazzaz *et al.* 1987).

Many orchids rely on a deceptive pollination strategy, a pollination strategy in which the flower provides floral cues indicating a food reward while not providing that reward (Faegri and van der Pijl 1971; Cozzolino and Widmer 2005). Deceptive pollination systems often show lower visitation and pollination relative to rewarding relatives (Nilsson 1980, 1984). To account for this reduced visitation and pollination, some have hypothesized that deception reduces geitonogamous pollination by causing pollinators to flee non-rewarding patches (Smithson 2002; Johnson *et al.* 2004; Schiestl 2005; Kindlmann and Jersáková 2006; Sun *et al.* 2009), referred to as the outcrossing hypothesis (Jersáková *et al.* 2006). Orchids

relying on food deception often depend on newly emergent or otherwise inexperienced insects for pollinator services (Jersáková *et al.* 2006). In *Dactylorhiza lapponica*, Sletvold *et al.* (2010) demonstrated strong pollinator-mediated selection on spur length and plant height in the open lawn community of a Norwegian fen. This and other studies of deceptive species (O'Connell and Johnston 1998; Johnson and Nilsson 1999; Gigord *et al.* 2001) indicate that a variety of floral traits (plant height, flower number, petal colour) may be targets of selection for increasing female reproductive success through pollinator visitation.

Pollen limitation, defined as the difference in seed production between open pollinated (natural pollination with no supplementation) and supplemental pollination treatments, occurs when the average open-pollinated seed production is significantly less than the average seed production of individuals receiving supplemental pollen (Knight 2003). While increased floral display size may reduce pollinator limitation, it may also substantially increase geitonogamy (de Jong *et al.* 1993; Barrett and Harder 1996; Snow *et al.* 1996; Galloway *et al.* 2002). Geitonogamy reduces female function by reducing the number or quality of the offspring, but also impacts male function by reducing the quantity of pollen available for export to other plants, also known as pollen discounting (Harder and Barrett 1995; Barrett 2003; Johnson *et al.* 2004).

After pollination, female reproductive success may be eroded by seed predation. Pre-dispersal seed predation can play an important role in determining fecundity and long-term population persistence (Louda and Potvin 1995; Russell *et al.* 2010). Chronic seed predation rates can limit population growth by reducing fecundity, while stochastic predation rates can play a more diffuse, but equally important role in population dynamics (Kolb *et al.* 2007). Just as pollinators are often attracted to large floral displays, seed predators may be attracted to the accompanying large ovule resource (Brody and Mitchell 1997; Kudoh and Whigham 1998; Galen and Cuba 2001; Irwin *et al.* 2003; Adler and Bronstein 2004; Shimamura *et al.* 2005; Stephens and Myers 2012) as many seed predators rely on ovule development to feed their offspring (Cariveau *et al.* 2004). Therefore, although dense floral resources may attract mutualist pollinators, the accompanying dense floral and fruit resources may simultaneously attract antagonist herbivores and seed predators, creating conflicting selective pressures (Louda and Potvin 1995; Strauss and Irwin 2004; Ågren *et al.* 2008). Similar to the influence of co-flowering species on plant–pollinator interactions, rates of pre-dispersal seed predation also can be influenced by community context through their attraction to other species that share seed predators (Recart *et al.* 2013).

Cypripedium species are deceptive, deciduous, terrestrial orchids with growth emerging from a subterranean rhizome (Stoutamire 1967). *Cypripedium candidum* Muhlenberg ex. Willdenow, the Small White Lady's Slipper, with yellow-green lateral sepals and petals with a white, purple-spotted labellum (Stoutamire 1967), occurs in calcareous prairies as well as fens and limestone barrens (Cusick 1980). The plants occur as single plants or large clumps (1–12 vegetative stems) containing a single large flower per vegetative stem. This floral architecture, combined with a short flowering period, severely limits the probability of reproduction by restricting opportunities for pollination. *Cypripedium candidum* flowers are pollinated by small (4–6 mm long) adrenid and halictid bees (Catling and Knerer 1980; Bowles 1983; Wake 2007). While pollen transport distances have not been specifically studied for adrenid and halictid bees, previous studies have found pollinia transport distances to vary widely (Nilsson *et al.* 1992; Kropf and Renner 2008). In the larger (8–17 mm long) *Andrena* bee species, Kropf and Renner (2008) found a maximum transport distance of 6.9 m. Data collected in a separate study (Walsh 2013) in Ohio *C. candidum* found pollen dispersal to be limited to 1 m within the focal plant.

The primary antagonist for *C. candidum*, *Stethobaris ovata* (Family: Curculiondae; subfamily: Baridinae), is a known weevil seed predator of *Cypripedium* spp. and other temperate orchid genera, with reports of adults in Canadian populations feeding on emerging shoots and flower buds (Light and MacConaill 2011). Adult weevils emerge in early spring along with *Cypripedium* shoots, and oviposit in developing fruits and, possibly stems, resulting in either fruit abortion or near total loss of the developing embryos (Light and MacConaill 2011). Predation rates on *C. parviflorum* in Canada vary from 32 to 53 % among plants in a population, depending on climate and availability of fruit resources (Light and MacConaill 2002). Little is known about the life history of the weevil, although they may complete two life cycles within a growing season (M. Light, pers. comm.).

The goal of this study was to assess the effects of plant and floral display sizes on both pollination and seed predation, and understand how these factors influence female reproductive success in the long-lived, highly specialized deceptive orchid, *C. candidum*. *Cypripedium candidum* is highly dependent on full sun in open areas and, as with many prairie species, populations begin to decline with the invasion of woody plants (Curtis 1946). In addition to the potential shading effects of encroaching woody vegetation, increased heterospecific stem density has been shown to reduce pollination and population recruitment (Wake 2007). Although the major challenges in orchid conservation research reside

in understanding their symbiotic associations with fungi and improving survival following propagation (Krupnick *et al.* 2013), effective management and restoration will require a more mechanistic understanding of how habitat changes influence all biotic interactions limiting recruitment. We hypothesized that an increased floral display size would (i) increase pollinator visitation as indicated by increasing fruit set, (ii) increase geitonogamy, resulting in increased fruit abortion, decreasing fruit maturation and offspring fitness, and (iii) increase attraction of antagonists by providing an attractive resource concentration for seed predators. To address these hypotheses, we conducted two pollen limitation experiments over the course of the 2009 and 2011 field seasons in two separate sites, examining the effects of plant size on pollinator limitation and seed predation in 2009, and the effect of pollen quality in a seed predator exclusion experiment in 2011.

Methods

The primary field site for this study was located in Northern Ohio (GPS coordinates available upon request). Historically *C. candidum* existed in at least seven Ohio counties (herbarium records OSU and BGSU). However, the Northern Ohio site is now one of the only two locations in the state where *C. candidum* remains (ODNR, pers. comm.). This site has a large, actively managed prairie area (~900 ha) with wooded areas intertwined. Vegetative cover at this location is dominated by *Andropogon gerardii*, *Sorghastrum nutans* and *Silphium terebinthinaceum*, with *Viola* spp., *Sisyrinchium montanum* and *Fragaria virginiana* co-flowering with *C. candidum*. The prairie is maintained through controlled burns in early spring approximately every 3 years (J. Windus, ODNR, pers. comm.), producing large, thriving populations (total $N \sim 6000$) in the calcareous soil, although our access to the area was restricted to a subset of the total population by the Ohio Division of Natural Resources.

2009 Study

In early May 2009, we established three randomly selected 60-m line transects through a patch of *C. candidum* ($n > 250$). The population density of *C. candidum* at the prairie was previously estimated at 3.26 plants m^{-2} (range = 1–9 plants m^{-2}; SD = 2.68) (Walsh 2013). At 5-m intervals, the plant closest to a transect was selected and a second plant of equal size (number of stems and flowers) was chosen within 0.5 m of the transect on the opposite side (total $N = 72$). Any plants with flowers that had already opened, or had any pollinia removed or deposited prior to set up of the experiment were excluded from the study, resulting in the exclusion of one sample point. The standard method of assessing pollen limitation compares the female fitness of open-pollinated plants with that of plants that have had all of their flowers hand pollinated (Wesselingh 2007) to avoid redirection of resources from non-pollinated flowers, which may bias information on fruit or seed set (Knight et al. 2006; Wesselingh 2007). One of the paired plants was randomly chosen to receive a hand pollination treatment on all its flowers (mean = 2.06, SE = 0.14, range 1–7) with pollinia from a different population from a site at least 100 m away, while the other member of the pair was open pollinated. The numbers of flowering stems, total stems, leaves per stem and the height of each flowering stem (to the nearest 0.1 cm) were also recorded. Because the site was burned in 2009 ca. 1 month prior to sampling, surrounding vegetation was sparse during the flowering period, precluding the collection of surrounding vegetation data.

Each flower received a single pollinium from a mixed batch of pollinia gathered earlier the same day from a different population at least 100 m away. After hand pollination, flowers were bagged with a mesh (mesh opening size 3 mm × 3 mm) to prevent accidental removal of the pollinium. Mesh bags were removed after all flowers in the experiment had dehisced (ca. 2 weeks) to allow weevil predation. Capsule development was recorded 1 month after floral dehiscence (June), as well as at maturity in August when all capsules were collected and scored for insect damage and seed production. Fruit abortion was scored as the number of fruits initiated in May minus the number of fruits matured in August. We saw no evidence of browsing or herbivory other than weevil damage during the experimental period. Mature capsules were dried at 60 °C for 3 days prior to separate weighing of capsule and seed masses. Mature capsules were scored as predated when circular insect exit holes, ca. 1 mm in diameter, weevil body parts (possibly molts) and a lack of mature seeds were observed.

2011 Study

In May 2011 prior to flower opening, we set up a second pollen limitation study on a different nearby population of similar size to the 2009 study on the same property. In this experiment, a hand self-pollination treatment was added, stem and flower number were controlled and fruits were protected from weevil predation to obtain enough fruits for analysis of the effect of pollen quality on fruit set and seed mass. Five 50-m transects were randomly located across the population. For this study, only three-flowered, three-stemmed individuals were chosen to control for plant size, apply the three treatment types and to limit resource reallocation issues that might arise if non-experimental flowers set fruit. At 5-m intervals along each transect, we tagged the nearest three-flowered, three-stemmed individual and wrapped unopened flowers in a mesh to prevent visitation and weevil predation. Each stem was randomly assigned to one of three treatments: hand self-pollinated, hand-outcross pollinated (from a population >100 m away) or open pollination. Plants were checked daily for open flowers and treatments were applied when the stigma became receptive. Open-pollinated stems had mesh bags removed as soon as flowers opened, while hand-outcross and hand-selfed flowers were re-bagged after receiving appropriate treatments. Following floral dehiscence, the pollinator exclusion bags were removed. Initial fruit set was scored 2 weeks after flower dehiscence, with green, enlarged fruits scored as pollinated and pale, shrunken or missing fruits scored as a failed pollination. All fruits at this time were covered in dialysis tubing and secured at both ends to exclude insect damage. Fruit abortion was scored 4 weeks after flower dehiscence as well as at the end of the study (August 2011). We collected fruits 3 months after flower dehiscence, dried them at 65 °C for 48 h and then weighed dissected seed mass to the nearest hundred-thousandth of a gram on a Mettler AE-240 scale (Mettler-Toledo Inc.). We estimated the effect of inbreeding depression (δ) on female reproductive success by calculating the mean per-family seed production for each maternal family that matured capsules on both the self and outcross pollination treatments as $\delta = 1 - (\omega_s/\omega_x)$ where ω_s is the seed mass produced by selfing and ω_x is the seed mass produced by hand-outcross (Johnston and Schoen 1994).

Data analysis. Analyses were performed using JMP v.9.0.2 (SAS Institute, Cary, CA, USA, 1989–2013). A generalized linear model (GLM) with an identity link function was used to assess the effects of treatment, average flowering stem height, number of flowers, number of stems and the interaction between treatment and average flowering stem height on percentage of initial capsule development (number of capsules produced/number of flowers), proportion of capsules matured and proportion of capsules preyed upon. The number of flowers and number of stems were not significant variables in any initial models and therefore we used pooled error terms to test for other effects in the final models. The effects of number of flowers, number of stems, average height of flowering stems, treatment and percentage of capsules produced on the probability of abortion, final fruit and seed mass were analysed using a GLM. The proportions of fruit set and predation were arcsin square root transformed and all count data were log transformed. To test for proportionality of increase in response to size variables we used an ln–ln regression, testing for a slope of one (Klinkhamer and de Jong 2005; Karron and Mitchell 2012). For the 2011 study, a GLM with an identity link function examined the effect of pollination treatments on initial fruit set, final fruit set and seed mass, followed by one-way ANOVAs to test for differences among treatment groups on fruit set, abortion rates and final seed mass. Post hoc analyses on the one-way ANOVA's were performed using a Tukey–Kramer HSD.

Results

2009 Study

Plants in the study had an average of 2.06 flowers (SE = 0.14, range 1–7) and 3.05 stems (SE = 0.23, range 1–9) per individual. Mature flowering stems had an average height of 22.6 cm (SE = 0.47), ranging between 15.1 and 40.9 cm. The average number of leaves on a plant varied little, with an average of 3.2 leaves per stem (SE = 0.01, range 3–4). Leaf length and width were not measured in this study based on prior work (Walsh 2008) that showed little between-individual variation.

In these experiments, C. candidum showed strong pollen limitation. Initial fruit set, measured 1 month after floral dehiscence, significantly increased in plants receiving supplemental pollen compared with open-pollinated flowers (pollination treatment $P < 0.0001$; whole model $P < 0.0001$, df = 5, AIC = 73.6, $R^2 = 0.33$; Fig. 1). Plants that received supplemental pollen had an initial fruit set of 87 % (SE = 0.037), while plants only serviced by pollinators had substantially lower initial fruit set (mean = 46 %, SE = 0.06, Fig. 1). Number of flowers and number of stems did not significantly predict the

percentage of initial or final capsules produced and were omitted from further analysis. In the subsequent GLM analysis (whole model: $P < 0.0001$; df = 3; AIC = 71.2; $R^2 = 0.29$), initial fruit set was similarly affected by pollination treatment ($P < 0.0001$), but was significantly influenced by flowering stem height only for open-pollinated plants (interaction effect of treatment × the average flowering stem height, $P = 0.02$; main overall effect of height $P = 0.11$).

Final fruit set, measured 3 months post-floral dehiscence, also differed between pollination treatments (model $P = 0.0004$; df = 1; AIC = 121.14; $R^2 = 0.16$; Fig. 1). Study plants receiving supplemental pollen matured capsules 87 % (SE = 0.87) of the time, while only 46 % (SE = 0.057) of open-pollinated plants fully developed fruit (Fig. 1). However, no measured plant size traits (number of flowers, number of stems, height of flowering stems and number of leaves) or interaction effects explained variation in final fruit set (Fig. 2).

Of the flowering stems that set fruit, 73 % were preyed upon. Surprisingly, predation rates were not influenced by availability of food for the weevils (numbers of flowers or fruits) or by plant size (number of stems) (whole model $P = 0.0024$; df = 5; AIC = 93.6; $R^2 = 0.23$). The average height of flowering stems was the only size variable to significantly explain the probability of predation ($P = 0.002$; whole model: $P = 0.0020$; df = 1; AIC = 98.5; $R = 0.14$), with taller stems less likely to be attacked ($F_{1,61} = 10.1, P = 0.002, R^2 = 0.14$; Fig. 3). Fruits suffering predation had poor seed production, as the seed mass of predated capsules was 89 % lower than the seed mass of capsules without predation. A total of 22 plants in the study aborted at least one fruit between initial and final fruit set measurements. Seed mass, measured as the dried and extracted seeds from the capsules, increased

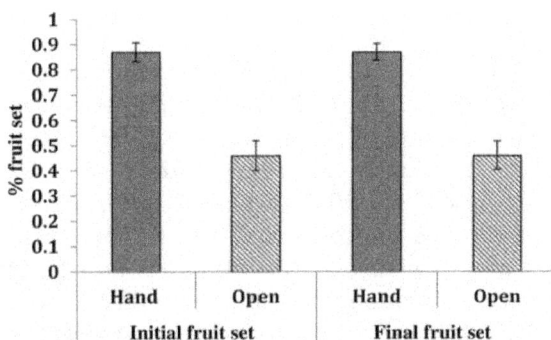

Figure 1. Effect of hand and open pollination treatments on the mean per cent of initial and final fruit set. Plants receiving supplemental pollen produced higher fruit set for both the initial time period (Hand mean = 0.87, SE = 0.037; Open mean = 0.46, SE = 0.06; $t = -4.11$, $n = 36$, $P < 0.001$) and final time period (Hand mean = 0.87, SE = 0.033; Open mean = 0.46, SE = 0.057, $t = -4.30$, $n = 36$, $P < 0.001$).

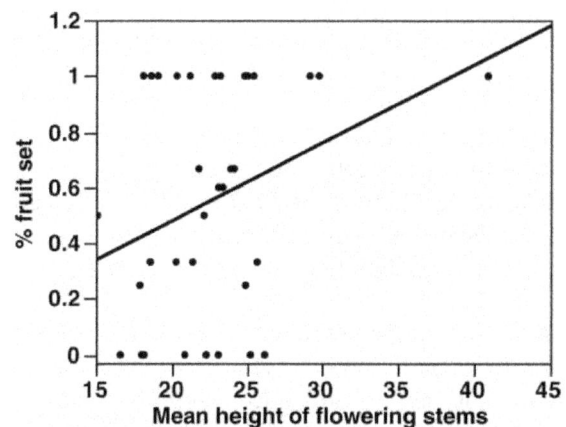

Figure 2. Relationship between per cent fruit set (number of fruits/number of flowers) of open-pollinated plants and the mean height of flowering stems. Linear regression, % fruit set $= -0.34 + 0.037 \times$ avg. height flowering stems, $F_{1,35} = 7.0$, $P = 0.012$, $R^2 = 0.17$.

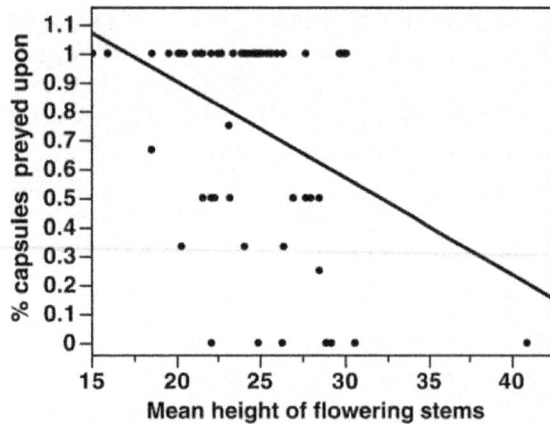

Figure 3. Relationship between the percentage of matured capsules preyed upon (number of capsules preyed upon/number of capsules produced) and the mean height of flowering stems of all plants in the study. Linear regression, % capsules preyed upon = 1.57 − 0.033 × avg. height flowering stems, $F_{1,61} = 10.12$, $P = 0.023$, $R^2 = 0.144$.

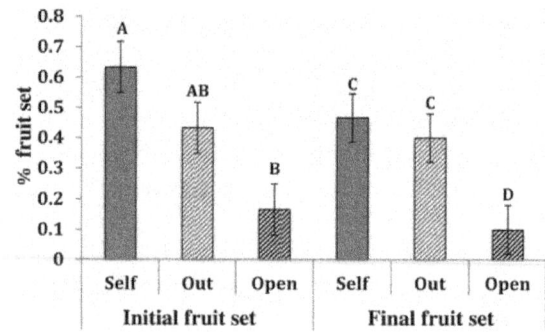

Figure 4. Effect of pollination treatments on initial and final fruit set in the 2011 study. $N = 30$. Mean fruit set for the initial (Self = 0.633, SE = 0.084; Outcross = 0.433, SE = 0.084; Open = 0.166, SE = 0.084) and final fruit maturation times (Self = 0.466, SE = 0.08; Outcross = 0.40, SE = 0.08; Open = 0.1, SE = 0.08) were significantly different as indicated by Tukey–Kramer HSD. Treatments separated by letter indicate significant difference. ANOVA: initial fruit set: $F_{2,89} = 7.73$, $P < 0.001$; final fruit set: $F_{2,89} = 5.73$, $P = 0.004$.

with increased fruit production (numbers of capsules, $P = 0.0161$) and decreased with the proportion of capsules predated ($P < 0.0001$) (whole model $P < 0.001$; df = 6; AIC = 320.5; $R^2 = 0.36$).

2011 Study

When fruits were protected from weevil predation and plant size and flower number were controlled, initial fruit set varied with pollen source. Plants receiving hand pollination had higher fruit set 2 weeks after floral dehiscence, with plants receiving self-pollen setting 63 % of their capsules (SE = 0.08) and plants receiving outcrossed pollen setting 43 % of their capsules (SE = 0.09). In contrast, fruit set on open-pollinated plants was considerably lower, with only 16 % (SE = 0.06) of open-pollinated flowers initially setting fruit. Initial fruit set differed among pollination treatments ($F_{2,89} = 7.73$, $P = 0.0008$; Fig. 4). Plants that received hand self-pollinations had a significantly higher initial fruit set than the open-pollinated plants ($P < 0.05$, Fig. 4).

Final fruit set, scored 12 weeks after floral dehiscence, was affected by the pollination treatments ($F_{2,89} = 5.73$, $P = 0.0046$; Fig. 4). As in the 2009 study, plants receiving natural pollinator service were pollen limited, as self and outcross hand-pollinations matured more fruits (self mean fruits per flower = 0.46, $P = 0.0058$; outcross = 0.4, $P = 0.0292$) compared with open-pollinated stems (mean fruits per flower = 0.1; SE = 0.081). However, pollen quality did not influence the probability of fruit maturation, as the fruit set of selfed and outcrossed hand-pollination treatments were similar ($P = 0.83$; Fig. 4). However, pollen quality of the pollination treatment did significantly influence final seed mass

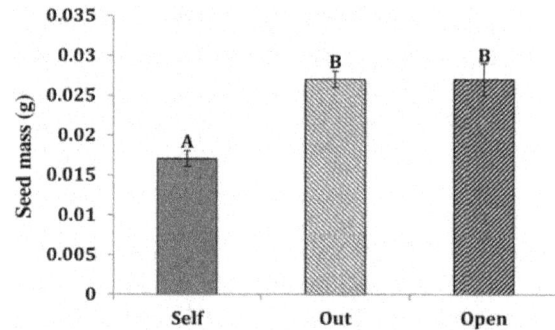

Figure 5. Effect of pollination treatment on seed mass (g) per fruit. The outcross (mean = 0.027, SE = 0.001) and open-pollinated plants (mean = 0.027, SE = 0.002) produced significantly higher seed mass than the self-pollinated plants (mean = 0.017, SE = 0.001) as indicated by Tukey–Kramer HSD, $N = 29$. Treatments separated by a different letter are significantly different. ANOVA: $F_{2,28} = 18.72$, $P < 0.001$, $R^2 = 0.59$.

($P < 0.0001$, $F_{2,28} = 18.72$, Fig. 5). The open and outcross treatments produced significantly larger seed mass per capsule than the self-pollinated treatments (open = 0.027 g, SE = 0.002; outcross = 0.026 g, SE = 0.001; self = 0.017 g, SE = 0.001), while seed masses of the open pollinated and outcross hand-pollination treatments were similar (Fig. 5). There was no significant difference in abortion rates among treatments ($F_{2,89} = 1.79$, $P = 0.172$). A power analysis indicated that 81 replicates would have been needed to reach a significance level of $P < 0.05$.

When the 12 individuals that matured fruit on both hand-pollination treatments were used to estimate the effect of inbreeding depression on seed production, most individuals showed increased seed mass when pollinated with outcrossed pollen (mean $\delta = 0.463$; SD = 0.31). However, the effect of selfing was variable across families. Nine families had an 11–67 % reduction in seed mass when selfed, two families matured a fruit but failed to produce any selfed seed within the fruit, while one family produced similar masses of seed in both the outcross and self hand-pollination treatments.

Discussion

Orchid species have been routinely shown to demonstrate consistent pollen limitation across multiple years (Snow and Whigham 1989; Ackerman and Montalvo 1990; Calvo 1990; Primack and Hall 1990; Dudash and Fenster 1997). Furthermore, deceptive orchids often produce only half as many fruits as their non-deceptive counterparts (Johnson and Bond 1997; Neiland and Wilcock 1998; Tremblay et al. 2004; Jersáková et al. 2006). This study provides strong evidence of pollen limitation in a deceptive orchid over two flowering seasons. In 2009, we observed a moderate fruit set from open pollination (46 %), while fruit set was greatly reduced (16.6 %) for C. candidum plants in 2011. This number closely parallels open-pollinated fruit set seen across multiple non-burn years at this site (16.5 %; Walsh 2013), as well as levels reported in other relatives, such as 10.5 % in C. calceolus (Kull 1998) and 5–13 % in C. acaule (O'Connell and Johnston 1998). Hand-pollinated fruit set was consistent between study years and burn/non-burn years, with at least 40 % of flowers setting fruit when supplemental pollen was provided. The mean temperature for the flowering month did not differ between 2009 and 2011 (15.5 °C); however, the 2011 study year received more than double the precipitation (18.5 cm) compared with 2009 (8.8 cm).

Pollen limitation, in principle, has two components, pollen quantity and pollen quality (Aizen and Harder 2007). In plant systems producing normal, dust-like pollen and large numbers of ovules, inadequate saturation of the stigmatic surface may result in only partial pollination. Orchids produce pollen aggregated into sac-like pollinia containing large amounts of pollen, although misplacement of the pollinia on the stigmatic surface by pollinators may result in incomplete pollination. Studies in the deceptive Dactylorhiza orchid have shown that flowers may need multiple visits in order to receive enough pollination for complete seed set (Sletvold and Ågren 2010; Sletvold et al. 2010). Although Aizen and Harder (2007) argue that pollen supplementation often involves high-quality outcross pollen that could inflate pollen deposition estimates, our 2011 study found that pollen quality manipulation in this system did not significantly increase fruit production, although it did significantly increase

seed mass. Different experiments from both years show no significant change in seed mass/fruit between open-pollinated and hand-outcrossed flowers, suggesting that open-pollinated flowers are usually outcrossed.

Although numerous studies have shown that an increase in floral display and plant size increases pollen receipt and fruit maturation (Peakall 1989; Meléndez-Ackerman and Ackerman 2001; Aragón and Ackerman 2004; Mitchell et al. 2004; Li et al. 2011), we saw no effect of number of flowers, number of stems or number of leaves on pollen receipt or overall fruit maturation. The population studied was relatively dense for an orchid population, 3.26 plants m^{-2} (range = 1–9 plants m^{-2}; SD = 2.68), and the large numbers of closely spaced individuals may have limited the ability to detect any effect of floral display size on fruiting success. However, we found strong evidence that greater flowering stem height increases initial fruit set, suggesting that taller plants were more likely to attract pollinators. Others have shown previously that the height and density of surrounding vegetation affects pollination and fruit production in deceptive orchids. Wake (2007) found increased seed set in C. candidum when surrounding vegetation was experimentally reduced, while the height of the flowering stem also increased pollination and fruit production in the closely related species C. acaule (O'Connell and Johnston 1998). Similarly, Sletvold et al. (2013) found strong pollinator-mediated selection for taller plants in the presence of taller vegetation in the deceptive D. lapponica. Given the tall grass prairie vegetation in which C. candidum occurs, a taller flower would be more visible to pollinators through the vegetation and therefore be more likely to receive pollinator servicing. Compared with other years (measured in a concurrent demographic survey; Walsh 2013), fruit set was unusually high in the 2009 study, which took place immediately after a controlled burn. This would suggest that both increased visibility to pollinators and increased nutrients from the burn may have contributed to this relatively high fruit set. Furthermore, taller flowering stems with increased sun exposure may offer a warmer microenvironment for the small bee pollinators in early spring, as well as greater opportunity for photosynthesis by developing fruits. Future studies on the effects of flowering stem height on seed dispersal in response to local and landscape variation in vegetation density and height may provide additional insights into the functional role of selection on plant traits affecting floral displays.

The presence of a deceptive pollination system may explain why our results with C. candidum are contrary to reports in the literature involving non-orchids. While other plants with larger floral displays attract pollinators from greater distances (Sih and Baltus 1987; Hessing

1988), the absence of a reward may discourage further foraging on the same plant, limiting any increase in fitness that would otherwise occur in a large, multi-flowered rewarding plant. Jersáková *et al.* (2006) cite numerous examples of deceptive orchids with reduced geitonogamy, while nectar addition experiments in deceptive orchids have found dramatic increases in self-pollination when reward is added (Johnson *et al.* 2004; Jersáková *et al.* 2006; Walsh 2013). In a deceptive system, a taller stem may increase the probability of pollinator attraction, but the visitor is expected to quickly depart after receiving no compensation for its efforts. Although food deceptive systems may increase pollen limitation compared with rewarding ones, it may be more advantageous to produce fewer, but higher quality fruits than producing additional lower quality (selfed) offspring, as seen in our 2011 data (in which seed mass substantially decreased with hand self-pollination) and several others (e.g. Tremblay *et al.* 2004; Jersáková *et al.* 2006).

These data describe a potential mechanism driving the classical outcrossing hypothesis, which explains the benefits of deceptive orchid pollination via increased outcrossing (Dafni and Ivri 1979; Nilsson 1983; Ackerman 1986; Johnson and Nilsson 1999; Jersáková *et al.* 2006). In their review of published estimates of inbreeding depression in orchids, Sletvold *et al.* (2012) noted that mean inbreeding depression for seed production was 33 %, regardless of mating system. While most individuals in this study produced greater seed mass on average when receiving outcrossed pollen (leading to a mean inbreeding depression of $\delta = 0.46$), this outcome was variable across all families, with one individual producing equal seed mass in both self and outcross treatments. The average number of flowers per plant within this population over a 4-year observation period was 1.75, with ~51 % of flowers setting fruit (Walsh 2013), indicating that although inbreeding depression in *C. candidum* might be overcome by setting an additional fruit, the floral display architecture (single flower/stem) of *C. candidum* makes this highly unlikely. Furthermore, a true estimate of inbreeding depression would require data on the germination and future growth and reproduction of the offspring and is likely to depend on environmental conditions (Cheptou and Donohue 2011; Murren and Dudash 2012). Although seed packets have been previously used to quantify germination in the field in some orchids (Rasmussen and Whigham 1993; Sletvold *et al.* 2012), attempts to germinate *C. candidum* using this method produced no seedlings over a 2-year study period (Walsh 2013).

In this study seed predators preferentially preyed on fruit with shorter flowering stems, exerting a strong concordant selective pressure reinforcing that of the pollinators. In total, seed predation heavily reduced total reproductive output of the population, with 73 % of all capsule-bearing stems attacked by a seed predator. All capsules appeared to be damaged by the same insect, most likely the weevil in the *Stethobaris* genus previously reported to prey upon *Cypripedium* fruit (Light and MacConaill 2011). Weevil predators in this genus are known to feed upon the leaves, flowers and developing capsules of many orchids, destroying most seeds by ovipositing in the maturing capsules. Contrary to our predictions, neither predation rates nor pollination success was related to other size variables such as numbers of flowers, stems and leaves. Our data suggest these weevil predators may prefer to forage on resources closer to the ground, where a lower predation risk may be associated with cover from litter and canopy vegetation, or where less energy is required to climb shorter stems to reach capsules for oviposition. Although our analyses do not indicate any response to food resource abundance such as the number of fruits on a plant, Recart *et al.* (2013) found a significant increase in abundance of another orchid weevil, *Sethobaris polita*, along with increased floral damage and reduced fruit set on a native Puerto Rican orchid, *Bletia patula*, in sites where an invasive orchid *Spathoglottis plicata* co-occurs. Complex plant–pollinator–seed predator interactions have been documented by others (Strauss and Irwin 2004), although instances of conflicting pressures seem to greatly outnumber instances of concurrent pressure.

Conclusions

Our study quantifies a three-way interaction between plants, pollinators and seed predators in a deceptive orchid system in which mutualists and antagonists are exerting concordant, reinforcing selection on a plant trait, reproductive stem height. Pollinators visited plants with taller stems more often, while another trait often associated with increasing pollinator visitation, floral display size, had no effect on pollination. Furthermore, measures of plant size, such as numbers of stems and leaves that might influence resources available for fruit maturation, did not affect female reproductive success. Seed predators may be attracted to more easily reached resources that are sheltered by surrounding vegetation and less apparent to their invertebrate predators (Marquis 1992; Palo *et al.* 1993). Additionally, the greater seed mass from both outcrossed hand and open pollination events compared with hand self-pollinations suggests that deceit pollination effectively prevents geitonogamy, so that most pollinations in *C. candidum* arise predominately from outcrossing. Although variable in magnitude and based on a small sample, inbreeding depression reduced seed mass by 11–67 % in all but one family, suggesting that conservation of small, at-risk populations that may be vulnerable to decreased pollination opportunities and increased geitonogamy should focus on facilitating outcrossing to increase recruitment.

In addition to the strong directional selective pressure on flowering stem height from both antagonists and mutualists, the height and density of the surrounding hetero-specific vegetation matrix likely enhances this selection for taller flowering stems. The evolution of this complex interaction may hold an important lesson in the conservation of *Cypripedium* spp. and other deceptive plants. Management activity that controls surrounding hetero-specific vegetation density and height during the flowering period may increase pollination success and fruit maturation, functioning as a cost-effective method to potentially increase rare plant offspring recruitment by modifying the pre-existing natural selective pressures on the biotic interactions of the system. However, while there have been several studies examining the effect of nectar addition on deceptive orchids (e.g. Johnson and Nilsson 1999; Johnson *et al.* 2004; Jersáková and Johnson 2006), there has been no study to date explicitly examining the demographic consequences of the deceptive pollination strategy in any plant. Although further research across multiple years and populations is needed to quantify the relative importance of pollination limitation and seed predation, orchid restoration efforts may benefit from research on whether management can be specifically targeted to ameliorate chronic pollination limitation or seed predation in orchids. As recently argued by Sletvold and Ågren (2014), spatial and temporal variation in selection mediated by the biotic environment strongly affects the extent of pollinator-mediated selection. We propose that the surrounding vegetation context that provides the arena for plant–pollinator–predator interactions, as well as seed dispersal, is an important, relatively understudied component that should be considered for both managers of species of conservation concern as well as biologists seeking greater understanding of the evolution and functional significance of floral traits in these complex interactions.

Sources of Funding

Our work was funded in part by an undergraduate research grant through the Science, Engineering and Technology Gateway Ohio (SETGO)-NSF program.

Contributions by the Authors

This research was completed as part of the R.P.W.'s dissertation studying under H.J.M. P.M.A. participated in the 2011 study as part of the SETGO undergraduate research program.

Acknowledgements

We thank the Ohio Department of Natural Resources for access to the field site, and particularly Jennifer Windus for streamlining of the permit process and providing valuable data on the *C. candidum* population. We also thank the members of the Michaels lab for their input and development of this research as well as Randy Mitchell, Karen Root, Moira van Staaden and Timothy Murnen for serving on the dissertation committee and providing critical feedback. We are indebted to two anonymous reviewers for their insights and excellent suggestions for improvements in the manuscript.

Literature Cited

Abdala-Roberts L, Parra-Tabla V, Salinas-Peba L, Herrera CM. 2009. Noncorrelated effects of seed predation and pollination on the perennial herb *Ruellia nudiflora* remain spatially consistent. *Biological Journal of the Linnean Society* **96**:800–807.

Ackerman JD. 1986. Mechanisms and evolution of food-deceptive pollination systems in Orchids. *Lindleyana* **1**:108–113.

Ackerman JD, Montalvo AM. 1990. Short- and long-term limitations to fruit production in a tropical Orchid. *Ecology* **71**:263–272.

Adler LS, Bronstein JL. 2004. Attracting antagonists: does floral nectar increase leaf herbivory? *Ecology* **85**:1519–1526.

Ågren J, Ehrlén J, Solbreck C. 2008. Spatio-temporal variation in fruit production and seed predation in a perennial herb influenced by habitat quality and population size. *Journal of Ecology* **96**: 334–345.

Ågren J, Hellström F, Toräng P, Ehrlén J. 2013. Mutualists and antagonists drive among-population variation in selection and evolution of floral display in a perennial herb. *Proceedings of the National Academy of Sciences of the USA* **110**:18202–18207.

Aizen MA, Harder LD. 2007. Expanding the limits of the pollen-limitation concept: effects of pollen quantity and quality. *Ecology* **88**:271–281.

Aragón S, Ackerman JD. 2004. Does flower color variation matter in deception pollinated *Psychilis monensis* (Orchidaceae)? *Oecologia* **138**:405–413.

Barrett SCH. 2003. Mating strategies in flowering plants: the out-crossing–selfing paradigm and beyond. *Philosophical Transactions of the Royal Society B: Biological Sciences* **358**:991–1004.

Barrett SCH, Harder LD. 1996. Ecology and evolution of plant mating. *Trends in Ecology and Evolution* **11**:73–79.

Bazzaz FA, Chiariello NR, Coley PD, Pitelka LF. 1987. Allocating resources to reproduction and defense. *BioScience* **37**:58–67.

Bowles ML. 1983. The tallgrass prairie orchids *Platanthera leucophaea* (Nutt.) Lindl. and *Cypripedium candidum* Muhl. ex Willd.: some aspects of their status, biology, and ecology, and implications toward management. *Natural Areas Journal* **3**:14–37.

Brody AK, Mitchell RJ. 1997. Effects of experimental manipulation of inflorescence size on pollination and pre-dispersal seed predation in the hummingbird pollinated plant *Ipomopsis aggregate*. *Oecologia* **110**:86–93.

Burd M. 1995. Pollinator behavioural responses to reward size in *Lobellia deckenii*: no escape from pollen limitation of seed set. *Journal of Ecology* **83**:865–872.

Burkhardt A, Delph LF, Bernasconi G. 2009. Benefits and costs to pollinating, seed-eating insects: the effect of flower size and fruit abortion on larval performance. *Oecologia* **161**:87–98.

Calvo RN. 1990. Inflorescence size and fruit distribution among individuals in three Orchid species. *American Journal of Botany* **77**: 1378–1381.

Cariveau D, Irwin RE, Brody AK, Garcia-Mayeya LS, von der Ohe A. 2004. Direct and indirect effects of pollinators and seed predators to selection on plant and floral traits. *Oikos* **104**:15–26.

Carlson JE, Holsinger KE. 2010. Natural selection on inflorescence color polymorphisms in wild Protea populations: the role of pollinators, seed predators, and intertrait correlations. *American Journal of Botany* **97**:934–944.

Catling PM, Knerer G. 1980. Pollination of the small white lady's-slipper (*Cypripedium candidum*) in Lambton County, southern Ontario. *The Canadian Field-Naturalist* **94**:435–438.

Cheptou PO, Donohue K. 2011. Environment-dependent inbreeding depression: its ecological and evolutionary significance. *New Phytologist* **189**:395–407.

Cozzolino S, Widmer A. 2005. Orchid diversity: an evolutionary consequence of deception? *Trends in Ecology and Evolution* **20**: 487–494.

Crone EE, Lesica P. 2004. Causes of synchronous flowering in *Astragalus scaphoides*, an iteroparous perennial plant. *Ecology* **85**: 1944–1954.

Crone EE, Miller E, Sala M. 2009. How do plants know when other plants are flowering? Resource depletion, pollen limitation and mast-seeding in a perennial wildflower. *Ecology Letters* **12**: 1119–1126.

Curtis JT. 1946. Use of mowing in management of white ladyslippers. *Journal of Wildlife Management* **10**:303–306.

Cusick AW. 1980. *Cypripedium candidum* Muhl ex Willd. Ohio Heritage Database.

Dafni A, Ivri Y. 1979. Pollination ecology of, and hybridization between, *Orchis coriophora* L. and *O. collina* Sol. ex Russ. (Orchidaceae) in Israel. *New Phytologist* **83**:181–187.

De Jong TJ, Waser NM, Klinkhamer PGL. 1993. Geitonogamy: the neglected side of selfing. *Trends in Ecology and Evolution* **8**: 321–325.

Dudash MR, Fenster CB. 1997. Multiyear study of pollen limitation and cost of reproduction in the iteroparous *Silene virginica*. *Ecology* **78**:484–493.

Faegri K, van der Pijl L. 1971. *The principles of pollination ecology*. Oxford, NY: Pergamon Press.

Galen C, Cuba J. 2001. Down the tube: pollinators, predators, and the evolution of flower shape in the alpine skypilot, *Polemonium viscosum*. *Evolution* **55**:1963–1971.

Galloway LF, Cirigliano T, Gremski K. 2002. The contribution of display size and dichogamy to potential geitonogamy in *Campanula americana*. *International Journal of Plant Sciences* **163**:133–139.

Gigord LDB, Macnair MR, Smithson A. 2001. Negative frequency dependent selection maintains a dramatic flower color polymorphism in the rewardless orchid *Dactylorhiza sambucina* (L.) Soo. *Proceedings of the National Academy of Sciences of the USA* **98**:6253–6255.

Gómez J. 2003. Herbivory reduces the strength of pollinator-mediated selection in the Mediterranean herb *Erysimum medio-hispanicum*: consequences for plant specialization. *The American Naturalist* **162**:242–256.

Grindeland JM, Sletvold N, Ims RA. 2005. Effects of floral display size and plant density on pollinator visitation rate in a natural population of *Digitalis purpurea*. *Functional Ecology* **19**:383–390.

Harder LD, Barrett SCH. 1995. Mating cost of large floral display in hermaphrodite plants. *Nature* **373**:512–515.

Hessing MB. 1988. Geitonogamous pollination and its consequences in *Gernium caespitosum*. *American Journal of Botany* **75**: 1324–1333.

Irwin RE, Brody AK. 2011. Additive effects of herbivory, nectar robbing and seed predation on male and female fitness estimates of the host plant *Ipomopsis aggregata*. *Oecologia* **166**:681–692.

Irwin RE, Strauss SY, Storz S, Emerson A, Guibert G. 2003. The role of herbivores in the maintenance of a flower color polymorphism in wild radish. *Ecology* **84**:1733–1743.

Jersáková J, Johnson SD. 2006. Lack of floral nectar reduces self-pollination in a fly-pollinated orchid. *Oecologia* **147**:60–68.

Jersáková J, Johnson SD, Kindlmann P. 2006. Mechanisms and evolution of deceptive pollination in orchids. *Biological Reviews of the Cambridge Philosophical Society* **81**:219–235.

Johnson SD, Bond WJ. 1997. Evidence for widespread pollen limitation of fruiting success in Cape wildflowers. *Oecologia* **109**: 530–534.

Johnson SD, Nilsson LA. 1999. Pollen carryover, geitonogamy, and the evolution of deceptive pollination systems in orchids. *Ecology* **80**:2607–2619.

Johnson SD, Peter CI, Agren J. 2004. The effects of nectar addition on pollen removal and geitonogamy in the non-rewarding orchid *Anacamptis morio*. *Proceedings of the Royal Society B: Biological Sciences* **271**:803–809.

Johnston MO, Schoen DJ. 1994. On the measurement of inbreeding depression. *Evolution* **48**:1735–1741.

Karron JD, Mitchell RJ. 2012. Effects of floral display size on male and female reproductive success in *Mimulus ringens*. *Annals of Botany* **109**:563–570.

Kindlmann P, Jersáková J. 2006. Effect of floral display on reproductive success in terrestrial orchids. *Folia Geobotanica* **41**:47–60.

Klinkhamer PGL, de Jong T. 2005. *Evolutionary ecology of plant reproductive strategies*. Cambridge, UK: Cambridge University Press.

Knight TM. 2003. Floral density, pollen limitation, and reproductive success in *Trillium grandiflorum*. *Oecologia* **137**:557–563.

Knight TM, Steets JA, Ashman T-L. 2006. A quantitative synthesis of pollen supplementation experiments highlights the contribution of resource reallocation to estimates of pollen limitation. *American Journal of Botany* **93**:271–277.

Kolb A, Ehrlén J. 2010. Environmental context drives seed predator-mediated selection on a floral display trait. *Evolutionary Ecology* **24**:433–445.

Kolb A, Ehrlen J, Eriksson O. 2007. Ecological and evolutionary consequences of spatial and temporal variation in pre-dispersal seed predation. *Perspectives in Plant Ecology, Evolution and Systematics* **9**:79–100.

Kropf M, Renner SS. 2008. Pollinator-mediated selfing in two deceptive orchids and a review of pollinium tracking studies addressing geitonogamy. *Oecologia* **155**:497–508.

Krupnick GA, McCormick MK, Mirenda T, Whigham DF. 2013. The status and future of orchid conservation in North America. *Annals of the Missouri Botanical Garden* **99**:180–198.

Kudoh H, Whigham DF. 1998. The effect of petal size manipulation on pollinator/seed-predator mediated female reproductive success of *Hibiscus moscheutos*. *Oecologia* 117:70–79.

Kull T. 1998. Fruit-set and recruitment in populations of *Cypripedium calceolus* L. in Estonia. *Botanical Journal of the Linnean Society* 126:27–38.

Li P, Huang BQ, Pemberton RW, Luo YB, Cheng J. 2011. Floral display influences male and female reproductive success of the deceptive orchid *Phaius delavayi*. *Plant Systematics and Evolution* 296:21–27.

Light MHS, MacConaill MC. 2002. Climatic influences on flowering and fruiting of *Cypripedium parviflorum* var. *pubescens*. In: Kindlmann P, Willems JH, Whigham DH, eds. *Trends and fluctuations and underlying mechanisms in terrestrial orchid populations*. Leiden, The Netherlands: Backhuys Publishers, 85–97.

Light MHS, MacConaill MC. 2011. Potential impact of insect herbivores on orchid conservation. *European Journal of Environmental Sciences* 1:115–124.

Louda SM, Potvin MA. 1995. Effect of inflorescence-feeding insects on the demography and lifetime of a native plant. *Ecology* 76:229.

Marquis RJ. 1992. Selective impact of herbivores. In: Fritz RS, Simms EL, eds. *Plant resistance to herbivores and pathogens, ecology, evolution and genetics*. University of Chicago Press, 301–325.

Meléndez-Ackerman EJ, Ackerman JD. 2001. Density-dependent variation in reproductive success in a terrestrial orchid. *Plant Systematics and Evolution* 227:27–36.

Mitchell RJ, Karron JD, Holmquist KG, Bell JM. 2004. The influence of *Mimulus ringens* floral display size on pollinator visitation patterns. *Functional Ecology* 18:116–124.

Murren CJ, Dudash MR. 2012. Variation in inbreeding depression and plasticity across native and non-native field environments. *Annals of Botany* 109:621–632.

Neiland MRM, Wilcock C. 1998. Fruit set, nectar reward, and rarity in the Orchidaceae. *American Journal of Botany* 85:1657–1671.

Nilsson LA. 1980. The pollination ecology of *Dactylorhiza sambucina* (Orchidaceae). *Botaniska Notiser* 133:367–385.

Nilsson LA. 1983. Mimesis of bellflower (*Campanula*) by the red helleborine orchid *Cephalanthera rubra*. *Nature* 305:799–800.

Nilsson LA. 1984. Anthecology of *Orchis morio* (Orchidaceae) at its outpost in the north. *Nova Acta Regiae Societatis Scientiarum Upsaliensis* 3:167–179.

Nilsson LA, Rabakonandrianina E, Pettersson B. 1992. Exact tracking of pollen transfer and mating in plants. *Nature* 360:666–668.

O'Connell LM, Johnston MO. 1998. Male and female pollination success in a deceptive orchid, a selection study. *Ecology* 79:1246–1260.

Palo TR, Gowda J, Högberg P. 1993. Species height and root symbiosis, two factors influencing antiherbivore defense of woody plants in East African savanna. *Oecologia* 93:322–326.

Parachnowitsch AL, Caruso CM. 2008. Predispersal seed herbivores, not pollinators, exert selection on floral traits via female fitness. *Ecology* 89:1802–1810.

Peakall R. 1989. The unique pollination of *Leporella fimbriata* (Orchidaceae): pollination by pseudocopulating male ants (*Myrmecia urens*, Formicidae). *Plant Systematics and Evolution* 167:137–148.

Primack RB, Hall P. 1990. Costs of reproduction in the Pink Lady's Slipper Orchid: a four-year experimental study. *The American Naturalist* 136:638–656.

Rasmussen HN, Whigham DF. 1993. Seed ecology of dust seeds insitu; a new study technique and its application in terrestrial orchids. *American Journal of Botany* 80:1374–1378.

Recart W, Ackerman JD, Cuevas AA. 2013. There goes the neighborhood: apparent competition between invasive and native orchids mediated by a specialist florivorous weevil. *Biological Invasions* 15:283–293.

Russell FL, Rose KE, Louda SM. 2010. Seed availability and insect herbivory limit recruitment and adult density of native tall thistle. *Ecology* 91:3081–3093.

Schiestl FP. 2005. On the success of a swindle: pollination by deception in orchids. *Naturwissenschaften* 92:255–264.

Shimamura R, Kachi N, Kudoh H, Whigham DF. 2005. Visitation of a specialist pollen feeder *Althaeus hibisci* Olivier (Coleoptera: Bruchidae) to flowers of *Hibiscus moscheutos* L.(Malvaceae) 1. *The Journal of the Torrey Botanical Society* 132:197–203.

Sih A, Baltus MS. 1987. Patch size, pollinator behavior, and pollinator limitation in catnip. *Ecology* 68:1679–1690.

Sletvold N, Ågren J. 2010. Pollinator-mediated selection on floral display and spur length in the orchid *Gymnadenia conopsea*. *International Journal of Plant Sciences* 171:999–1009.

Sletvold N, Ågren J. 2011. Nonadditive effects of floral display and spur length on reproductive success in a deceptive orchid. *Ecology* 92:2167–2174.

Sletvold N, Ågren J. 2014. There is more to pollinator-mediated selection than pollen limitation. *Evolution* 68:1907–1918.

Sletvold N, Grindeland JM, Ågren J. 2010. Pollinator-mediated selection on floral display, spur length and flowering phenology in the deceptive orchid *Dactylorhiza lapponica*. *New Phytologist* 188:385–392.

Sletvold N, Grindeland JM, Zu P, Ågren J. 2012. Strong inbreeding depression and local outbreeding depression in the rewarding orchid *Gymnadenia conpsea*. *Conservation Genetics* 13:1305–1315.

Sletvold N, Grindeland JM, Ågren J. 2013. Vegetation context influences the strength and targets of pollinator-mediated selection in a deceptive orchid. *Ecology* 94:1236–1242.

Smithson A. 2002. The consequences of rewardlessness in orchids: reward-supplementation experiments with *Anacamptis morio* (Orchidaceae). *American Journal of Botany* 89:1579–1587.

Snow AA, Whigham DF. 1989. Costs of flower and fruit production in *Tipularia discolor* (Orchidaceae). *Ecology* 70:1286–1293.

Snow AA, Spira TP, Simpson R, Klips RA. 1996. The ecology of geitonogamous pollination. In: Lloyd DG, Barrett SCH, eds. *Floral biology*. New York: Chapman & Hall, 191–216.

Stephens AEA, Myers JH. 2012. Resource concentration by insects and implications for plant populations. *Journal of Ecology* 100:923–931.

Stoutamire WP. 1967. Flower biology of the Lady's-Slippers (Orchidaceae: *Cypripedium*). *The Michigan Botanist* 6:159–175.

Strauss SY, Irwin RE. 2004. Ecological and evolutionary consequences of multispecies plant-animal interactions. *Annual Review of Ecology, Evolution, and Systematics* 35:435–466.

Sun H-Q, Cheng J, Zhang F-M, Luo Y-B, Ge S. 2009. Reproductive success of non-rewarding *Cypripedium japonicum* benefits from low spatial dispersion pattern and asynchronous flowering. *Annals of Botany* **103**:1227–1237.

Tremblay RL, Ackerman JD, Zimmerman JK, Calvo RN. 2004. Variation in sexual reproduction in orchids and its evolutionary consequences: a spasmodic journey to diversification. *Biological Journal of the Linnean Society* **84**:1–54.

Wake CM. 2007. Micro-environment conditions, mycorrhizal symbiosis, and seed germination in *Cypripedium candidum*: strategies for conservation. *Lankesteriana* **7**:423–426.

Walsh RP. 2008. *Cypripedium* (Orchidaceae) Hybridization along a Prairie/Woodland ecotone: evidence for hybridization using morphology and genetics. MS Thesis, Bowling Green State University, USA.

Walsh RP. 2013. Pollination ecology and demography of a deceptive orchid. PhD Dissertation, Bowling Green State University, USA.

Wesselingh RA. 2007. Pollen limitation meets resource allocation: towards a comprehensive methodology. *New Phytologist* **174** 26–37.

6

Host tree phenology affects vascular epiphytes at the physiological, demographic and community level

Helena J. R. Einzmann[1]*, Joachim Beyschlag[1], Florian Hofhansl[2], Wolfgang Wanek[2] and Gerhard Zotz[1,3]

[1] Department of Biology and Environmental Sciences, Carl von Ossietzky University of Oldenburg, Carl-von-Ossietzky-Straße 9-11, D-26111 Oldenburg, Germany
[2] Department of Microbiology and Ecosystem Science, University of Vienna, Althanstrasse 14, A-1090 Vienna, Austria
[3] Smithsonian Tropical Research Institute, Apartado Postal 0843-03092, Balboa, Ancon, Panamá, República de Panamá

Associate Editor: Markus Hauck

Abstract. The processes that govern diverse tropical plant communities have rarely been studied in life forms other than trees. Structurally dependent vascular epiphytes, a major part of tropical biodiversity, grow in a three-dimensional matrix defined by their hosts, but trees differ in their architecture, bark structure/chemistry and leaf phenology. We hypothesized that the resulting seasonal differences in microclimatic conditions in evergreen vs. deciduous trees would affect epiphytes at different levels, from organ physiology to community structure. We studied the influence of tree leaf phenology on vascular epiphytes on the Island of Barro Colorado, Panama. Five tree species were selected, which were deciduous, semi-deciduous or evergreen. The crowns of drought-deciduous trees, characterized by sunnier and drier microclimates, hosted fewer individuals and less diverse epiphyte assemblages. Differences were also observed at a functional level, e.g. epiphyte assemblages in deciduous trees had larger proportions of Crassulacean acid metabolism species and individuals. At the population level a drier microclimate was associated with lower individual growth and survival in a xerophytic fern. Some species also showed, as expected, lower specific leaf area and higher $\delta^{13}C$ values when growing in deciduous trees compared with evergreen trees. As hypothesized, host tree leaf phenology influences vascular epiphytes at different levels. Our results suggest a cascading effect of tree composition and associated differences in tree phenology on the diversity and functioning of epiphyte communities in tropical lowland forests.

Keywords: Community assembly; Crassulacean acid metabolism (CAM); diversity; microclimate; specific leaf area; water-use efficiency.

Introduction

Tropical forests are characterized by an unmatched diversity of organismal forms. More than 300 tree species may be found in a single hectare of forest (Balslev *et al.* 1998); and a single tree may in the extreme case be inhabited by almost 200 species of vascular epiphytes (Catchpole and Kirkpatrick 2010). Currently, studies on processes that shape tropical plant communities show a very strong bias towards a single life form, i.e. trees (e.g. Kraft and Ackerly 2010; Metz 2012; Baldeck *et al.* 2013; Brown *et al.* 2013). However, other life forms, e.g. understorey herbs and shrubs, lianas and epiphytes, can make up a substantial proportion of local plant diversity (Gentry and Dodson 1987). Indeed, in some humid

* Corresponding author's e-mail address: helena.einzmann@uni-oldenburg.de

montane forests, epiphytes alone may account for a similar fraction of local vascular plant diversity as all other plant life forms combined (Kelly *et al.* 2004).

It is hardly conceivable that vascular epiphyte assemblages made up of structurally dependent, mostly herbaceous plants arranged in a three-dimensional matrix of available substrate supplied by trees are structured by the same biotic and abiotic factors as their hosts. For example, interspecific competition and the impact of pathogens and herbivores, which are very important for trees, seem to be rather irrelevant for epiphytes (Zotz and Hietz 2001). One important biotic factor affecting community assembly of these structurally dependent plants could be host tree identity itself. Arguably, narrow host tree specificity is found only in exceptional cases, whereas a certain degree of host preference is not uncommon (ter Stege and Cornelissen 1989; Laube and Zotz, 2006; Martinez-Melendez *et al.*, 2008). Mechanistically, such preferences are probably related to differences in tree architecture, bark structure and chemistry or leaf phenology of the host. Furthermore, there is also stratification within individual trees (Freiberg 1996; Freiberg, 1999; Krömer *et al.*, 2007). ter Stege and Cornelissen (1989), for example, found that epiphyte assemblages on lower canopy branches (Johansson zone III; Johansson 1974) differed from those on middle and outer canopy branches (Johansson zones IV and V) in a lowland rainforest in Guyana. Several studies have investigated the impact of a variety of tree characteristics on epiphyte distribution, such as tree size (Zotz and Vollrath 2003), branch diameter (Zimmerman and Olmsted 1992) or bark water-holding capacity (WHC, Callaway *et al.* 2002).

Differing tree characteristics create a variety of microclimates within a tree crown that should directly impact epiphyte distributions (Pittendrigh 1948; Benzing, 1995). The changing light regime along the vertical axis of the canopy is one important factor. Lower and more central crown parts are more humid than the outer portions and exposed parts within a tree are usually the driest (e.g. Freiberg 1997; Wagner *et al.* 2013). Stem diameter, tree height and other architectural features of tree species seem to influence stem flow (Hölscher *et al.* 2003), and the stratification of the crown structure influences the amount and the chemical composition of rainwater reaching the epiphytes (Hofhansl *et al.* 2012). However, gradients are not stable in time but are affected by phenological changes of the tree usually associated with wet and dry seasons. Seasonality may be an important factor for the establishment of epiphytes; especially when the host tree is deciduous (Zotz and Winter 1994). Indeed, it has been found that differing microenvironments in evergreen and deciduous trees can lead to significant differences in epiphyte cover (Cardelus

2007). The situation is complex, however, because other traits, e.g. bark characteristics, can aggravate or, alternatively, counteract the possible effects of phenology since epiphyte growth is significantly higher in trees with bark with high WHC (Callaway *et al.* 2002).

The work presented here was initiated to understand the effect of tree leaf phenology on vascular epiphytes at different levels, from leaf traits of individual epiphytes and demographic parameters to community composition, both at a taxonomic and a functional level. To this end, we documented the differences in environmental conditions in the crowns of five tree species on Barro Colorado Island (BCI), which were deciduous, semi-deciduous and evergreen. We hypothesized that:

(i) deciduous trees host less diverse epiphyte communities,

(ii) traits usually interpreted as adaptation to drought such as Crassulacean acid metabolism (CAM) are more prevalent in epiphyte communities in deciduous trees,

(iii) individual growth is lower in epiphyte populations growing in deciduous trees,

(iv) trait differences associated with drought are also detected at an intraspecific level, thus epiphytes growing in deciduous trees have different trait values such as lower specific leaf area (SLA) and higher $\delta^{13}C$ values compared with conspecifics growing in evergreen trees.

Methods

Study site and survey

The study was conducted on BCI, Republic of Panama. This biological reserve in the Gatun Lake of the Panama Canal, which is administered by the Smithsonian Tropical Research Institute (STRI), is covered by a semi-deciduous lowland forest with varying canopy heights of up to 50 m (Leigh *et al.* 2004). The annual precipitation averages 2600 mm with a pronounced dry season from January through April. During these 4 months, both total rainfall and its frequency are strongly reduced. Rainless periods regularly expand to around 10 days, the maximum rainless period on record being 31 days (based on data from 2006 to 2012 measured by the Physical Monitoring Program of STRI—http://biogeodb.stri.si.edu/physical_monitoring/research/barrocolorado, 10 November 2014).

Five tree species were selected for the study. Selection criteria were emergent crowns, thus deciduousness would have a strong effect on microclimate; more or less even distribution across the island; accessibility with the single rope technique. The following species were selected: *Anacardium excelsum* (Bertero ex Kunth) Skeels (Anacardiaceae) and *Brosimum alicastrum* Sw. (Moraceae; both evergreen), *Ceiba pentandra* (L.) Gaertn. (Malvaceae) as a semi-deciduous species and *Pseudobombax septenatum* (Jacq.) Dugand and *Cavanillesia platanifolia* (Humb. & Bonpl.) Kunth (both deciduous Malvaceae). In the following, species are addressed by their genus names. The deciduous trees are leafless during the entire dry season (Fig. 1). Semi-deciduous *Ceiba* trees usually lose their leaves only for a few weeks, but every 4 to 5 years when flowering and fruiting the leafless phase lasts for up to 20 weeks (Windsor, unpublished). All chosen individuals (four to five per species, average height 40 m) were large canopy trees lacking lianas and were distributed over the entire island **[see Supporting Information]**. Tree crown diameter was estimated via stride length and diameter at breast height (dbh) determined with a tape measure.

All trees were accessed with the single rope technique (Perry 1978). Epiphytes on the trunk and on branches in the inner canopy could be accessed directly; those in the outer canopy were recorded with the help of binoculars from the central crown. Abundances of all epiphytes were quantified in a 90° sector of each tree crown and then extrapolated. The remaining crown was screened for additional species. Nomadic vines and hemiepiphytes (as defined by Zotz 2013) were recorded but excluded from statistical analyses unless being truly epiphytic, i.e. without contact to the ground. Species names follow Croat (1978) and Tropicos (2012). Voucher specimens were deposited in the Herbarium of the University of Panama. Since the delimitation of individual plants was often difficult, Sanford's (1968) definition of a stand was followed: a group of rhizomes, stems and leaves belonging to one species, which is clearly separated from conspecifics. Small juveniles were only included when exceeding 20 % of the maximum size of a given species.

Microclimate

The trees we studied were canopy trees with at least partly emergent crowns. Microclimatic differences should be most pronounced within crowns, while stems are protected by both crown and surrounding vegetation leading to more homogeneous microclimatic conditions. Thus, we confined microclimate measurements to tree crowns. A crown was defined as the upper part of the tree with all major limbs, encompassing Johansson zones III–V (Johansson 1974). Total radiation in the inner crown (analogous to Johansson zone III) of each studied tree was documented for about 1 year from 22 March 2011 to 6 March 2012 with one HOBO pendant data loggers (Onset Computer Corporation, Pocasset, USA) per tree. Loggers were placed in a southerly direction within Johansson zone III. The pendant loggers measure ambient luminance in lux within a broad wavelength spectrum

(ca. 150–1200 nm). Thus, the data are not numerically comparable with photosynthetic active radiation, but allow a relative comparison. As such we defined the highest measured daily mean as 100 % and expressed all other readings relative to this extreme. Readings were taken every 40 min. Loggers were calibrated against each other. For statistical analyses we calculated daily means.

Evaporation was estimated after Didham and Lawton (1999). Per tree four millilitre-scaled test tubes were filled with 10 mL of water each, a filter paper was added to enlarge the evaporating surface. The tubes were left in the crown (oriented in the four cardinal directions in Johansson zone III) for 2–6 h around noon, allowing the calculation of hourly evaporation. Because daily variation in evaporation was large and measurements were done on different days from 19 February to 24 March 2010 and from 5 March to 21 March 2011 (in *Cavanillesia*), we standardized our data. Evapotranspiration is continuously measured with an ET gauge (Spectrum Technologies, Inc., Aurora, USA) on top of a walk-up tower at a height of 48 m by the Physical Monitoring Program of STRI. In March 2011, we compared the readings obtained with our method on the tower with readings of the ET gauge on 4 days. This allowed us to standardize our data and to express evaporation measured in tree crowns as a proportion of the same day's evapotranspiration measured at the tower.

To allow comparisons of the microclimatic conditions within tree crowns, the study period was divided into dry and wet seasons. The two dry seasons covered by this study were quite distinct, but differences within the tree species were consistent. Thus, we combined the data of both dry seasons. For regions with pronounced dry periods, dry season duration is commonly defined as the period to accumulate 10 % of annual total rainfall (Reiser and Kutiel 2008). This percentage also represents the average amount reported by the Physical Monitoring Programme of STRI. The dry season on the BCI is not only characterized by lower absolute rainfall but also by a lower frequency of rainfall events. Therefore, we defined the start and the end of dry seasons according to a combination of frequency and amount of rainfall. The first dry season ended on 23 May 2011, which was the last day of a series of at least 5 days with <0.01 % of the year's total rainfall. The start of the second dry season was 24 December 2011, which was the first day of a series of at least 5 days with <0.01 % of the year's total rainfall. As expected, the total rainfall in the dry seasons was ca. 10 % of the annual total rainfall. Rainfall data from the study period are given in Fig. 2.

Tree bark

Samples from major branches in the inner crown of four trees per species were collected by cutting off pieces of

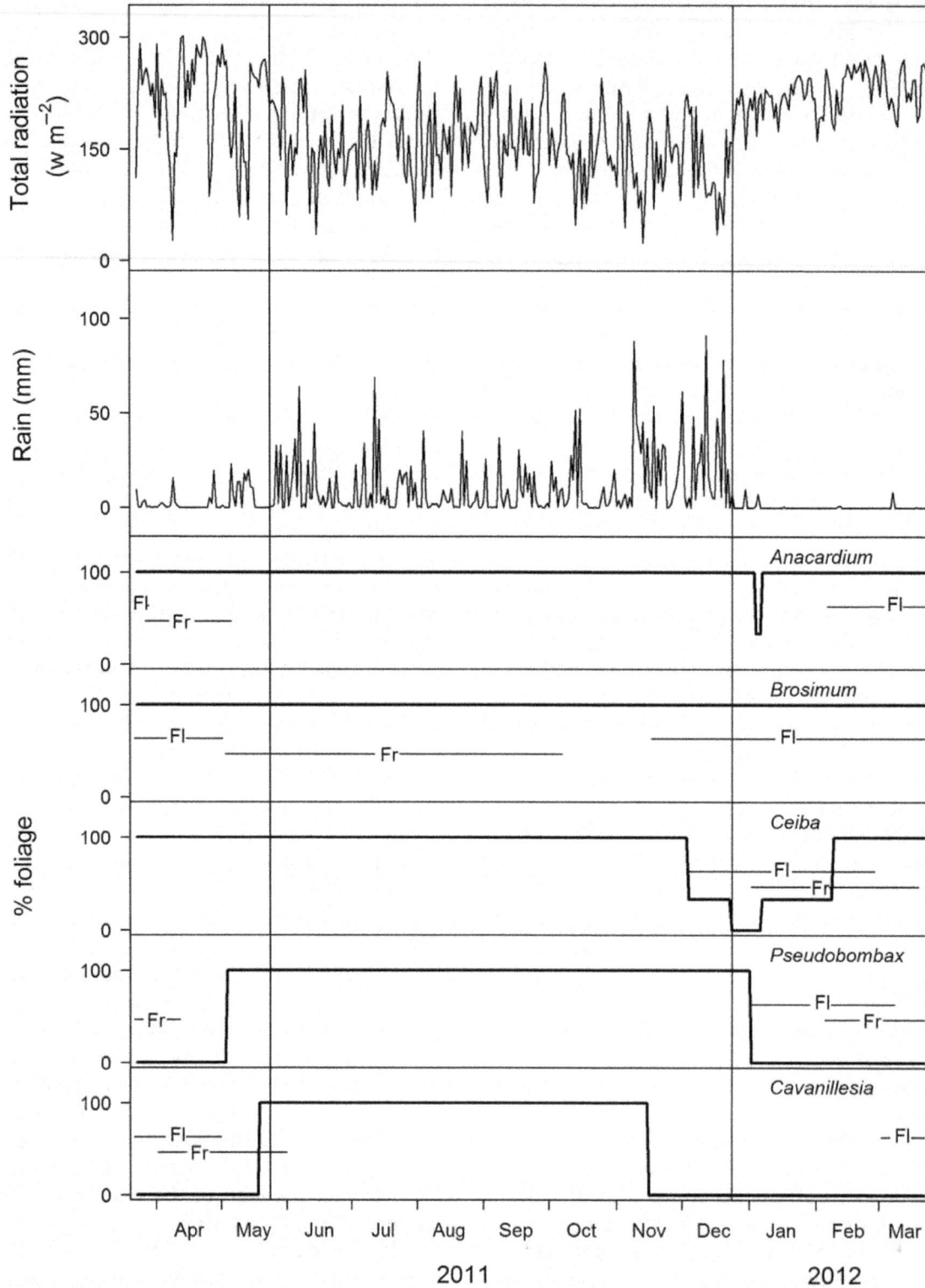

Figure 1. Seasonal changes in climate during the study period and long-term average phenology of the trees studied. The upper plots show daily mean radiation and daily precipitation data measured at 48 m, above the forest canopy. Thin vertical lines indicate the limits of dry and wet seasons—typically, the dry season ends earlier than in 2011. In the five tree plots: thick lines indicate leaf status, 'Fl' time of flowering and 'Fr' fruiting. Leaves of *A. excelsum* are exchanged within days, whereas *B. alicastrum* is never leafless. *Ceiba pentandra* may lose its leaves for about three months, which happens only every 4 to 5 years. *Pseudobombax septenatum* and *C. platanifolia* are drought-deciduous trees. Climate data are from a walk-up tower of the Physical Monitoring Program of STRI. Phenology data are from the Panama species database provided by the STRI website and Condit *et al.* (2011).

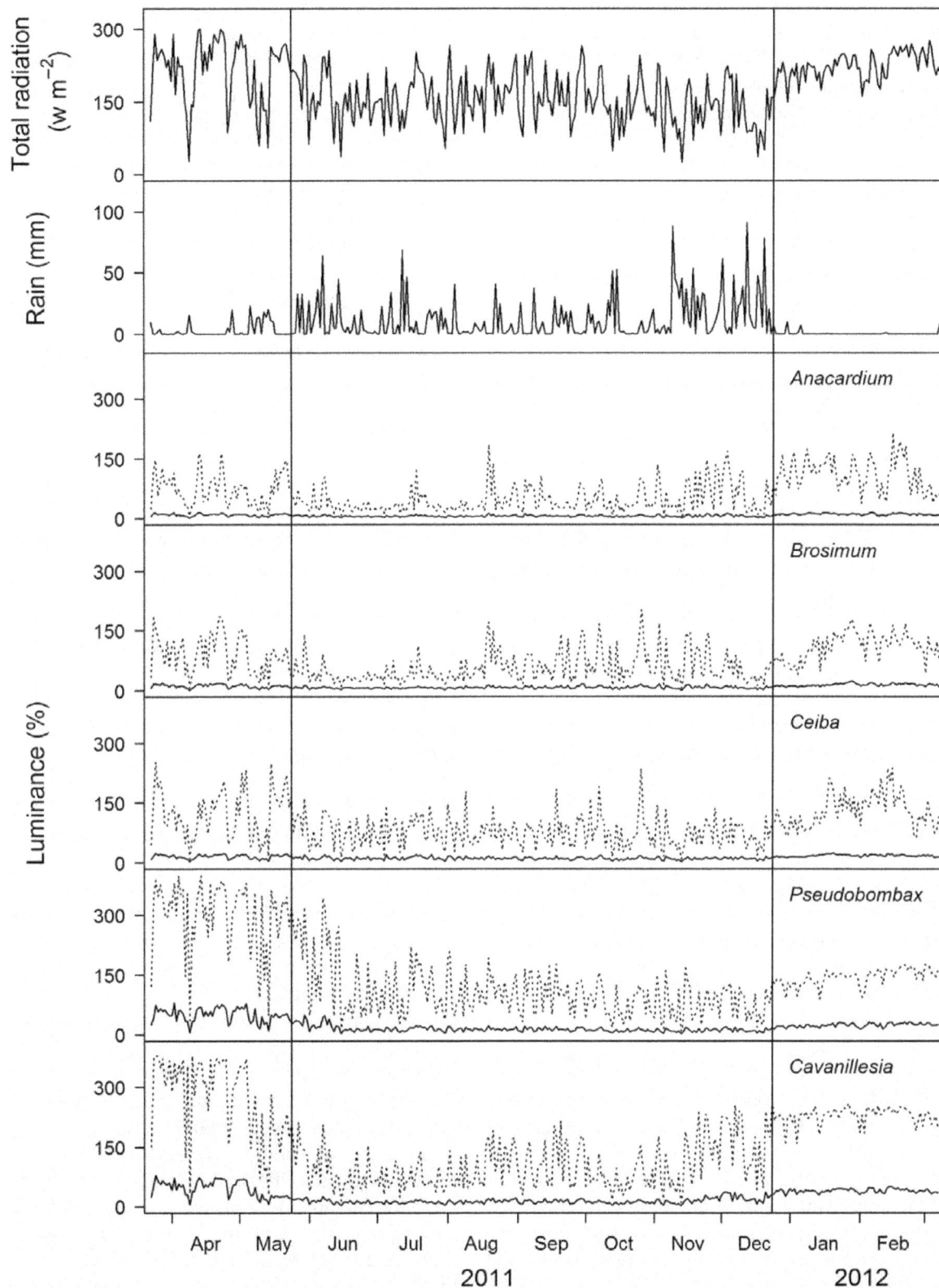

Figure 2. Seasonal changes in radiation and rainfall. Data for radiation and rain data were obtained from the Physical Monitoring Program of STRI. Light is presented as daily mean, while rain represents daily integrals. Daily means (solid line) of total radiation in percentage of the highest daily mean measured in the tree crowns of the five species studied (*A. excelsum*, *n* = 4 trees; *B. alicastrum*, *n* = 4; *C. pentandra*, *n* = 4; *P. septenatum*, *n* = 3; *C. platanifolia*, *n* = 4) over the course of 1 year. The dotted lines depict the mean daily maximum. The vertical lines mark the limits of dry and wet seasons (see MM for definition).

the phellem. This material was used to determine bark WHC (compare with Callaway et al. 2002). Briefly, samples were cleaned of any moss, oven-dried (50 °C) and their dry weight determined. Water-holding capacity was determined after soaking the samples in water for half an hour. Excess water was shaken off in a consistent manner and samples were weighed again after a 5-min pause (=maximum WHC per area). Subsequently, water loss rates were determined by keeping the samples at a constant temperature of ~22 °C for 8 h on a lab bench and weighing them at hourly intervals and expressing the loss as percentage of the maximum WHC.

Epiphyte growth

We selected two abundant species for growth analysis, which formed accessible populations in a reasonable subset of the study trees. Growth of individuals in populations of Dimerandra emarginata and Niphidium crassifolium was estimated between March 2011 and March 2012. These two species are generally abundant on BCI, are relatively well adapted to drought and grew in almost all surveyed trees. However, plants were not always accessible. We were able to include seven populations of Dimerandra in one Brosimum, three Ceiba and three Pseudobombax trees and eight populations of Niphidium in two Brosimum, three Ceiba, two Pseudobombax and one Cavanillesia trees. Populations from conspecific tree individuals were pooled for the analyses.

Initial size, i.e. the length of the longest shoot of a plant (Dimerandra) and length of the longest leaf (Niphidium), was measured; plants were marked and mapped. The original size was defined as 100 %. The following year the difference in the length of the longest shoot/leaf to first year's length was calculated as the relative positive or negative growth in per cent and analysed with initial size as a covariate. Species differences in survival and survival of the same species on different trees were analysed with Fisher's exact test for count data.

Epiphyte physiology

A total of 509 leaf samples of 59 species were analysed (usually six replicates) for foliar nitrogen (N) concentration and δ^{13}C- and δ^{15}N values. δ^{13}C values were used (i) to distinguish CAM- and C_3-species, using the generally accepted value of -20 ‰ as the cut-off between these photosynthetic pathways (Zotz 2004) and (ii) among the C_3 plants they were used as a proxy for long-term water-use efficiency in intraspecific comparisons (Van de Water et al. 2002) (see the next paragraph).

A large proportion (271 samples) represented five abundant species that were used to analyse intraspecific variation in SLA, δ^{13}C value, δ^{15}N value and leaf N concentration in relation to host tree species and microclimate.

For SLA the leaf surface was estimated via pixel counting of photographs using an image processing program (ImageJ 1.45, Rasband 1997–2012) and subsequent determination of leaf dry weight. Physiological parameters in epiphytes tend to vary substantially with plant size (Zotz et al. 2001). Hence, we recorded the plant size and used it as a covariate in our analyses. For the orchids their stem length, and for the two ferns the length of their longest leaf were taken as a proxy for size. All five abundant epiphyte species were C_3 plants (Zotz and Ziegler 1997): D. emarginata (G. Mey) Hoehne, Maxillaria uncata Lindl., Scaphyglottis behrii (Rchb. f.) Benth.& Hook. f. ex Hemsl. (all three Orchidaceae), N. crassifolium (L.) Lellinger (Polypodiaceae) and Vittaria lineata (L.) J. Sm. (Vittariaceae). These five species were selected for their high abundance that allowed us to sample them on most tree species. Dimerandra and Niphidium grew on all five study tree species and Niphidium could be sampled on at least one of each. Dimerandra could not be sampled on Anacardium. Vittaria grew in all but the Cavanillesia trees, but could not be sampled on Pseudobombax. Maxillaria and Scaphyglottis were not found on the two deciduous tree species.

Data analysis

Statistical tests were conducted with R 2.15.2 (R Development Core Team 2012). All data was related to tree species as factor variables: microclimate, bark maximum WHC and water loss rates, epiphyte diversity (Shannon Index), percentage of CAM species within tree crowns and the foliar traits SLA, δ^{13}C value, δ^{15}N value and N concentration. These data were tested for normal distribution with 'Lillifors (Kolmogorov-Smirnov) test for normality', and homogeneity of variances was tested with the 'Fligner-Killeen Test of Homogeneity of Variances'. When prerequisites were fulfilled we used parametric tests ANOVA and ANCOVA. Otherwise, we used the Kruskal–Wallis rank sum test (KW). Post-hoc tests were Tukey HSD for parametric data sets and Nemenyi-Damico-Wolfe-Dunn test for non-parametric data sets. Count data of survival from the growth study were analysed with Fisher's exact test for count data. The Shannon diversity index of abundance data was generated with EstimateS 8.20 (Colwell 2009). To assess the compositional similarity of epiphyte assemblages within the tree crowns (intra-specific and all species pooled) we calculated the multiple-assemblage similarity profile following Jost et al. (2011). This profile illustrates how differentiation of the assemblages varies between all species ($q = 0$), the typical species ($q = 1$) and the dominant species ($q = 2$). For the canonical correspondence analyses, we used Canoco 4.5 (ter Braak and Šmilauer 1997). This multivariate technique detects variations in the species assemblage caused by environmental variables. Tree species identity was incorporated in the analysis as supplementary nominal variable. Other environmental variables tested were evaporation, luminance (separated for the dry and wet seasons) and dbh as a measure of tree size. The statistical significance of the environmental variables was tested using the Monte Carlo permutation test with manual forward selection (Monte Carlo test run 9999 times). The analyses were run with abundance data and binary data (presence/absence). Scaling of data was symmetric between inter-species and inter-sample distances. The scaling type was a biplot scaling and data were log-transformed and rare species were downweighted.

In all box plots the median is depicted as bold black bar. The box represents the inner quartile range (IQR), while whiskers extend to extreme values within the first quartile $-1.5 \times$ IQR and, respectively, within the third quartile $+1.5 \times$ IQR. Empty circles indicate values beyond this range.

Results

Abiotic and biotic conditions

Microclimate: Luminance was three to four times higher in deciduous vs. evergreen tree crowns in the dry season (deciduous: 36 and 43 % of highest daily mean, evergreen: 10 and 13 %; KW: $P < 0.001$, $\chi^2 = 1228.9$, df = 4; Table 1, Fig. 2). The lowest luminance was observed in *Anacardium* and the highest in *Cavanillesia* (Table 1). These differences between deciduous and evergreen trees persisted during the wet season but were less pronounced. Luminance in the inner crowns of drought-deciduous trees with full foliage exceeded that in evergreen trees by about 2-fold (deciduous: 14 % of highest

Table 1. Mean luminance and evaporation in the study trees during dry and wet seasons. Superscript lower case letters indicate significant differences between trees (luminance KW: $P < 0.001$, n per species = 785–1404; evaporation ANOVA: $P < 0.05$, $F_{(4,17)} = 4.0$). Luminance was calculated as percentage of the highest daily mean. Evaporation was calculated in relation to measurements with an ET gauge of STRI (see MM for details). [+] Evergreen, [°] semi-deciduous, [−] deciduous species.

Species	Dry season		Wet season
	Luminance (%)	Evaporation (%)	Luminance (%)
A. excelsum[+]	9.7[a]	58[a]	5.5[a]
B. alicastrum[+]	13.1[b]	76[ab]	7.1[b]
C. pentandra[°]	16.2[c]	69[ab]	10.1[c]
P. septenatum[−]	35.9[d]	101[ab]	13.5[d]
C. platanifolia[−]	42.5[e]	121[b]	13.8[d]

daily mean, evergreen: 6 and 7 %; KW: $P < 0.001$, $\chi^2 = 1418.7$, df = 4, Table 1).

Average evaporation, measured only in the dry season, was about twice as high in deciduous trees compared with evergreen ones, but means were only significantly different between *Cavanillesia* and *Anacardium* trees (ANOVA: $P < 0.05$, $F_{(4,17)} = 4.0$; Table 1).

Bark: Species differences in water storage capacity of the bark were unrelated to leaf phenology. Maximum water content ranged from 20 to 144 mg cm^{-2}. Each pair of evergreen and deciduous trees had one species with relatively high and one with relatively low WHC (Table 2). However, these trends were not significant (KW: $P = 0.08$, $\chi^2 = 8.3$, df = 4). On the other hand, rates of water loss tended to be highest in the deciduous species. Water loss after 8 h was significantly higher for *Ceiba* and *Cavanillesia* compared with *Brosimum* (ANOVA: $P < 0.05$, $F_{(4,15)} = 4.8$; Table 2).

Epiphyte flora and assemblages

In total, the 24 studied trees were host to an estimated 26 000 individual epiphytes (= stands) from 83 holoepiphyte species, two hemiepiphyte species and two nomadic vines, which had no contact with the ground [see **Supporting Information** for a complete list]. These represented 16 families (Table 3). Three families accounted for more than two-thirds of all species, i.e. Orchidaceae (40 %, 35 spp.), Araceae (13 %, 11 spp.) and Bromeliaceae (10 %, 9 spp.). Ranking families by abundance leads to relatively small changes. Orchidaceae was even more prominent (56 %), followed by four families with at least 5 % of all individuals: Polypodiaceae (11 %), Dryopteridaceae, Araceae (both 8 %) and Bromeliaceae (5 %).

Three epiphyte species have not been documented previously for BCI (*Elaphoglossum* cf. *doanense*, *E. peltatum* and *E. latum*—all Dryopteridaceae), while the fourth, *Maxillaria acervata* (Orchidaceae), had been collected twice by the last author in recent years, without publishing this finding.

The largest number of species was found on the five *Anacardium* trees (65 species), which represents more than two-thirds of all epiphyte species found in this study. In *Brosimum* and *Ceiba* trees, we still found 49 and 50 species, respectively. The deciduous *Pseudobombax* and *Cavanillesia* supported considerably fewer epiphytes with only 20 and 7 species, respectively. None of the epiphyte species found on the deciduous trees were restricted to these tree species.

The results shown so far represent the assemblages of the entire trees. Epiphytes growing on the trunks of emergent trees are less affected by tree phenology since they are protected by the crown and the surrounding vegetation. Therefore, we expected differences in the epiphyte

Table 2. Water-holding capacity and water loss of the bark of five tree species. Values are means $\pm$ SD ($n = 4$) for each tree species (*A. excelsum* (Ana), *B. alicastrum* (Bro), *C. pentandra* (Cei), *P. septenatum* (Pse), *C. platanifolia* (Cav)). Significant differences are indicated by superscript lower case letters (ANOVA: $P < 0.05$, Tukey HSD).

	Ana	Bro	Cei	Pse	Cav
Bark WHC (mL cm^{-2})	94.0 $\pm$ 37.7	36.0 $\pm$ 15.5	64.5 $\pm$ 22.1	90.3 $\pm$ 61.7	33.3 $\pm$ 7.5
Bark water loss (%)	62.0 $\pm$ 6.7^{ab}	52.8 $\pm$ 13.6^{a}	74.3 $\pm$ 5.2^{b}	65.8 $\pm$ 7.8^{ab}	75.8 $\pm$ 7.3^{b}

Table 3. Epiphyte abundance and diversity in the trees studied. Shown are families of epiphytes found in the five tree species studied: *A. excelsum* (Ana, $n = 5$), *B. alicastrum* (Bro, $n = 4$), *C. pentandra* (Cei, $n = 5$), *P. septenatum* (Pse, $n = 5$) and *C. platanifolia* (Cav, $n = 5$). Species numbers are given in parentheses. Prevalence of CAM was calculated for each column's total excluding nomadic vines and hemiepiphytes. Data summarize epiphytes on a given tree, except 'CAM species—crown' and 'Shannon index' that refer to crown data only. Significant differences are indicated by superscript lower case letters (ANOVA: $P < 0.05$, Tukey HSD). *Including two species of nomadic vines. †Hemiepiphyte.

	Individuals (species)	Ana	Bro	Cei	Pse	Cav
Total	25 592 (87)	10 963 (65)	3432 (49)	8826 (50)	2296 (20)	75 (7)
Araceae*	2050 (11)	1504 (10)	288 (8)	249 (7)	7 (4)	2 (2)
Aspleniaceae	564 (2)	554 (2)	5 (1)	5 (1)	0 (0)	0 (0)
Bromeliaceae	1398 (9)	1041 (7)	201 (2)	154 (3)	2 (2)	0 (0)
Cactaceae	84 (2)	5 (1)	53 (2)	26 (2)	0 (0)	0 (0)
Clusiaceae†	118 (1)	19 (1)	23 (1)	76 (1)	0 (0)	0 (0)
Cyclanthaceae	218 (1)	8 (1)	0 (0)	210 (1)	0 (0)	0 (0)
Davalliaceae	379 (1)	139 (1)	23 (1)	217 (1)	0 (0)	0 (0)
Dryopteridaceae	2135 (6)	1827 (6)	2 (1)	305 (2)	1 (1)	0 (0)
Gesneriaceae	377 (1)	0 (0)	377 (1)	0 (0)	0 (0)	0 (0)
Hymenophyllaceae	30 (2)	30 (2)	0 (0)	0 (0)	0 (0)	0 (0)
Lycopodiaceae	14 (1)	5 (1)	5 (1)	4 (1)	0 (0)	0 (0)
Orchidaceae	14 329 (35)	5128 (23)	1701 (19)	5486 (21)	1982 (10)	32 (2)
Piperaceae	630 (5)	177 (2)	87 (4)	363 (2)	0 (0)	3 (1)
Polypodiaceae	2778 (7)	325 (5)	560 (6)	1564 (5)	291 (2)	38 (2)
Rubiaceae†	132 (1)	99 (1)	5 (1)	28 (1)	0 (0)	0 (0)
Vittariaceae	356 (2)	102 (2)	102 (1)	139 (2)	13 (1)	0 (0)
CAM plants	1242 (14)	117 (6)	319 (11)	401 (10)	374 (3)	31 (1)
CAM individuals (%)	4.9	1.1	9.3	4.5	16.3	41.3
CAM species (%)	16.0	9.0	21.2	18.5	13.6	14.3
CAM species—crown (%)		6.6 $\pm$ 5.5^{a}	18.5 $\pm$ 15.7^{ab}	22.6 $\pm$ 12.0^{ab}	31.3 $\pm$ 6.4^{b}	8.3 $\pm$ 16.7^{ab}
Shannon index		2.1 $\pm$ 0.6^{a}	2.0 $\pm$ 0.7^{a}	1.9 $\pm$ 0.7^{a}	0.7 $\pm$ 0.3^{b}	0.4 $\pm$ 0.5^{b}

assemblages to be most pronounced or even to be confined to the emergent crowns. Indeed we did not find significant differences in epiphyte diversity on the trunk between tree species (ANOVA: $P = 0.5$, $F_{(4,14)} = 0.9$), and excluded all epiphytes growing on the main trunks from further analysis. The average epiphyte diversity in the crowns of the evergreen and semi-deciduous species was ~4-fold higher compared with the two deciduous tree species (ANOVA: $P < 0.001$, $F_{(4,19)} = 9.8$; Table 3). Likewise, differences in epiphyte abundance among tree crowns were also significant (KW: $P < 0.01$, $\chi^2 = 13.6$, df = 4; data not shown). Deciduous trees hosted fewer individuals than evergreen and semi-deciduous trees, although the difference was only significant between deciduous *Cavanillesia* and both evergreen *Anacardium* and semi-deciduous *Ceiba*.

The consistency of the most abundant species between individual tree crowns of a species is indicated by the sensitivity parameter $q = 2$. *Anacardium* showed the least congruence (12 %) while most abundant species' identity was highly consistent in *Pseudobombax* (83 %, Fig. 3). Congruence of abundant species increased from *Brosimum* (34 %) to *Ceiba* (47 %). Deciduous *Cavanillesia* showed very low congruence with 14 % unlike its counterpart *Pseudobombax*. However, they were similar in respect to the high evenness in the species occurrence among their tree crowns, which is reflected in the less steep slopes of their similarity profile compared with the evergreen and semideciduous trees. All tree individuals pooled shared ~28 % of the most abundant species in their crowns.

The differentiation of epiphyte assemblages in evergreen and deciduous tree crowns was also reflected by a canonical correspondence analysis (Fig. 4). When environmental variables were added, luminance during the dry season and evaporation contributed significantly to the explanation of the variation in species data. Diameter at breast height as a proxy for tree size/age was also significant. An analysis with binary data yielded very similar results although evaporation had no significant influence anymore (data not shown).

Crassulacean acid metabolism

We compared the percentages of CAM-performing species and individuals in different tree crown assemblages

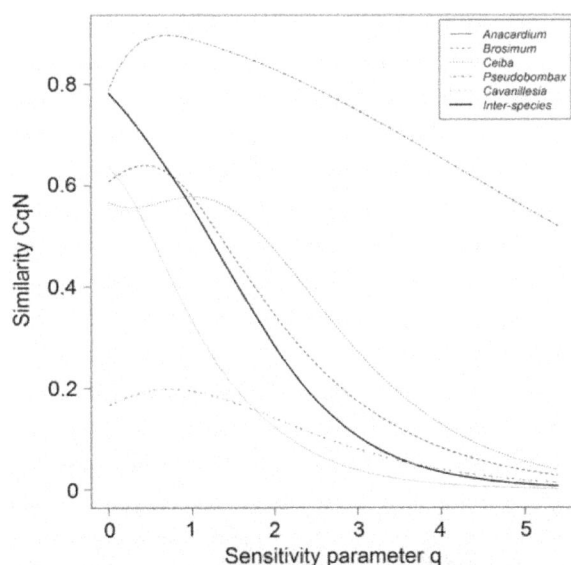

Figure 3. The multiple-assemblage similarity profile CqN (for $q = 0 - 5$) of epiphyte assemblages in the tree crowns of *A. excelsum* (5 trees/60 epiphyte species), *B. alicastrum* (4/43), *C. pentandra* (5/47), *P. septenatum* (5/7), *C. platanifolia* (4/7) and of epiphyte assemblages in tree crowns between all trees (23/79).

to address functional changes apart from floristic ones. Disregarding *Cavanillesia*, there was a gradual increase in CAM species from evergreen, semi-deciduous to deciduous tree crowns. The difference in the proportion of CAM species between deciduous *Pseudobombax* and evergreen *Anacardium* was significant (ANOVA: $P < 0.05$, $F_{(4,18)} = 3.7$; Table 3). The low percentage of CAM epiphytes in *Cavanillesia* crowns (33 % in one tree) is probably a sampling artefact, since *Cavanillesia* hosts very few individuals and species in general: we never found more than eight individuals from three species in any tree (Table 3). Xerophytic C_3 species such as *N. crassifolium* or *Polypodium polypodioides* were the only epiphytes in three of five *Cavanillesia* crowns. There was no significant difference in the abundance of CAM epiphytes with tree species although the percentage of CAM-performing individuals could reach >50 % in deciduous tree crowns, while evergreen and semi-deciduous trees never hosted >20 % in their crowns (data not shown).

Individual performance and microclimatic differences

Growth and survival were studied in *Dimerandra* and *Niphidium*. Survival was significantly lower for *Dimerandra* than for *Niphidium* (Fisher's exact test: $P < 0.05$). Tree identity did not affect the survival or growth of *Dimerandra* (Fisher's exact test: $P = 0.6$; KW: $P = 0.08$, $\chi^2 = 5.1$, df $= 2$) but survival and growth of *Niphidium* differed significantly among trees (Fisher's exact test: $P < 0.05$; ANOVA: $P < 0.001$, $F_{(3,40)} = 12.6$; [see Supporting Information]).

Dimerandra's mean annual relative growth rate (RGR) ranged from 8 % in *Ceiba* to 42 % in *Pseudobombax*. This coincides with the prevalence of *Dimerandra* in *Pseudobombax* crowns where it accounted for 70 % of all epiphytes. Survival of *Niphidium* was much lower in *Pseudobombax* (only 5 of 13 individuals were still alive in 2012) compared with that in both *Brosimum* (24 of 25 individuals survived) and *Ceiba* (22 of 26 individuals survived). Also indicative of harsh conditions in deciduous *Pseudobombax* were the slow RGRs of *Niphidium*.

Intraspecific differences in SLA, $\delta^{13}C$, $\delta^{15}N$ and N concentration of leaves in relation to microclimatic differences

Possible differences among populations of the same species associated with microclimatic differences were assessed for several leaf traits (SLA, $\delta^{13}C$, $\delta^{15}N$ and leaf N concentration) in *D. emarginata*, *Maxillaria uncata*, *N. crassifolium*, *Scaphyglottis behrii* and *V. lineata*.

We used size as a covariate in our analyses as it has been shown to influence physiological parameters of epiphytes (Zotz et al. 2001). Our results, however,

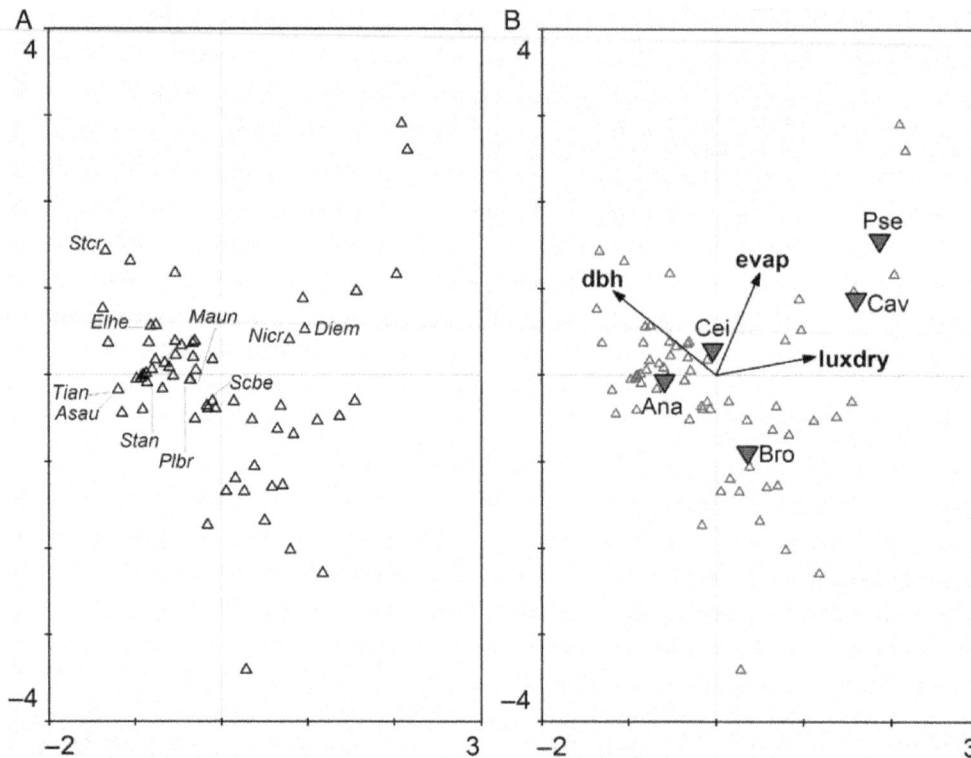

Figure 4. Species-conditional triplot (divided into two graphs for easy interpretation) based on a canonical correspondence analysis of the log-transformed epiphyte species abundance data (rare species were down-weighted) displaying 18 % of the inertia (=weighted variance) in the abundances and 84 % of variance in the weighted averages and class totals of species with respect to environmental variables. The eigenvalues of Axis 1 (horizontally) and Axis 2 (vertically) are 0.26 and 0.15, respectively. Displayed are species (small triangles) and tree centroids (down triangles). Quantitative environmental variables are indicated by arrows in (B) (dbh of trees, evaporation and luminance measured in dry season). In the analysis 79 true epiphyte species were included; only the 10 most abundant species are labelled in (A). Epiphyte species names are abbreviated with the first two letters of the genus and epitheton: *Asplenium auritum, Dimerandra emarginata, Elaphoglossum herminieri, Maxillaria uncata, Niphidium crassifolium, Pleurothallis brighamii, Scaphyglottis behrii, Stenospermation angustifolium, Stelis crescentiicola* and *Tillandsia anceps.* Tree species are: *A. excelsum, B. alicastrum, C. pentandra, P. septenatum* and *C. platanifolia.*

were very inconsistent between the various species regarding whether the influence was significant for a given parameter and occasionally also regarding the direction of the correlation when a significant influence was found **[see Supporting Information]**.

There was a trend towards lower SLA in deciduous trees for *Niphidium* (ANCOVA: $P < 0.001$, $F_{(4,73)} = 5.8$, Fig. 5). Specific leaf area of *Dimerandra* and *Vittaria* also showed significant differences between tree species but without a clear pattern and the result was blurred by a significant interaction with size.

In deciduous trees, $\delta^{13}C$ was generally less negative in *Scaphyglottis* (Fig. 6). *Niphidium* also showed the same trend, but interpretation is difficult because of a significant interaction with relative plant size. *Maxillaria* and *Vittaria* were not found in deciduous trees, while *Dimerandra* showed no significant differences in $\delta^{13}C$ values between the trees included.

The only species with significant differences in $\delta^{15}N$ values between trees was *Scaphyglottis*, which showed a

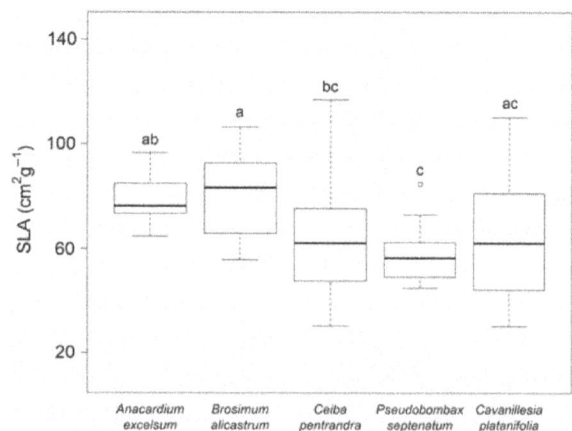

Figure 5. Specific leaf area of *N. crassifolium* individuals (83) sampled on different tree species. Letters indicate significant differences (ANCOVA with plant size as the covariate: $P < 0.05$, $F_{(4,73)} = 5.8$).

tendency towards more negative values when growing in deciduous tree crowns (Fig. 7). *Dimerandra* and *Niphidium* showed consistent results, with higher $\delta^{15}N$ values

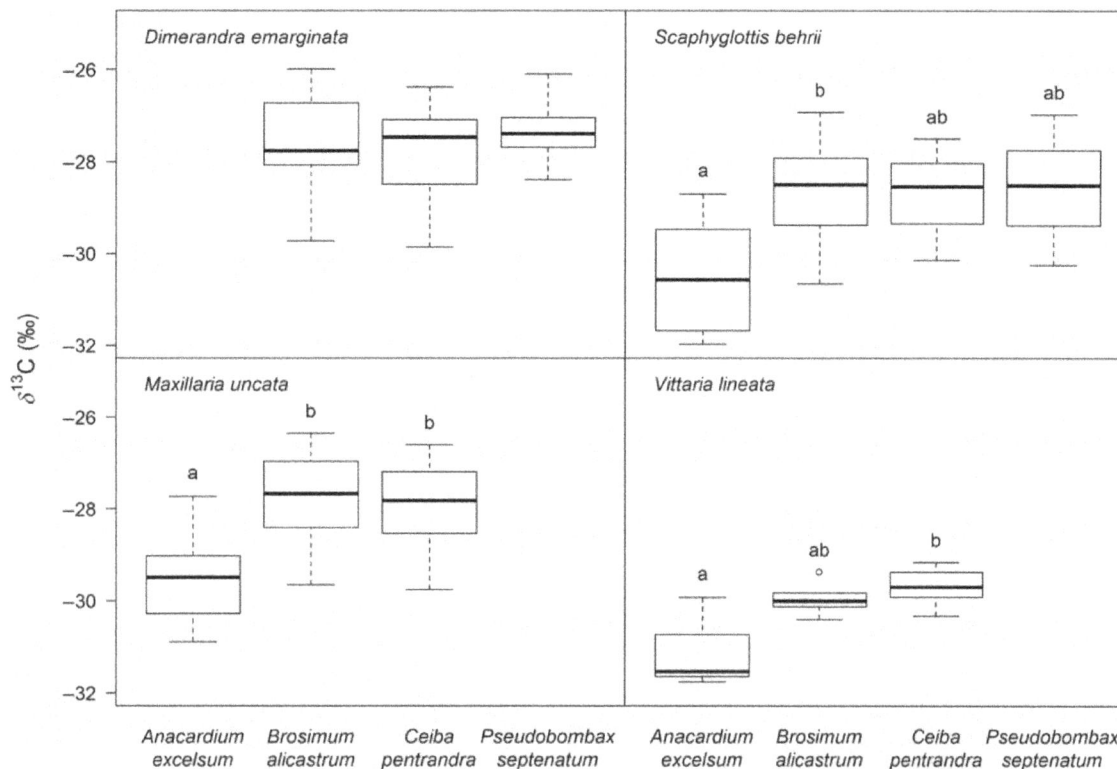

Figure 6. $\delta^{13}C$ values for *Dimerandra* (66), *Maxillaria* (40), *Scaphyglottis* (27) and *Vittaria* (15) sampled on different tree species. Letters indicate significant differences (ANCOVA with plant size as the covariate: *Maxillaria* $P < 0.05$, $F_{(1,36)} = 6.2$ and *Scaphyglottis* $P < 0.05$, $F_{(3,22)} = 5.4$; ANOVA: *Vittaria* $P < 0.05$, $F_{(2,11)} = 5.5$).

in evergreen *Brosimum* than in semi-deciduous *Ceiba* and deciduous *Pseudobombax*. $\delta^{15}N$ values for *Niphidium* in *Anacardium* and *Cavanillesia* did not vary from those of plants growing in *Ceiba*, however. We were not able to obtain samples of *Dimerandra* from *Anacardium* or *Cavanillesia*. Yet there was a significant interaction of host tree species with relative plant size of *Dimerandra* and *Niphidium*. *Maxillaria* was not found in deciduous trees.

The only species with significant differences in leaf N concentrations between trees were *Niphidium* (ANOVA: $P < 0.001$, $F_{(4,74)} = 8.1$) and *Dimerandra*. But for the latter there was a significant interaction with relative size and for *Niphidium* there was no clear tendency or grouping with leaf phenology [**see Supporting Information**]. No size-related variation was found.

Discussion

This study is based on the understanding that the composition of a community is strongly influenced by abiotic and biotic factors and tries to identify general rules to explain distributional patterns of vascular epiphytes in time and space (Díaz et al. 1999). This approach contrasts with the neutral theory (Hubbell 2001), which emphasizes the importance of stochastic processes. From a deterministic

point of view the environment acts as a filter that governs the final community composition at a given location (Keddy 1992). Zobel's (1997) distinction of three different species pools (actual, local and regional) provides a useful conceptual framework for the comparison of epiphyte assemblages found on each of our five study species (= actual species pool) vs. those reported for BCI (=regional species pool). The 92 species (including five nomadic vines that had ground contact) occurring on the 24 studied trees represent almost half of the entire epiphyte flora of BCI (15.6 km², Croat 1978). This large number is comparable with the number of species found on 12 similarly large trees at the San Lorenzo Canopy Crane site (83 species, Zotz and Bader 2011) located ~20 km as the crow flies from BCI in a much wetter climatic setting. Additionally, we found three ferns and an orchid species, which are not included in the flora of BCI by Croat (1978). Although we found a large number of species on the few study trees, more than 100 epiphyte species of the regional species pool were not covered by our study. Some of these 'missing' epiphyte species would probably be found with more extensive sampling on other conspecific trees, but others may indeed be excluded by particular traits of the studied tree species. Leaf phenological patterns can arguably act as a potent filter for the local

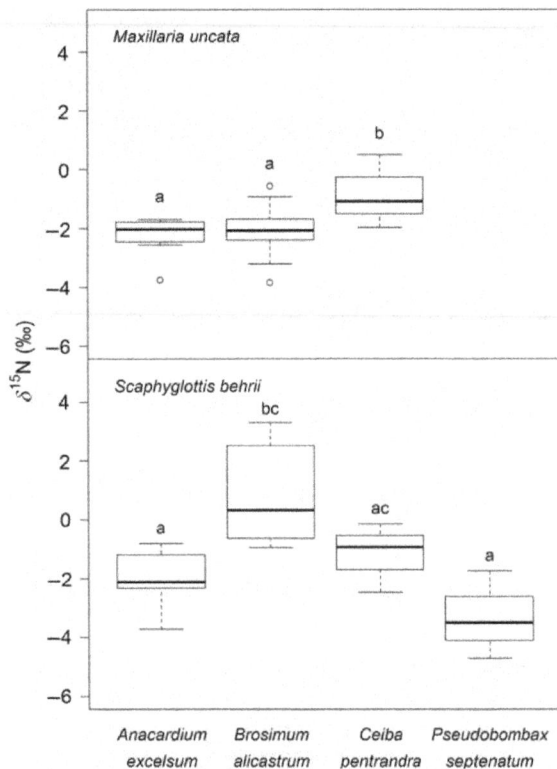

Figure 7. $\delta^{15}N$ values for *Maxillaria* (38) and *Scaphyglottis* (28) sampled in different tree crowns. Letters indicate significant differences (ANCOVA with plant size as the covariate: *Maxillaria* $P < 0.05$, $F_{(1,33)} = 15.3$; ANOVA: *Scaphyglottis* $P < 0.05$, $F_{(3,23)} = 13.9$).

species pool. We analysed its influence at different levels. In the composition of epiphyte assemblages, we found more drought-adapted species in deciduous trees. At the level of physiological plasticity some species showed adjustment to drier conditions and reduced growth performance in populations of individual species growing in deciduous trees. Besides documenting these effects, we also identified the mechanism: luminance and evaporation showed the influence of host tree leaf phenology on the microclimate in tree crowns, while bark characteristics seem unrelated.

As expected, we found marked differences in microclimatic conditions during the dry season between evergreen and deciduous trees, e.g. much higher luminance and much higher evaporative demand in deciduous tree crowns. The number of epiphyte species, which thrive under such demanding conditions, is obviously limited. Thus, both species diversity and epiphyte abundance were significantly reduced in deciduous trees. A multiple-assemblage similarity profile (Fig. 3) indicated that epiphyte assemblages of the deciduous trees (*Pseudobombax*) change less in respect to their most dominant species, and epiphyte abundances were also more even. In contrast, epiphyte assemblages of evergreen trees

shared fewer abundant species, and had a lower percentage of abundant species. This is typical for species-rich communities with few abundant but many rare species. Averaged over all tree individuals trees shared more abundant species than did *Anacardium* individuals at an intra-specific level.

In agreement with our results, the percent cover of vascular epiphytes in an evergreen tree species (*Hyeronima alchorneoides*) was significantly higher compared with a deciduous tree species in La Selva, Costa Rica (*Lecythis ampla*, Cardelus 2007). Both studies consistently indicate that epiphytes growing in deciduous trees are also generally found in evergreen trees (Cardelus 2007). An effect of deciduousness on epiphyte assemblages in the wet forest at La Selva is somewhat surprising, because there precipitation exceeds 100 mm per month even during the short dry season (Cardelus and Chazdon 2005). Moreover, the studied tree species were deciduous during the rainy season. This suggests that the persistent differences between deciduous and evergreen trees in luminance, and possibly other microclimatic variables in the rainy season on BCI are relevant as well.

Tree phenology did not only affect the floristic composition of epiphyte assemblages (Figs 3 and 4). At a functional level, there was a general increase in the proportion of species using the water-conserving CAM-pathway, from evergreen *Anacardium* over semi-evergreen *Ceiba* to drought-deciduous *Pseudobombax* (Table 3). The seeming deviation of *Cavanillesia* is probably a sampling artefact due to the low number of species found in the crowns. At the level of the entire *Cavanillesia* trees, >40 % of all individuals used CAM, compared with an average of ca. 5 % in the two evergreen species (Table 3). In summary, the drought-deciduous tree species offer drier growth conditions for epiphytes throughout the year, which leads to impoverished epiphyte assemblages with a large proportion of CAM species.

The studied components of microclimate are clearly not the sole factors affecting the diversity of these epiphyte assemblages, e.g. *Pseudobombax* supported more diverse assemblages compared with *Cavanillesia*. Bark water storage capacity is generally assumed to be important for epiphytes and, Barkman (1958) listed it as a key trait apart from bark pH, bark electrolyte concentration and bark buffer capacity. The highest maximum water content we measured exceeded that reported by Callaway *et al.* (2002) by more than a third, but maximum WHC of tree bark did not correlate with either epiphyte diversity or abundance. In another study communities of epiphytic bryophytes could be related to some extent to the water storage capacity of the host (Studlar 1982). However, water storage capacity greatly varies even within single tree species as the bark structure varies with age and position in the tree, and our data thus have to be interpreted with caution. Water loss rates of the bark samples in our study, which were within the range of the values reported by Callaway et al. (2002), may have more explanatory power: semi-deciduous and deciduous trees had consistently higher water loss rates compared with evergreen trees, which should reinforce any effect associated with phenology. We have recently shown that small differences in water supply may be crucial during germination under the demanding conditions of tree canopies (Wester and Zotz 2011). Contrary to Callaway et al. (2002) and Studlar (1982), we did not find an association of water storage capacity of host tree bark and epiphyte diversity. As there was no consistent correlation of the water storage of tree bark and phenology, this aspect did not enhance the phenology effect whereas differences in water loss rates arguably did. We conclude that water storage capacity is a minor filter in our system.

We also expected differences in functional attributes of conspecifics growing on trees with varying phenology. ^{13}C in tissues can be a good proxy for whole plant water-use efficiency (Cernusak et al. 2008), more positive $\delta^{13}C$ values being generally associated with drier microclimates (e.g. Lajtha and Getz 1993; Van de Water et al. 2002). Leaf structure, i.e. SLA, is also known to respond to growing conditions (Cornelissen et al. 2003). Expectations only partly bore out. A slight trend towards lower SLA was found in only one of five species, but $\delta^{13}C$ values of most species were indeed more positive in deciduous tree crowns. Thus, although not entirely consistent, there are detectable intraspecific differences in functional traits of epiphyte populations in trees with different leaf phenology.

Plant size has been identified as an important covariate in physiological studies with epiphytes (Schmidt et al. 2001). Since an increase in plant size almost inevitably leads to a decrease in surface/volume ratios, plant water relations must be affected, e.g. because identical transpiration rates would deplete stored water more rapidly in smaller plants under otherwise identical conditions (Schmidt et al. 2001). This would immediately result in a less favourable water status in smaller plants. Possible adaptive responses are reduced SLA or reduced stomatal conductance in small plants. However, we found no consistent size dependence, neither for SLA nor for $\delta^{13}C$, other local factors obviously being of overriding importance.

The mean $\delta^{15}N$ value of all tested epiphytes of our study was in the range of a study including epiphytes from Brazil, Australia and Solomon Islands (Stewart et al. 1995) and slightly lower $\delta^{15}N$ values were reported for Costa Rican and Mexican epiphytes (Hietz et al. 2002; Hietz and Wanek 2003). Epiphytes are supplied with N from various sources, e.g. via dry and wet deposition of reactive N from the atmosphere, biological N_2 fixation and decomposition of dead canopy organic material (Benzing 1989). Since different sources can differ in their $\delta^{15}N$ signatures (Robinson 2001), the $\delta^{15}N$ values of epiphyte tissue can be used to infer their main N source. Atmospheric N has, by definition, a $\delta^{15}N$ value of 0‰ and biologically fixed N therefore commonly has isotope signatures close to this. The isotopic signatures of N inputs from dry and wet deposition are highly variable (e.g. Lindberg et al. 1986; Heaton et al. 1997; Högberg, 1997) but $\delta^{15}N$ values tend to be higher in dry than in wet deposition (Moore 1977; Garten 1992; Heaton et al. 1997). Low $\delta^{15}N$ signatures in epiphytes have therefore been explained by efficient use of atmospheric wet deposition (Stewart et al. 1995). Tree foliage captures dry deposition of reactive N forms and channels it into the inner crown to the epiphytes during rain events. This scenario should result in a larger yearly portion of dry deposition in evergreen trees than in deciduous trees. Consequently, we expected epiphytes in evergreen trees to show less negative $\delta^{15}N$ values compared with deciduous trees. Less atmospheric N deposition could also result in lower leaf N concentrations of epiphytes in deciduous trees. However, any such effect could easily be confounded by reduced growth due to drought stress. Of the five studied epiphyte species only one showed the expected trend, which can hardly be taken as strong support for the scenario outlined above. Regarding N, our results showed no consistent pattern either. This might partly be due to drought-induced growth inhibition, which we actually documented for one species in Pseudobombax. Overall, our data do not allow us to make strong conclusions on possible effects of leaf phenology on the N cycle in tree canopies.

Individual growth performance integrates all physiological processes that change in response to the environment and thus integrates the environmental influence. This has been e.g. shown in a 7-year study of the population dynamics of the epiphytic orchid Aspasia principissa (Zotz and Schmidt 2006). Intraspecific variation at the level of individual growth was studied in two species. Differences in growth were inconsistent. While Niphidium was hampered in its growth in deciduous trees, the opposite trend was observed for Dimerandra. Species that can cope with the drier conditions in deciduous trees can take advantage of increased radiation. Dimerandra is a good example for this. This C_3 species copes well with drier microclimate and obviously prefers more exposed sites. Although it does grow in evergreen trees, it almost always grew in the outermost sections (Johansson zone V) of Anacardium crowns. Tolerance to severe drought of this light-demanding species also explains its abundance in many *Pseudobombax* crowns. Therefore, its overall poor survival compared with *Niphidium* is rather surprising. No differences regarding the survival or growth rate among populations were found between different tree species. In contrast, survival and growth of *Niphidium* were lowest in deciduous trees. Thus, it was possible to relate microclimatic conditions to intraspecific growth differences at least for one species. Species that dominated in the drier climate of deciduous trees were not limited by more humid climate of evergreen trees, as they did occur in evergreen trees as well. We did observe that these species grew preferentially in the drier and lighter parts of the evergreen tree crowns. Thus, the drier climate of deciduous tree crowns poses a hard filter for all species that cannot cope with these conditions. Similarly, Larrea and Werner (2010) and Toledo-Aceves *et al.* (2012) found impoverished epiphyte communities in remnant or shade trees drier compared with fragments of humid montane forest. In contrast, the more humid climate of evergreen trees poses only a partial filter to drought-tolerant epiphytes.

We included *C. pentandra* in this study because a semi-deciduous tree should offer intermediate growth conditions. Consistent with this expectation, many aspects studied in this report, from species composition to physiological traits, were intermediate between evergreen and deciduous trees. In many regards, the combination of a rather open, only briefly leafless, crown with large limbs makes *Ceiba* a relatively good host for vascular epiphytes (Valdivia 1977).

Conclusions

To conclude, we found that tree phenology imposes an imprint on epiphyte assemblages at different levels of integration, from physiology to species composition of epiphytes on individual trees. The effect of host tree phenology was not always clear-cut and entirely consistent, but is arguably modified by many other factors, only some of which could be studied. Neither dismissing the importance of these additional filters nor dismissing the role of chance our study describes one important environmental filter that structures the local communities of vascular epiphytes.

Sources of Funding

Financial support from the Smithsonian Tropical Research Institute for field work in Panama is acknowledged.

Contributions by the Authors

G.Z. conceived the study design. H.J.R.E. and J.B. performed field work and data analyses. F.H. and W.W. did isotope analyses. H.E. and G.Z. wrote the paper, which was reviewed and revised by all other authors.

Acknowledgements

The authors acknowledge the permission to work in the 50-ha plot on Barro Colorado Island by Rick Condit (STRI). They are grateful to Osvaldo Calderón (STRI) for pointing out several study trees to them. Katherine Araúz, Yader Sageth and Eduardo Sánchez (all STRI) are all acknowledged for their assistance in the field. Further, the authors are grateful to Glenda Mendieta Leiva (University of Oldenburg) for her help in the determination of species and for her advice on climbing techniques.

Supporting Information

The following Supporting Information is available in the online version of this article –

Figure S1. Map of Barro Colorado Island (BCI, modified after Kinner & Paton 1998) showing the approximate locations of the study trees. Evergreen trees are triangles: *Anacardium excelsum* (blue) and *Brosimum alicastrum* (green); circles represent the semi-evergreen *Ceiba pentandra*; quadrats represent the drought-deciduous species: *Pseudobombax septenatum* (yellow) and *Cavanillesia platanifolia* (purple). The large yellow rectangle represents the 50-ha plot.

Figure S2. Growth of *Niphidium crassifolium* in four tree species measured from 2011 to 2012 (ANOVA: $P < 0.001$, $F_{(3,40)} = 12.6$). *Brosimum* ($n = 2$ trees), *Ceiba* ($n = 4$), *Pseudobombax* ($n = 3$) and *Cavanillesia* ($n = 1$).

Figure S3. Leaf nitrogen (N) concentration of *N. crassifolium* (81) sampled in different tree crowns. Letters indicate significant differences (ANOVA: $P < 0.001$, $F_{(4,74)} = 8.1$).

Table S1. Species abundances of epiphytes and hemiepiphytes. The studied species were *A. excelsum* ($n = 5$), *B. alicastrum* ($n = 4$), *C. pentandra* ($n = 5$), *P. septenatum* ($n = 5$) and *C. platanifolia* ($n = 5$). Nutrient concentrations are usually averages of six replicates, exceptional cases with a single sample are shown in italics. Some $\delta^{13}C$ values were taken from other publications ([1] Zotz & Ziegler 1997; [2] Zotz 2004). $\delta^{13}C$ and $\delta^{15}N$ in ‰, C and N in %. * Hemiepiphyte or nomadic vine *sensu* Zotz (2013). † Nomadic vines with established ground contact via roots (excluded from further analyses).

Table S2. Summary of ANCOVA results for size dependence of four leaf traits. Analysed were size and its significant interaction (*) with 'host species' of the studied leaf traits (SLA, $\delta^{13}C$, $\delta^{15}N$, N) for the focal epiphyte species (all studied host species, *A. excelsum*, *B. alicastrum*, *C. pentandra P. septenatum C. platanifolia*, other = all other except the specified). ↑ Positive correlation, ↓ negative correlation, – no correlation.

Literature Cited

Baldeck CA, Harms KE, Yavitt JB, John R, Turner BL, Valencia R, Navarrete H, Bunyavejchewin S, Kiratiprayoon S, Yaacob A, Supardi MNN, Davies SJ, Hubbell SP, Chuyong GB, Kenfack D, Thomas DW, Dalling JW. 2013. Habitat filtering across tree life stages in tropical forest communities. *Proceedings of the Royal Society B Biological Sciences* **280**:1–8.

Balslev H, Valencia R, Paz y Miño G, Christensen H, Nielsen I. 1998. Species count of vascular plants in one hectare of humid lowland forest in Amazonian Ecuador. In: Dallmeier F, Comiskey JA, eds. *Forest biodiversity in North, Central and South America, and the Caribbean.* Paris: Parthenon Publishing Group, 585–594.

Barkman JJ. 1958. *Phytosociology and ecology of Cryptogamic epiphytes.* Assen: Van Gorcum & Co.

Benzing DH. 1989. The mineral nutrition of epiphytes. In: Lüttge U, ed. *Vascular plants as epiphytes.* Berlin, Germany: Springer.

Benzing DH. 1995. The physical mosaic and plant variety in forest canopy. *Selbyana* **16**:159–168.

Brown C, Burslem DFRP, Illian JB, Bao L, Brockelman W, Cao M, Chang LW, Dattaraja HS, Davies S, Gunatilleke CVS, Gunatilleke IAUN, Huang J, Kassim AR, LaFrankie JV, Lian J, Lin L, Ma K, Mi X, Nathalang A, Noor S, Ong P, Sukumar R, Su SH, Sun IF, Suresh HS, Tan S, Thompson J, Uriarte M, Valencia R, Yap SL, Ye W, Law R. 2013. Multispecies coexistence of trees in tropical forests: spatial signals of topographic niche differentiation increase with environmental heterogeneity. *Proceedings of the Royal Society B Biological Sciences* **280**:1–8.

Callaway RM, Reinhart KO, Moore GW, Moore DJ, Pennings SC. 2002. Epiphyte host preferences and host traits: mechanisms for species-specific interactions. *Oecologia* **132**:221–230.

Cardelus CL. 2007. Vascular epiphyte communities in the inner-crown of *Hyeronima alchorneoides* and Lecythis ampla at La Selva Biological Station, Costa Rica. *Biotropica* **39**:171–176.

Cardelus CL, Chazdon RL. 2005. Inner-crown microenvironments of two emergent tree species in a lowland wet forest. *Biotropica* **37**:238–244.

Catchpole DJ, Kirkpatrick JB. 2010. The outstandingly speciose epiphytic flora of a single stranger fig (*Ficus crassiuscula*) in a Peruvian montane cloud forest. In: Bruijnzeel LA, Scatena FN, Hamilton LS, eds. *Tropical montane cloud forests.* New York: Cambridge University Press, 142–146.

Cernusak LA, Winter K, Aranda J, Turner BL. 2008. Conifers, angiosperm trees, and lianas: growth, whole-plant water and nitrogen use efficiency, and stable isotope composition (delta C-13 and delta O-18) of seedlings grown in a tropical environment. *Plant Physiology* **148**:642–659.

Colwell RK. 2009. Estimate S: statistical estimation of species richness and shared species from samples. Version 8.2. <http://viceroy.eeb.uconn.edu/estimates/index.html> (10 November 2014).

Condit R, Pérez R, Daguerre N. 2011. *Trees of Panama and Costa Rica.* Princeton and Oxford: Princeton University Press.

Cornelissen JHC, Lavorel S, Garnier E, Diaz S, Buchmann N, Gurvich DE, Reich PB, ter Steege H, Morgan HD, van der Heijden MGA, Pausas JG, Poorter H. 2003. A handbook of protocols for standar-

dised and easy measurement of plant functional traits worldwide. *Australian Journal of Botany* **51**:335–380.

Croat TB. 1978. *Flora of Barro Colorado Island.* Stanford: Stanford University Press.

Díaz S, Cabido M, Casanoves F. 1999. Functional implications of trait-environment linkages in plant communites. In: Weiher E, Keddy P, eds. *Ecological assembly rules—perspectives, advances, retreats.* UK: Cambridge University Press, 338–362.

Didham RK, Lawton JH. 1999. Edge structure determines the magnitude of changes in microclimate and vegetation structure in tropical forest fragments. *Biotropica* **31**:17–30.

Freiberg M. 1996. Spatial distribution of vascular epiphytes on three emergent canopy trees in French Guiana. *Biotropica* **28**:345–355.

Freiberg M. 1997. Spatial and temporal pattern of temperature and humidity of a tropical premontane rain forest tree in Costa Rica. *Selbyana* **18**:77–84.

Freiberg M. 1999. The vascular epiphytes on a *Virola michelii* tree (Myristicaceae) in French Guiana. *Ecotropica* **5**:75–81.

Garten CT. 1992. Nitrogen isotope composition of ammonium and nitrate in bulk precipitation and forest throughfall. *International Journal of Environmental Analytical Chemistry* **47**:33–45.

Gentry AH, Dodson CH. 1987. Contribution of non-trees to species richness of a tropical rain forest. *Biotropica* **19**:149–156.

Heaton THE, Spiro B, Madeline S, Robertson C. 1997. Potential canopy influences on the isotopic composition of nitrogen and sulphur in atmospheric deposition. *Oecologia* **109**:600–607.

Hietz P, Wanek W. 2003. Size-dependent variation of carbon and nitrogen isotope abundances in epiphytic bromeliads. *Plant Biology* **5**:137–142.

Hietz P, Wanek W, Wania R, Nadkarni NM. 2002. Nitrogen-15 natural abundance in a montane cloud forest canopy as an indicator of nitrogen cycling and epiphyte nutrition. *Oecologia* **131**:350–355.

Hofhansl F, Wanek W, Drage S, Huber W, Weissenhofer A, Richter A. 2012. Controls of hydrochemical fluxes via stemflow in tropical lowland rainforests: effects of meteorology and vegetation characteristics. *Journal of Hydrology* **452–453**:247–258.

Högberg P. 1997. Tansley review No. 95 15N natural abundance in soil–plant systems. *New Phytologist* **137**:179–203.

Hölscher D, Köhler L, Leuschner C, Kappelle M. 2003. Nutrient fluxes in stemflow and throughfall in three successional stages of an upper montane rain forest in Costa Rica. *Journal of Tropical Ecology* **19**:557–565.

Hubbell SP. 2001. *The unified neutral theory of biodiversity and biogeography.* Princeton, New Jersey: Princeton University Press.

Johansson D. 1974. Ecology of vascular epiphytes in West African rain forest. *Acta Phytogeographica Suecica* **59**:136–267.

Jost L, Chao A, Chazdon RL. 2011. Compositional similarity and β (beta) diversity. In: Magurran AE, McGill BJ, eds. *Biological diversity—frontiers in measurement and assessment.* New York: Oxford University Press Inc, 66–84.

Keddy PA. 1992. Assembly and response rules: two goals for predictive community ecology. *Journal of Vegetation Science* **3**:157–164.

Kelly DL, O'Donnovan G, Feehan J, Murphy S, Drangeid SO, Marcano-Berti L. 2004. The epiphyte communities of a montane rain forest in the Andes of Venezuela: patterns in the distribution of the flora. *Journal of Tropical Ecology* **20**:643–666.

Kinner D, Paton S. 1998. Barro Colorado Island Image Map, version

1.0. Version 1.0 ed. Smithsonian Tropical Research Institution. <http://mapserver.stri.si.edu/geonetwork/srv/en/main.home> (10 November 2014).

Kraft NJB, Ackerly DD. 2010. Functional trait and phylogenetic tests of community assembly across spatial scales in an Amazonian forest. *Ecological Monographs* **80**:401–422.

Krömer T, Kessler M, Gradstein SR. 2007. Vertical stratification of vascular epiphytes in submontane and montane forest of the Bolivian Andes: the importance of the understory. *Plant Ecology* **189**: 261–278.

Lajtha K, Getz J. 1993. Photosynthesis and water-use efficiency in Pinyon-Juniper communites along an elevation gradient in northern New Mexico. *Oecologia* **94**:95–101.

Larrea ML, Werner FA. 2010. Response of vascular epiphyte diversity to different land-use intensities in a neotropical montane wet forest. *Forest Ecology and Management* **260**:1950–1955.

Laube S, Zotz G. 2006. Neither host-specific nor random: vascular epiphytes on three tree species in a Panamanian lowland forest. *Annals of Botany* **97**:1103–1114.

Leigh EG, Loo de Lao S, Condit R, Hubbell SP, Foster RB, Pérez R. 2004. Barro Colorado Island forest dynamics plot, Panama. In: Losos EC, Leigh JEG, eds. *Tropical forest diversity and dynamism: findings from a large-scale plot network*. Chicago: University of Chicago Press, 451–463.

Lindberg SE, Lovett GM, Richter DD, Johnson DW. 1986. Atmospheric deposition and canopy interactions of major ions in a forest. *Science* **231**:141–145.

Martinez-Melendez N, Perez-Farrera MA, Flores-Palacios A. 2008. Vertical stratification and host preference by vascular epiphytes in a Chiapas, Mexico, cloud forest. *Revista De Biologia Tropical* **56**: 2069–2086.

Metz MR. 2012. Does habitat specialization by seedlings contribute to the high diversity of a lowland rain forest? *Journal of Ecology* **100**: 969–979.

Moore H. 1977. The isotopic composition of ammonia, nitrogen dioxide and nitrate in the atmosphere. *Atmospheric Environment (1967)* **11**:1239–1243.

Perry DR. 1978. Method of access into crowns of emergent and canopy trees. *Biotropica* **10**:155–157.

Pittendrigh CS. 1948. The bromeliad–anopheles–malaria complex in Trinidad. I. The bromeliad flora. *Evolution* **2**:58–89.

R Development Core Team. 2012. *R: a language and environment for statistical computing*. Vienna, Austria: R Foundation for Statistical Computing.

Rasband W. 1997–2012. *ImageJ*. 1.44 edn. Bethesda, Maryland, USA: US National Institute of Health.

Reiser H, Kutiel H. 2008. Rainfall uncertainty in the Mediterranean: time series, uncertainty, and extreme events. *Theoretical and Applied Climatology* **104**:357–375.

Robinson D. 2001. Delta N-15 as an integrator of the nitrogen cycle. *Trends in Ecology and Evolution* **16**:153–162.

Sanford WW. 1968. Distribution of epiphytic orchids in semi-deciduous tropical forest in southern Nigeria. *Journal of Ecology* **56**:697–705.

Schmidt G, Stuntz S, Zotz G. 2001. Plant size—an ignored parameter in epiphyte ecophysiology. *Plant Ecology* **153**:65–72.

Stewart GR, Schmidt S, Handley LL, Turnbull MH, Erskine PD, Joly CA. 1995. [15]N natural abundance of vascular rainforest epiphytes: implications for nitrogen source and acquisition. *Plant, Cell and Environment* **18**:85–90.

Studlar SM. 1982. Host specificity of epiphytic bryophytes near mountain-lake, Virginia. *Bryologist* **85**:37–50.

ter Braak CJF, Šmilauer P. 1997. *CANOCO reference manual and Cano-Draw for Windows user's guide: software for canonical community ordination version 4.5*. Ithaka, NY, Microcomputer Power.

ter Stege H, Cornelissen JHC. 1989. Distribution and ecology of vascular epiphytes in lowland rain forest of Guyana. *Biotropica* **21**:331–339.

Toledo-Aceves T, Garcia-Franco JG, Hernandez-Rojas A, MacMillan K. 2012. Recolonization of vascular epiphytes in a shaded coffee agroecosystem. *Applied Vegetation Science* **15**:99–107.

Tropicos. 2012. Missouri Botanical Garden. <http://www.tropicos.org> (10 November 2014).

Valdivia PE. 1977. Estudio botánico y ecológico de la región del Río Uxpanapa, Veracruz. N°4. Las epífitas. *Biotica* **2**:55–81.

Van de Water PK, Leavitt SW, Betancourt JL. 2002. Leaf δ^{13}C variability with elevation, slope aspect, and precipitation in the southwest United States. *Oecologia* **132**:332–343.

Wagner K, Bogusch W, Zotz G. 2013. The role of the regeneration niche for the vertical stratification of vascular epiphytes. *Journal of Tropical Ecology* **29**:277–290.

Wester S, Zotz G. 2011. Seed comas of bromeliads promote germination and early seedling growth by wick-like water uptake. *Journal of Tropical Ecology* **27**:115–119.

Zimmerman JK, Olmsted IC. 1992. Host tree utilization by vascular epiphytes in a seasonally inundated forest (Tintal) in Mexico. *Biotropica* **24**:402–407.

Zobel M. 1997. The relative role of species pools in determining plant species richness: an alternative explanation of species coexistence? *Trends in Ecology and Evolution* **12**:266–269.

Zotz G. 2004. How prevalent is Crassulacean acid metabolism among vascular epiphytes? *Oecologia* **138**:184–192.

Zotz G. 2013. Hemiepiphyte: a confusing term and its history. *Annals of Botany* **111**:1015–1020.

Zotz G, Bader MY. 2011. Sampling vascular epiphyte diversity—species richness and community structure. *Ecotropica* **17**: 103–112.

Zotz G, Hietz P. 2001. The physiological ecology of vascular epiphytes: current knowledge, open questions. *Journal of Experimental Botany* **52**:2067–2078.

Zotz G, Schmidt G. 2006. Population decline in the epiphytic orchid *Aspasia principissa*. *Biological Conservation* **129**:82–90.

Zotz G, Vollrath B. 2003. The epiphyte vegetation of the palm *Socratea exorrhiza*—correlations with tree size, tree age and bryophyte cover. *Journal of Tropical Ecology* **19**:81–90.

Zotz G, Winter K. 1994. A one-year study on carbon, water and nutrient relationships in tropical C$_3$-CAM hemi-epiphytes, *Clusia uvitana* Pittier. *New Phytologist* **127**:45–60.

Zotz G, Ziegler H. 1997. The occurrence of Crassulacean acid metabolism among vascular epiphytes from Central Panama. *New Phytologist* **137**:223–229.

Zotz G, Hietz P, Schmidt G. 2001. Small plants, large plants: the importance of plant size for the physiological ecology of vascular epiphytes. *Journal of Experimental Botany* **52**:2051–2056.

7

Influence of sulfur and cadmium on antioxidants, phytochelatins and growth in Indian mustard

Humayra Bashir[1], Mohamed M. Ibrahim[2,3], Rita Bagheri[1], Javed Ahmad[1], Ibrahim A. Arif[2,3], M. Affan Baig[1] and M. Irfan Qureshi[1]*

[1] Proteomics and Bioinformatics Lab, Department of Biotechnology, Jamia Millia Islamia, New Delhi 110025, India
[2] Department of Botany and Microbiology, Science College, King Saud University, PO Box 2455, Riyadh, Saudi Arabia
[3] Department of Botany and Microbiology, Faculty of Science, Alexandria University, PO Box 21511, Alexandria, Egypt

Associate Editor: Wen-Hao Zhang

Abstract. Soils in many parts of the world are contaminated with heavy metals, leading to multiple, deleterious effects on plants and threats to world food production efficiency. Cadmium (Cd) is one such metal, being toxic at relatively low concentrations as it is readily absorbed and translocated in plants. Sulfur-rich compounds are critical to the impact of Cd toxicity, enabling plants to increase their cellular defence and/or sequester Cd into vacuoles mediated by phytochelatins (PCs). The influence of sulfur on Cd-induced stress was studied in the hyperaccumulator plant Indian mustard (*Brassica juncea*) using two sulfur concentrations (+S, 300 μM SO_4^{2-} and S-deficient $-$S, 30 μM SO_4^{2-}) with and without the addition of Cd (100 μM $CdCl_2$) at two different time intervals (7 and 14 days after treatment). Compared with control plants (+S/$-$Cd), levels of oxidative stress were higher in S-deficient ($-$S/$-$Cd) plants, and greatest in S-deficient Cd-treated ($-$S/+Cd) plants. However, additional S (+S/+Cd) helped plants cope with oxidative stress. Superoxide dismutase emerged as a key player against Cd stress under both $-$S and +S conditions. The activity of ascorbate peroxidase, glutathione reductase and catalase declined in Cd-treated and S-deficient plants, but was up-regulated in the presence of sulfur. Sulfur deficiency mediated a decrease in ascorbate and glutathione (GSH) content but changes in ascorbate (reduced : oxidized) and GSH (reduced : oxidized) ratios were alleviated by sulfur. Our data clearly indicate that a sulfur pool is needed for synthesis of GSH, non-protein thiols and PCs and is also important for growth. Sulfur-based defence mechanisms and the cellular antioxidant pathway, which are critical for tolerance and growth, collapsed as a result of a decline in the sulfur pool.

Keywords: Antioxidants; cadmium; growth; oxidative stress; phytochelatins; sulfur.

Introduction

In recent years, sulfur deficiency has become a major problem for agricultural productivity, reducing both crop quality and yields. Legitimate approaches to reducing emissions of sulfur into the atmosphere have resulted in a concomitant decrease in atmospheric deposition of sulfur on agricultural land. Based on crop demand, fertilizer-use efficiency and current inputs, the worldwide sulfur-deficit has been estimated to be 10.4 million tonnes annually (Haneklaus *et al.* 2007). Increased food production will further raise sulfur requirements, elevating this sulfur-deficit to 12.5 million tonnes by 2015 (Randazzo 2009).

* Corresponding author's e-mail address: mirfanq@gmail.com; miqureshi@jmi.ac.in

As an essential macronutrient required for the proper growth and development of plants, sulfur plays a critical role in a number of cellular processes, such as in protein disulfide bridges (Saito 2000), mediating electron transport in iron–sulfur (Fe–S) clusters, the redox cycle, detoxification of heavy metals and xenobiotics, vitamin co-factors (Hell and Hillebrand 2001) and the metabolism of secondary products (Hell 1997; Saito 2000). Sulfur is thus a major plant nutrient (Takahashi et al. 1997; Leustek et al. 2000; Saito 2000; Lee and Korban 2002) that contributes to an increase in crop yield, directly adding nutritional value and improving the efficiency of use of other essential plant nutrients (Salvagiotti et al. 2009), particularly nitrogen and phosphorus. Research on plant adaptations to S-related stresses has shifted from an emphasis on excessive inputs and acidification to how deficiencies impact crop production. Sulfur deficiency not only impacts crop quality and yield, but it also raises the demand for adequate fertilization to resist biotic and abiotic stresses. Exposure of plants to excessive toxic metals like cadmium (Cd) may affect the uptake and metabolism of S and negatively impact the yield and plant resistance to abiotic stresses (Gill and Tuteja 2011).

Cadmium is a non-essential heavy metal that is readily absorbed and rapidly translocated in plants, which makes it highly bio-available and thus toxic, even at relatively low concentrations. Cadmium exerts its phytotoxicity by interfering with several basic events of plant growth, development and physiology (Qadir et al. 2004). Cadmium induces oxidative stress leading to the overproduction of harmful reactive oxygen species (ROS) (Zhang et al. 2010). These ROS may cause damage to cell membranes, proteins, DNA replication and repair. A major effect of this metal, observed in most plants studied to date, is the inhibition of photosynthesis by altering chlorophyll synthesis (Pietrini et al. 2003; Maksymiec and Krupa 2006) and the light-harvesting Chl a/Chl b protein complex II (Qureshi et al. 2010), and interfering with RuBisCO activation (Prasad 1995). Therefore, Cd-mediated reductions in photosynthetic activity lead to declines in crop yield (Ouzounidou et al. 1997). Cadmium also interferes with processes such as carbohydrate and nitrogen metabolism (Sanità di Toppi and Gabbrielli 1999), enzyme catalysis (Van Assche and Clijsters 1990) and water balance (Perfus-Barbeoch et al. 2002).

Sulfur plays a critical role in allowing plants to protect themselves against heavy metal toxicity, especially Cd toxicity (Gill and Tuteja 2011). Numerous studies have shown that S is involved in the biosynthesis of heavy metal detoxification agents (Anjum et al. 2012), such as non-protein thiols (NPTs) including phytochelatins (PCs), glutathione (GSH) (Noctor et al. 2011) and Cd -sulfide crystallites (Robinson et al. 1993). Harada et al. (2002)

reported that Cd induces the production of thiol compounds and transcripts of the S-assimilation pathway. It is noteworthy that a PC-deficient *Arabidopsis thaliana* mutant, cad1, exhibits Cd, As and Zn hypersensitivity (Cobbett and Goldsbrough 2002; Tennstedt et al. 2009). Cadmium activates the S-assimilation pathway responsible for the synthesis of cysteine (Cys), a precursor of GSH biosynthesis. Glutathione (an NPT) acts as an important antioxidant in mitigating Cd-induced stress (Meuwly and Rauser 1992). It also plays an important role in PC synthesis, which has a proven role in Cd detoxification (Park et al. 2012).

Plants are also provided with an efficient mechanism for protection against ROS and peroxidation reactions. The majority of ROS-scavenging pathways in plants involve superoxide dismutase (SOD), which is found in almost all cellular compartments, the water–water cycle in chloroplasts and the ascorbic acid (AsA)–GSH cycle in chloroplasts, cytosol, mitochondria, apoplast and peroxisomes, in which ascorbate peroxidase (APX) and GR play crucial roles. Catalase (CAT) removes H_2O_2 in peroxisomes (Mittler 2002).

A varying degree of antioxidative response and tolerance is exhibited by different plant species. Several studies have demonstrated that most of the main S-responsive genes involved in sulfate assimilation are induced by Cd, suggesting the existence of a general adaptive response to an increase in cellular demand for reduced S (Nocito et al. 2006). Hence the availability of S to plants is crucial.

Brassica juncea (L.) Czern. (Indian Mustard) belongs to the Brassicaceae family and is an important oilseed crop that is known to tolerate considerable amounts of Cd (Qadir et al. 2004). In this study, we investigated the effects of Cd, S-deficiency and their combination on S-assisted defence against oxidative stress, and changes in cellular antioxidant activity, the contents of NPTs, PCs and photosynthetic pigments, and growth parameters in *B. juncea*.

Methods

Experimental design

Seeds of mustard (*B. juncea* cv. Pusa Jaikisan), procured from the Indian Agriculture Research Institute (IARI), were treated with 1 % (v/v) sodium hypochlorite solution for 10 min followed by thorough washing in deionized water. The seeds were germinated on moist soil (Soilrite®; 1 kg per pot). For S-nutrition, conditions were as described in Abdallah et al. (2010). One set of plant cultures, considered as the control, received Hoagland nutrient solution (Hoagland and Arnon 1950) with 300 μM SO_4^{2-} (herein referred to as 'S-sufficient') and was designated as (+S). A second set received the same Hoagland nutrient solution with an S concentration 10 times lower ($30\,\mu M\,SO_4{}^{2-}$, herein referred to as 'S-deficient') than that of the control, and was designated as ($-S$). After 10 days of growth, both sets were divided into two further sets. Cadmium ($100\,\mu M\,CdCl_2$) prepared in corresponding Hoagland nutrient media was supplied to one set of $+S$ and $-S$ daily according to the water-holding capacity (WHC) of the soil. The plants were grown in a growth chamber: 16/8 h light/dark period, photon flux density of $150 \pm 10\,\mu mol$ photons $m^{-2}\,s^{-1}$, 25/20 °C (day/ night) temperature and 75 % relative humidity. There were four replicates of each treatment: ($T_1 = +S/-Cd$), ($T_2 = -S/-Cd$), ($T_3 = -S/+Cd$), ($T_4 = +S/+Cd$). Leaves were harvested from the plants after 7 and 14 days of Cd treatment and immediately used for biochemical parameter, leaf area and also fresh and dry weight (DW) analysis.

Thiobarbituric acid reactive substances (TBARS)

The magnitude of oxidative stress was measured by estimating the content of TBARS following the method of Heath and Packer (1968). Fresh tissue was ground to a powder using liquid nitrogen in a pre-chilled mortar and pestle. The powder was mixed into a paste in 1 % (w/v) trichloroacetic acid (TCA; 10 mL g^{-1} fresh weight, FW). The extract was centrifuged at $9660 \times g$ for 5 min. A 1.0-mL aliquot of supernatant was placed in a separate tube, to which 4.0 mL of 0.5 % (w/v) thiobarbituric acid (TBA) was added. The mixture was heated at 99 °C for 30 min. It was then quickly cooled in an ice bath and centrifuged at $2817 \times g$ for 5 min to clarify the reaction mixture. The absorbance of the supernatant at 532 nm was measured and corrected for unspecific turbidity by subtracting the value at 600 nm.

Superoxide dismutase assay

The SOD assay was performed using the method of Dhindsa et al. (1981), based on the ability of SOD to inhibit photochemical reduction of nitroblue tetrazolium (NBT). The assay mixture, consisting of 1.5 mL reaction buffer (containing 0.1 M sodium phosphate buffer, pH 7.5, 1 % w/v PVP), 13 mM of L-methionine, 0.1 mL of enzyme extract with equal amounts of 1 M Na_2CO_3, 2.25 mM NBT solution, 3 mM EDTA and $60\,\mu M$ riboflavin and 1.0 mL of double-distilled water (DDW), was incubated under a 15-W fluorescent lamp at 28 °C. The absorbance of the irradiated reaction mixtures at 560 nm was compared with the non-irradiated mixture and per cent inhibition of colour was plotted as a function of the volume of enzyme extract corresponding to 50 % reduction of NTB, which was considered as one unit of enzyme activity and expressed as mg^{-1} protein min^{-1}.

Ascorbate peroxidase assay

The APX extraction and assay were performed as described by Qureshi et al. (2007). Frozen leaf (0.3 g) was homogenized in 3 mL cold extraction buffer (0.5 M phosphate buffer containing 1 % (w/v) polyvenylpyrrolidone (PVP), 1 % (v/v) Triton-X 100, 100 mM EDTA, pH 7.8) in an ice bath. The crude extract was centrifuged at $6708 \times g$ for 15 min at 4 °C. The supernatant was used to measure APX activity. All reagents were prepared fresh before the assay. The assay reaction mixture contained 0.1 M potassium phosphate buffer (pH 7.4), 0.5 mM ascorbate, 0.3 % (v/v) H_2O_2 and $100\,\mu L$ enzyme extract in a total volume of 1 mL. The assay was allowed to equilibrate at 25 °C for 1 min before the addition of hydrogen peroxide, which initiated the reaction. Ascorbate peroxidase assay activity was determined by monitoring the rate of ascorbate oxidation as indicated by a reduction in the absorbance at 290 nm for 3 min at 25 °C. A control reaction was prepared by replacing the ascorbate with reaction buffer. A unit of APX is defined as the amount required to oxidize $1\,\mu mol$ of ascorbate min^{-1} at 25 °C (290 nm extinction coefficient of $2.8\,mmol^{-1}\,cm^{-1}$).

Glutathione reductase assay

To prepare the crude enzyme extract, 0.5 g leaf material was ground to a powder in liquid nitrogen using a pre-chilled mortar and pestle. The powder was homogenized in 2 mL of cold 0.1 M potassium phosphate buffer (pH 7.2) in an ice bath. The crude extract was centrifuged at $5433 \times g$ for 15 min at 4 °C. The supernatant was collected and used for assay. The glutathione reductase (GR) assay was modified from Anderson (1985). The 1-mL assay contained 0.02 mM oxidized glutathione (GSSG) and 0.2 mM NADPH in a buffer (0.1 M potassium phosphate buffer, pH 7.2). The assay was initiated with the addition of 0.2 mL of enzyme extract and activity was monitored by a decrease in absorbance at 340 nm for 3 min at 25 °C. A unit of activity is the amount of enzyme that catalyses the reduction of $1\,\mu mol$ of GSSG min^{-1} at 25 °C.

Catalase assay

Catalase (CAT) activity was determined using the method of Aebi (1984). Fresh leaf material (0.5 g), ground in 5 mL of extraction buffer (0.5 M Na phosphate, pH 7.3, 3 mM EDTA, 1 % w/v PVP, 1 % v/v Triton X-100) was centrifuged at $13\,148 \times g$ for 20 min at 4 °C. Catalase activity in the supernatant was determined by monitoring the disappearance of H_2O_2, according to the decrease in absorbance at 240 nm. The reaction was run in a final volume of 2 mL of reaction buffer (0.5 M Na phosphate, pH 7.3) containing 0.1 mL of 3 mM EDTA, 0.1 mL of enzyme extract and 0.1 mL of 3 mM H_2O_2 for 5 min. Catalase activity was calculated by using a coefficient of absorbance of $0.036 \, mM^{-1} \, cm^{-1}$. One unit of enzyme determines the amount necessary to decompose 1 μmol of H_2O_2 per min.

Ascorbate content

Ascorbate (AsA) was estimated using a modification of Law et al.'s method (1983). Fresh leaf material (0.1 g) was ground to a powder in a mortar and pestle using liquid nitrogen. The powder was homogenized in 2 mL of extraction buffer and was centrifuged at 6708 × g for 10 min. To 400 μL of supernatant, 10 % (w/v) TCA (200 μL) was added. The mixture was vortex-mixed and cooled in ice for 5 min and 10 μL of 5 M NaOH was added. The supernatant mixture was centrifuged at 822 × g for 5 min.

The supernatant fraction was divided into two separate centrifuge tubes (200 μL each) to measure total (AsA + DHA) and reduced (AsA) ascorbate. To estimate total ascorbate, 100 μL of dithiothreitol (DTT) and 200 μL of reaction buffer (150 mM potassium phosphate buffer) were added and the assay mixture was thoroughly mixed and incubated for 15 min at 25 °C; 100 μL of 0.5 % (w/v) N-ethylmaleimide was then added. The contents of the other tube were mixed with 200 μL of reaction buffer (150 mM potassium phosphate buffer) and 200 μL of double-distilled water. Both samples were vortex-mixed and incubated at room temperature for 30 s. To each tube was then added 400 μL of 10 % (w/v) TCA, 400 μL of 44 % (v/v) H_3PO_4, 444 μL of 4 % (w/v) bipyridyl and 200 μL of 3 % (w/v) $FeCl_3$. After vortex-mixing, samples were incubated at 37 °C for 60 min and the absorbance was recorded at 525 nm on a UV–VIS spectrophotometer (Model DU 640, Beckman, USA). A calibration curve was prepared from different concentrations of AsA. The amount was expressed in nmol g^{-1} FW.

Glutathione content

Glutathione content was determined using the method of Anderson (1985), which measures total GSH [reduced glutathione (GSH) and oxidized glutathione (GSSG)] using a GR-catalyzed reaction. Glutathione was extracted by homogenizing 0.5 g of frozen fresh leaves in 2 mL of 5 % (w/v) 5-sulfosalicylic acid to reduce the oxidation of GSH. The powder was homogenized in 2 mL of cold 0.5 M potassium phosphate buffer (pH 7.8) in an ice bath. The crude extract was centrifuged at 5433 × g for 15 min at 4 °C and used for subsequent assay.

The 1.0-mL assay for oxidized GSH contained 40 μL of 0.15 % (w/v) 5-5′-dithio-bis(2-nitrobenzoic acid) (DTNB) and 0.5 mL plant extract in 0.1 M potassium phosphate buffer (pH 7.8). The assay mixture (minus GR and leaf extract) was allowed to equilibrate at 30 °C for 5 min. After

the addition of GSH-containing extract, the absorbance at 412 nm was monitored. The reaction blank was prepared by replacing the plant extract with 5 % (w/v) 5-sulfosalicylic acid. To the same tube 0.2 units per assay GR from yeast and 50 μL of 0.4 % (w/v) NADPH were added and the reaction was allowed to run for 30 min at 25 °C, after which the absorbance at 412 nm was measured. The concentration of GSH was determined through comparison with a standard curve of the reaction rate as a function of the concentration of GSSG and was expressed as nmol g^{-1} FW.

Non-protein thiols

Non-protein thiols (NPTs) were estimated with the method described by Howe and Merchant (1992), using Ellman's reagent. A calibration curve was prepared using GSH (Sigma, MO, USA) to estimate NPTs in samples. The amount of NPT was expressed as nmol g^{-1} FW.

Phytochelatins

Phytochelatin content (PCs) in the tissue was calculated by subtracting the amount of GSH from the amount of total NPTs, and expressed as nmol g^{-1} FW.

$$PCs \, (nmol \, g^{-1} \, FW) = NPT - total \, GSH.$$

Chlorophyll estimation and growth parameters

Chlorophyll content was estimated using the method of Hiscox and Israelstam (1979). Fresh leaves were collected, washed with deionized water and kept in vials. Ten millilitres of DMSO were added to the vials, which were then kept in an oven at 65 °C for 1 h. Absorbance was recorded at 480, 645, 520 and 663 nm on a Beckman DU 640B spectrophotometer. The chlorophyll concentration was calculated with the help of formulae given by Arnon (1949) and expressed as mg g^{-1} FW.

Plants were uprooted carefully from the soil, washed with double-distilled water to remove soil and kept between moist filter paper to avoid desiccation. The leaf area was measured with the help of a portable leaf area meter (Model LI 3000A, LI-COR, USA) and expressed in cm^2 per plant. To determine the leaf dry matter, the plants were dissected at the stem and petiole junction and the leaves were dried in an oven at 65 °C for 2 days. The dried samples were then weighed to determine the plant DW.

Statistical analysis

All the estimates of sample variability are given in terms of the standard error (SE). A Student's t-test was used to identify statistical differences between pairs of means at a confidence level of ≥95 % for each set of data using ANOVA. The data are means ± SE from 10 samples in four replicates ($n = 4$, $P \leq 0.05$, significant at 5 % level, $P \leq 0.01$, significant at 1 % level).

Results

Thiobarbituric acid reactive substances

To assess S-deficiency and Cd-induced oxidative cell damage, the content of TBARS was determined. Under S-deficiency, a 32 % increase in TBARS was observed over the control; a further 84 % increase occurred following the application of Cd to these plants. However, the presence of S (+S/+Cd) lowered the oxidative threat by 33 % at 7 DAT and 32 % at 14 DAT during Cd treatment (Fig. 1).

Response of enzymatic and non-enzymatic antioxidants

To study the stress response in plants exposed to Cd and the protection offered in the presence of S, changes in the activities of enzymatic and non-enzymatic antioxidants involved in the AsA-GSH cycle were analysed in the leaves of mustard.

Antioxidative enzymes

The activity of antioxidative enzymes viz. SOD (Fig. 2A), APX (Fig. 2B), GR (Fig. 2C) and CAT (Fig. 2D) was determined over an experimental period of 7 DAT and 14 DAT of Cd exposure in both S-sufficient and S-deprived plants. Sulfur deficiency significantly suppressed the activities of SOD (42 and 38 %), APX (35 and 28 %), GR (30 and 50 %)

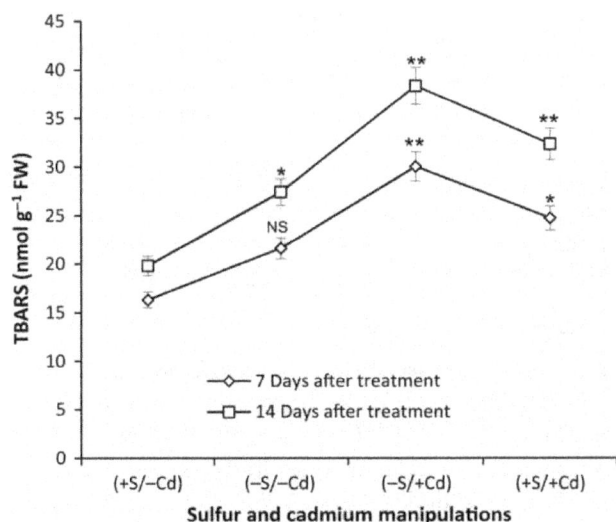

Figure 1. Effect of S-deficiency and Cd stress on the magnitude of oxidative stress in mustard leaf. The values are mean and standard error of the mean (mean $\pm$ SE) of 10 samples and four replicates ($n = 4$, *$P \leq 0.05$; **$P \leq 0.01$; NS, non-significant). +S = 300 μM SO_4^{2-}, $-S = 30$ μM SO_4^{2-}, $-Cd =$ No $CdCl_2$, +Cd = 100 μM $CdCl_2$.

and CAT (10 and 28 %) as noted at 7 DAT and 14 DAT, respectively. However, SOD activity was up-regulated in $-S/+Cd$ plants both at 7 DAT (32 %) and 14 DAT (21 %). The activity of SOD further increased to 136 % (7 DAT) and 153 % (14 DAT) under Cd stress when in the presence of sufficient S (+S/+Cd).

Cadmium treatment of S-deprived plants proved the most deleterious, leading to a decline in the activity of APX (47 and 45 %), GR (55 and 54 %) and CAT (36 and 30 %) at 7 DAT and 14 DAT, respectively. However, S-sufficient plants showed significant resistance to Cd stress (+S/+Cd). The activity of APX (13 and 24 %), GR (26 and 46 %) and CAT (16 and 53 %) was increased at 7 DAT and 14 DAT, respectively.

Non-enzymatic antioxidants

Significant changes in the content of non-enzymatic antioxidants including ascorbates (ascorbate, AsA; dehydroascorbate, DHA; total ascorbate, AsA + DHA) and their ratios (AsA/DHA) were seen under S-deficiency and Cd stress. Sulfur-deprived ($-S/-Cd$) plants showed a 23 % (7 DAT) and 47 % (14 DAT) increase in DHA content. Cadmium stress during S-deprivation ($-S/+Cd$) further increased the DHA content by 40 % at 7 DAT and 58 % at 14 DAT. Sulfur-sufficient but Cd-treated (+S/+Cd) plants showed a 33 % (7 DAT) and 60 % (14 DAT) increase in the DHA content.

In contrast to DHA, the AsA content suffered a decline under every treatment and at both time points. AsA declined by 37 and 22 % ($-S/-Cd$), 42 and 57 % ($-S/+Cd$) and 33 and 22 % (+S/+Cd) at 7 DAT and 14 DAT, respectively. Similarly, the total ascorbate (AsA + DHA) content suffered a decline, but of a lesser magnitude, in all treatments and at all time points. The total ascorbate content declined by 24 and 10 % ($-S/-Cd$), 24 and 38 % ($-S/+Cd$) and 19 and 8 % (+S/+Cd) at 7 DAT and 14 DAT, respectively. The AsA/DHA ratio showed significant alterations; compared with the control (+S/$-Cd$) value of 3.66 (7 DAT) and 4.95 (14 DAT), the AsA/DHA ratio dropped to 1.87 and 2.63, 1.50 and 1.33 and 1.83 and 2.41 under $-S/-Cd$, $-S/+Cd$ and +S/+Cd, respectively (Table 1).

The GSH contents (reduced form, GSH; oxidized form, GSSG; total GSH, GSH + GSSG) and their ratios (GSH/GSSG) also showed significant alterations. Sulfur-deprived ($-S/-Cd$) plants showed a decrease of 23 % (7 DAT) and 12 % (14 DAT) in the GSSG content. Cadmium stress during Sulfur deficiency ($-S/+Cd$) further decreased the GSSG content to 12 % at 7 DAT and 23 % at 14 DAT. Sulfur-sufficient but Cd-treated (+S/+Cd) plants, however, showed a 160 % (7 DAT) and 124 % (14 DAT) increase in the GSSG content.

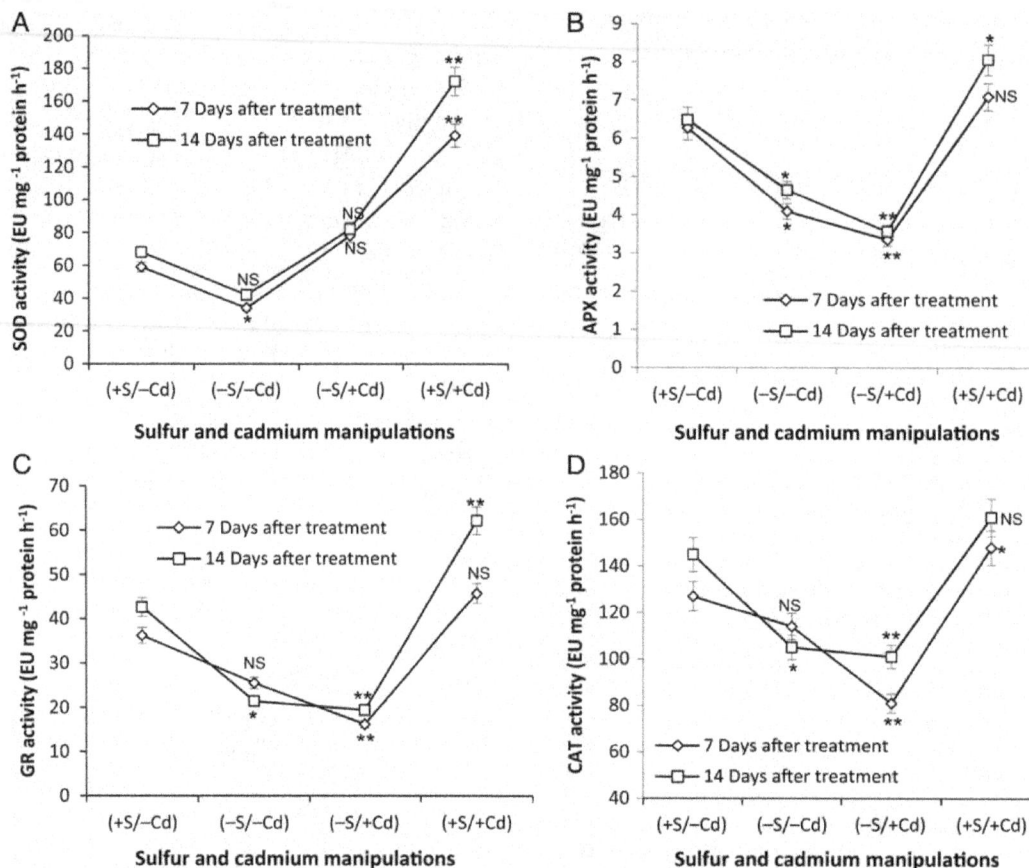

Figure 2. Effect of S-deficiency and Cd stress on the activity of antioxidant enzymes. (A) Superoxide dismutase, (B) APX, (C) GR and (D) catalase (CAT) in mustard leaf. The values are mean and standard error of the mean (mean $\pm$ SE) of 10 samples and four replicates ($n = 4$, $*P \leq 0.05$; $**P \leq 0.01$; NS, non-significant). $+S = 300 \mu M$ SO_4^{2-}, $-S = 30 \mu M$ SO_4^{2-}, $-Cd = No$ $CdCl_2$, $+Cd = 100 \mu M$ $CdCl_2$.

A similar trend was exhibited by the GSH content in all treatments and at both time points. Glutathione content declined by 28 and 18 % ($-S/-Cd$), 26 and 34 % ($-S/+Cd$) at 7 DAT and 14 DAT, respectively. A significant increase of 110 % (7 DAT) and 81 % (14 DAT) in the GSH content over the control was observed in S-sufficient but Cd-treated ($+S/+Cd$) plants. However, the total GSH content (GSH + GSSG) declined in all treatments, except $+S/+Cd$, and at both time points. The total GSH content declined by 16 and 17 % ($-S/-Cd$), 27 and 32 % ($-S/+Cd$) at 7 DAT and 14 DAT, respectively. However, the presence of sufficient S during Cd stress ($+S/+Cd$) helped the plant to accumulate 119 % (7 DAT) and 89 % (14 DAT) more total GSH over the control. The GSH/GSSG ratio showed significant alterations; compared with the control ($+S/-Cd$) value of 4.36 (7 DAT) and 4.50 (14 DAT), the GSH/GSSG ratio dropped to 4.06 and 4.23, 3.66 and 3.81 and 3.52 and 3.63 under $-S/-Cd$, $-S/+Cd$ and $+S/+Cd$, respectively (Table 2).

NPTs and PCs

The NPT content declined by 65 and 69 % ($-S/-Cd$), 49 and 51 % ($-S/+Cd$) at 7 DAT and 14 DAT, respectively.

However, S-sufficient but Cd-treated ($+S/+Cd$) plants showed a 32 % (7 DAT) and 79 % (14 DAT) increase in the NPT content over the control (Table 2).

Similarly, the PC content (PCs = NPTs $-$ total glutathione) declined under each treatment, except $+S/+Cd$, at both time points. The PC content declined by 80 and 92 % ($-S/-Cd$), 59 and 60 % ($-S/+Cd$) at 7 DAT and 14 DAT, respectively. However, the presence of sufficient S during Cd stress ($+S/+Cd$) helped the plant to almost maintain (-2 % at 7 DAT) or increase ($+75$ % at 14 DAT) the PC content over the control (Table 2).

Chlorophyll and growth

Sulfur deficiency resulted in a significant decrease in chlorophyll content; Chl a decreased by 21 % (7 DAT) and 23 % (14 DAT). Moreover, application of Cd to S-deprived ($-S/+Cd$) plants resulted in a drastic decrease of 53 % at 7 DAT and 55 % at 14 DAT. However, plants with sufficient S were able to resist damage with a decline of 32 and 38 % at 7 DAT and 14 DAT, respectively. The Chl b content decreased by 25 % (7 DAT) and 16 % (14 DAT) in $-S/-Cd$, 39 % (7 DAT) and 39 % (14 DAT) in $-S/+Cd$ and 16 % (7 DAT) and 26 % (14 DAT) in Cd-treated plants

Table 1. Changes in different attributes of ascorbates (reduced ascorbate, AsA; oxidized ascorbate, DHA; total ascorbate, AsA+DHA and AsA/DHA ratio) under different combinations of Sulfur (S) and Cadmium (Cd) in mustard leaf. Units are expressed in nmol g⁻¹ FW. The values are mean and standard error of the mean (mean ± SE) of 10 samples and four replicates ($n = 4$; *$P \leq 0.05$; **$P \leq 0.01$; NS, non-significant). Figures in parentheses are per cent variations with respect to control (+S/−Cd). +S = 300 µM SO_4^{2-}, −S = 30 µM SO_4^{2-}, −Cd = No $CdCl_2$, +Cd = 100 µM $CdCl_2$.

Treatment/parameter	Days after treatment (DAT)							
	7				14			
	+S/−Cd Control	−S/−Cd	−S/+Cd	+S/+Cd	+S/−Cd Control	−S/−Cd	−S/+Cd	+S/+Cd
DHA (nmol g⁻¹ FW)	67.3 ± 3.2 (00)	82.9 ± 3.85 (+23 %)NS	94.6 ± 4.18 (+40 %)**	89.9 ± 4.17 (+33 %)*	61.6 ± 2.6 (00)	90.6 ± 4.8 (+47 %)**	97.4 ± 5.10 (+58 %)**	98.4 ± 5.33 (+60 %)**
AsA (nmol g⁻¹ FW)	246.6 ± 12.2 (00)	155.3 ± 7.5 (−37 %)**	142.3 ± 7.2 (−42 %)**	164.6 ± 7.9 (−33 %)*	304.9 ± 15.3 (00)	238.4 ± 12.7 (−22 %)NS	129.4 ± 4.9 (−57 %)**	237.2 ± 9.15 (−22 %)NS
AsA + DHA (nmol g⁻¹ FW)	313.9 ± 13.4 (00)	238.2 ± 10.6 (−24 %)*	236.9 ± 10.2 (−24 %)*	254.5 ± 12.1 (−19 %)NS	366.5 ± 12.3 (00)	329 ± 11.7 (−10 %)NS	226.8 ± 9.2 (−38 %)*	335.6 ± 13.1 (−8 %)NS
AsA/DHA	3.66	1.87	1.50	1.83	4.95	2.63	1.33	2.41

supplied with a sufficient amount of S (+S/+Cd). Total Chl ($a + b$) content decreased by 22 % (7 DAT) and 21 % (14 DAT) in −S/−Cd, 49 % (7 DAT) and 50 % (14 DAT) in −S/+Cd and, 28 % (7 DAT) and 35 % (14 DAT) in Cd-treated plants supplied with a sufficient amount of S (+S/+Cd). In fact, compared with the control, the ratio of Chl a to Chl b tended to decrease in plants treated with Cd particularly under Sulfur deficiency (Table 3).

Leaf area was also reduced by ∼28 % (7 DAT) and 31 % (14 DAT) in S-deprived plants and the reduction was more pronounced (48 % (7 DAT) and 54 % (14 DAT)) when S-deficient plants were exposed to Cd (Fig. 3A). The presence of S in Cd-stressed plants, however, had a lesser effect of −35 % (7 DAT) and −36 % (14 DAT). Sulfur deficiency led to a reduction of leaf DW by 27 % (7 DAT) and 40 % (14 DAT). Upon exposure to Cd, S-deficient plants further decreased DW, by 40 and 50 % at 7 DAT and 14 DAT, respectively. However, the DW of leaves in S-sufficient plants treated with Cd decreased by 21 % (7 DAT) and 38 % (14 DAT) (Fig. 3B).

Discussion

Cadmium stress as well as S deprivation leads to oxidative stress in plants (Pilon et al. 2006; Bashir et al. 2013) due to peroxidation of biomolecules including lipids. In this study, an increase in the magnitude of oxidative stress was observed in response to S-deficiency and Cd stress. Cadmium induced more oxidative stress, particularly in Sulfur deficient conditions. This indicated the formation of more ROS in leaves, not only under S-deficiency but also under Cd stress, with the most formed under dual stress (−S/+Cd). However, the presence of S reduced the magnitude of oxidative stress (Fig. 1), indicating that Cd becomes more deleterious under Sulfur deficiency. This could be because antioxidant systems in plants might suffer a limitation of S for the synthesis of antioxidant enzymes and other peptides/proteins, and ROS quenching molecules such as GSH. Thus, it is evident that Cd severely impairs the plant's ability to counter-attack ROS during S-deficiency. Hu et al. (2015) also found that a source of S (SO_2) helped wheat to resist oxidative stress.

Antioxidative enzymes

Cadmium influences the ascorbate-GSH antioxidant system (Markovska et al. 2009) and metabolism of essential elements (Dong et al. 2006), including non-enzymatic (GSH; AsA; α-tocopherol and carotenoids) and enzymatic (SOD, APX, GR, CAT, etc.) antioxidants. In the present study, SOD, APX, GR and CAT activities were studied. The activity of all these enzymes was lower under S-deficiency, which may be attributed to the limitation

Table 2. Changes in different attributes of thiols (−SH) including reduced GSH, oxidized glutathione (GSSG), NPTs and PCs, and soluble protein content under different combinations of Sulfur (S) and Cd in mustard leaf. Values for GSH, GSH and GSH + GSSG are expressed in nmol g^{-1} FW, NPT and PCs in nmol SH mg^{-1} protein and soluble protein in mg g^{-1} FW. The values are the mean and standard error of the mean (mean ± SE) of 10 samples and four replicates ($n = 4$; *$P \leq 0.05$; **$P \leq 0.01$; NS, non-significant). Numbers in parentheses are per cent variations with respect to control (+S/−Cd). +S = 300 μM SO_4^{2-}, −S = 30 μM SO_4^{2-}, −Cd = No $CdCl_2$, +Cd = 100 μM $CdCl_2$.

Treatment/parameter	Days after treatment (DAT)							
	7				14			
	+S/−Cd Control	−S/−Cd	−S/+Cd	+S/+Cd	+S/−Cd Control	−S/−Cd	−S/+Cd	+S/+Cd
GSSG	23.2 ± 1.8 (00)	17.9 ± 1.2 (−23 %)*	20.5 ± 1.4 (−12 %)NS	60.4 ± 2.8 (+160 %)**	27.8 ± 1.8 (00)	24.3 ± 1.9 (−12 %)NS	21.5 ± 1.8 (−23 %)NS	62.4 ± 4.2 (+124 %)**
GSH	101.2 ± 6.1 (00)	72.8 ± 6.2 (−28 %)	75.1 ± 7.1 (−26 %)	212.7 ± 7.2 (+110 %)**	125.2 ± 3.9 (00)	102.8 ± 4.8 (−18 %)NS	82.1 ± 3.4 (−34 %)*	226.7 ± 9.8 (+81 %)**
GSH + GSSG	124.4 ± 6.7 (00)	90.7 ± 4.9 (−16 %)NS	95.6 ± 5.4 (−27 %)*	273.1 ± 9.6 (+119 %)**	153 ± 4.5 (00)	127.1 ± 5.2 (−17 %)NS	103.6 ± 5.2 (−32 %)*	289.1 ± 12.1 (+89 %)**
GSH/GSSG	4.36	4.06	3.66	3.52	4.50	4.23	3.81	3.63
NPTs	445.3 ± 31.3 (00)	155.8 ± 10.1 (−65 %)**	227.8 ± 16.2 (−49 %)*	586.1 ± 36.4 (+32 %)*	495.2 ± 40.2 (00)	154.7 ± 7.4 (−69 %)**	241.6 ± 15.3 (−51 %)*	887.5 ± 100.2 (+79 %)**
PCs	320.9 ± 18.4 (00)	65.1 ± 4.6 (−80 %)**	132.2 ± 9.3 (−59 %)**	313 ± 22.6 (−2 %)NS	342.2 ± 24.4 (00)	27.6 ± 2.6 (−92 %)**	138 ± 8.4 (−60 %)**	598.4 ± 44.7 (+75 %)**

Table 3. Changes in chlorophyll contents (mg g^{-1} FW) and chlorophyll a/b ratio in mustard leaf treated with different combinations of Sulfur (S) and Cd. The values are mean and standard error of the mean (mean ± SE) of 10 samples and four replicates (n = 4; *P ≤ 0.05; **P ≤ 0.01; NS, non-significant). Numbers in parentheses are per cent variations with respect to control (+S/−Cd). +S = 300 µM SO$_4^{2-}$, −S = 30 µM SO$_4^{2-}$, −Cd = No CdCl$_2$, +Cd = 100 µM CdCl$_2$.

Treatment/ parameter	Days after treatment (DAT)							
	7				14			
	+S/−Cd Control	−S/−Cd	−S/+Cd	+S/+Cd	+S/−Cd Control	−S/−Cd	−S/+Cd	+S/+Cd
Chl a	2.38 ± 0.10 (00)	1.87 ± 0.09 (−21 %)^{NS}	1.12 ± 0.05 (−53 %)**	1.62 ± 0.07 (−32 %)*	2.52 ± 0.14 (00)	1.93 ± 0.14 (−23 %)*	1.14 ± 0.05 (−55 %)**	1.56 ± 0.06 (−38 %)*
Chl b	0.85 ± .04 (00)	0.64 ± 0.03 (−25 %)*	0.52 ± 0.03 (−39 %)*	0.71 ± 0.04 (−16 %)^{NS}	0.92 ± 0.06 (00)	0.77 ± 0.05 (−16 %)^{NS}	0.56 ± 0.05 (−39 %)*	0.68 ± 0.04 (−26 %)^{NS}
Chl a + b	3.23 ± 0.17 (00)	2.51 ± 0.14 (−22 %)^{NS}	1.64 ± 0.10 (−49 %)**	2.33 ± 0.08 (−28 %)*	3.44 ± 0.21 (00)	2.70 ± 0.16 (−21 %)*	1.70 ± 0.12 (−50 %)**	2.24 ± 0.15 (−35 %)*
Chl a/b ratio	2.80	2.92	2.15	2.28	2.74	2.51	2.03	2.29

in the amount of S-containing amino acids and hence lower synthesis of enzymes. Under dual (−S/+Cd) stress the activity of APX, GR and CAT further decreased as a result of S deficiency. Interestingly, Cd increased the activity of SOD despite limited S availability, which perhaps is due to the channelling of infrastructural S towards the synthesis of more SOD, to counter ROS. A similar observation of lesser degree was made in Arabidopsis by Bashir et al. (2013). Unlike SOD, APX and GR activities did not increase under dual (−S/+Cd) stress but decreased further, indicating that Cd-tolerant mustard prefers the strengthening of SOD rather than the ascorbate-GSH antioxidant system. These results also clearly show that due to the lower activity of the ascorbate-GSH antioxidant system accumulation of H$_2$O$_2$ caused high damage to chlorophylls. The maximum decrease in APX, GR and CAT activities occurred under Cd stress during S-deficiency, which could be attributed to the binding of Cd metal with the thiol groups of these enzymes (Romero-Puertas et al. 2007). SOD showed a different response perhaps due to Cd entrapment by peptides, GSH/GSH oligomers or antioxidant enzymes, helping up-regulate SOD activity. Sulfur deficiency significantly suppressed APX and GR activities. The decline in APX and GR activities may be due to GSH depletion and a subsequent reduction in the ascorbate-GSH cycle (Gomes-Junior et al. 2006) as shown in Fig. 3B and C. The recovery of CAT activity at 14 DAT in S-deficient plants might be due to the presence of comparatively more Fe (metal ligand of CAT) in the absence of S and that the defence system acts better against oxidative stress and/or compensates for the decrease in other antioxidant enzymes such as APX and GR but at non-chloroplast locations in the cell. The scenario was totally different in S-sufficient plants exposed to Cd. The activities of all four enzymes (SOD, APX, GR and CAT) were up-regulated; the maximum SOD was 136–153 %.

Ascorbate, glutathione and sulfur-rich compounds

The ascorbate-GSH cycle appears to be of great importance in controlling the cellular redox status, especially upon exposure to Cd. Both ascorbate and GSH are essential for scavenging ROS (Anjum et al. 2008) and are important in controlling metal homeostasis. The ratio of AsA in the reduced form (AsA) to that in the oxidized state (DHA) is considered an important indicator of the redox status of the cell and the degree of oxidative stress experienced. Data (Table 1) show that there was an overall increase in DHA and a decrease in AsA and total ascorbate content, indicating inhibition of ascorbate synthesis. However, the AsA/DHA ratio varied; from 4.95 (control) it dropped to 2.63 (−S/−Cd), 1.33 (−S/+Cd) and 2.41 (+S/+Cd) at 14 DAT. The AsA/DHA ratios can be attributed

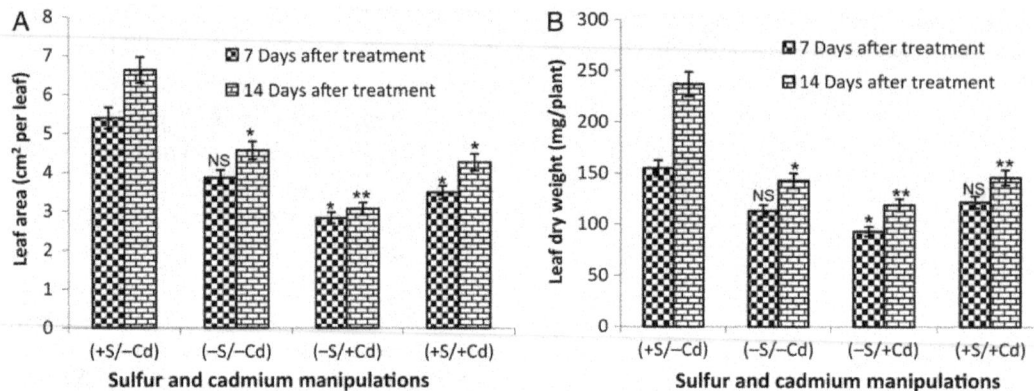

Figure 3. Effect of S-deficiency and Cd stress on (A) leaf area (cm^2 $leaf^{-1}$) and (B) dry weight (mg $plant^{-1}$) of mustard leaf. The values are mean and standard error of the mean (mean $\pm$ SE) of 10 samples and four replicates ($n = 4$; $*P \leq 0.05$; $**P \leq 0.01$; NS, non-significant). $+S = 300$ μM SO_4^{2-}, $-S = 30$ μM SO_4^{2-}, $-Cd =$ No $CdCl_2$, $+Cd = 100$ μM $CdCl_2$.

to APX activity, which helps in AsA regeneration and increases the AsA/DHA ratio.

Glutathione is the main S-storage compound and an important antioxidant in plants. The content and redox level of GSH was measured to determine the effect of S-deficiency on the amount of GSH. In S-deficient plants, the GSH content decreased compared with the control. This decrease was more pronounced when S-deficient mustard was exposed to Cd. However, when S was supplied to these plants, there was a marked increase in GSH levels (Table 2). Thus, S plays a key role in limiting the cellular damage and inducing defence mechanisms against the ROS in response to Cd treatment through the accumulation of GSH. The combination of Cd stress with S-deficiency provided novel data in the present study, but otherwise the results were consistent with the results obtained by Smeets et al. (2005), who reported that heavy metals such as Cd bind to GSH, forming metal-thiolate compounds. This suggests that GSH might be involved in the synthesis of PCs, which could detoxify Cd ions. It also plays an indirect role in protecting membranes by maintaining α-tocopherol and zeaxanthin in the reduced form. Glutathione levels and GSH/GSSG ratios are often indicative of the stress faced by the plant (Tausz et al. 2004). The response of GSH and redox state has varied in different studies. In this study, S-deficiency, Cd stress under S-deficiency and Cd stress during sufficient S decreased the GSH/GSSG ratio, from 4.50 in the control to 4.23 ($-S/-Cd$), 3.81 ($-S/+Cd$) and 3.63 ($+S/+Cd$) at 14 DAT, showing a ratio-maintaining capability of plant only in the presence of sufficient S. The data suggest a direct correlation between GR activity and GSH/GSSG ratio.

The addition of sufficient S tended to recover GSH levels and maintain the redox status of the cells during short- and long-term exposure to Cd. Under Cd stress, the formation of Cd-GSH and Cd-PC complexes reduces the

free Cd concentration in the cytoplasm and helps suppress activation of the stress-related responses in plant metabolism (Metwally et al. 2005). It has been shown that S-nutrition status is associated with plant response to Cd, at least at the plasma membrane H^+ATPase level (Astolfi et al. 2005), with a concomitant uptake of sulfate at a higher rate (Nocito et al. 2002). Interestingly, the levels of NPTs and PCs fell far below those of control plants, indicating that the cell utilizes reserves of S-rich compounds that are found in relatively high concentrations in Indian mustard, perhaps making it Cd-tolerant and a hyperaccumulator. Adaptation of sulfate uptake and assimilation is assumed to be a crucial determinant for plant survival in a wide range of adverse environmental conditions since different S-containing compounds are involved in plant responses to both biotic and abiotic stresses (May et al. 1998; Rausch and Wachter 2005). Cysteine provided to plants through sulfate assimilation pathways acts as a source of reduced S for the biosynthesis of a number of S-containing compounds, including GSH. Cysteine synthesis has previously been shown to be improved with the addition of S sources, and a Cys deficiency limits the synthesis of GSH (Nikiforova et al. 2006; Chekmeneva et al. 2011). Therefore, GSH synthesis in plant cells as observed in the present study relies on and is regulated by the plant's S supply. A sufficient amount of S helped plants not only to maintain, but also to accumulate GSH, NPTs and PCs by 90, 79 and 57 %, respectively, over the control at 14 DAT. The level of different types of PCs was shown to increase with 100 μM $CdCl_2$ in mustard (D'Allessandro et al. 2013) using a proteomic approach, strongly supporting the protective role of PCs against Cd. Phytochelatin binds to toxic metal and then the metal-PC complex is sequestered in the vacuole and perhaps induces specific transporters that mediate Cd tolerance to Arabidopsis (Park et al. 2012). However, the efficiency of PC-based Cd-detoxification is subject to the availability of other nutrients such as iron (Astolfi *et al.* 2011).

Chlorophyll and growth

Cadmium may directly inhibit the process of photosynthesis by its interaction with enzymes (Siedlecka and Krupa 1999). Indirectly, Cd may inhibit the synthesis of photosynthetic pigments or cause their degradation; S-deficiency further worsens the scenario (Bashir *et al.* 2013). The effects of Cd on N and S assimilation have been studied in several plants. The current results showed that both Cd and S-deficiency significantly reduced the chlorophyll content and the Chl *a* to Chl *b* ratio in a hyper-accumulator plant, Indian mustard. Both S and Cd, alone or in combination, cause leaf chlorosis (Bashir *et al.* 2013). Several studies have suggested that Cd-induced leaf chlorosis might be due to impairment of the Mg^{2+} insertion into protoporphyrinogen (Gillet *et al.* 2006) or direct Chl destruction as a consequence of Mg ion substitution in both Chl *a* and *b* (Küpper *et al.* 1998). Our results showed clear and very rapid inhibition of Chl *a*, rather than Chl *b*, mainly in response to Cd stress, with or without S-deficiency. As an early response to S-deficiency, the Chl *a*/Chl *b* ratio increased slightly (− S/ − Cd at 7 DAT) but in the remaining treatments there was a significant decrease in the Chl *a*/Chl *b* ratio. Cadmium may also impair the S uptake, leading to leaf chlorosis and a disrupted pigment ratio. It has also been proposed that a reduction in chlorophyll might be caused by direct interference by Cd with enzymes of the Chl biosynthesis pathways or by Cd interference with the correct assembly of the pigment–protein complexes of the photosystems (Baryla *et al.* 2001; Qureshi *et al.* 2010). It is also possible that the Chl decrease may be due to strong oxidation of the photochemical apparatus and a reduction in chloroplast density and size. Sulfur deficiency in combination with Cd further alleviated chlorophyll inhibition. However, in the presence of sufficient S, we observed lower destruction of chlorophyll. S is required to repair Fe-S-containing protein complexes, including the incorporation of Fe-S clusters into apoproteins and the stabilization of biomolecules with sulfolipids. It has also been proposed that Cd may influence the biosynthesis of chlorophyll by affecting protochlorophyllide reductase, which, however, contains oxygen-tolerant Fe-S clusters (Nomata *et al.* 2008), indicating the role of S in protection.

When the overall impact of S-deficiency and Cd stress on leaf area and leaf biomass was analysed, S showed a positive role under Cd stress. Sulfur deficiency and Cd caused a significant decrease in leaf area and DW, at least with sufficient S-supply. In the current study, growth inhibition was more severe in − S/+Cd (up to 54 % in leaf area and 50 % in DW) plants than in S-deficiency (up to 31 % in leaf area and 40 % in DW) or Cd stress (up to 36 % in leaf area and 38 % in DW), indicating that the effect of Cd treatment and Sulfur deficiency is synergistic. Moreover, leaf area was found to be more sensitive in response to Cd stress under S-deficiency. Ouzounidou *et al.* (1997) suggested that the inhibitory action of heavy metals on root-shoot and leaf growth seems principally to be due to chromosomal aberrations and abnormal cell divisions. It may also be correlated with Cd-induced inhibition of photosynthetic processes and other enzymes involved in leaf expansion. Decreased plant growth caused by heavy metals is a collective consequence of inhibition of photosynthesis, translocation of photosynthetic products and cell division (Drążkiewicz and Baszyński 2005).

Conclusions

In conclusion, we confirmed that S-deficiency increases the susceptibility of plants to Cd-generated oxidative damage and modulates the AsA-GSH cycle. We propose that S helps the accumulation of GSH and other S-rich compounds to detoxify metals. In turn, SOD expression is up-regulated to reduce the concentration of superoxide radicals. This study demonstrates that SOD is not the first line of defence against metal stress and that S-rich compounds play a prime role. Further, S-deprived plants lack S-defence and efficient antioxidative mechanisms, making Cd more dangerous. Under S-deficiency, S-containing defence metabolites (GSH, PCs, etc.) as well as enzyme co-factors (Fe-S clusters) decline. The clear decrease in Fe-S clusters, under S-deficiency, is crucial for the maintenance of photosynthesis-related molecules and activities, but the presence of Cd under such conditions might exacerbate oxidative stress, with a highly adverse effect on growth.

Sources of Funding

The authors extend their appreciation to the Deanship of Scientific Research at King Saud University for funding this work through research group no. RGP-VPP RG-1435-042.

Contributions by the Authors

The research design and preparation of the manuscript are credited to M.I.Q., H.B. and M.M.I. H.B., R.B., J.A. and M.A.B. contributed to data collection in replicates. M.I.Q., H.B. and I.A.A. analysed the data.

Literature Cited

Abdallah M, Dubousset L, Meuriot F, Etienne P, Avice JC, Ourry A. 2010. Effect of mineral sulphur availability on nitrogen and sulphur uptake and remobilization during the vegetative growth of *Brassica napus* L. *Journal of Experimental Botany* **61**:2635–2646.

Aebi H. 1984. Catalase in vitro. *Methods in Enzymology* **105**:121–126.

Anderson ME. 1985. Determination of glutathione and glutathione disulfides in biological samples. *Methods in Enzymology* **113**:548–570.

Anjum N, Umar S, Ahmad A, Iqbal M, Khan N. 2008. Sulphur protects mustard (*Brassica campestris* L.) from cadmium toxicity by improving leaf ascorbate and glutathione. *Plant Growth Regulation* **54**:271–279.

Anjum NA, Ahmad I, Mohmood I, Pacheco M, Duarte AC, Pereira E, Umar S, Ahmad A, Khan NA, Iqbal M, Prasad MNV. 2012. Modulation of glutathione and its related enzymes in plants' responses to toxic metals and metalloids—a review. *Environmental and Experimental Botany* **75**:307–324.

Arnon DI. 1949. Copper enzymes in isolated chloroplasts: polyphenol oxides in *Beta vulgaris*. *Plant Physiology* **24**:1–15.

Astolfi S, Zuchi S, Passera C. 2005. Effect of cadmium on H+ATPase activity of plasma membrane vesicles isolated from roots of different S-supplied maize (*Zea mays* L.) plants. *Plant Science* **169**:361–368.

Astolfi S, Zuchi S, Neumann G, Cesco S, Sanità di Toppi L, Pinton R. 2011. Response of barley plants to Fe deficiency and Cd contamination as affected by S starvation. *Journal of Experimental Botany* **63**:1241–1250.

Baryla A, Carrier P, Franck F, Coulomb C, Sahut C, Havaux M. 2001. Leaf chlorosis in oilseed rape plants (*Brassica napus*) grown on cadmium-polluted soil: causes and consequences for photosynthesis and growth. *Planta* **212**:696–709.

Bashir H, Ahmad J, Bagheri R, Nauman M, Qureshi MI. 2013. Limited sulfur resource forces *Arabidopsis thaliana* to shift towards non-sulfur tolerance under cadmium stress. *Environmental and Experimental Botany* **94**:19–32.

Chekmeneva E, Gusmão R, Díaz-Cruz JM, Ariño C, Esteban M. 2011. From cysteine to longer chain thiols: thermodynamic analysis of cadmium binding by phytochelatins and their fragments. *Metallomics* **3**:838–846.

Cobbett C, Goldsbrough P. 2002. Phytochelatins and metallothioneins: roles in heavy metal detoxification and homeostasis. *Annual Review of Plant Biology* **53**:159–182.

D'Allessandro A, Taamalli M, Gevi F, Timperio AM, Zolla L, Ghnaya T. 2013. Cadmium stress response in *Brassica juncea*: hints from proteomics and metabolomics. *Journal of Proteome Research* **12**:4979–4997.

Dhindsa RH, Plumb-Dhindsa P, Thorpe TA. 1981. Leaf senescence correlated with increased level of membrane permeability, lipid peroxidation and decreased level of SOD and CAT. *Journal of Experimental Botany* **32**:93–101.

Dong J, Wu F, Zhang G. 2006. Influence of cadmium on antioxidant capacity and four microelement concentrations in tomato seedlings (*Lycopersicon esculentum*). *Chemosphere* **64**:1659–1666.

Drążkiewicz M, Baszyński T. 2005. Growth parameters and photosynthetic pigments in leaf segments of *Zea mays* exposed to cadmium, as related to protection mechanisms. *Journal of Plant Physiology* **162**:1013–1021.

Gill SS, Tuteja N. 2011. Cadmium stress tolerance in crop plants—probing the role of sulfur. *Plant Signaling and Behavior* **6**:215–222.

Gillet S, Decottignies P, Chardonnet S, Maréchal P. 2006. Cadmium response and redoxin targets in *Chlamydomonas reinhardtii*: a proteomic approach. *Photosynthesis Research* **89**:201–211.

Gomes-Junior RA, Moldes CA, Delite FS, Pompeu GB, Gratão PL, Mazzafera P, Lea PJ, Azevedo RA. 2006. Antioxidant metabolism of coffee cell suspension cultures in response to cadmium. *Chemosphere* **65**:1330–1337.

Haneklaus S, Bloem E, Schnug E, de Kok LJ, Stulen I. 2007. Sulfur. In: Barker AV, Pilbeam DJ, eds. *Handbook of plant nutrition*. Boca Raton: CRC Press, 183–238.

Harada E, Yamaguchi Y, Koizumi N, Hiroshi S. 2002. Cadmium stress induces production of thiol compounds and transcripts for enzymes involved in sulfur assimilation pathways in *Arabidopsis*. *Journal of Plant Physiology* **159**:445–448.

Heath RL, Packer L. 1968. Photoperoxidation in isolated chloroplasts. I. Kinetics and stoichiometry of fatty acid peroxidation. *Archives of Biochemistry and Biophysics* **125**:189–198.

Hell R. 1997. Molecular physiology of plant sulfur metabolism. *Planta* **202**:138–148.

Hell R, Hillebrand H. 2001. Plant concepts for mineral acquisition and allocation. *Current Opinion in Biotechnology* **12**:161–168.

Hiscox JD, Israelstam GF. 1979. A method for the extraction of chlorophyll from leaf tissue without maceration. *Canadian Journal of Botany* **57**:1332–1334.

Hoagland DR, Arnon DI. 1950. The water culture method for growing plants without soil. *California Agriculture Experimental Station Circular* **347**:461–465.

Howe G, Merchant S. 1992. Heavy metal activated synthesis of peptides in *Chlymodomonas reinhartii*. *Plant Physiology* **98**:127–136.

Hu K-D, Bai G-S, Li W-J, Yan H, Hu L-Y, Li Y-H, Zhang H. 2015. Sulfur dioxide promotes germination and plays an antioxidant role in cadmium-stressed wheat seeds. *Plant Growth Regulation* **75**:271–280.

Küpper H, Küpper F, Spiller M. 1998. In situ detection of heavy metal substituted chlorophylls in water plants. *Photosynthesis Research* **58**:123–133.

Law MY, Charles SA, Halliwell B. 1983. Glutathione and ascorbic acid in spinach (*Spinacia oleracea*) chloroplasts. The effect of hydrogen peroxide and paraquat. *Biochemical Journal* **210**:899–903.

Lee SL, Korban SK. 2002. Transcriptional regulation of *Arabidopsis thaliana* phytochelatin synthase by cadmium during early stages of plant development. *Planta* **215**:689–693.

Leustek T, Martin MN, Bick JA, Davies JP. 2000. Pathways and regulation of sulfur metabolism revealed through molecular and genetic studies. *Annual Review of Plant Physiology and Plant Molecular Biology* **51**:141–165.

Maksymiec W, Krupa Z. 2006. The effects of short-term exposition to Cd, excess Cu ions and jasmonate on oxidative stress appearing in *Arabidopsis thaliana*. *Environmental and Experimental Botany* **57**:187–194.

Markovska YK, Gorinova NI, Nedkovska MP, Miteva KM. 2009. Cadmium-induced oxidative damage and antioxidant responses in *Brassica juncea* plants. *Biologia Plantarum* **53**:151–154.

May MJ, Vernoux T, Sánchez-Fernández R, Van Montagu M, Inzé D. 1998. Evidence for posttranscriptional activation of γ-glutamylcysteine synthetase during plant stress responses. *Proceedings of the National Academy of Sciences of the USA* **95**: 12049–12054.

Metwally A, Safronova VI, Belimov AA, Dietz K-J. 2005. Genotypic variation of the response to cadmium toxicity in *Pisum sativum* L. *Journal of Experimental Botany* **56**:167–178.

Meuwly P, Rauser WE. 1992. Alteration of thiol pools in roots and shoots of maize seedlings exposed to cadmium: adaptation and developmental cost. *Plant Physiology* **99**:8–15.

Mittler R. 2002. Oxidative stress, antioxidants and stress tolerance. *Trends in Plant Science* **7**:405–410.

Nikiforova VJ, Bielecka M, Gakiere B, Krueger S, Rinder J, Kempa S, Morcuende R, Scheible W-R, Hesse H, Hoefgen R. 2006. Effect of sulfur availability on the integrity of amino acid biosynthesis in plants. *Amino Acids* **30**:173–183.

Nocito FF, Pirovano L, Cocucci M, Sacchi GA. 2002. Cadmium-induced sulfate uptake in maize roots. *Plant Physiology* **129**:1872–1879.

Nocito FF, Lancilli C, Crema B, Fourcroy P, Davidian JC, Sacchi GA. 2006. Heavy metal stress and sulfate uptake in maize roots. *Plant Physiology* **141**:1138–1148.

Noctor G, Queval G, Mhamdi A, Chaouch S, Foyer CH. 2011. Glutathione. *The Arabidopsis Book* **9**:e0142.

Nomata J, Ogawa T, Kitashima M, Inoue K, Fujita Y. 2008. NB-protein (BchN–BchB) of dark-operative protochlorophyllide reductase is the catalytic component containing oxygen-tolerant Fe–S clusters. *FEBS Letters* **582**:1346–1350.

Ouzounidou G, Moustakas M, Eleftheriou EP. 1997. Physiological and ultrastructural effects of cadmium on wheat (*Triticum aestivum* L.) leaves. *Archives of Environmental Contamination and Toxicology* **32**:154–160.

Park J, Song W-Y, Ko D, Eom Y, Hansen TH, Schiller M, Lee TG, Martinoia E, Lee Y. 2012. The phytochelatin transporters AtABCC1 and AtABCC2 mediate tolerance to cadmium and mercury. *The Plant Journal* **69**:278–288.

Perfus-Barbeoch L, Leonhardt N, Vavasseur A, Forestier C. 2002. Heavy metal toxicity: cadmium permeates through calcium channels and disturbs the plant water status. *The Plant Journal* **32**:539–548.

Pietrini F, Iannelli MA, Pasqualini S, Massacci A. 2003. Interaction of cadmium with glutathione and photosynthesis in developing leaves and chloroplasts of *Phragmites australis* (Cav.) Trin. ex Steudel. *Plant Physiology* **133**:829–837.

Pilon M, Abdel-Ghany S, Ye H, Van Hoewyk D, Pilon-Smits EAH. 2006. Biogenesis of iron-sulfur cluster proteins in plastids. In: Setlow JK, ed. *Genetic engineering principles and methods.* New York: Wiley, 101–117.

Prasad MNV. 1995. Inhibition of maize leaf chlorophylls, carotenoids and gas exchange functions by cadmium. *Photosynthetica* **31**: 635–640.

Qadir S, Qureshi MI, Javed S, Abdin MZ. 2004. Genotypic variations in phytoremediation potential of *Brassica juncea* cultivars exposed to Cd stress. *Plant Science* **167**:1171–1181.

Qureshi MI, Abdin MZ, Qadir S, Iqbal M. 2007. Lead-induced oxidative stress and metabolic alterations in *Cassia angustifolia* Vahl. *Biologia Plantarum* **51**:121–128.

Qureshi MI, D'Amici GM, Fagioni M, Rinalducci S, Zolla L. 2010. Iron stabilizes thylakoid protein–pigment complexes in Indian mustard during Cd-phytoremediation as revealed by BN-SDS-PAGE and ESI-MS/MS. *Journal of Plant Physiology* **167**: 761–770.

Randazzo CA. 2009. Sulphur—essential to the fertilizer industry as a raw material, plant nutrient, and soil amendment. In: *15th AFA International Annual Fertilizers Forum and Exhibition*, Cairo, Egypt.

Rausch T, Wachter A. 2005. Sulfur metabolism: a versatile platform for launching defence operations. *Trends in Plant Science* **10**: 503–509.

Robinson NJ, Tommy AM, Kushe C, Jackson PJ. 1993. Plant metallothioneins. *Biochemical Journal* **295**:1–10.

Romero-Puertas MC, Corpas FJ, Rodríguez-Serrano M, Gómez M, del Río LA, Sandalio LM. 2007. Differential expression and regulation of antioxidative enzymes by cadmium in pea plants. *Journal of Plant Physiology* **164**:1346–1357.

Saito K. 2000. Regulation of sulfate transport and synthesis of sulfur-containing amino acids. *Current Opinion in Plant Biology* **3**:188–195.

Salvagiotti F, Castellarin JM, Miralles DJ, Pedrol HM. 2009. Sulfur fertilization improves nitrogen use efficiency in wheat by increasing nitrogen uptake. *Field Crops Research* **113**:170–177.

Sanità di Toppi L, Gabbrielli R. 1999. Response to cadmium in higher plants. *Environmental and Experimental Botany* **41**:105–130.

Siedlecka A, Krupa Z. 1999. Cd/Fe interaction in higher plants—its consequences for the photosynthetic apparatus. *Photosynthetica* **36**:321–331.

Smeets K, Cuypers A, Lambrechts A, Semane B, Hoet P, Van Laere A, Vangronsveld J. 2005. Induction of oxidative stress and antioxidative mechanisms in *Phaseolus vulgaris* after Cd application. *Plant Physiology and Biochemistry* **43**:437–444.

Takahashi H, Yamazaki M, Sasakura N, Watanabe A, Leustek T, de Almeida Engler J, Engler G, Van Montagu M, Saito K. 1997. Regulation of sulfur assimilation in higher plants: a sulfate transporter induced in sulfate-starved roots plays a central role in *Arabidopsis thaliana*. *Proceedings of the National Academy of Sciences of the USA* **94**:11102–11107.

Tausz M, Sircelj H, Grill D. 2004. The glutathione system as a stress marker in plant ecophysiology: is a stress-response concept valid? *Journal of Experimental Botany* **55**:1955–1962.

Tennstedt P, Peisker D, Böttcher C, Trampczynska A, Clemens S. 2009. Phytochelatin synthesis is essential for the detoxification of excess Zinc and contributes significantly to the accumulation of Zinc. *Plant Physiology* **149**:938–948.

Van Assche F, Clijsters H. 1990. Effects of metals on enzyme activity in plants. *Plant, Cell and Environment* **13**:195–206.

Zhang ZC, Chen BX, Qiu BS. 2010. Phytochelatin synthesis plays a similar role in shoots of the cadmium hyperaccumulator *Sedum alfredii* as in non-resistant plants. *Plant, Cell and Environment* **33**: 1248–1255.

8

Germination and ultrastructural studies of seeds produced by a fast-growing, drought-resistant tree: implications for its domestication and seed storage

Helene Fotouo-M.*, Elsa S. du Toit* and Petrus J. Robbertse

Department of Plant Production and Soil Science, University of Pretoria, Pretoria 0002, South Africa

Associate Editor: John Runions

Abstract. Seed ageing during storage is one of the main causes of reduction in seed quality and this results in loss of vigour and failure to thrive. Finding appropriate storage conditions to ameliorate deterioration due to ageing is, therefore, essential. Ultrastructural changes in cellular organelles during storage and seed germination rates are valuable indices of damage that occurs during seed ageing. There is increasing interest in *Moringa oleifera* Lam. because of its multiple uses as an agroforestry crop. Seeds of this species lose their viability within 6–12 months of harvest but no scientific information is available on the longevity of seed stored in the fruit (capsules). In most undeveloped countries, seeds are still stored inside the fruit by traditional methods in special handmade structures. In this experiment we tried to simulate these traditional storage conditions. Capsules of *Moringa* were stored at ambient room temperature for 12, 24 and 36 months. The ultrastructure, solute leakage and viability of seed were investigated. The ultrastructure of 1-year-old seed showed no sign of deterioration. It was evident, however, that some cells of the 3-year-old seed had deteriorated. The remnants of the outer and inner two integuments that remain tightly attached to the cotyledons probably play a role in seed dormancy. No significant difference was found between germination percentage of fresh and 1-year-old seed. The germination percentage decreased significantly from 2 years of storage onward. The decrease in seed viability during storage was associated with a loss in membrane integrity which was evidenced by an increase in electrolyte leakage. Our findings indicate that the longevity of *M. oleifera* seeds can be maintained if they are stored within their capsules.

Keywords: Deterioration; dormancy; lipid bodies; membrane leakage; protein bodies; seed; storage.

Introduction

In seeds, age-induced deterioration results from various internal changes. Reserve substances may be altered so that they no longer supply the nutritional requirements of the embryo (Simola 1974). Membrane aberrations are said to increase with seed ageing (Berjak and Villiers 1970) and result in increased leakage of metabolites and ions (Roberts and Ellis 1982; Ouyang et al. 2002).

As seed deterioration increases, the rate of germination decreases, and production of weak seedlings with loss of vigour increases progressively. According to Garcia de Castro and Martinez-Honduvilla (1984), seed ageing is a complex process, so it is essential to investigate this process at the subcellular level in order to understand the best conditions for seed storage. Despite the publication of numerous papers on seed ultrastructure at various phases of development during the last two decades

* Corresponding author's e-mail address: fotouoh@yahoo.fr; elsa.dutoit@up.ac.za

(Berjak *et al.* 1994; Algan and Bakar 1997; Wang and Berjak 2000; Wiśniewska *et al.* 2006; Teixeira and Machado 2008; Egorova *et al.* 2010; Moura *et al.* 2010; Chan and Belmonte 2013; Hu *et al.* 2013; and citations therein) knowledge of the ultrastructural changes that occur during seed storage is still insufficient. The complexity of seed tissues and the difficulties in preparing samples for microscopy have until now precluded ultrastructural studies on *Moringa oleifera* Lam. seeds.

Moringa oleifera, or miracle tree, is known in many parts of the world for its multiple uses as an agroforestry crop. The leaves, flowers and immature fruits are edible (Anwar and Bhanger 2003). The leaves are a good source of protein, vitamin A, B and C and minerals such as calcium and iron (Dahot 1988). A number of medicinal properties have been ascribed to various parts of this tree (Kumar *et al.* 2010). *Moringa* seeds contain ~35–40 % oil, commercially known as 'Ben oil' (Anwar and Rashid 2007). *Moringa* domestication and commercialization is still a challenge as its agronomic properties have not been well elucidated. It is a perennial tree that in most cases is grown as an intercrop or in an agroforestry set-up and only bears fruit (capsule) once a year. The crop is mainly propagated by seed (Radovich 2012). Many studies (Bezerra *et al.* 2004; Madinur 2007; De Oliveira *et al.* 2009) have found that *M. oleifera* seeds lose their viability and vigour within 6–12 months depending on the conditions in which they are stored. High-quality seed is essential for most crops including agroforestry crops like *Moringa*. Seed ageing is one of the main causes of reduction in seed quality (FAO/IPGRI 1994). Finding appropriate storage conditions to ameliorate deterioration is essential.

Simple techniques such a hessian sacks, cotton bags, paper containers, cardboard, aluminium cans and foil, glass jars and plastic film stored at ambient temperature have been used to maintain seed viability in both domesticated and wild sources since early times (Willan 1985; Ellis *et al.* 1996; Akai 2009). In sub-Saharan Africa and other developing countries around the world, traditional methods of seed storage have been used for many years with little or no modification (Mannan and Tarannum 2011). Traditional storage facilities such as underground pits and different types of solid wall bins (made of timber, earth and stone) are usually used by subsistence or small-scale farmers to keep part of their products to be used as seed in the following planting season (Ali Ahmed and Ahmed Alama 2010). Farmers must also be able to store their products until the next successful harvest and this might be more than a year in the case of crop failure (Blum and Bekele 2000).

Seed stored within the pods of some legumes can withstand infestation by storage pests and tend to extend seed viability. This relatively positive feature is often used by farmers to minimize crop damage by storing in pods. The method can be applied to a variety of edible legumes, and is particularly valid for bambara nuts, groundnuts and cow peas (Odogola 1994). Cow pea pods are usually hand-picked when mature, bagged and then hauled to a place where they are stored for a variable period until they are threshed. After threshing, the seed becomes more exposed to post-harvest insect pests and is vulnerable to these insects throughout subsequent storage (Murdock *et al.* 2003). It is with this background in mind that this study was initiated.

No study has investigated the effects of traditional methods of storage, which are used primarily by most small and poor farmers. The aim of this study was to determine the germination percentage of *M. oleifera* seed that was left within the fruit (capsule) and stored at ambient room temperature and to provide details on ultrastructural changes that occur during seed ageing.

Methods

Seed material

Fruits (capsules) of *M. oleifera* were harvested during 2009 and 2011 from an orchard growing on the Experimental Farm of the University of Pretoria (25°45S, 28°16E). They were bagged in open poly mesh bags and stored at ambient room temperature (annual average temperature: 23–25 °C) for 12, 24 and 36 months. Seeds from the 2011 harvest were assessed after 12 and 24 months and seeds from the 2009 harvest were assessed after 36 months. After threshing, only seeds with the physical characteristics of maturity were selected for the study.

Seed morphology and ultrastructural studies

The outer part of the seed coat of fresh, 12-month-old seeds and 36-month-old seeds was removed. Seeds were imbibed for 24 h after which the samples were divided into two sub-samples: one for light microscopic and another for transmission electron microscopic examination.

Seed morphology. Morphological characteristics of the seed were observed using a dissecting microscope. The cotyledons and the embryo were carefully separated. Photographs were taken with the Zeiss Discovery V20 stereo microscope (Jena, Germany).

Light microscopy. The cotyledons of the seeds were separated and fixed with FAA (formaldehyde, acetic acid and ethanol) inside polytops. Thereafter, samples were dehydrated in an ethanol series (30, 50, 70 and 100 %) and each concentration was replaced after 24 h, except for 100 % alcohol that was repeated. Ethanol was extracted from the samples with a series of xylene

(30, 50, 70 and 100 %). After gradual wax infiltration, the samples were embedded in pure wax and mounted on stubs. A microtome (2040 Autocut Sterea Star Zoom, Reichert Jung-0.7× to 42 × 570, Leica, Vienna, Austria) was used to cut samples at ~10 μm. Sections were stained with saffranin, counterstained with fast green and mounted in DPX mountant. Pictures were taken with a digital camera (Nikon DXM 1200) mounted on a Zeiss Discovery V20 stereo microscope and light microscope (Nikon/SMZ-1, Japan).

Transmission electron microscopy (TEM). Small samples (2 × 2 mm) were excised from the cotyledons to include the epidermis and sub-epidermal layers as well as samples from the central part of the cotyledons and then prepared for TEM, according to Coetzee and Van Der Merwe (1996). The samples were fixed for 3 days in 2.5 % glutaraldehyde in 0.075 M phosphate buffer (pH 7.4), after which they were rinsed three times (10 min each) in 0.075 M phosphate buffer. Samples were further post-fixed in 0.5 % aqueous osmium tetroxide for 2 h and thereafter rinsed three times with distilled water. This was followed by dehydration in an ethanol series (30, 50, 70, 90 and 100 %), infiltrated with 30 and 60 % quetol for 1 h each and pure quetol for 4 h and then polymerized at 60 °C for 39 h. Ultrathin sections were prepared using a Reichert Ultracut E ultramicrotome (Vienna, Austria). The sections were stained with 4 % aqueous uranyl acetate and lead citrate (Reynolds 1963) for viewing and photographing with a JEOL JEM-2100F transmission electron microscope (JEOL, Tokyo, Japan).

Seed germination

Seeds of all the treatments were germinated using the same procedure. Each treatment consisted of four replicates of 50 seeds. Prior to germination, the outer part of the seed coat was removed to minimize fungal attack. The inner part of the seed coat containing remnants of the inner integument remained tightly attached to the cotyledons and could not be removed without damaging the cotyledons. The seeds were then disinfected in 1 % of sodium hypochlorite for 25 min and rinsed three times in sterile distilled water in a the laminar airflow hood. Seeds were germinated according to ISTA (2006) procedure. The 50 seeds of each replicate were distributed on paper towel rolls soaked with 70 mL of water and incubated in controlled temperature chambers at alternative temperatures of 20/30 °C. Germinated seeds were counted for the first time after 7 days and the last count after 14 days.

Tetrazolium test (TTA). The TTA was conducted on seeds that did not germinate to test for possible lack of viability. The test was done according to ISTA (ISTA 2006). Tissues necessary for development of seed into seedling should stain red.

Electrolyte leakage

Solute leakage of seeds was estimated by placing one seed gram (i.e. without seed coat) into 10 mL of distilled water for 24 h at 25 °C (Rao et al. 2006). This was replicated at least four times ($\geq$20 seeds/treatments). Electrical conductivity of the water containing the leakage was measured with a conductivity meter (Mettler Toledo, 8603. Switzerland).

Statistical analysis

Germination percentage and electrical conductivity were statistically analysed using STATISCA software (STATISCA 12, Statsoft 2013). A one-way analysis of variance (ANOVA) was performed to determine the statistical difference between germination percentage and electrolyte leakage as influenced by storage duration. The Duncan post-hoc test ($P < 0.05$) was used to check the significance between groups.

Results

Seed morphology and ultrastructure

Seed morphology. The general morphology of *M. oleifera* seed is represented in Fig. 1. The seeds are round, protected with a brownish seed coat containing three white wings (Fig. 1A). When trying to remove the seed coat, the inner part containing remnants of the outer integument and the inner integument remains tightly attached to the cotyledons as a creamy white layer with three stripes containing vascular bundles below the wings (Fig. 1B and C). The diminutive embryonic axis (1.5–2 mm) is located in the small cavity between the two cotyledons near the micropilar region (Fig. 1D). The embryonic axis has a distinguished radicle (whitish) and plumule that is slightly split into two ends, representing the primordial of the first leaves (Fig. 1E).

Light microscopy. The transverse section of a mature *M. oleifera* seed reveals cotyledons enclosed by inner endotesta and inner integument composed of different layers of non-living cells (Fig. 2A and B). As described by Muhl (2014), the endotesta originates from parts of the outer integument and is made up of layers of elongated, thickly reticulate cells. Below the endotesta is the compressed inner integument with prominent epidermis cells having a thick cell wall. Cells of the cotyledons are isometric or roundish in shape and filled

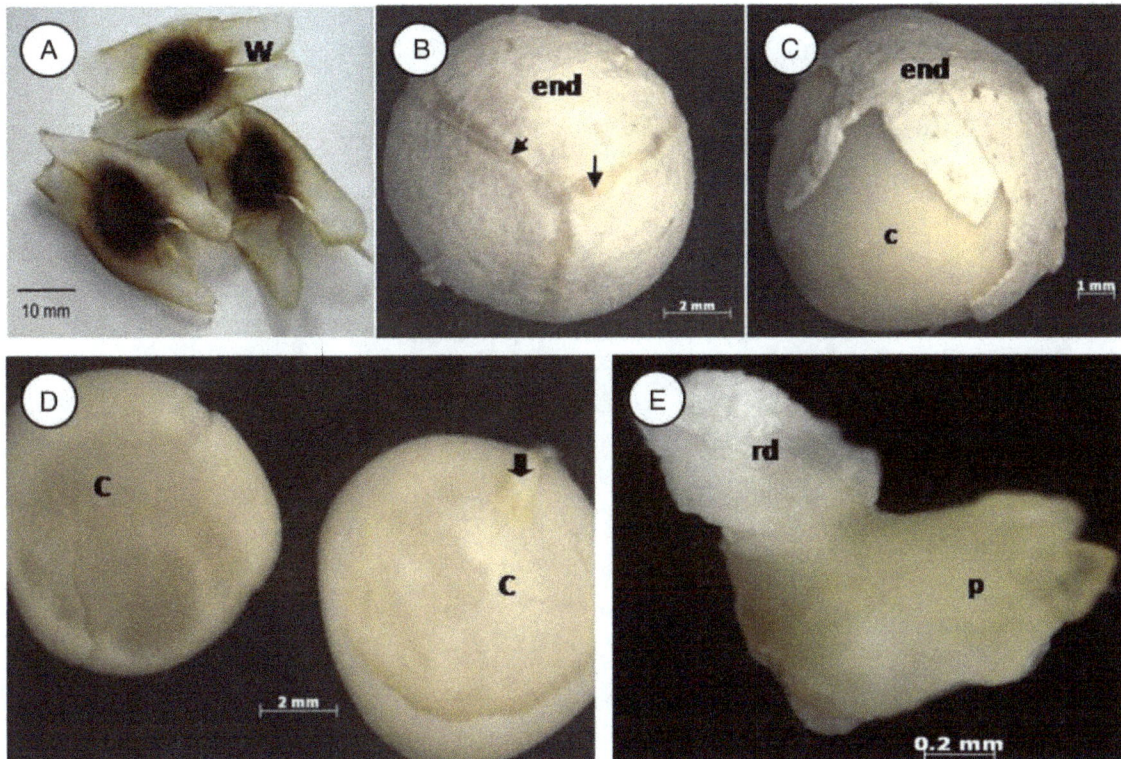

Figure 1. *Moringa* seed morphology. (A) Seed with seed coat and wings. (B) Seed with internal seed coat and vascular bundle. (C) Seed with partially removed internal seed coat. (D) Separated cotyledon. (E) Embryo axis. Wing (w), vascular bundle (arrowhead), cotyledon (c), endotesta (end), embryonic axis (black arrow), radical (rd), plumule (p).

with storage material. Vascular bundles are absent in sub-epidermal tissue (Fig. 2C) and present in the central layers (Fig. 2D). No difference was visible between seed of different ages with the light microscope.

Transmission electron microscope (TEM)

Cotyledons. As seen with the light microscope, the epidermal cells of the inner integument have thick cell walls (Fig. 4A). A small nucleus is present at the centre of some cells (picture not shown). Below the inner integument are the epidermal cells of the cotyledon (Figs 3A, 4B and 7A). They are occupied by lipid bodies and a conspicuous nucleus. Protein bodies are seldom present and plastids are absent. No difference was found between epidermal cells of seeds from different storage periods.

The structure of sub-epidermal cells of the cotyledon was uniform. Lipid bodies surrounded the protein bodies and filled most of the remaining space in all healthy cells, irrespective of the age (Figs 3C, 4C–D and 5C–D). The lipid bodies are numerous, round to oval in shape, small in size and have a uniform grey interior. Protein bodies occupy most of the cytoplasm volume. Globoid inclusions were present occasionally (Fig. 4D). Most organelles were not noticeable in the cotyledon as the entire cytoplasm was filled with storage materials, except for the irregularly

shaped nucleus with a well-defined nuclear membrane (Fig. 3D).

Sub-epidermal cells of the 1-year cotyledon showed no sign of deterioration (Fig. 4). In the 3-year-old cotyledon, the deterioration was not uniform. Some cells seemed to be still perfectly healthy (Fig. 5C) while damage was noticeable in others (Fig. 5D and E). The membrane of the protein body was deteriorated, causing lipid bodies to enter the protein body (Fig. 5D). Cell deterioration was also marked by the collapse of the cell wall adjacent to the intercellular cavity (Fig. 5E). A degraded Golgi apparatus next to the broken cell wall (Fig. 5E) was observed.

In the 3-year-old seeds, the central cells of the cotyledons were almost completely filled with scattered oil bodies (Fig. 6A and B). The protein bodies were disrupted and smaller in size compared with those found in sub-epidermal cells. No sign of deterioration was evident in central cells of cotyledons of the 1- and 3-year-old seeds.

Embryo axis. The main storage material in the embryo axis was protein bodies. They occupied more of the cell volume compared with the sub-epidermal cells of the cotyledons (Fig. 7A–C). Globoid inclusions were small and numerous when present. Lipid bodies formed a wall lining along the plasma membrane and built a single layer around protein bodies. The cells of the embryo

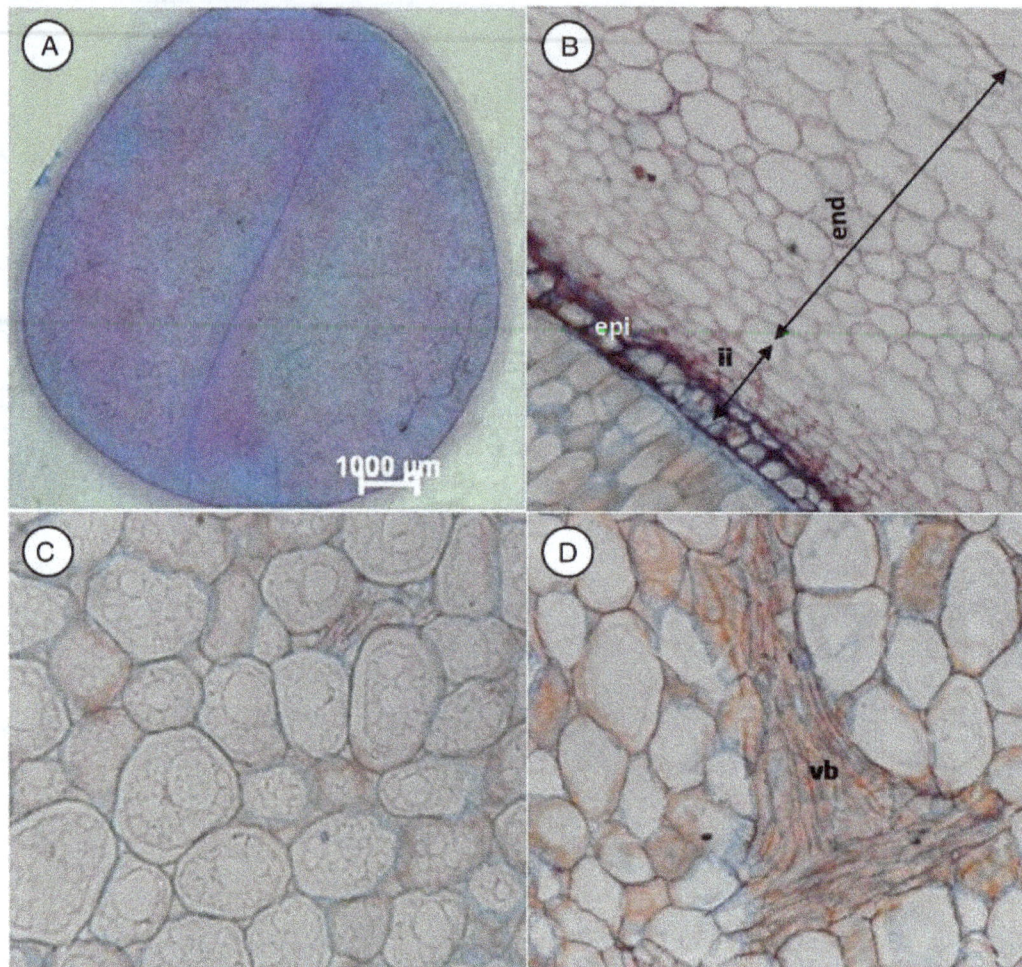

Figure 2. Light micrographs of parts of *M. oleifera* seed. (A) Transverse section of cotyledons and remnant of seed coat. (B) Endotesta, inner integument (ii) covering the outer cell layer of the cotyledon. (C) Sub-epidermal cell layers of cotyledon. (D) Central tissue of cotyledon. Endotesta (end), epidermis of inner integument (epi), vascular bundle (vb).

axis were not totally filled with storage material as was the case with cotyledon cells; the remaining space was filled with ground substances or cytoplasm and various organelles. A prominent nucleus with well-defined nucleolus was present in most cells. Mitochondria with cristae and a double membrane (Fig. 7D) were present in cells of all ages. Golgi apparatus was spotted in cells of fresh embryo (Fig. 7E).

As was the case with sub-epidermal cells of the cotyledon, some cells of the embryo axis still maintained the ultrastructure of healthy cells while others showed some anomalies (Fig. 7F). In some cells, the cytoplasm had shrunk and detached from the cell wall. The extent of the detachment varied from one cell to another; some were moderate and others severe.

Seed viability and electrical conductivity

The mean germination percentage and the electrolyte conductivity of seed stored for different periods are represented in Table 1. The germination percentage was determined to be 67 % for the fresh seeds and decreased to 16 %, while the seed viability (germination percentage + TTA test) was initially 80 % and declined to 18 % after the fruits were stored at ambient temperature for 3 years. Seeds maintained a high germination percentage until the end of 1 year of storage; after 2 years seed viability declined significantly ($P < 0.05$). The electrical conductivity of seed increased significantly as the storage duration increased.

Discussion

Morphology and ultrastructural studies

Cotyledons. The ultrastructure of cotyledonary tissues of *M. oleifera* is similar to those of other oily seeds (Muller *et al.* 1975; Young *et al.* 2004; Kuang *et al.* 2006; Donadon *et al.* 2013) and other seeds in general (Cecchifiordi *et al.* 2001; Pinzón-Torres *et al.* 2009). The seeds are filled with storage material that hampers the visualization of organelles, except for the lobed

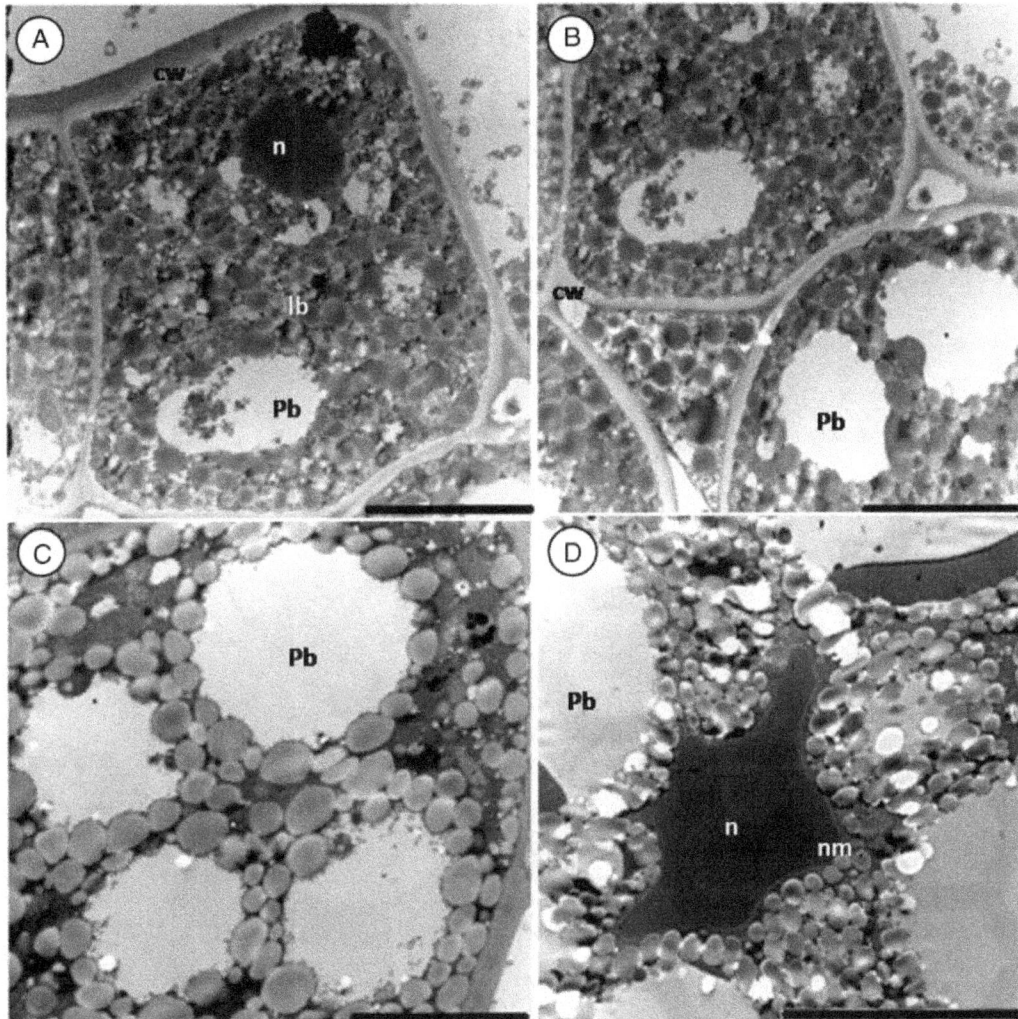

Figure 3. Sections of a fresh (control) *M. oleifera* seed (cotyledons). (A) Epidermis with fewer protein bodies and a thick external cell wall. (B) Epidermal and sub-epidermal cell. (C and D) Sub-epidermal cells filled with large protein and lipid bodies. Note the electron-dense nucleus (n in D) of irregular shape with well-defined membrane squeezed between storage materials. Lipid bodies (lb), protein bodies (pb), nuclear membrane (nm). Scale bars: A–C (5 μm); D (10 μm).

nucleus and plastids that were spotted in some cells. The scarcity of organelles may suggest low metabolic activity and the restricted function of cotyledons as storage organs (Pinzón-Torres *et al.* 2009). The organization of storage material in different tissues is probably determined by the role they play and the order in which they are used during germination. Starch grains are totally absent in the cotyledons.

The ultrastructure of the 1-year-old cotyledons was similar to that of the fresh cotyledons. In the sub-epidermal cells of the 3-year-old seed, the membrane of the protein bodies had ruptured, leading to the coalescence of protein bodies into a large confluent masse and the engulfment of lipid bodies by the protein substances. Similar observations have been reported by Smith (1978) and Dawidowicz-Grzegorzewska and Podstolski (1992), in

artificially aged lettuce and *Brassica napus* seeds. Protein bodies contain several hydrolytic enzymes, among them phosphatases. The release of their contents into the cytosol as a result of membrane deterioration can cause localized cellular autolysis (Dawidowicz-Grzegorzewska and Podstolski 1992). The breakdown of the cell wall and the disintegration of the Golgi apparatus were also evident in severely deteriorated cells. Breakages of cell wall have not been reported in previous studies. One would be tempted to assign this breakage to sample preparation but with the presence of disintegrated Golgi apparatus next to the broken cell wall, the breakage was likely caused by deterioration during ageing. Cellular membrane damage is mediated by oxidative attack, which promotes phospholipid degradation and the loss of membrane organization (Pukacka and Ratajczak 2007). The decrease in size of lipid

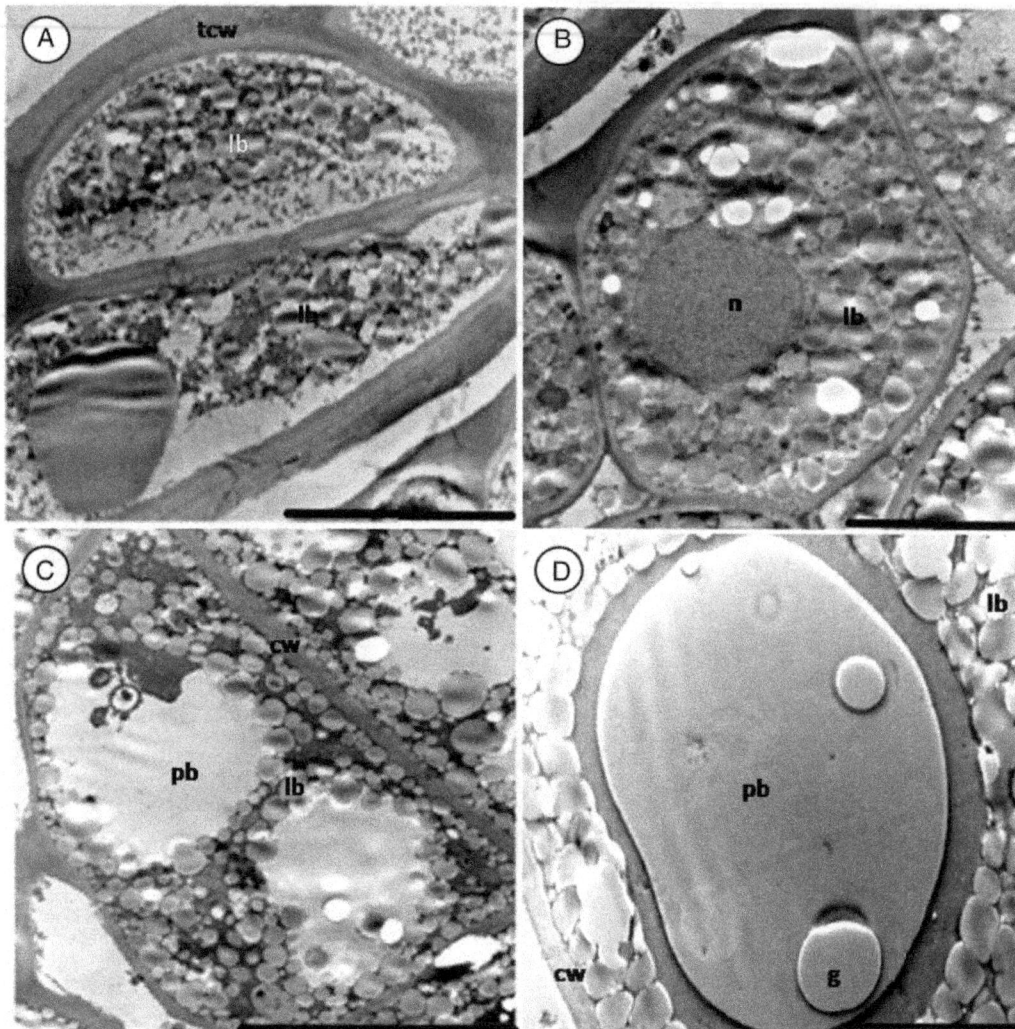

Figure 4. Section of 1-year-old seed (cotyledon) of *M. oleifera*. (A) Epidermis cells of the inner integument. Note the thick cell wall. (B) Epidermal cells of cotyledons. (C) Sub-epidermal cell filled with protein and lipid bodies. (D) Protein bodies embedded in the cytoplasm. Note globoid (g) inclusions in protein bodies. Scale bars: A (10 μm); B–D (5 μm).

bodies and their coalescence have been reported in aged and non-viable seed by other authors (Garcia de Castro and Martinez-Honduvilla 1984; Neya *et al.* 2004; Walters *et al.* 2005; Donadon *et al.* 2013). Lipid bodies of *M. oleifera* remained intact after 3 years of storage.

Embryo axis. As found for *M. oleifera*, lipid and protein bodies in cells of the embryo axis have been reported in the seed of many species such as *B. napus* (Kuraś 1984; Dawidowicz-Grzegorzewska and Podstolski 1992), *Origananum majorana* (Wiśniewska *et al.* 2006), *Amaranthus hypocondriacus* (Ciombra and Salema 1994), *Picea mariana* (Wang and Berjak 2000) and carrot (Dawidowicz-Grzegorzewska 1997). Germination is initiated in the embryo. Protein content in the embryo is higher than that in the cotyledon. The protein reserve function is to supply amino acids for the formation of enzymes

during germination, which are used by the cell for hydrolysis of storage materials (Pernollet 1978). Lipid bodies in embryo axis cells of a number of species are thought to serve as reservoirs during germination (Pinzón-Torres *et al.* 2009).

Damage observed in the embryo was different from that of the cotyledons. The separation of the plasmalemma from the cell wall in non-viable seeds has been observed in many species (Anderson 1970; Garcia de Castro and Martinez-Honduvilla 1984) but has also been linked to imbibition damage (Hoekstra *et al.* 1999). Thus membrane withdrawal can occur either as a result of ageing or during sample preparation (fixation). In this study, the storage may have played a huge role as the leaching of cellular constituents in the 3-year-old seeds was found to be significantly higher than in other seeds. The nuclei maintained their morphology in the old seed even when

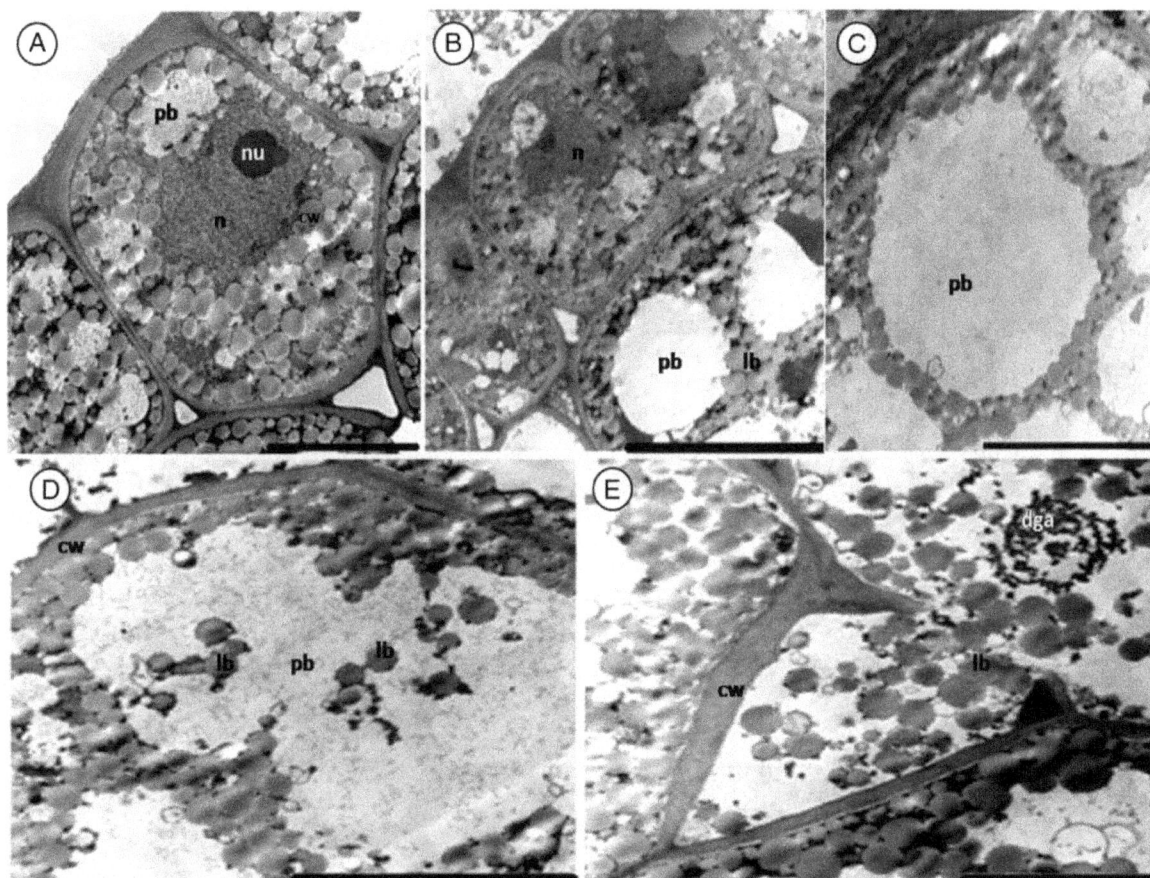

Figure 5. Sections of 3-year-old seed (cotyleton) of *M. oleifera*. (A) Epidermal cells with few protein bodies between lipid bodies. (B) Epidermal and sub-epidermal cells. (C) Sub-epidermal cells. Note the presence of protein body membrane. (D and E) Deteriorated sub-epidermal cells. (D) Membrane has deteriorated and protein appears coalescent. (E) Broken cell wall allowing merging between the plasmalemma of intercellular space and adjacent cells. Note the presence of a deteriorated Golgi apparatus (dga). Scale bars: A and E (5 μm); B–D (10 μm).

some parts of the cells were damaged. The resistance of nucleus to deterioration during senescence has also been reported by Simola (1974).

Seed viability and electrical conductivity

Nouman *et al.* (2012) observed that dehulling of *Moringa* seed did not significantly increase the germination rate while Mubvuma *et al.* (2013) reported that they scarify the seed before planting in order to increase water uptake. According to our observations (Figs 3B and 5A–B), it is not possible to remove the entire seed coat without damaging the seed, which suggests that for any germination test, research should be more specific about the removal of the seed coat.

Initial germination percentages of 84 and 93 % for *M. oleifera* have been reported by Bezerra *et al.* (2004) and Madinur (2007), respectively. The germination percentages obtained in the present study compare well with dehulled seeds of Nouman *et al.* (2012). According to the TTA test 13 % of the seeds were dormant, probably as a result of the remnants of the seed coat remaining

attached to the cotyledons. This finding is supported by Nouman *et al.* (2012), who reported that priming the seed with *Moringa* leaf extract improved the emergence of *Moringa* seedlings. Similar observations were made by Mubvuma *et al.* (2013); they found an increase in germination percentages after storing seeds for 60 days at 25 and 35 °C, but they argue that *Moringa* seeds are non-dormant because fresh seeds can readily germinate after exposure to favourable conditions and that the improvement after exposure to high temperature may be explained by genetic adaptation of seed within their centres of origin where the temperature ranges between 30 and 35 °C. By the same logic it is likely that *Moringa* seeds acquire dormancy when they are grown in an environment with an average annual temperature below 30 °C. High temperatures and priming treatment are some of the methods used to break seed dormancy.

De Oliveira *et al.* (2009) reported that *M. oleifera* seeds retain their viability after 6 months of storage at ambient temperature, irrespective of the type of storage. This was supported by a previous study done by Bezerra *et al.*

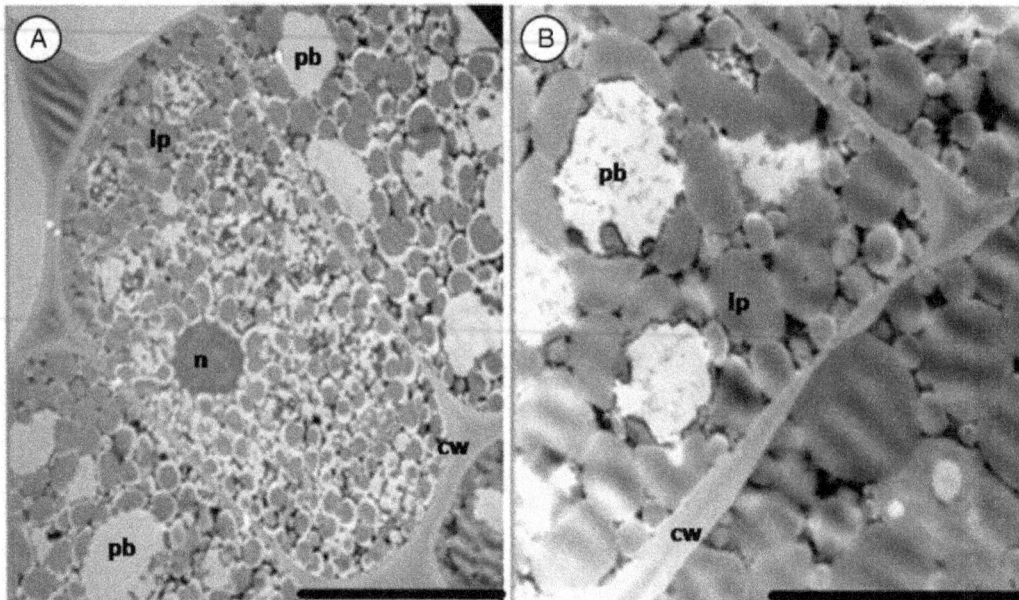

Figure 6. Section of central cotyledon cells of 3-year-old *M. oleifera*. (A) Scattered lipid bodies occupy most of the cell volume. (B) Protein bodies are small and disrupted. Scale bars: A and B (10 μm).

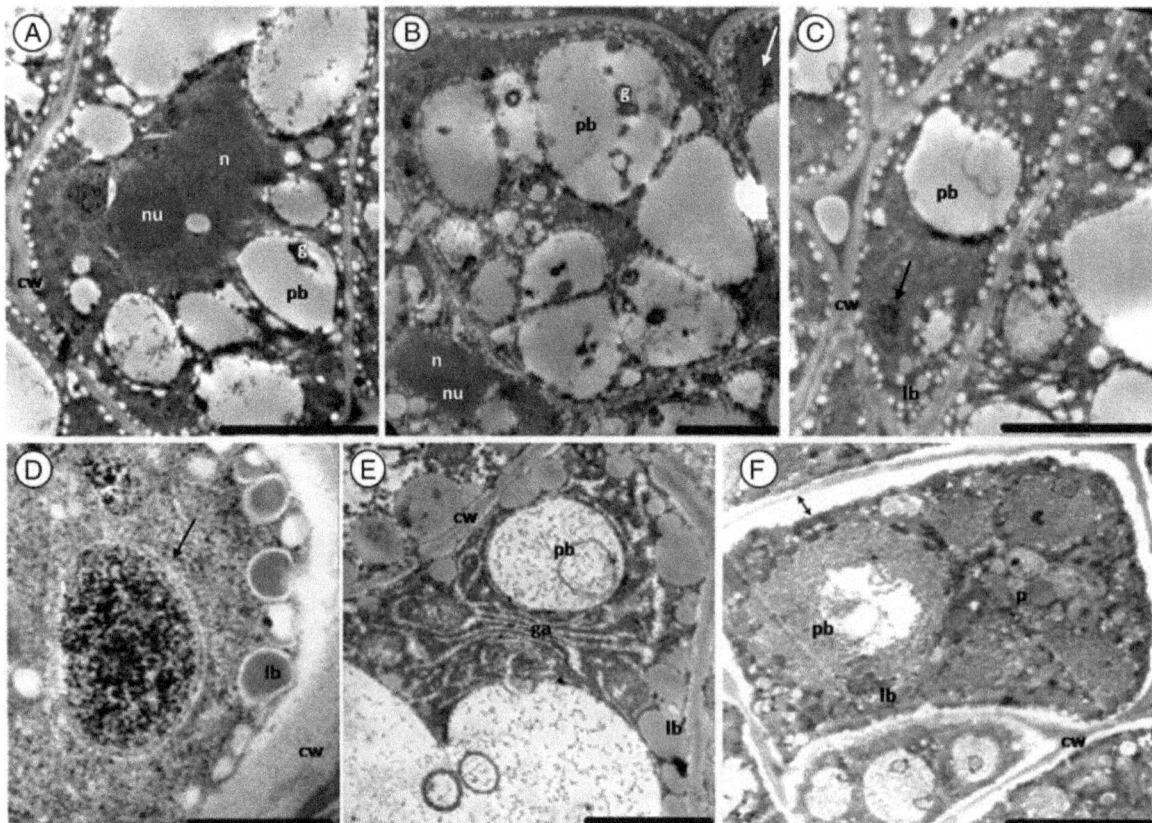

Figure 7. Section of embryo axis of *M. oleifera*. (A–C) Healthy cells of fresh, 1- and 3-year-old seeds. (D) Mitochondria with double membrane (arrow). (E) Golgi apparatus (ga) occupying part of the cytoplasm. (F) Deteriorated cells of embryo axis of 3-year-old seed. Note the withdrawal of the plasmalemma (double arrow). Scale bars: A–C and F (5 μm); D (1 μm); E (2 μm).

Table 1. Average germination percentage, with or without tetrazolium test (TTA) and electrical conductivity of seed, left in capsules (fruit) for different storage periods. Data shown are means of four replicates $\pm$ SE. Averages with different letters are significantly different ($P \leq 0.05$).

Storage duration (years)	Germination		Electrical conductivity (μS/cm)
	Germ %	Germ % +TTA	
Fresh (0)	67 ± 1.4^a	80 ± 0.96^a	21.83 ± 0.7^a
1	64 ± 1.7^a	72 ± 1.31^a	32.945 ± 0.5^b
2	40 ± 3.1^b	46 ± 3.6^b	88.8 ± 3.01^c
3	16 ± 0.85^c	18 ± 0.62^c	101.88 ± 1.76^c

(2004). The authors also observed a 65 % decrease in germination percentage after 12 months and a complete loss of viability after 24 months. Madinur (2007) reported a continuous significant decrease from 2- to 12-month storage at room temperature. In the present study, the seed viability remained almost unchanged after 12 months and maintained 46 % viability after 24 months of storage. These values are considerably higher than those found in studies where seeds were separated from the capsules before storage. Seed viability decreases as a result of loss of membrane integrity (Fig. 6D), which is supported by the increase of solute leakage (Table 1). Higher leachates have also been recorded in other stored high oil content seed such as *Carthamus tinctorius* (Alivand *et al.* 2012) and sunflower (Kallapa 1982). According to these authors the higher electrolyte leakage in oily seed is a result of the leaching out of free fatty acid during storage.

Conclusions

The remnants of the outer integument and the inner integument that remain tightly attached to the cotyledons when trying to remove the seed coat probably play a role in seed dormancy. The decrease in seed viability during storage is associated with the loss in membrane integrity and it is confirmed by the increase in electrolyte leakage. The longevity of *M. oleifera* seeds can be extended if they are stored within their capsules (fruit). Capsule storage does not require additional equipment and can be adopted by small farmers. *Moringa oleifera* seed are high in oil content, making fixative infiltration very difficult and therefore affecting the quality of the micrographs.

Sources of Funding

This work was funded by the National Research Foundation, South Africa.

Contributions by the Authors

All authors contributed extensively to the work presented in this paper. H.F.-M. designed, performed experiments, analysed data and wrote the manuscript. E.S.d.T. and P.J.R. were involved in designing and supervising data analysis, and edited the manuscript.

Acknowledgements

We are thankful to Mr C. van der Merwe and Mrs A. Buys from the Laboratory for Microscopy and Microanalysis.

Literature Cited

Akai CY. 2009. Effect of storage environment on the quality of the physic nut (*Jatropha curcas* L.) seed. Master dissertation. Department of Horticulture, Kwame Nkrumah University of Science and Technology, Kumasi, Ghana.

Algan G, Bakar HN. 1997. The ultrastructure of the mature embryo sac in the natural tetraploid of red clover (*Trifolium pratense* L.): that has a very low rate of seed formation. *Acta Societatis Botanicorum Poloniae* **66**:13–20.

Ali Ahmed EE, Ahmed Alama SH. 2010. Sorghum (*Sorghum bicolor* (L.) *Moench.*) seed quality as affected by type and duration of storage. *Agriculture and Biology Journal of North America* **1**:1–8.

Alivand R, Afshari RT, Sharifzadeh F. 2012. Storage effects on electrical conductivity and fatty acid composition of *Carthamus tinctorius* seed. In: *International Conference of Agriculture Engineering*, 8–12 July 2012, Valencia, Spain.

Anderson JD. 1970. Metabolic changes in partially dormant wheat seeds during storage. *Plant Physiology* **46**:605–608.

Anwar F, Bhanger MI. 2003. Analytical characterization of *Moringa oleifera* seed oil grown in temperate regions of Pakistan. *Journal of Agricultural and Food Chemistry* **51**:6558–6563.

Anwar F, Rashid U. 2007. Physico-chemical characteristics of *Moringa oleifera* seeds and seed oil from a wild provenance of Pakistan. *Pakistan Journal of Botany* **39**:1443–1453.

Berjak P, Villiers TA. 1970. Ageing in plant embryos. I. The establishment of the sequence of development and senescence in the root cap during germination. *New Phytologist* **69**:929–938.

Berjak P, Bradford KJ, Kovach DA, Pammenter NW. 1994. Differential effects of temperature on ultrastructural responses to dehydration in seeds of *Zizania palustris*. *Seed Science Research* **4**: 111–121.

Bezerra AME, Filho SM, Freitas JBS, Teófilo EM. 2004. Avaliação da qualidade das sementes de *Moringa oleifera* Lam. durante o armazenamento (Evaluation of quality of the drumstick seeds during the storage). *Ciência e Agrotecnologia* **28**: 1240–1246.

Blum AA, Bekele A. 2000. The use of indigenous knowledge by farmers in Ethiopia when storing grains on their farms. In: *Paper submitted to the 16th Symposium of the International Farming Systems Association*, Santiago, Chile, 27–29 November 2000.

Cecchifiordi A, Palandri M, Turicchia S, Gabrieletani, Di Falco P. 2001. Characterization of the seed reserves in *Tillandsia* (Bromeliaceae) and ultrastructural aspects of their use at germination. *Caryologia* **54**:1–16.

Chan A, Belmonte MF. 2013. Histological and ultrastructural changes in canola (*Brassica napus*) funicular anatomy during the seed life-cycle. *Botany* **91**:671–679.

Ciombra S, Salema R. 1994. *Amaranthus hypochondriacus*: seed structure and localization of seed reserves. *Annals of Botany* **74**:373–379.

Coetzee J, Van Der Merwe CF. 1996. *Preparation of biological material for the electron microscopy*. South Africa: Unit of Electron Microscopy, University of Pretoria.

Dahot MU. 1988. Vitamin contents of flowers and seeds of *Moringa oleifera*. *Pakistan Journal of Biochemistry* **21**:21–24.

Dawidowicz-Grzegorzewska A. 1997. Ultrastructure of carrot seeds during matriconditioning with Micro-Cel E. *Annals of Botany* **79**:535–545.

Dawidowicz-Grzegorzewska A, Podstolski A. 1992. Age-related changes in the ultrastructure and membrane properties of *Brassica napus* L. seeds. *Annals of Botany* **69**:39–46.

De Oliveira LM, Ribeiro MCC, Maracaja PB, Carvalho GS. 2009. Qualidade fisiologica de sementes de Moringa emfunção do tipo de embalagem, ambiente e tempo de armazenamento 1 (In Portuguese with English Abstract). *Revista Caatinga Mossoró* **22**:70–75.

Donadon JR, Resende O, Teixeira SP, dos Santos JM, Moro FV. 2013. Effect of hot air drying on ultrastructure of crambe seeds. *Drying Technology* **31**:269–276.

Egorova VP, Zhao Q, Lo Y, Jane W, Cheng N, Hou S, Dai H. 2010. Programmed cell death of the mung bean cotyledon during seed germination. *Botanical Studies* **51**:439–449.

Ellis RH, Hong TD, Astley D, Pinnegar AE, Kraak HL. 1996. Survival of dry and ultra-dry seeds of carrot, groundnut, lettuce, oilseed rape, and onion during five years' hermetic storage at two temperatures. *Seed Science and Technology* **24**:347–358.

FAO/IPGRI Genebank Standards. 1994. Food and Agriculture Organization of the United Nations, Rome, International Plant Genetic Resources Institute, Rome. ftp://ftp.fao.org/docrep/fao/meeting/015/aj680e.pdf.

Garcia de Castro MF, Martinez-Honduvilla CJ. 1984. Ultrastructural changes in naturally aged *Pinus pinea* seeds. *Physiologia Plantarum* **62**:581–588.

Hoekstra FA, Golovina EA, Van Aelst AC, Hemminga MA. 1999. Imbibitional leakage from anhydrobiotes revisited. *Plant, Cell and Environment* **22**:1121–1131.

Hu Z, Hua W, Zhang L, Deng L, Wang X, Liu G, Hao W, Wang H. 2013. Seed structure characteristics to form ultrahigh oil content in rapeseed. *PLoS One* **8**:e62099.

ISTA. 2006. *International rules for seed testing, rules 1993*, 2006 edn. International Seed Testing Association.

Kallappa PK. 1982. Prediction of potential of seed lots of sunflower BSH-1. *Seed Technical News* **32**:146.

Kuang A, Blasiak JS, Chen S, Bingham G, Mary E. 2006. Modification of reserve deposition in wheat and brassica seeds by synthetic atmospheres and microgravity. *Gravitational and Space Biology Bulletin* **19**:161–162.

Kumar PS, Mishra D, Ghosh G, Panda CS. 2010. Medicinal uses and pharmacological properties of *Moringa oleifera*. *International Journal of Phytomedicine* **2**:210–216.

Kuraś MW. 1984. Activation of rape (*Brassica napus* L.) embryo during seed germination. III. Ultrastructure of dry embryo axis. *Acta Societatis Botanicorum Poloniae* **53**:171–186.

Madinur NI. 2007. Seed viability in drumstick (*Moringa oleifera* Lamk.). Master Thesis. University of Agricultural Sciences, Dwarwad.

Mannan MA, Tarannum N. 2011. Assessment of storage losses of different pulses at farmers' level in Jamalpur region of Bangladesh. *Bangladesh Journal of Agricultural Research* **36**:205–212.

Moura EF, Ventrella MC, Motoike SY. 2010. Anatomy, histochemistry and ultrastructure of seed and somatic embryo of *Acrocomia aculeata* (Arecaceae). *Scientia Agricola* **67**:399–407.

Mubvuma MT, Mapanda S, Mashonjowa E. 2013. Effect of storage temperature and duration on germination of moringa seeds (*Moringa oleifera*). *Greener Journal of Agricultural Sciences* **3**:427–432.

Muhl QE. 2014. Flowering, fruitgrowth and intracellular storage component formation in developing *Moringa oleifera* Lam. seed as influenced by irrigation. PhD Thesis, University of Pretoria, Pretoria.

Muller LL, Hensarling TP, Jacks TJ. 1975. Cellular ultrastructure of jojoba seed. *Journal of the American Oil Chemists' Society* **52**:164–165.

Murdock LL, Seck D, Ntoukam G, Kitch L, Shade RE. 2003. Preservation of cowpea grain in sub-Saharan Africa—Bean/Cowpea CRSP contributions. *Field Crops Research* **82**:169–178.

Neya O, Golovina EA, Nijsse J, Hoekstra FA. 2004. Ageing increases the sensitivity of neem (*Azadirachta indica*) seeds to imbibitional stress. *Seed Science Research* **14**:205–217.

Nouman W, Siddiqui MT, Basra SMA, Afzal I, Rehman H. 2012. Enhancement of emergence potential and stand establishment of *Moringa oleifera* Lam. by seed priming. *Turkish Journal of Agriculture and Forestry* **36**:227–235.

Odogola WR. 1994. Postharvest management and storage legumes. AGROTEC, UNDP/OPS Regional Programme, RAF/92/R51. Harare, Zimbabwe. http://www.academia.edu/2658013/SUSTAINABLE_HORTICULTURAL_PRODUCTION_IN_THE_TROPICS_PRODUCTION_IN_THE_TROPICS.

Ouyang XR, van Voorthuysen T, Toorop PE, Hilhorst HWM. 2002. Seed vigour, aging, and osmopriming affect anion and sugar leakage during imbibition of maize (*Zea mays* L.) caryopses. *International Journal of Plant Sciences* **163**:107–112.

Pernollet JC. 1978. Protein bodies of seeds: ultrastructure, biochemistry, biosynthesis and degradation. *Phytochemistry* **17**:1473–1480.

Pinzón-Torres JA, dos Santos VR, Schiavinato MA, Maldonado S. 2009. Biochemical, histochemical and ultrastructural characterization of *Centrolobium robustum* (Fabaceae) seeds. *Hoehnea* **36**:149–160.

Pukacka S, Ratajczak E. 2007. Age-related biochemical changes during storage of beech (*Fagus sylvatica* L.) seeds. *Seed Science Research* **17**:45–53.

Radovich T. 2012. Farm and Forestry production and marketing profile for Moringa (*Moringa oleifera*). Specialty crops for pacific Island agroforestry. http://agroforestry.net/scps.

Rao RGS, Singh PM, Rai M. 2006. Storability of onion seeds and effects of packaging and storage conditions on viability and vigour. *Scientia Horticulturae* **110**:1–6.

Reynolds ES. 1963. The use of lead citrate at high pH as an electron-opaque stain in electron microscopy. *Journal of Cell Biology* **17**:208–212.

Roberts EH, Ellis RH. 1982. Physiological, ultrastructural and metabolic aspects of seed viability. In: Khan AA, ed. *The physiology and biochemistry of seed dormancy and germination*. Amsterdam: Elsevier Biomedical Press, 465–485.

Simola LK. 1974. Ultrastructural changes in the seeds of *Pinus sylvestris* L. during senescence. *Studia Forestalia Suecica*, nr 119.

Smith MT. 1978. Cytological changes in artificially aged seeds during imbibition.

Teixeira SP, Machado SR. 2008. Storage sites in seeds of *Caesalpinia echinata* and *C. ferrea* (Leguminosae) with considerations on nutrients flow. *Brazilian Archives of Biology and Technology* **51**: 127–136.

Walters C, Landré P, Hill L, Corbineau F, Bailly C. 2005. Organization of lipid reserves in cotyledons of primed and aged sunflower seeds. *Planta* **222**:397–407.

Wang BSP, Berjak P. 2000. Beneficial effects of moist chilling on the seeds of black spruce (*Picea mariana* [Mill.] B.S.P.). *Annals of Botany* **86**:29–36.

Willan RL. 1985. A guide to forest seed handling. FAO Forestry Paper 20/2. Rome, Italy: FAO. Chapter 7. http://www.fao.org/docrep/006/ad232e/ad232e00.htm.

Wiśniewska M, Lotocka B, Suchorska-Tropiło K, Dąbrowska B. 2006. Embryo ultrastructure in *Origanum majorana* L. (Lamiaceae) after seed conditioning. *Acta Biologica Cracoviensia Botanica* **48**:105–116.

Young CT, Schadel WE, Pattee HE, Sanders TH. 2004. The microstructure of almond (*Prunus dulcis* (Mill.) D.A.Webb cv. 'Nonpareil') cotyledon. *LWT—Food Science and Technology* **37**: 317–322.

9

Effects of a native parasitic plant on an exotic invader decrease with increasing host age

Junmin Li[1,2]*, Beifen Yang[1,2], Qiaodi Yan[1,2], Jing Zhang[2,3], Min Yan[3] and Maihe Li[4]

[1] Zhejiang Provincial Key Laboratory of Plant Evolutionary Ecology and Conservation, Taizhou 318000, China
[2] Institute of Ecology, Taizhou University, Taizhou 318000, China
[3] School of Life Science, Shanxi Normal University, Linfen 041004, China
[4] Ecophysiology Group, Forest Dynamics, Swiss Federal Research Institute WSL, 8903 Birmensdorf, Switzerland

Associate Editor: James F. Cahill

Abstract. Understanding changes in the interactions between parasitic plants and their hosts in relation to onto-genetic changes in the hosts is crucial for successful use of parasitic plants as biological controls. We investigated growth, photosynthesis and chemical defences in different-aged *Bidens pilosa* plants in response to infection by *Cuscuta australis*. We were particularly interested in whether plant responses to parasite infection change with changes in the host plant age. Compared with the non-infected *B. pilosa*, parasite infection reduced total host biomass and net photosynthetic rates, but these deleterious effects decreased with increasing host age. Parasite infection reduced the concentrations of total phenolics, total flavonoids and saponins in the younger *B. pilosa* but not in the older *B. pilosa*. Compared with the relatively older and larger plants, younger and smaller plants suffered from more severe damage and are likely less to recover from the infection, suggesting that *C. australis* is only a viable biocontrol agent for younger *B. pilosa* plants.

Keywords: Defence; deleterious effect; growth; invasive plant; parasitic plant.

Introduction

A parasitic plant is a type of angiosperm (flowering plant) that directly attaches to another plant via a haustorium (Press 1998). Over 4500 known plant species are parasitic to some extent and acquire some or all of their water, carbon and nutrients from a host (Press 1998; Li *et al.* 2014). Parasitic plants are classified as stem or root parasites including facultative, hemiparasitic and holoparasitic forms (Yoder and Scholes 2010).

Infection by parasitic plants has been considered as an effective method for controlling invasive plants because the parasites partially (hemiparasites) or completely (holoparasites) absorb water, nutrients and carbohydrates from their host plants, suppressing the vitality of the host (Parker *et al.* 2006; Yu *et al.* 2008, 2009; Li *et al.* 2012). For example, the holoparasite *Cuscuta australis*, native to China, can inhibit the growth of *Bidens pilosa*, an invasive plant in China, and thus serve as an effective biological control agent for controlling the invasive *B. pilosa* (Zhang *et al.* 2012, 2013). Compared with the effects of feeding by herbivores, the defence responses of plants infected by parasitic plants have rarely been studied (Runyon *et al.* 2006; Ranjan *et al.* 2014), even though such knowledge is important for the successful use of parasitic plants as enemies against invasive plants.

* Corresponding author's e-mail address: lijmtzc@126.com

It has been documented that plant defences to herbivore or pathogen damage vary with a plant's ontogenetic stages (Boege and Marquis 2006; Barton and Koricheva 2010; Tucker and Avila-Sakar 2010; Barton 2013). The ontogenetic patterns of plant defences were found to differ with plant life form (woody, herbaceous and grass, Barton and Koricheva 2010; Massad 2013), growth stage (seedlings, juveniles, mature plants, Barton and Koricheva 2010; Houter and Pons 2012; Barton 2013), development stage (flowering stage, fruiting stage, Tucker and Avila-Sakar 2010) and growth rate (slow-growing plant and fast-growing plant, Massad 2013). For herbaceous plants, for example, young plants are normally more heavily chemically defended than older ones (Cipollini and Redman 1999; Barton and Koricheva 2010; Massad 2013). However, as expected from the resource limitation hypothesis, the smaller reserves of resources stored in younger plants may negatively influence secondary metabolites in comparison with the larger resource reserves stored in mature plants, as expected from the resource limitation hypothesis (Bryant et al. 1991). Thus, younger plants may be less defended and less able to recover after herbivory or parasitic infestation (Hódar et al. 2008).

Similar to those defence reaction induced by herbivores and pathogens infection, plants may increase their chemical complexes to defend against parasitic infection through plant hormones, salicylic acid and jasmonic acid pathway (Runyon et al. 2010). Little attention has been paid to the ontogenetic changes of invasive plants inresponse to holoparasites. Wu et al. (2013) found that Cuscuta campestris seedlings cannot parasitize the invasive Mikania micrantha if the stem diameter of the host is ≥0.3 cm. Our field investigation found that the infestation rate of plants, such as B. pilosa, Solidago canadensis and Phytolacca americana, by C. australis decreased with increasing host age (data not shown). Accordingly, we conducted an experiment to understand the host defences in relation to host age in a holoparasite–host system. The growth characteristics and the concentrations of the main chemical defences were determined in different-aged invasive B. pilosa plants infected by C. australis to test the hypothesis that younger hosts are more easily damaged and less able to recover than the older ones because the younger plants with limited resource reserves have less capacity to produce chemical defences and to promote compensatory growth. We aimed to answer the following questions: (i) Do younger and older B. pilosa plants differ in their responses to infection by C. australis? (ii) Are these differences in responses are correlated with the growth of different-aged invasive host plants? The answers to these questions could provide basic scientific knowledge for using C. australis to manage the invasive plant B. pilosa.

Methods

Plant species

Bidens pilosa is native to the tropical America and has widely spread throughout China. It is an annual forb and can grow up to 1 m in height and produces numerous seeds every year, and it grows both in nutrient-rich and -poor soils. In November, 2009, seeds of B. pilosa were collected near Sanfeng temple (121°16′E, 28°88′N) in Linhai City, Zhejiang Province, China, and stored in a low-humidity storage cabinet (HZM-600, Beijing Biofuture Institute of Bioscience and Biotechnology Development) until use.

Cuscuta australis, a native annual holoparasitic plant species to South China, and is considered a noxious weed of agriculture (Yu et al. 2011). It can infect a wide range of herbs and shrubs (e.g. plants in the families of Fabaceae and Asteraceae), including the invasive plants M. micrantha, Ipomoea cairica, Wedelia trilobata, Alternanthera philoxeroides and Bidens (Yu et al. 2011; Wang et al. 2012; Zhang et al. 2012).

Experimental design

We conducted a greenhouse experiment at Taizhou University (121°17′E, 28°87′N) in Linhai City, Zhejiang Province, China. We sowed B. pilosa seeds in trays with sand to germinate in a greenhouse on 13 March, 22 March and 6 April 2011, to create three different-aged B. pilosa seedlings of three different ages. Approximately 20 days after sowing, B. pilosa seedlings (~10 cm in height) were transplanted into pots (28 cm in inner diameter and 38 cm deep; 1 seedling per pot) filled with 2.5 kg yellow clay soil mixed with sand in a 2 : 1 ratio (v : v). Plant materials and stones were removed from the yellow clay soil collected from fields in Linhai. The soil mixture had a pH of 6.64 ± 0.01, with an organic matter content of 15.74 ± 2.65 g kg^{-1}, available nitrogen of 0.27 ± 0.10 g kg^{-1}, available phosphorus of 0.026 ± 0.004 g kg^{-1} and available potassium of 0.049 ± 0.003 g kg^{-1}.

The pots were randomly placed in a greenhouse and irrigated with tap water twice daily. One week after transplantation, 2 g slow release fertilizer (Scotts Osmocote, N : P : K = 20 : 20 : 20, The Scotts Miracle-Gro Company, Marysville, OH, USA) was added to each pot.

On 5 June, when B. pilosa plants were of ages 59 days (mean height 32.0 cm and mean diameter 2.7 mm), 74 days (mean height 62.6 cm and mean diameter 5.1 mm) and 83 days old (mean height 93.3 cm and mean diameter 5.7 mm), plants were infected by C. australis manually. Three 15-cm long segments of parasitic C. australis stems collected from fields in Linhai were twined onto the stems of a B. pilosa plant to induce infection. After 24 h, most of C. australis successfully parasitized the host and died segments were substituted by new ones. For each age class, six individuals were infected and six plants were left intact as controls ($n = 6$). Six individuals were harvested, separated into shoots and roots, and then dried at 70 °C for 72 h, to determine the initial plant biomass (W_1) at the beginning of infection (t_1, i.e. 5 June).

Measurements

On 30 June 2011, i.e. after 26 days of infection, the net photosynthetic rate (P_n) of B. pilosa plants was determined on fully expanded, mature sun leaves in the upper canopy between 10:00 and 11:30 am, using a portable photosynthesis system (LI-6400/XT, LI-COR Biosciences, Lincoln, NE, USA). For each measurement, three leaves per plant were chosen, and six consecutive measurements were performed.

On 9 July 2011 (t_2), i.e. 35 days after infection, when C. australis was flowering and the host plants were 94, 109 and 118 days old, respectively, all plants were harvested. Cuscuta australis plants were separated from their hosts and dried at 70 °C for 72 h to determine the C. australis' biomass (B_c). The host plants were separated into leaves, stems and roots. Leaves, stems and roots of the host plants were dried at 70 °C for 72 h to determine their biomass (W_2). The relative growth rate (RGR) of biomass was calculated with the equation RGR = (ln W_2 − ln W_1)/(t_2 − t_1) (González-Santana et al. 2012; Li et al. 2012).

The dried stems of the host plants were ground using a universal high-speed grinder (F80, Xinkang Medical Instrument Co. Ltd, Jiangyan, Jiangsu). The powder was filtered through a 20-mesh sieve and stored in a drier until chemical analysis.

Approximately 0.1 g of powder was extracted three times with 70 % ethanol (v/v) under reflux at 90 °C and the aqueous extract was used to measure the concentration of total phenolics and total flavonoids. The concentration of total phenolics and total flavonoids was determined using the Folin–Denis method and AlCl$_3$ reaction method according to Cortés-Rojas et al. (2013) and Jin et al. (2007). Absorbance at 750 nm for total phenolics and 420 nm for total flavonoids was determined with a T6 UV–VIS spectrophotometer (Beijing Purkinje General Instrument Co. Ltd, Beijing, China). Gallic acid and rutin (purchased from National Institutes for Food and Drug Control, Beijing, China) were used as the standard for total phenolics and total flavonoids, respectively.

Approximately 0.1 g of powder was extracted three times with 70 % methanol under reflux at 70 °C, and the aqueous extract was used to measure the concentration of saponins and alkaloids. The concentration of total saponins and alkaloids was determined by a colourimetric method and bromocresol green reaction method, respectively, according to Li et al. (2006) and Jin et al. (2006). Absorbance at 560 nm (saponins) and 470 nm (alkaloids) was determined with a T6 UV–VIS spectrophotometer (Beijing Purkinje General Instrument Co. Ltd, Beijing, China). Ginsenosides-Re and berberin HCl (purchased from National Institutes for Food and Drug Control, Beijing, China) were used as the standard for saponins and alkaloids, respectively.

Approximately 0.1 g of powder was extracted three times with boiled water and the aqueous extract was used to measure the concentration of soluble tannin via the potassium permanganate redox titrations method according to Li et al. (2007). Gallic acid (purchased from National Institutes for Food and Drug Control, Beijing, China) was used as the standard.

Data analysis

The plastic responses (PR) of the host plants to parasite infection were calculated for all plant traits studied, using the equation PR = (VPP − MVC)/MVC (Barton 2008), where VPP is the value of a trait in a parasite-infected plant and MVC is the mean value of that trait in the same-aged controls. For example, if the mean biomass of the non-parasitized B. pilosa is A and the biomass of the parasitized B. pilosa is B, then PR = (B − A)/A. Such PR values reflect the relative changes in the host traits caused by parasites. A value of PR = 0 indicates no response. A value of PR < 0 indicates a negative response, whereas a value of PR > 0 indicates a positive response of the host to parasite infection.

The normality of the distribution and the homogeneity of the data were checked (Kolmogorov–Smirnov test) before any further statistical analysis. A two-way ANOVA was used to analyse the effects of parasitism and host age on the host traits studied, followed by a one-way ANOVA and Tukey's HSD analysis to test the difference in means within a parasite treatment and between the infected and non-infected plants. All tests were conducted at a significance level of $P < 0.05$ using SPSS (version 16.0).

Results

Effects of C. australis infection on host growth

Cuscuta australis infection decreased the root, stem, leaf and total biomass of the hosts compared with the controls within each age class (Fig. 1, upper panel), and this negative effect on host growth significantly decreased with increasing host age (Fig. 1, lower panel). Compared with the non-infected control hosts, the total plant biomass of the infected hosts decreased 84 % for the 59-day-old plants, 48 % for the 74-day-old plants and 21 % for the 83-day-old plants.

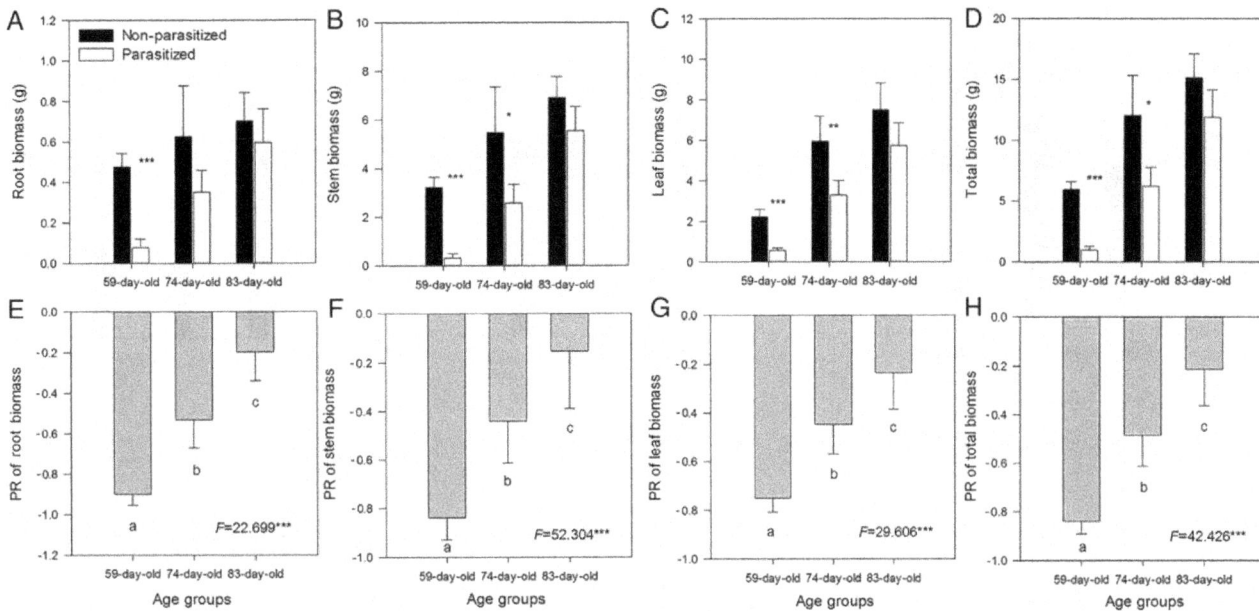

Figure 1. The root (A), stem (B), leaf (C) and total plant biomass (D) of different-aged invasive *B. pilosa* plants infected and not infected by *C. australis*, and the PR of the stem (E), root (F), leaf (G) and total plant biomass (H) of the infected *B. pilosa* plants. Values are given as means +1 SD (*n* = 6). Asterisks in the upper panel indicate significant difference in means between non-infected and infected plants within the same age class at *P < 0.05, **P < 0.01 and ***P < 0.001, respectively. Different letters in the lower panel indicate significant difference between PRs (*P* < 0.05).

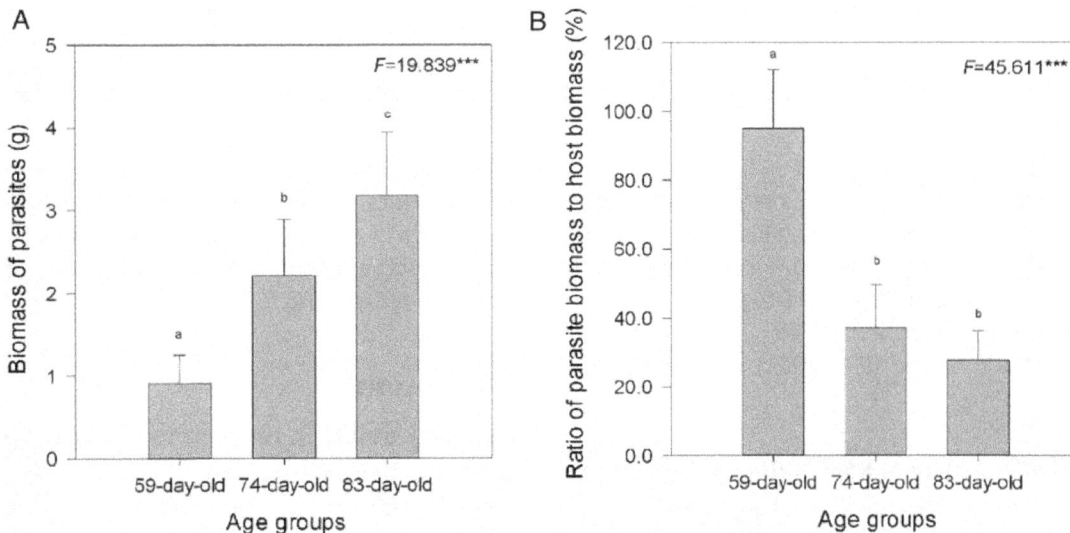

Figure 2. The biomass of parasites (A) of different-aged invasive *B. pilosa* plants, and the ratio of parasite biomass to host biomass (B). Values are given as means +1 SD (*n* = 6). Different letters indicate significant difference between host plants of different ages at *P* < 0.05. *F*-value and significance levels are given. ***Significant difference in means between plants within the same age class at *P* < 0.001.

The growth of the parasites significantly increased with host age (Fig. 2A). However, the biomass ratio of parasite to host significantly decreased with host age (Fig. 2B).

The infection of *C. australis* significantly decreased the growth rates in the younger (i.e. the 59- and 74-day-old hosts; both *P* < 0.001) but not in the 83-day-old hosts compared with those in the corresponding controls

(Fig. 3A and D). The infection significantly suppressed the net photosynthetic rates only in the younger (59- and 74-day-old) hosts but not in the older hosts (Fig. 3B and E). *Cuscuta australis* infection had no effects on the root/shoot ratio in different-aged hosts (Fig. 3C), and the root/shoot ratio tended to decrease with increasing host age (Fig. 3C). The negative effect of *C. australis*

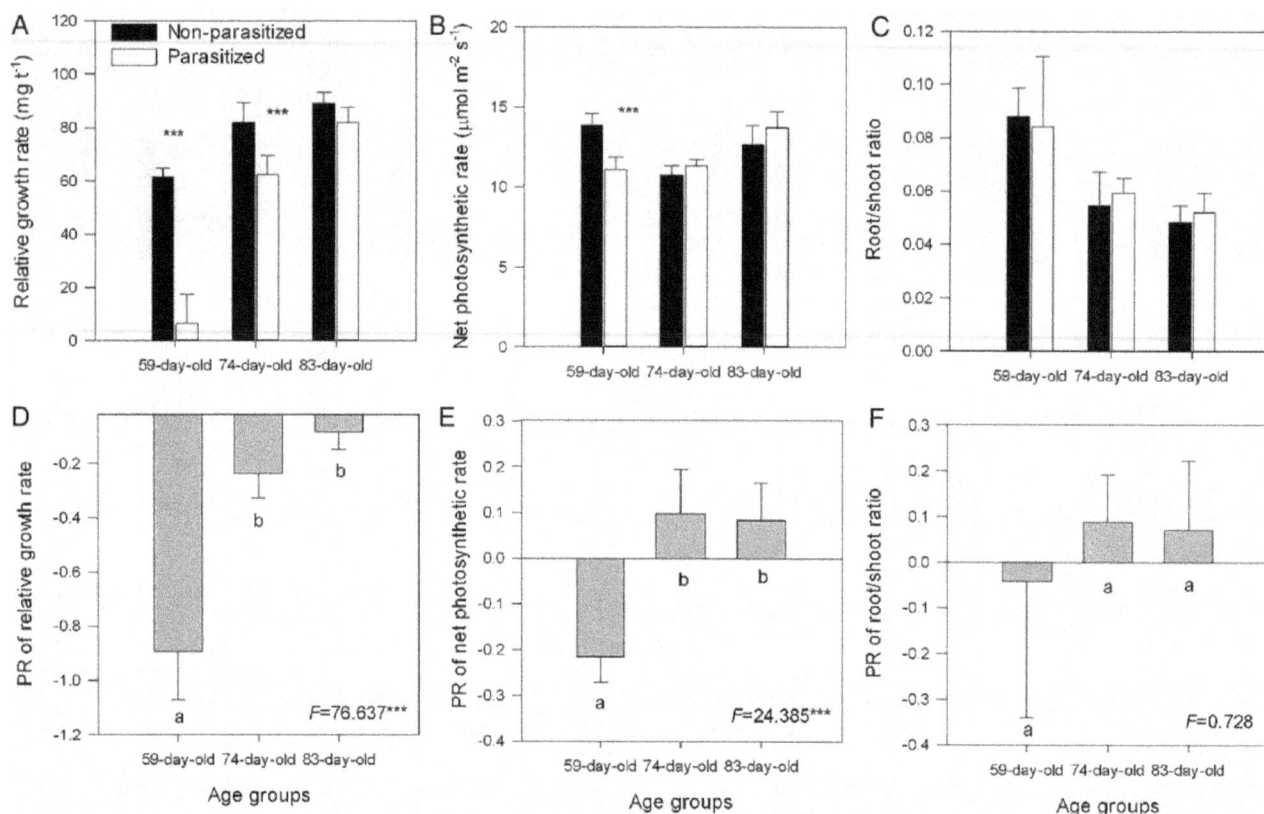

Figure 3. The RGR (A), net photosynthetic rate of leaves (B) and root/shoot ratio (C) of different-aged invasive *B. pilosa* plants infected and not infected by *C. australis*, and the PR of RGR of plant biomass (D), net photosynthetic rate of leaves (E) and ratio of root biomass to shoot biomass (F) of different-aged invasive *B. pilosa* to the parasitic *C. australis*. Values are given as means +1 SD ($n = 6$). Different letters indicate significant difference between host plants of different ages at $P < 0.05$. F-value and significance levels are given. Asterisks indicate significant difference in means between plants within the same age class at ***$P < 0.001$.

infection on *B. pilosa*'s RGRs (Fig. 3D), net photosynthetic rates (Fig. 3E) and root/shoot ratios (Fig. 3F) decreased with increasing host age (Fig. 3D–F). Host age (A) interacted with parasites (P) to affect the total biomass ($P < 0.01$ for $A \times P$ interaction), RGRs ($P < 0.001$) and net photosynthetic rates ($P < 0.001$) of the hosts (Table 1).

Effects of infection on host's secondary metabolites

The infection significantly decreased the concentrations of the phenolics in the younger plants (59 and 74 days old) but not in the older hosts (Fig. 4C), and the negative effect of *C. australis* decreased with increasing host age (Fig. 4D). The infection significantly decreased the concentrations of the terpenoids in the younger plants (59 days old) but significantly increased those concentrations in the older plants (74 and 83 days old) (Fig. 4G). Host age interacted with parasite infection to influence the levels of total phenols and saponins (Fig. 4C and G). No effects of host age, parasite infection and their interaction on the concentrations of tannin, total flavonoids and alkaloids in the hosts were observed (Fig. 4A, E and

I). The effect of infection on the concentration of total flavonoids in the younger plants was negative (59 days old), whereas that in the older plants was positive (74 and 83 days old) (Fig. 4F).

Discussion

Recovery ability in relation to the host age

The present study found that the deleterious effects of parasite infection on the older *B. pilosa* were significantly less severe than on the younger plants, indicating that the younger plants were more sensitive to parasite infection than the older ones. These results supported our hypothesis that the damage to younger *B. pilosa* caused by *C. australis* infection is greater than the damage to older *B. pilosa*.

Previous studies have shown that older hosts exhibit a defence mechanism that hampers the development of haustoria and thus mitigates parasite infection (Runyon et al. 2006; Meulebrouck et al. 2009; Lee and Jernstedt 2013). However, this conclusion was not supported by the present study because the parasite biomass did not

Table 1. Two-way ANOVA analysis of age and parasitism on the growth of invasive B. pilosa. Bold values indicate significant effects on the growth of B. pilosa.

	df	Leaf biomass		Stem biomass		Root biomass		Total biomass		Root/shoot ratio		Relative growth rate		Net photosynthetic rate	
		F	P	F	P	F	P	F	P	F	P	F	P	F	P
Age (A)	2	97.135	**<0.001**	59.992	**<0.001**	19.819	**<0.001**	121.986	**<0.001**	23.926	**<0.001**	175.610	**<0.001**	29.308	**<0.001**
Parasitism (P)	1	43.267	**<0.001**	51.293	**<0.001**	29.009	**<0.001**	116.349	**<0.001**	0.112	0.740	137.485	**<0.001**	131.144	**<0.001**
A × P	2	1.036	0.367	2.368	0.111	3.065	0.061	9.763	**<0.01**	0.342	0.713	38.004	**<0.001**	50.772	**<0.001**

decrease but increased with increasing host age in association with host size. Older hosts, with larger size and greater resource storage, could support greater growth of the attached parasites, leading to a mean increase in parasite biomass by 142 % for the 74-day-old hosts and 248 % for the 83-day-old hosts compared with that for the 59-day-old hosts.

Our study found that root, stem, leaf and total biomass, RGR and photosynthetic ability were significantly negatively affected by parasite infection in the younger plants (59 and 74 days old) but not in the older plants (83 days old), indicating that the young plants fail to compensate whereas the older plants do (Tan et al. 2004; Zhang et al. 2012). Herbivory can induce compensatory growth by stimulating photosynthesis, altering mass allocation and increasing growth rates (Markkola et al. 2004; Hódar et al. 2008). Therefore, the responses of younger B. pilosa to parasite infection is not similar to the responses of plants to damage by herbivores (Barton and Koricheva 2010; Barton 2013). Stout et al. (2002) have found that younger rice plants appeared to be less tolerant to herbivory than older rice plants, though Elger et al. (2009) have reported a higher sensitivity to herbivore attacks in young seedlings of British grassland species than in older conspecifics.

Parasite infection had no effects on the root/shoot ratio in different-aged hosts, i.e. parasite infection did not alter the mass allocation to roots and shoots. This result may be due primarily to changes in the light conditions caused by the attack behaviour of the parasite. Herbivores destroy parts of plants, resulting in increases in light quality and quantity within a plant and thus leading to increases in photosynthesis and growth rates. In contrast, parasite infection may shade a host and decrease the light intensity within a host plant, especially if the hosts plants are small or young, thus leading to decreases in photosynthesis and growth rates (Rijkers et al. 2000).

The decreases in photosynthesis induced by parasitic infection led to a limited availability of resources, which further resulted in lower growth rates in smaller and younger hosts. These results indicated that the resistance of hosts to parasitic infection is negatively correlated with the availability of the resources stored in a host plant (Shen et al. 2013). For woody forest plants, negative effects of mistletoe infection on host tree growth and mortality have been consistently and extensively reported (Shaw et al. 2008; Logan et al. 2013), and those negative growth effects have widely been considered to result from decreased photosynthetic production (Meinzer et al. 2004) caused by decreased leaf size and leaf N content (Ehleringer et al. 1986; Cechin and Press 1993; Logan et al. 1999; Mishra et al. 2007).

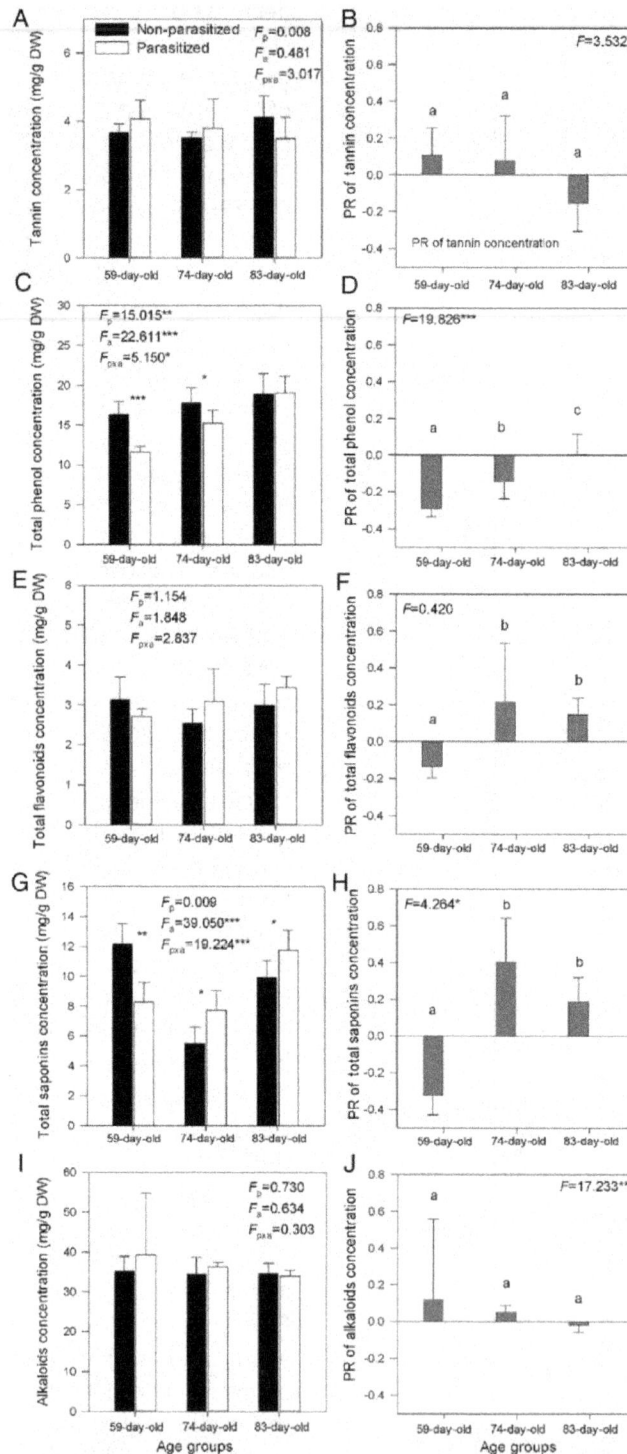

Figure 4. The concentrations (mean values + 1 SD, $n = 6$) of tannin (A), total phenolics (C), total flavonoids (E), saponins (G) and alkaloids (I) in different-aged invasive *B. pilosa* plants infected and not infected by *C. australis*, and the PR of tannin (B), total phenolics (D), total flavonoids (F), saponins (H) and alkaloids (J) in stems of the different-aged *B. pilosa* plants to the parasitic *C. australis*. Different letters indicate significant difference between host plants of different ages at $P < 0.05$. F-value and significance levels are given. Asterisks indicate significant difference in means between plants within the same age class at $*P < 0.05$, $**P < 0.01$ and $***P < 0.001$, respectively.

Chemical defence relative to host age

Little attention has been paid to changes in chemical defences, such as alkaloids, phenolics, flavonoids, cyanogenic glycosides (Elger *et al.* 2009; Quintero and Bowers 2013), induced by parasitic infection, whereas ontogenetic changes in chemical defences against herbivory are well documented (Van Zandt and Agrawal 2004; Barton 2008). The concentrations of cyanogenic glycoside (Schappert and Shore 2000), nicotine (Ohnmeiss and Baldwin 2000), alkaloids (Ohnmeiss and Baldwin 2000; Elger et al. 2009) and phenolics (Donaldson et al. 2006; Elger et al. 2009) have been found to be higher in older tissues/plants than in younger tissues/plants exposed to herbivory. The present study found that the concentrations of total phenolics, total flavonoids and saponins were significantly negatively affected by C. australis infection in younger B. pilosa but not in older plants. A positive response of total phenolics was found only in the 83-day-old plants; a positive response of total flavonoids and saponins occurred in both 74- and 83-day-old plants. These results supported our initial hypothesis that the younger plants with limited resource reserves have less leeway to produce chemical defences. Similarly, it has been reported that older plants that had accumulated resources over a long period were better able to maintain anti-herbivore defences than younger plants with limited resources (Boege 2005; Elger et al. 2009).

However, other studies have reported opposite or neutral responses of chemical defences to herbivory (Thomson et al. 2003; Barton and Koricheva 2010). Goodger et al. (2013) demonstrated that the levels of phenolics produced in response to herbivory were highest in seedlings compared with those in juveniles and mature trees. A meta-analysis based on data from 36 published studies also did not find any clear relationships between ontogenetic stage and chemical defences induced by herbivory (Barton and Koricheva 2010). Ontogenetic patterns of plant chemical responses to herbivory or parasite infection might vary with plant species, chemical compounds and disturbance (e.g. parasite or insect infection) (Webber and Woodrow 2009; Rohr et al. 2010). Further studies are needed to identify, clarify and quantify the chemical defences in different hosts and different-aged hosts infected by different parasites.

Plant resistance traits, including physical defences traits and chemical secondary compounds, and tolerance traits such as compensatory re-growth are detectable, when plants are damaged by insect, pathogen or parasite infection (Stowe et al. 2000; Quintero and Bowers 2013). Previous studies prove that trade-offs between resistance and tolerance traits of plants do occur following damage (Orians et al. 2010; Quintero and Bowers 2013). However, several studies reported that the trade-offs between resistance and tolerance traits are not detectable for all development stages (e.g. Quintero and Bowers 2013), for example, during the seedling development stage (Barton 2008). Our results showed that younger hosts had higher concentrations of tannin and alkaloids but lower concentrations of flavonoids, phenolics and terpenoids, as well as a lower re-growth ability; conversely, older hosts had lower concentration of tannin and alkaloids, but higher concentrations of flavonoids, phenolics and terpenoids, as well as a higher capability of compensatory growth. These results provided an indirect evidence for supporting the above-mentioned viewpoints that younger hosts with lower biomass invest fewer resources in growth but more resources in higher concentrations of tannin and alkaloids, whereas the older hosts have a higher biomass and higher concentrations of flavonoids, phenolics and terpenoids, but lower concentrations of tannin and alkaloids. Accordingly, the overall pattern of response might depend on the age or size of the host and even on the diversity of damage.

In conclusion, the intensity of damage caused by the parasite C. australis was dependent on the host age or size associated with the host biomass, showing that younger and smaller hosts were more easily damaged and less able to recover. This result supported the resource limitation hypothesis (Bryant et al. 1991) because hosts with increasing host age (or size) have more resources available to defend against or limit the establishment of haustoria (Runyon et al. 2010). Our field investigation found that B. pilosa plants with a height >90 cm were resistant because they were less infected by C. australis, whereas B. pilosa plants with a height <30 cm were effectively and successfully controlled by C. australis. Our results, therefore, indicate that the parasite C. australis could be successfully used as a potential biological agent in the control of invasive B. pilosa plants only in the early stages of development. However, as Boege (2005) indicated, further studies are needed to better and fully understand the mechanisms underlying the effects of host ontogeny on the responses of the host plants to herbivores or (holo) parasites.

Sources of Funding

This work was financially supported by the National Natural Science Foundation of China (No. 31270461; No. 30800133) and National Natural Science Foundation of Zhejiang Province (No. Y5110227).

Contributions by the Authors

J.L. and M.Y. conceived and designed research. B.Y., Q.Y. and J.Z. conducted experiments. J.L. analysed data. J.L. and M.L. wrote the manuscript. All authors read and approved the manuscript.

Literature Cited

Barton KE. 2008. Phenotypic plasticity in seedling defense strategies: compensatory growth and chemical induction. Oikos 117:917–925.

Barton KE. 2013. Ontogenetic patterns in the mechanisms of toler-
    ance to herbivory in *Plantago*. *Annals of Botany* **112**:711–720.

Barton KE, Koricheva J. 2010. The ontogeny of plant defense and her-
    bivory: characterizing general patterns using meta-analysis. *The
    American Naturalist* **175**:481–493.

Boege K. 2005. Influence of plant ontogeny on compensation to leaf
    damage. *American Journal of Botany* **92**:1632–1640.

Boege K, Marquis RJ. 2006. Plant quality and predation risk mediated
    by plant ontogeny: consequences for herbivores and plants.
    Oikos **115**:559–572.

Bryant JP, Provenza FD, Pastor J, Reichardt PB, Clausen TP, du Toit JT.
    1991. Interactions between woody plants and browsing
    mammals mediated by secondary metabolites. *Annual Review
    of Ecology, Evolution and Systematics* **22**:431–446.

Cechin I, Press MC. 1993. Nitrogen relations of the sorghum-Sfriga
    hermonthica hostparasite association: growth and photosyn-
    thesis. *Plant, Cell and Environment* **16**:237–247.

Cipollini DF Jr, Redman AM. 1999. Age-dependent effects of jasmonic
    acid treatment and wind exposure on foliar oxidase activity and
    insect resistance in tomato. *Journal of Chemical Ecology* **25**:
    271–281.

Cortés-Rojas DF, Chagas-Paula DA, da Costa FB, Souza CRF,
    Oliveira WP. 2013. Bioactive compounds in *Bidens pilosa*
    L. populations: a key step in the standardization of phytopharma-
    ceutical preparations. *Brazilian Journal of Pharmacognosy* **23**:
    28–35.

Donaldson JR, Kruger EL, Lindroth RL. 2006. Competition- and
    resource-mediated tradeoffs between growth and defensive
    chemistry in trembling aspen (*Populus tremuloides*). *New Phytol-
    ogist* **169**:561–570.

Ehleringer JR, Cook CS, Tieszen LL. 1986. Comparative water use and
    nitrogen relationships in a mistletoe and its host. *Oecologia* **68**:
    279–284.

Elger A, Lemoine DG, Fenner M, Hanley ME. 2009. Plant ontogeny and
    chemical defence: older seedlings are better defended. *Oikos*
    118:767–773.

González-Santana IH, Márquez-Guzmán J, Cram-Heydrich S,
    Cruz-Ortega R. 2012. *Conostegia xalapensis* (Melastomataceae):
    an aluminum accumulator plant. *Physiologia Plantarum* **144**:
    134–145.

Goodger JQD, Heskes AM, Woodrow IE. 2013. Contrasting ontogenet-
    ic trajectories for phenolic and terpenoid defences in *Eucalyptus
    froggattii*. *Annals of Botany* **112**:651–659.

Hódar JA, Zamora R, Castro J, Gómez JM, García D. 2008. Biomass
    allocation and growth responses of Scots pine saplings to simu-
    lated herbivory depend on plant age and light availability. *Plant
    Ecology* **197**:229–238.

Houter NC, Pons TL. 2012. Ontogenetic changes in leaf traits of
    tropical rainforest trees differing in juvenile light requirement.
    Oecologia **169**:33–45.

Jin ZX, Li JM, Zhu XY. 2006. The content analysis of total alkaloids in a
    rare and extincted plant *Sinocalycanthus chinensis*. *Journal of
    Fujian Forestry Science and Technology* **33**:7–10.

Jin ZX, Li JM, Zhu XY. 2007. Content of total flavonoids and total
    chlorogenic acid in the endangered plant *Sinocalycanthus
    chinensis* and their correlations with the environmental
    factors. *Journal of Zhejiang University (Science Edition)* **33**:
    454–457.

Lee KB, Jernstedt JA. 2013. Defense response of resistant host *Impa-
    tiens balsamina* to the parasitic angiosperm *Cuscuta japonica*.
    Journal of Plant Biology **56**:138–144.

Li JM, Jin ZX, Zhu XY. 2006. Analysis and determination of total sap-
    onin content in an endangered plant *Calycanthus chinensis*. *Jour-
    nal of Northwest Forestry University* **21**:147–150.

Li JM, Jin ZX, Zhu XY. 2007. Comparison of the total tannin in different
    organs of *Calycanthus chinensis*. *Guihaia* **27**:944–947.

Li J, Jin Z, Song W. 2012. Do native parasitic plants cause more dam-
    age to exotic invasive hosts than native non-invasive hosts? An
    implication for biocontrol. *PLoS ONE* **7**:e34577.

Li JM, Jin ZX, Hagedorn F, Li MH. 2014. Short-term parasite-infection
    alters already the biomass, activity and functional diversity of soil
    microbial communities. *Scientific Reports* **4**:6895–6902.

Logan BA, Demmig-Adams B, Rosenstiel TN, Adams WW III. 1999.
    Effect of nitrogen limitation on foliar antioxidants in relationship
    to other metabolic characteristics. *Planta* **209**:213–220.

Logan BA, Reblin JS, Zonana DM, Dunlavey RF, Hricko CR, Hall AW,
    Schmiege SC, Butschek RA, Duran KL, Emery RJN, Kurepin LV,
    Lewis JD, Pharis RP, Phillips NG, Tissue DT. 2013. Impact of east-
    ern dwarf mistletoe (*Arceuthobium pusillum*) on host white
    spruce (*Picea glauca*) development, growth and performance
    across multiple scales. *Physiologia Plantarum* **147**:502–513.

Markkola A, Kuikka K, Rautio P, Härmä E, Roitto M, Tuomi J. 2004.
    Defoliation increases carbon limitation in ectomycorrhizal sym-
    biosis of *Betula pubescens*. *Oecologia* **140**:234–240.

Massad TJ. 2013. Ontogenetic differences of herbivory on woody and
    herbaceous plants: a meta-analysis demonstrating unique
    effects of herbivory on the young and the old, the slow and the
    fast. *Oecologia* **172**:1–10.

Meinzer FC, Woodruff DR, Shaw DC. 2004. Integrated responses of
    hydraulic architecture, water and carbon relations of western
    hemlock to dwarf mistletoe infection. *Plant, Cell and Environment*
    27:937–946.

Meulebrouck K, Verheyen K, Brys R, Hermy M. 2009. Limited by the
    host: host age hampers establishment of holoparasite *Cuscuta
    epithymum*. *Acta Oecologica* **35**:533–540.

Mishra JS, Moorthy BTS, Bhan M, Yaduraju NT. 2007. Relative toler-
    ance of rainy season crops to field dodder (*Cuscuta campestris*)
    and its management in niger (*Guizotia abyssinica*). *Crop Protec-
    tion* **26**:625–629.

Ohnmeiss TE, Baldwin IT. 2000. Optimal defense theory predicts the
    ontogeny of an induced nicotine defense. *Ecology* **81**:1765–1783.

Orians CM, Hochwender CG, Fritz RS, Snäll T. 2010. Growth and
    chemical defense in willow seedlings: trade-offs are transient.
    Oecologia **163**:283–290.

Parker JD, Burkepile DE, Hay ME. 2006. Opposing effects of native and
    exotic herbivores on plant invasions. *Science* **311**:1459–1461.

Press MC. 1998. Dracula or Robin Hood? A functional role for root
    hemiparasites in nutrient poor ecosystems. *Oikos* **82**:609–611.

Quintero C, Bowers MD. 2013. Effects of insect herbivory on induced
    chemical defences and compensation during early plant
    development in *Penstemon virgatus*. *Annals of Botany* **112**:661–669.

Ranjan A, Ichihashi Y, Farhi M, Zumstein K, Townsley B,
    David-Schwartz R, Sinha NR. 2014. De novo assembly and charac-
    terization of the transcriptome of the parasitic weed dodder

identifies genes associated with plant parasitism. *Plant Physiology* **166**:1186–1199.

Rijkers T, Pons TL, Bongers F. 2000. The effect of tree height and light availability on photosynthetic leaf traits of four neotropical species differing in shade tolerance. *Functional Ecology* **14**:77–86.

Rohr JR, Raffel TR, Hall CA. 2010. Developmental variation in resistance and tolerance in a multi-host-parasite system. *Functional Ecology* **24**:1110–1121.

Runyon JB, Mescher MC, De Moraes CM. 2006. Volatile chemical cues guide host location and host selection by parasitic plants. *Science* **313**:1964–1967.

Runyon JB, Mescher MC, Felton GW, de Moraes CM. 2010. Parasitism by *Cuscuta pentagona* sequentially induces JA and SA defence pathways in tomato. *Plant, Cell and Environment* **33**:290–303.

Schappert PJ, Shore JS. 2000. Cyanogenesis in *Turnera ulmifolia* L. (Turneraceae). II. Developmental expression, heritability and cost of cyanogenesis. *Evolutionary Ecology Research* **2**:337–352.

Shaw DC, Huso M, Bruner H. 2008. Basal area growth impacts of dwarf mistletoe on western hemlock in an old-growth forest. *Canadian Journal of Forest Research* **38**:576–583.

Shen H, Xu SJ, Hong L, Wang ZM, Ye WH. 2013. Growth but not photosynthesis response of a host plant to infection by a holoparasitic plant depends on nitrogen supply. *PLoS ONE* **8**:e7555.

Stout MJ, Rice WC, Ring DR. 2002. The influence of plant age on tolerance of rice to injury by the rice water weevil, *Lissorhoptrus oryzophilus* (Coleoptera: Curculionidae). *Bulletin of Entomological Research* **92**:177–184.

Stowe KA, Marquis RJ, Hochwender CG, Simms EL. 2000. The evolutionary ecology of tolerance to consumer damage. *Annual Review of Ecology and Systematics* **31**:565–595.

Tan DY, Guo SL, Wang CL, Ma C. 2004. Effects of the parasite plant (*Cistanche deserticola*) on growth and biomass of the host plant (*Haloxylon ammodendron*). *Forest Research* **17**:472–478.

Thomson VP, Cunningham SA, Ball MC, Nicotra AB. 2003. Compensation for herbivory by *Cucumis sativus* through increased photosynthetic capacity and efficiency. *Oecologia* **134**:167–175.

Tucker C, Avila-Sakar G. 2010. Ontogenetic changes in tolerance to herbivory in *Arabidopsis*. *Oecologia* **164**:1005–1015.

Van Zandt PA, Agrawal AA. 2004. Specificity of induced plant responses to specialist herbivores of the common milkweed *Asclepias syriaca*. *Oikos* **104**:401–409.

Wang RK, Guang M, Li YH, Yang BF, Li JM. 2012. Effect of the parasitic *Cuscuta australis* on the community diversity and the growth of *Alternanthera philoxeroides*. *Acta Ecologica Sinica* **32**:1917–1923.

Webber BL, Woodrow IE. 2009. Chemical and physical plant defence across multiple ontogenetic stages in a tropical rain forest understorey tree. *Journal of Ecology* **97**:761–771.

Wu Z, Guo Q, Li MG, Jiang L, Li FL, Zan QJ, Zheng J. 2013. Factors restraining parasitism of the invasive vine *Mikania micrantha* by the holoparasitic plant *Cuscuta campestris*. *Biological Invasions* **15**:2755–2762.

Yoder JI, Scholes JD. 2010. Host plant resistance to parasitic weeds; recent progress and bottlenecks. *Current Opinion in Plant Biology* **13**:478–484.

Yu H, Yu FH, Miao SL, Dong M. 2008. Holoparasitic *Cuscuta campestris* suppresses invasive *Mikania micrantha* and contributes to native community recovery. *Biological Conservation* **141**:2653–2661.

Yu H, He WM, Liu J, Miao SL, Dong M. 2009. Native *Cuscuta campestris* restrains exotic *Mikania micrantha* and enhances soil resources beneficial to natives in the invaded communities. *Biological Invasions* **11**:835–844.

Yu H, Liu J, He WM, Miao SL, Dong M. 2011. *Cuscuta australis* restrains three exotic invasive plants and benefits native species. *Biological Invasions* **13**:747–756.

Zhang J, Yan M, Li JM. 2012. Effect of differing levels parasitism from native *Cuscuta australis* on invasive *Bidens pilosa* growth. *Acta Ecologica Sinica* **32**:3137–3143.

Zhang J, Li JM, Yan M. 2013. Effects of nutrients on the growth of the parasitic plant *Cuscuta australis* R.Br. *Acta Ecologica Sinica* **33**:2623–2631.

10

Investigating the impacts of recycled water on long-lived conifers

Lloyd L. Nackley, Corey Barnes and Lorence R. Oki*

Department of Plant Sciences, University of California, Davis, CA 95616, USA

Associate Editor: Abad Chabbi

Abstract. Recycled wastewater is a popular alternative water resource. Recycled water typically has higher salinity than potable water and therefore may not be an appropriate water source for landscapes planted with salt-intolerant plant species. Coast redwoods (*Sequoia sempervirens*) are an important agricultural, horticultural and ecological species assumed to be salt intolerant. However, no studies have analysed how salinity impacts coast redwood growth. To determine salt-related growth limitations, as well as susceptibility to particular salt ions, we divided 102 *S. sempervirens* 'Aptos Blue' saplings evenly into 17 salinity treatments: a control and four different salts (sodium chloride, calcium chloride, sodium chloride combined with calcium chloride, and sodium sulfate). Each salt type was applied at four different concentrations: 1.0, 3.0, 4.5 and 6.0 dS m^{-1}. Trees were measured for relative growth, and leaves were analysed for ion accumulation. Results showed that the relative stem diameter growth was inversely proportional to the increase in salinity (electrical conductivity), with R^2 values ranging from 0.72 to 0.82 for different salts. Analysis of variance tests indicated that no particular salt ion significantly affected growth differently than the others ($P > 0.1$). Pairwise comparisons of the means revealed that moderately saline soils (4–8 dS m^{-1}) would decrease the relative height growth by 30–40 %. Leaf tissue analysis showed that all treatment groups accumulated salt ions. This finding suggests reduced growth and leaf burn even at the lowest ion concentrations if salts are not periodically leached from the soil. Regardless of the specific ions in the irrigation water, the results suggest that growth and appearance of coast redwoods will be negatively impacted when recycled water electrical conductivity exceeds >1.0 dS m^{-1}. This information will prove valuable to many metropolitan areas faced with conserving water while at the same time maintaining healthy verdant landscapes that include coast redwoods and other long-lived conifers.

Keywords: California; drought; Mediterranean climate; reclaimed water; urban forestry; urban horticulture.

Introduction

Water used to irrigate important verdant, social landscapes (e.g. arboreta, public parks and golf courses) faces competition with other uses of fresh water including increasing agricultural and urban demands (Hamilton *et al.* 2005). Recycled wastewater has been highlighted as one of the most affordable alternative resources for agricultural, industrial and urban non-potable purposes in arid and semi-arid regions like California, where current fresh water reserves are at a critical limit (Lazarova *et al.* 2001). In California natural prolonged periods of summer drought have been exacerbated in recent years by low winter rainfall. California's 2014 Water Year, which ended

* Corresponding author's e-mail address: lroki@ucdavis.edu

30 September 2014 was the third driest in 199 years of record; and was the warmest year on record (USGS 2015). In addition, California's population is estimated to increase by 15.4 million residents (a 39 % increase) over the next 50 years (Palmer and Schooling 2013). Both the rise in population and the uneven distribution of these new inhabitants will cause an increase in water demand (Hanak and Davis 2006). To mitigate the effects of increased competition for limited potable water, horticulturalists and municipalities in California and in arid and semi-arid climates around the world are developing sources of recycled wastewater (Hamilton *et al.* 2005; Miller 2006; Toze 2006).

Types of wastewaters used for recycling include treated and untreated sewage effluent, storm water runoff, domestic greywater and industrial wastewater (Toze 2006). Although recycled water meets many social and environmental objectives by reducing competition for fresh water, there are some drawbacks that make it less suitable than potable water for horticultural applications. Primarily, recycled water often has a greater salt concentration than potable water. Although the salinity of recycled water is not usually high enough to make it unsuitable for irrigation (Vartanian 2008), it can contain 10 times more salt (e.g. $1.0-2.0$ dS m^{-1}) than potable water (~ 0.1 dS m^{-1}). Thus, recycled water can be harmful to salt-intolerant plants (Maas 1986).

Salinity in low concentrations (<2.0 dS m^{-1}) has been shown to have adverse effects on growth and physiology of many plants (Kozlowski 1997; Chaves *et al.* 2009). Salinity impacts plant growth by decreasing the osmotic potential of the soil and imposing physiological drought, or through toxic effects from high concentrations of particular ions, such as sodium or chloride that can injure the plant (Chaves *et al.* 2009). Although there is an extensive literature on the negative effect of salt on plant growth for many agricultural crops (Sohan *et al.* 1999; Sultana *et al.* 1999; Katerji *et al.* 2003; Zheng *et al.* 2008), there is a limited amount of information quantifying growth responses to salt for important horticultural species. In particular, there is only one report published about the salt tolerance of the coast redwood tree (*Sequoia sempervirens*) (Wu and Guo 2006), which is surprising given this species' important ecological and horticultural value.

The coast redwood is emblematic of western US conifers known for its towering height (>100 m) and longevity (>1500 years). This charismatic tree species' native range extends along the fog-belt of the Pacific coast from southern Oregon to central California. The coast redwood is an important timber species, prized in building for its burnt-sienna coloured wood that is naturally decay resistant. Coast redwoods are also used extensively in Pacific horticulture (CA, OR, WA), in public parks, golf courses, highways and private landscapes; and are popular horticultural specimens used throughout the USA and in temperate climates around the world. Although the coast redwood is indigenous within a Mediterranean climate, which is typified by long periods of summer drought, coast redwoods thrive in areas with significant summertime moisture, typically derived from abundant marine fog. Moisture input from fog drip in the summer can constitute 30 % or more of the total water input each year (Dawson 1998). The coast redwood is characterized as having low to moderate drought tolerance (Sunset Books 2000) and requires supplemental irrigation where fog or summer precipitation events are lacking. Without natural precipitation (rain or fog) or supplemental irrigation, dry summer conditions may inhibit the performance of mature individuals of coast redwood in urban settings where signs of water stress often include leaf senescence and stem die back (Litvak *et al.* 2011) (Fig. 1).

The work presented herein was initiated to fill a knowledge gap by determining the level of tolerance of coast redwood to sodium and chloride. The research was designed in response to reports from water districts in the San Francisco Bay Area, which claimed that coast redwoods within public parks had shown signs of decline or death after irrigation with recycled water. To determine the effects of sodium and chloride ions on the growth and health of redwoods, *Sequoia sempervirens* 'Aptos Blue' specimens were placed in a greenhouse and irrigated daily with one of 17 treatments represented by a non-saline nutrient solution that was used as the control treatment plus four different salt solutions at four different concentrations. We hypothesized that redwoods

Figure 1. A photograph of coast redwoods (*Sequoia sempervirens*) that are growing in a public park in the San Francisco Bay Area, CA. The coast redwoods are the tall, slender conifers, exhibiting characteristic leaf senescence (browning) and stem dieback. The cause of stress of these trees was identified as water stress. Water stress can be caused by lack of soil moisture, or physiological drought from increased soil salinity.

would be classifiable as a 'salt-sensitive' species, demonstrated by declines in growth at soil salinity concentrations <3.0 dS m^{-1}. Further, we hypothesized that different salt solutions would be more toxic than others, represented by statistically different growth responses.

Methods

Experimental design

The experiment was conducted in a glasshouse at the UC Davis Environmental Horticulture Complex (Davis, CA, USA). Greenhouse daytime low and high temperatures were maintained between 21 and 24 °C, and night-time low and high temperatures were maintained between 13 and 17 °C. No artificial lighting was supplied to the plants. The glasshouse was divided into two blocks to control for natural gradients of sunlight, temperature and humidity. Pots were placed 1 m apart throughout the two blocks. One hundred and two *Sequoia sempervirens* 'Aptos Blue' saplings in 8 L pots (21 cm tall, with a 21 cm diameter tapering to 18.5 cm) were obtained from Generation Growers, Modesto, CA, USA. Potting media contained a mix of humus and sand in a 4:1 volumetric ratio, 6.0 kg m^{-3} dolomite, 0.6 kg m^{-3} calcium nitrate, 1.2 kg m^{-3} ferrous sulfate heptahydrate, 3.0 kg m^{-3} nitroform, 2.4 kg m^{-3} double super phosphate and 1.2 kg m^{-3} oyster shell lime.

The salinity treatments consisted of a control, as well as four different salts: sodium chloride (NaCl), calcium chloride (CaCl$_2$), sodium chloride and calcium chloride (NaCl + CaCl$_2$) and sodium sulfate (Na$_2$SO$_4$). Each salt was applied at four different concentrations represented by electrical conductivity (EC) of 1.0, 3.0, 4.5 and 6.0 dS m^{-1}. NaCl was selected because it is the most common salt in recycled water. Na$_2$SO$_4$ was used to isolate Na symptoms, whereas CaCl$_2$ served to isolate Cl symptoms. The combination of NaCl and CaCl$_2$ provided a treatment simulating environmental conditions, where combinations of monovalent and multivalent cations would be present in the irrigation water and/or soil. Each salt type was added to a one-quarter strength Hoagland's fertilizer 'Solution 2' which had an EC of 0.5 dS m^{-1} (Epstein and Bloom 2005). The control treatment received only the modified Hoagland's, without additional salt. Six trees were replicated in each of 17 treatments. Treatments were initialized on 15 October 2005.

Dosatron® DI-16 injectors (Dosatron USA, Clearwater, FL, USA) were used to mix the salinity treatments into the irrigation water. Three Netafim® Woodpecker pressure-compensating emitters (Netafim Irrigation, Fresno, CA, USA, rated 4 L h^{-1}) at each pot produced an average total flow rate of 12.8 L h^{-1} (SE = 0.08, $n = 9$). Multiple emitters at each pot allowed for uniform saturation of the container medium. Daily irrigations were

Table 1. Mean ($\pm$ SE) cumulative treatment and leachate EC values from the testing period 12 July 2005 to 1 September 2007. A leaching fraction of 0.4–0.5 was applied to all treatments independently. Irrigation treatment salinity concentrations were evaluated weekly by collecting solute from an emitter tube at each tree prior to the day's irrigation cycle.

Treatment	Cumulative mean treatment EC (dS m^{-1}) $\pm$ 1 SE	Cumulative mean leachate EC (dS m^{-1}) $\pm$ 1 SE
Control 0.5 dS m^{-1}	0.57 ± 0.01	0.66 ± 0.01
NaCl 1.0 dS m^{-1}	1.05 ± 0.01	1.67 ± 0.05
NaCl 3.0 dS m^{-1}	3.12 ± 0.03	4.52 ± 0.11
NaCl 4.5 dS m^{-1}	4.32 ± 0.05	5.71 ± 0.11
NaCl 6.0 dS m^{-1}	5.72 ± 0.08	7.08 ± 0.12
CaCl$_2$ 1.0 dS m^{-1}	1.06 ± 0.01	1.54 ± 0.02
CaCl$_2$ 3.0 dS m^{-1}	2.95 ± 0.02	5.08 ± 0.13
CaCl$_2$ 4.5 dS m^{-1}	4.52 ± 0.04	7.10 ± 0.16
CaCl$_2$ 6.0 dS m^{-1}	6.12 ± 0.04	8.83 ± 0.17
NaCl + CaCl$_2$ 1.0 dS m^{-1}	1.09 ± 0.01	1.61 ± 0.03
NaCl + CaCl$_2$ 3.0 dS m^{-1}	2.94 ± 0.03	4.60 ± 0.11
NaCl + CaCl$_2$ 4.5 dS m^{-1}	4.59 ± 0.03	6.83 ± 0.16
NaCl + CaCl$_2$ 6.0 dS m^{-1}	6.10 ± 0.04	8.40 ± 0.15
Na$_2$SO$_4$ 1.0 dS m^{-1}	1.09 ± 0.01	1.73 ± 0.05
Na$_2$SO$_4$ 3.0 dS m^{-1}	3.10 ± 0.04	4.68 ± 0.11
Na$_2$SO$_4$ 4.5 dS m^{-1}	4.71 ± 0.01	6.08 ± 0.09
Na$_2$SO$_4$ 6.0 dS m^{-1}	6.10 ± 0.02	7.37 ± 0.11

scheduled with a Hunter® ICC irrigation timer (Hunter Industries Inc., San Marcos, CA, USA). A leaching fraction of 0.4–0.5 was applied to all treatments independently. The leaching fraction is defined as the ratio of the quantity of water draining past the root zone to that infiltrated into the soil's surface. This fraction was used to isolate symptoms related to the salt treatments by eliminating stress due to both insufficient water and increasing container EC due to evapotranspiration. Further, this leaching fraction was designed to provide sufficient irrigation treatment volume to allow for uniform saturation of the container medium. Irrigation treatment salinity concentrations were evaluated weekly by collecting solute from the emitter tube at each tree during the day's irrigation cycle. After the irrigation cycle, a portable meter was used to test the EC and pH of each sample leachate (Table 1).

Data collection

Stem diameter and stem length (i.e. tree height) were measured every second week starting on 25 September

2005 and ending 3 January 2007. A set of digital calipers (Fisher Scientific, Pittsburgh, PA, USA) was placed around the trunk at a height of 3 cm above the potting medium in a constant orientation for each tree. The trunk was marked to indicate the points of contact for the calipers and the diameter was measured across these points each time. Tree height was evaluated every third week starting 15 September 2005 and ending 8 January 2007. Height was measured with a tape from an indicated point on the pot rim to the apex of the central leader of the tree.

The concentration of salt ions accumulated in the leaves was determined from analyses of leaves sampled from the previous flush of growth. These leaves were identified as originating from lignifying stem segments occurring directly behind the youngest, light green leaves on solid green stems. Consistency of tissue maturity has been shown to be an important characteristic for obtaining comparable results (Mills and Jones 1996). Leaf tissue-sampling events occurred on 17 October 2005, 9 January 2006, 18 May 2006, 22 September 2006 and 15 January 2007. The experiment was terminated shortly after the fifth sampling. Both proximal (P) and distal (D) leaf blade sections were collected on each date. The distal portions of leaves were removed first; the halfway cut point was determined visually. Then the basal sections of the cut leaves were removed by cutting them as closely to the stem as possible. A minimum of 1.5 g dry weight (3.8 g fresh weight, 39 % dry: fresh weight ratio) was collected for each sample. The dried samples were analysed for % Ca^+, % Cl^- and % Na^+ by using the 'Nitric/Perchloric Wet Ashing Open Vessel' (P – 3.10) technique, and Cl was analysed using the '2 % Acetic Acid Extraction' (P – 4.20) technique by Dellavalle® Laboratory, Inc. (Fresno, CA, USA). Ion accumulation rates were evaluated for the different concentrations within each salinity treatment type (e.g. 1.0 dS m^{-1} NaCl vs. 6.0 dS m^{-1} NaCl), as well as within treatment concentration level between the various salinity treatment types (e.g. 1.0 dS m^{-1} NaCl vs. 1.0 dS m^{-1} CaCl$_2$).

Statistical analysis

The experimental design represented a randomized complete block (RCB) with every treatment appearing in either block. Blocking of experimental groups in the greenhouse was used to control for variances that may have originated from the environmental conditions. The treatment and block effects were treated as random variables. Analysis of variance (ANOVA) tests were used to evaluate if differences in stem (height) growth and stem diameter growth could be confidently attributed ($P < 0.05$) to blocking, salinity type and salinity concentration. For each salt type, a linear regression was used to determine the

influence increasing salinity concentrations (EC) had on height and diameter growth responses (Fig. 2). Tukey's HSD were used as a post-hoc pairwise comparison of means within treatments when statistical differences were attributable to the treatment variables (salt or concentration). Relative comparisons of the mean ion concentrations, collected in proximal and distal tissues, were used to understand the differences attributable to salinity type and salinity concentration (Figs 3–5).

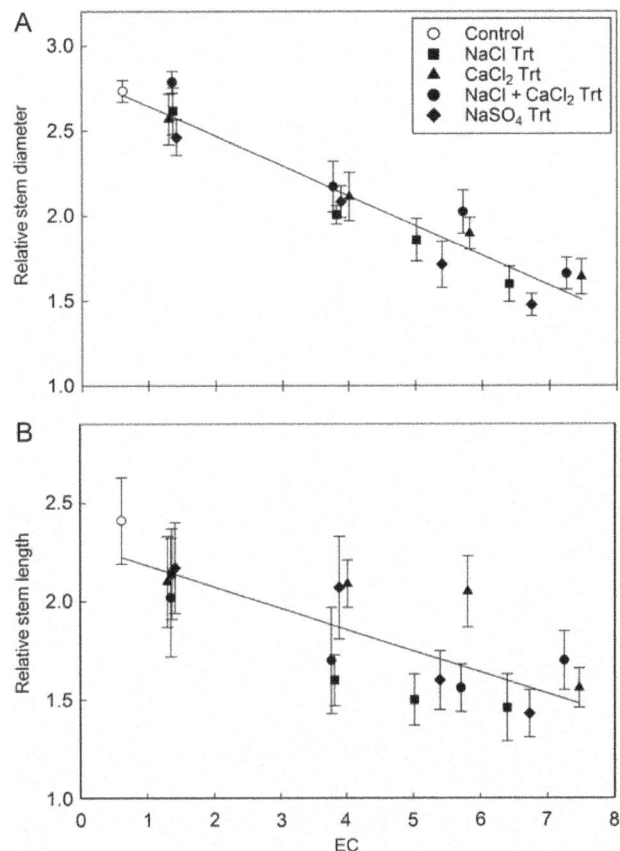

Figure 2. (A) Irrespective of salt ion the relative diameter growth was inversely proportional to the increase in salinity ($R^2 = 0.75$), which was measured as the mean of the irrigation and leachate EC. The R^2 relationship between EC and relative diameter growth values for the control and four concentrations within each salinity type ranged from 0.72 for the CaCl$_2$ treatments to 0.82 for the Na$_2$SO$_4$ treatments. Analysis of variance tests indicated that the specific salt (or ion combination) did not have a significantly greater effect on the diameter growth of the stem ($P > 0.1$). (B) The specific ion (or ion combination) did not have a significantly greater effect on the vertical growth of the stem ($P > 0.1$). The relative stem growth was poorly explained as a response to increasing salinity (EC) ($R^2 = 0.27$). Yet, there were significant differences in height ($P < 0.001$) between the different concentration levels within the salt treatment. Pairwise multiple comparisons of the means suggest that the final relative height in the control group and the lowest ion concentration (1.0 dS m^{-1}) were significantly greater ($P < 0.05$) when compared with the two higher ion concentrations (4.5 and 6.0 dS m^{-1}).

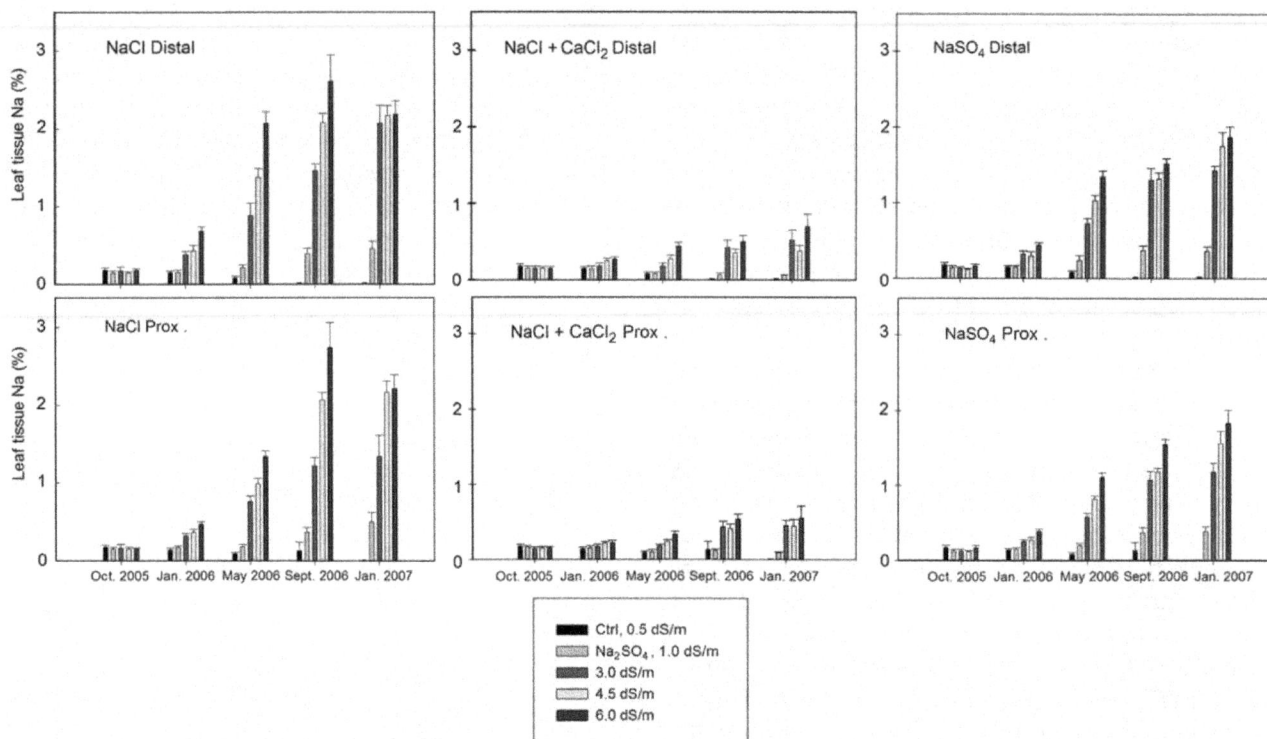

Figure 3. Na$^+$ ion concentration in the distal and proximal ends of the leaf tips from all treatment groups (excluding the CaCl$_2$ treatments). At every harvest the greatest % Na$^+$ in leaf tissues (distal and proximal) were detected in the 6.0 dS m^{-1} group in NaCl treatment. Over time, the % Na tissue concentration increased in all salinity concentrations across all ion treatments. The per cent concentration did not differ greatly between distal and proximal ends of the leaf for any treatment combination. The distal and proximal leaf tissues harvested from plants treated with NaCl had 2–3 times greater Na$^+$ when compared with the NaCl + CaCl$_2$ group, and 25 % greater than the NaSO$_4$ group.

Results

Growth

The stem length (height) and diameter measurements were highly dependent on the natural variability of initial starting size rather than the treatment effects. To elucidate the treatment effects we investigated the relative growth for each tree, rather than the absolute growth. Relative growth was calculated by dividing the recorded height or diameter measurements by the original (9 September 2005) height or diameter measurements. Analysis of variance results suggest no significant differences ($P > 0.1$) attributable to the experimental blocks. The magnitude of salinity effects, measured as mean EC from the cumulative irrigation EC and the cumulative leachate EC, was similar for all four salt treatments. Trees grown at 6.0 dS m^{-1} increased ~1.5 times the original diameter. Trees grown at 1.0 dS m^{-1} increased ~2.5 times the original diameter (Fig. 2A), which was similar to the growth by the control group. The proportion of the variation in stem diameter growth responses explained by increasing salinity (EC) ranged from $R^2 = 0.72$ for the CaCl$_2$ treatments, $R^2 = 0.74$ for NaCl + CaCl$_2$, $R^2 = 0.77$ NaCl, to $R^2 = 0.82$ for the Na$_2$SO$_4$ treatment. Analysis of

variance tests indicated that the specific salt (or ion combination) did not have a significantly greater effect on the diameter growth of the stem ($P > 0.1$). An ANOVA testing the response of height growth to the salt treatments suggested significant differences ($P < 0.001$) between the different concentration levels within the salt treatments.

Means of all the treatments showed an inverse relationship between stem length growth and increasing EC (Fig. 2B). Yet, the mean relative stem length and the corresponding standard errors overlapped greatly. Intermediate salinity concentrations caused the most variation within and between treatments in relative tree height (Fig. 2B). Relationships between EC and relative height was poorly described by linear modelling with R^2 values for each salinity type ranging from 0.18 for the NaCl + CaCl$_2$ treatments, $R^2 = 0.2$ CaCl$_2$, $R^2 = 0.34$ NaSO$_4$, to 0.38 for the NaCl treatment. A Tukey HSD pairwise comparison revealed that the final relative height (15 January 2007) of the control group (2.4 ± 0.2) was not significantly greater ($P > 0.1$) than the final relative height of any of the lowest salinity concentration (1.0 dS m^{-1}) (NaCl: 2.1 ± 0.2; CaCl$_2$: 2.1 ± 0.2, NaCl + CaCl$_2$: 2.0 ± 0.3; Na$_2$SO$_4$: 2.2 ± 0.2). Yet Tukey HSD pairwise comparisons suggest that the final relative height in the control group and the lowest

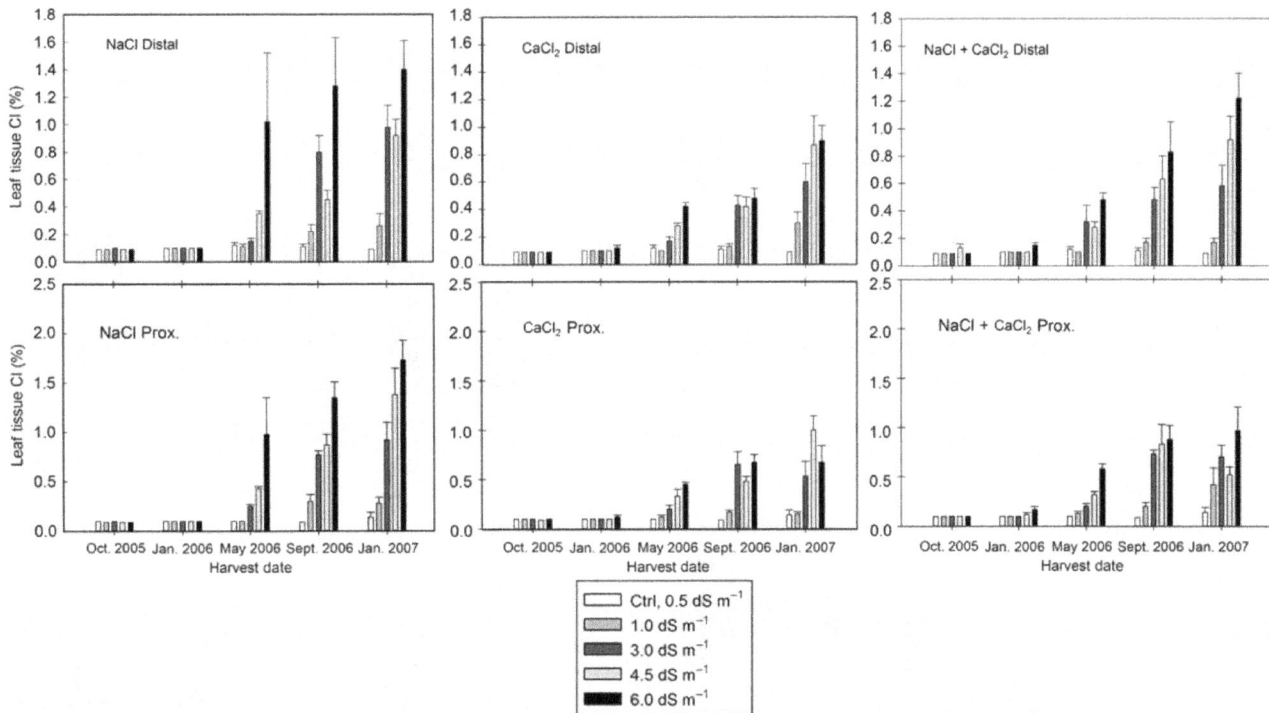

Figure 4. The rate of Cl^- increase was relative to the salinity concentration rate in the irrigation solution, with the greatest tissue concentrations in the plants in treatment groups with the greatest ions. The % Cl^- concentration did not differ greatly between distal and proximal ends of the leaf for any treatment combination. The Cl^- concentration did not initially increase (October 2005–January 2006). However, subsequent harvests exhibited increasing concentration of Cl^- ions in the leaf tissues.

ion concentration (1.0 dS m^{-1}) were significantly greater ($P < 0.05$) when compared with the two higher ion concentrations (4.0 and 6.0 dS m^{-1}). The significant differences between the high and low salinity groups suggest that if recycled water generates moderately saline soils (4–8 dS m^{-1}) that the relative height growth of coast redwoods may decrease by 30–40 %.

Tissue analysis

At every harvest, the greatest % Na^+ in leaf tissues (D or P) was detected in the 6.0 dS m^{-1} group in NaCl treatment (Fig. 3). The per cent concentration did not differ greatly between D or P leaf sections for any treatment combination and hereafter will not be differentiated. Over time, the % Na^+ tissue concentration increased in all salinity concentrations across all ion treatments (excluding $CaCl_2$, which did not contain Na^+ ions). The leaf tissues harvested from plants treated with NaCl had 2–3 times greater Na^+ when compared with the $NaCl + CaCl_2$ group, and 25 % greater than the $NaSO_4$ group.

Similar to the Na^+ tissue analyses, the rate of Cl^- increase was relative to the salinity concentration in the irrigation solution with the greatest tissue concentrations occurring in the plants in treatment groups with the greatest ions concentration (Fig. 4). Like Na^+, the % Cl^- concentration did not differ greatly between D and P

ends of the leaf for any treatment combination. The leaf Cl^- concentration did not initially increase (October 2005–January 2006). However, subsequent harvests exhibited increasing concentrations of Cl^- ions in the leaf tissues for all treatments containing Cl^- ions.

Calcium ion concentration in the leaf tissues followed the same pattern as Na^+ and Cl^-, with similar concentrations being found in D and P segments and the greatest concentrations being relative to the treatment ion concentrations, which increased with exposure over time (Fig. 5). A notable difference between the % Na^+ (Fig. 3), % Cl^- (Fig. 4) and % Ca^+ (Fig. 5) analyses is that the % Ca^+ in the control group was nearly an order of magnitude greater than either of the other two ions (1 % compared with 0.1 %). Corresponding to this relatively increased baseline, the final concentrations of Ca^+ ions found in the leaf tissues at January 2007 were >2 % for all Ca^+ containing salt treatments. Conversely, no treatment had >2 % Cl^-, and only the 6.0 NaCl treatments had >2 % Na.

Discussion

Growth responses

The impetus of the study was to quantify the toxicity of Na^+ and Cl^- to coast redwood, since horticultural

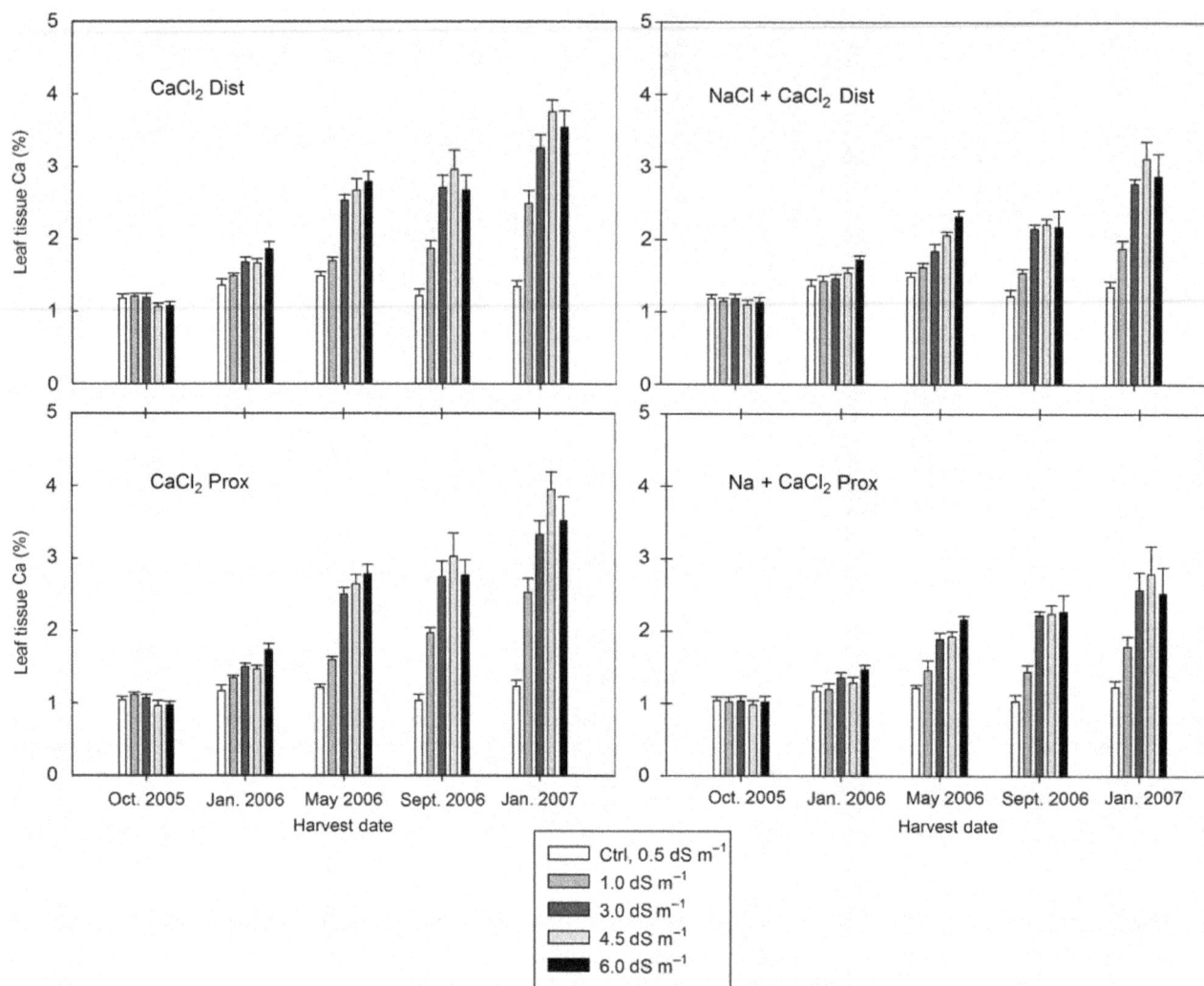

Figure 5. Calcium ion concentration in the leaf tissues followed the same pattern as Na^+ and Cl^-, with similar concentrations being found in distal and proximal segments and the greatest concentrations being relative to the treatment concentrations, which increased with exposure over time. The final concentrations of Ca^+ ions found in the leaf tissues in January 2007 were >2 % for all Ca^+ containing salt treatments. Conversely, no treatment had >2 % Cl^-, and only the 6.0 NaCl treatments had >2 % Na.

problems had been observed in coast redwoods irrigated with recycled water salinity in the Santa Clara, CA, USA area, and NaCl is the primary salt in recycled water in the Santa Clara, CA, USA, area. The salinity of recycled water may be the single most important parameter in determining its suitability for irrigation (EPA 2012). Salinity limits vegetative and reproductive growth of plants by inducing severe physiological dysfunctions and causing widespread direct and indirect harmful effects, even at low salt concentrations (Kozlowski 1997). Although the salts present in recycled water can vary greatly depending on the water source, the more common elemental ions include B^{3+}, Ca^{2+}, Cl^-, Na^+ and SO_4^- (Martinez and Clark 2012).

The treatments were designed to isolate particular ion effects: NaCl, the salt of interest, $CaCl_2$ to isolate the chloride effect, Na_2SO_4 to isolate the sodium effect, and the combination of NaCl + $CaCl_2$ to represent a mixture of mono- and divalent cations that would be closer to a 'real' exposure. The similar growth patterns and lack of statistical differences ($P > 0.1$) between salt ions at each respective concentration level (Fig. 2) suggest that none of the ions or ion combinations were more toxic than the others for coast redwood growth. Growth was significantly decreased when salinity in irrigation water was >3 dS m^{-1} when compared with water with 1.0 dS m^{-1} levels of salt ($P < 0.05$) (Fig. 2). The diameter growth responses to increasing salinity were fairly uniform with nearly three-quarters or more of the variation ($R^2 = 0.72 - 0.82$) being explained by increasing salinity. The height (stem length) responses were also negatively correlated with increasing salinity. Yet scarcely a third, at best, of the variation was attributable to increasing salinity ($R^2 = 0.18-0.38$). At intermediate salinity concentrations (EC 3.5 and 5.5) it appears (Fig. 2B) that $CaCl_2$ was less inhibiting on stem growth than other salts (e.g. NaCl). Yet differences in salt type were not found to be statistically significant determinant of stem growth ($P > 0.2$). The coast redwood is a hexaploid—each of its cells containing six sets of chromosomes, with 66 chromosomes. Natural genotypic variation is incredibly complex in this species and long-lived tree species in general. Additional analyses with greater sample sizes over longer duration could potentially improve the clarity of response between salt type and stem elongation. Nonetheless, the lack of statistical differences in growth reduction between specific salt ions agrees with previous research that did not find specific ion effects on redwood leaf responses to salt spray (Wu and Guo 2006). Confirming that overall salinity reduces growth, more so than specific ion toxicity, is an important finding because it suggests that a variety of sources of recycled water may be suitable for irrigation water in areas where coast redwoods are growing. However, these effects may change over time depending on the soil and climate where the trees are growing.

Salt may become progressively concentrated in the root zone because the plant roots absorb water but very little salt (Kozlowski 1997). In semi-arid and arid environments, saline soils can be problematic due to the lack of adequate rainfall to leach ions from the surface soil layers. The concentration of salt ions can have different effects in different soils. For instance, Na cations are known to interact with anionic clay soils resulting in swelling dispersion of the clay particles, which can reduce soil permeability (Halliwell et al. 2001). The buildup of salts in any type of soil will reduce the hydraulic conductivity, which can impact on the ability of water to infiltrate into the soil profile, and thus reduce the water availability to irrigated plants (Toze 2006). Plants will accumulate non-toxic osmolytes in the leaves to counter the change in soil osmotic potential (Greenway and Munns 1980; Munns and Tester 2008). In plants that are unable to exclude the salt, of salt ions, overabundant salt ions will also accumulate in leaf tissue vacuoles (Greenway and Munns 1980; Chaves et al. 2009). While this concentration of salt ions in the vacuoles can provide temporary osmotic balance and allow for water uptake, it can also become toxic over time as the ions begin to degrade the chloroplasts (Chaves et al. 2009). The statistical differences in growth across the given concentrations (Fig. 2) are even more clearly represented by the visual effects of leaf burn and necrosis (Fig. 6). The continually increasing salt ion concentrations in the leaf tissues (Figs 3–5) suggests that there is the possibility for reduced growth and leaf burn even at the lowest ion concentrations. These results demonstrate that salts must be periodically leached from the soil to prevent the accumulation of greater ionic concentrations in the leaves over time. Salt accumulation in the soil may not be problematic in areas where winter rainfall is sufficient to leach the salt ions. However, this will depend in part on the soil chemistry and structural properties including infiltration and drainage rates.

There has only been one previous study published regarding salt tolerance of coast redwoods (Wu and Guo 2006). Wu and Guo (2006) reported that coast redwood variety Aptos Blue should be placed in the salt sensitive category, because leaves were significantly chlorotic when foliar sprays were applied at a concentration of >2.0 dS m^{-1}. The experimental design of their study differed considerably from the work presented herein. Wu and Guo's (2006) salt tolerance categorization was based on the quantity of leaf yellowing after the application of a foliar spray containing salts. Yellowing is an important consideration for horticultural aesthetics, however, because the tolerable level of leaf chlorosis will depend on the management objectives, it is a more subjective measurement compared with the growth responses reported herein. Another important difference is that Wu and Guo's (2006) study focussed on foliar contact, which is important in situations where small trees may be sprayed by overhead irrigation. Yet salt spray on leaves may not be relevant in landscape situations where the spray from an overhead sprinkler is well below the crown of a mature coast redwood. We believe that our results, which analysed impacts of soil salinity, compliment the previous research on foliar responses to salinity. This is especially so considering that if recycled water is being utilized as a means of water conservation, in which case it should be used in conjunction with efficient irrigation methods (e.g. drip or micro spray emitters) (Pereira et al. 2002). Foliar spray on mature trees would be unlikely from these more efficient forms of irrigation.

Management considerations for recycled water in urban landscapes and horticulture

Sodium and other forms of salt are difficult to remove and may be persistent in recycled water (Toze 2006). Reverse osmosis, the most commonly used desalination technology (Fritzmann et al. 2007), is typically reserved for the production of high-quality recycled water (i.e. potable water) and would not be practical or economical for landscape irrigation (Toze 2006). Considering that long-term accumulation is a concern for coast redwoods, if salts were present in recycled water, irrigation management would have to rotate between recycled water and fresh water to leach salt ions, utilize recycled water blended with higher quality water to reduce salinity or depend on natural precipitation events. In a container nursery

Figure 6. Photographs of the leaves of individual trees taken on 22 September, 2006, ~1 year after the salinity treatments had been initiated. The visual evidence presented in these photos is characteristic of the leaf burn that was typical for all plants within each particular salt/concentration group. The photos display how little the leaves were affected at 1.0 dS m^{-1}, with progressively more burn at higher concentrations.

setting—where the soil media is well drained and leached regularly—recycled water with low salinity (<2.0 dS m^{-1}) could also be applied safely to irrigate coast redwoods as long as the containers were leached periodically with non-saline water to inhibit accumulation overtime.

The responses demonstrated in our study by the coast redwood, including decreased growth at moderate salinity levels and mortality at high salinity, were similar to responses reported for other conifers (Croser et al. 2001). Similar to our finding of increased salt

accumulation over time (Figs 3–5), needle burn in ponderosa pine (*Pinus ponderosa* L.) irrigated with recycled wastewater was largely correlated with needle Na$^+$ concentration (Qian et al. 2005). However, it has been reported that loblolly pine (*Pinus taeda* L.) irrigated with untreated laundry wastewater had improved shoot growth, which the authors attributed to the CaCl$_2$ found in the wastewater (Warren et al. 2004). Considering our results, these findings may not translate to all conifers. Further, the presence of CaCl$_2$ was found to decrease nutrient uptake in Norway spruce (*Picea abies* (L.) Karst.) (Bogemans *et al.* 1989). In general, there are very few studies of the impacts of salinity on conifers, which presents a challenge for mangers using recycled water and an opportunity for future studies. We conclude that *Sequoia sempervirens* 'Aptos Blue' was sensitive to salinity regardless of salt composition, particularly as the EC of the irrigation water exceeded 1.0 dS m^{-1} (Figs 2 and 6).

Conclusions

Fresh water is an essential resource—integral to all ecological and societal activities (Gleick 2012). Growing global human populations must manage finite fresh water resources to meet basic human needs while also ensuring that the extraction of water from natural sources (e.g. rivers, lakes, aquifers etc.) does not deleteriously affect other important ecosystem services derived from freshwater ecosystems (Millennium Ecosystem Assessment 2005).Water reuse for non-potable (i.e. irrigation, industrial) or indirect potable (e.g. discharge into drinking water reservoirs or supply) purposes has been considered across the USA, but particularly in drier or drought-ridden communities such as Arizona, California, Colorado and Texas; or communities experiencing substantial population and economic growth that place a strain on water supplies (e.g. Georgia and Florida) (Hartley 2006).

Unlike monoculture agriculture that manages a single species most landscape plantings include a diverse assemblage of species. Therefore, salt concentrations in recycled water must be acceptable for the most sensitive landscape plant species. The growth responses of the coast redwood to drip irrigation with a variety of salt ions across a range of salinity concentrations had never been reported until this work. This information will prove valuable to the many metropolitan areas faced with conserving water while at the same time maintaining healthy, verdant landscapes that include coast redwoods and other long-lived conifers. For instance, recycled water is a necessity in many metropolitan communities in California, including the San Francisco suburb Redwood City. Redwood City delivers ~2500 ML year^{-1} of recycled water for municipal irrigation, which is a portion of the 40 150 ML year^{-1} of recycled water provided by the South Bayside System Authority for municipal uses (primarily urban landscapes) (Ingram *et al.* 2006). The EC of this recycled water typically ranges from 1.0 to 1.5 dS m^{-1} (Santa Clara Valley Water District 2007). As the name Redwood City suggests, the coast redwood is an important component of the urban forest, and is the 'City Tree' of nearby Palo Alto, CA, USA, a city named for a charismatic *Sequoia sempervirens* 'El Palo Alto' (Spanish for tall stick). The results from this study show that like other long-lived conifers, coast redwoods may be able to tolerate the low levels of salt that are typically found in recycled water. In landscapes where interaction between clay soils and saline recycled water may exacerbate salt stress, more drought tolerant/salt tolerant conifers such as pines (e.g. *Pinus palustris* Mill., *P. pinea* L., *P. pinaster* Aiton, and *P. radiata* D. Don) (Antonellini and Mollema 2010) could be suitable replacements to coast redwoods.

Sources of Funding

Our work was funded by the Santa Clara Valley Water District (CA, USA); and the Cities of Palo Alto (CA, USA) and Mountain View (CA, USA).

Contributions by the Authors

L.L.N. and C.B. primarily collaborated in the drafting of the manuscript. L.L.N. primarily oversaw the analysis and interpretation of the data. C.B. and L.R.O. primarily collaborated in the design and acquisition of data. L.R.O. also provided critical revisions throughout the drafting of the manuscript.

Acknowledgements

We thank Drs Richard Evans and James Oster for their guidance and advice.

Literature Cited

Antonellini M, Mollema PN. 2010. Impact of groundwater salinity on vegetation species richness in the coastal pine forests and wetlands of Ravenna, Italy. *Ecological Engineering* **36**: 1201–1211.

Bogemans J, Neirinckx L, Stassart JM. 1989. Effect of deicing NaCl and CaCl$_2$ on spruce (*Picea abies* (L.) sp.). *Plant and Soil* **120**: 203–211.

Chaves MM, Flexas J, Pinheiro C. 2009. Photosynthesis under drought and salt stress: regulation mechanisms from whole plant to cell. *Annals of Botany* **103**:551–560.

Croser C, Renault S, Franklin J, Zwiazek J. 2001. The effect of salinity on the emergence and seedling growth of *Picea mariana*, *Picea glauca*, and *Pinus banksiana*. *Environmental Pollution* **115**:9–16.

Dawson TE. 1998. Fog in the California redwood forest: ecosystem inputs and use by plants. *Oecologia* **117**:476–485.

EPA. 2012. Guidelines for water reuse. EPA/600/R-12/618. Washington, DC.

Epstein E, Bloom AJ. 2005. *Mineral nutrition of plants: principles and perspectives*. Sunderland, MA: Sinauer Associates Inc.

Fritzmann C, Löwenberg J, Wintgens T, Melin T. 2007. State-of-the-art of reverse osmosis desalination. *Desalination* **216**: 1–76.

Gleick PH. 2012. *The world's water: the biennial report on freshwater resources*, Vol. 7 (Pacific Institute for Studies in Development and Security and Environment, Ed.). Washington, DC: Island Press.

Greenway H, Munns R. 1980. Mechanisms of salt tolerance in nonha-lophytes. *Annual Review of Plant Physiology* **31**:149–190.

Halliwell DJ, Barlow KM, Nash DM. 2001. A review of the effects of wastewater sodium on soil physical properties and their implications for irrigation systems. *Australian Journal of Soil Research* **39**:1259–1267.

Hamilton AJ, Boland A-M, Stevens D, Kelly J, Radcliffe J, Ziehrl A, Dillon P, Paulin B. 2005. Position of the Australian horticultural industry with respect to the use of reclaimed water. *Agricultural Water Management* **71**:181–209.

Hanak E, Davis M. 2006. California economic policy: lawns and water demand in California. San Francisco, CA, USA: Public Policy Institute of California.

Hartley TW. 2006. Public perception and participation in water reuse. *Desalination* **187**:115–126.

Ingram PC, Young VJ, Millan M, Chang C, Tabucchi T. 2006. From controversy to consensus: the Redwood City recycled water experience. *Desalination* **187**:179–190.

Katerji N, van Hoorn JW, Hamdy A, Mastrorilli M. 2003. Salinity effect on crop development and yield, analysis of salt tolerance according to several classification methods. *Agricultural Water Management* **62**:37–66.

Kozlowski TT. 1997. Responses of woody plants to flooding and salinity. *Tree Physiology* **17**:490.

Lazarova V, Levine B, Sack J, Cirelli G, Jeffrey P, Muntau H, Salgot M, Brissaud F. 2001. Role of water reuse for enhancing integrated water management in Europe and Mediterranean countries. *Water Science and Technology: A Journal of the International Association on Water Pollution Research* **43**:25–33.

Litvak E, McCarthy HR, Pataki DE. 2011. Water relations of coast redwood planted in the semi-arid climate of southern California. *Plant, Cell and Environment* **34**:1384–1400.

Maas E. 1986. Salt tolerance of plants. *Applied Agricultural Research* **1**:12–26.

Martinez CJ, Clark MW. 2012. *Using reclaimed water for landscape irrigation*. Gainesville, FL, USA: Florida Cooperative Extension Service, University of Florida.

Millennium Ecosystem Assessment. 2005. *Ecosystems and human well-being: biodiversity synthesis*. Washington, DC: Millennium Ecosystem Assessment.

Miller WG. 2006. Integrated concepts in water reuse: managing global water needs. *Desalination* **187**:65–75.

Mills HA, Jones JB. 1996. *Plant analysis handbook*. Athens, GA: Micro-Macro Publishing.

Munns R, Tester M. 2008. Mechanisms of salinity tolerance. *Annual Review of Plant Biology* **59**:651–681.

Palmer HD, Schooling B. 2013. *New population projections: California to surpass 50 million in 2049*. Sacramento, CA, USA: California Department of Finance.

Pereira LS, Oweis T, Zairi A. 2002. Irrigation management under water scarcity. *Agricultural Water Management* **57**:175–206.

Qian Y, Fu J, Klett J, Newman S. 2005. Effects of long-term recycled wastewater irrigation on visual quality and ion concentrations of ponderosa pine. *Journal of Environmental Horticulture* **23**:185–189.

Santa Clara Valley Water District, the City of Mountain View and South Bay Water Recycling. 2007. Solutions Project Report.

Sohan D, Jasoni R, Zajicek J. 1999. Plant–water relations of NaCl and calcium-treated sunflower plants. *Environmental and Experimental Botany* **42**:105–111.

Sultana N, Ikeda T, Itoh R. 1999. Effect of NaCl salinity on photosynthesis and dry matter accumulation in developing rice grains. *Environmental and Experimental Botany* **42**:211–220.

Sunset Books. 2000. *Sunset western garden book*. Menlo Park, CA: Sunset Publishing Corporation.

Toze S. 2006. Reuse of effluent water—benefits and risks. *Agricultural Water Management* **80**:147–159.

USGS. 2015. *The California Drought*. US Department of the Interior. Webpage Last Modified 22 December 2014, Accessed 3 January 2015.

Vartanian GM. 2008. *Managing salinity of recycled water for landscape irrigation: The link between plants, soils, salts, and recycled water*. Sacramento, CA, USA: California Department of Water Resources.

Warren SL, Amoozegar A, Robarge WP, Niewoehner CP, Reece WM. 2004. Effect of graywater on growth and appearance of ornamental landscape plants. In: Mankin KR, ed. *On-site wastewater treatment X*. St. Joseph, MI: American Society of Agricultural and Biological Engineers.

Wu L, Guo X. 2006. Response of two coast redwood (*Sequoia sempervirens* Endl.) varieties to moderate levels of salt and boron spray measured by stress symptoms: implications for landscape irrigation using recycled water. *Environmental and Experimental Botany* **58**:130–139.

Zheng Y, Wang Z, Sun X, Jia A, Jiang G, Li Z. 2008. Higher salinity tolerance cultivars of winter wheat relieved senescence at reproductive stage. *Environmental and Experimental Botany* **62**:129–138.

11

Effect of temperature and nutrients on the growth and development of seedlings of an invasive plant

Hana Skálová[1]*, Lenka Moravcová[1], Anthony F. G. Dixon[2,3], P. Kindlmann[3,4] and Petr Pyšek[1,5]

[1] Department of Invasion Ecology, Institute of Botany, The Czech Academy of Sciences, CZ-252 43, Průhonice, Czech Republic
[2] School of Biological Sciences, University of East Anglia, Norwich NR4 7TJ, Norfolk, UK
[3] Department of Biodiversity Research, Global Change Research Centre, The Czech Academy of Sciences, Bělidla 4a, CZ-602 00, Brno, Czech Republic
[4] Institute for Environmental Studies, Charles University in Prague, Benátská 2, CZ-128 01, Prague, Czech Republic
[5] Department of Ecology, Faculty of Science, Charles University in Prague, Viničná 7, CZ-128 44, Prague, Czech Republic

Associate Editor: Colin M. Orians

Abstract. Plant species distributions are determined by the response of populations to regional climates; however, little is known about how alien plants that arrive in central Europe from climatically warmer regions cope with the temperature conditions at the early stage of population development. *Ambrosia artemisiifolia* (common ragweed) is an invasive annual plant causing considerable health and economic problems in Europe. Although climate-based models predict that the whole of the Czech Republic is climatically suitable for this species, it is confined to the warmest regions. To determine the factors possibly responsible for its restricted occurrence, we investigated the effects of temperature and nutrient availability on its seedlings. The plants were cultivated at one of seven temperature regimes ranging from 10 to 34 °C, combined with three nutrient levels. The data on the rate of leaf development were used to calculate the lower developmental threshold (LDT, the temperature, in °C, below which development ceases), the sum of effective temperatures (SET, the amount of heat needed to complete a developmental stage measured in degree days above LDT) and width of the thermal window. The rate of development decreased with decrease in temperature and nutrient supply. Besides this, the decrease in the availability of nutrients resulted in decreased LDT, increased SET and wider thermal window. The dependence of LDT and SET on the availability of nutrients contradicts the concept that thermal constants do not vary. Our results highlight temperature as the main determinant of common ragweed's distribution and identify nutrient availability as a factor that results in the realized niche being smaller than the fundamental niche; both of these need to be taken into account when predicting the future spread of *A. artemisiifolia*.

Keywords: *Ambrosia artemisiifolia*; common ragweed; invasive species; non-indigenous plants; nutrient limitation; plant nutrition; rate of development; thermal time.

Introduction

Plant species distributions are determined by the response of populations to regional climates. Alien plants, many of which arrive in regions from areas with a warmer climate, provide a suitable model for studying the effect of temperature on determining their distribution. Since

* Corresponding author's e-mail address: hana.skalova@ibot.cas.cz

the early stage of population development is crucial for determining whether or not a species can successfully establish in a given region (Richardson *et al.* 2000; Blackburn *et al.* 2011), knowing the response of seedlings to temperature can greatly improve our understanding of invasion and provide knowledge necessary for the efficient management of invasive species.

Common ragweed (*Ambrosia artemisiifolia*) is an invasive wind-pollinated annual plant causing considerable health and economic problems (Rybníček and Jäger 2001; Chauvel *et al.* 2006; Kömives *et al.* 2006; Fumanal *et al.* 2007; Essl *et al.* 2009; Šikoparija *et al.* 2009). It was introduced in Europe in the 19th century from its native range in North America, where it occurs in eastern and central USA (Kartesz and Meacham 1999; Brandes and Nitzsche 2007), and at the beginning of 2000s it was reported as a neophyte by 36 countries, in the majority of which it was naturalized (Lambdon *et al.* 2008). After an extended lag phase, it has spread since the 1940s via transportation networks and contaminated crop seed, and recently rapidly invaded central Europe (Chauvel *et al.* 2006; Brandes and Nitzsche 2007; Dullinger *et al.* 2009; Essl *et al.* 2009; Richter *et al.* 2013a, b; Martin *et al.* 2014). It is predicted that it will spread further in Europe as the species is assumed to be favoured by ongoing global warming (Richter *et al.* 2013b; Storkey *et al.* 2014). In its global invaded range, which includes all continents and some islands (New Zealand, Hawaii, Madagascar, Mauritius; Brandes and Nitzsche 2007), it occurs in a wide range of open and nutrient-rich, disturbed ruderal habitats and arable land (Chauvel *et al.* 2006; Essl *et al.* 2009; Pinke *et al.* 2011; Pyšek *et al.* 2012a).

Climate-dependent phenological models suggest that the distribution of *A. artemisiifolia* in Europe is limited by low temperatures in the North where plants are prevented from completing reproduction by autumn frosts, which kill adult plants (Chapman *et al.* 2014), and drought in the South, which inhibits seed germination and seedling emergence (Storkey *et al.* 2014). On the other hand, other environmental variables such as habitat type, land use, crop type and soil nutrients also play a role in the occurrence and abundance of *A. artemisiifolia* at a regional scale (Essl *et al.* 2009; Pinke *et al.* 2011, 2013). At a local scale, soil disturbance and removal of vegetation enhance seedling recruitment of ragweed from the soil-seed bank (Fumanal *et al.* 2008), by favouring the growth of juveniles as this species is not a good competitor (Bazzaz 1974; Leskovšek *et al.* 2012a). Its success as an invasive species is associated with its high production of seed (1200–2500 seeds per plant; Fumanal *et al.* 2007; Moravcová *et al.* 2010), which form a long-term persistent soil-seed bank, with seeds remaining viable for up to 40 years (Baskin and Baskin 1980).

Ambrosia artemisiifolia was introduced with clover seed from North America and first recorded in the Czech Republic in 1883, and is classified as an invasive neophyte (term used for species introduced after 1500 A.D.) in that country (Pyšek *et al.* 2012b). Naturalized populations occur only in the warmest parts of the country, i.e. in the Elbe region, and southern and northern Moravia, with casual populations scattered throughout the rest of the country (Pyšek *et al.* 2012a). This species prefers open dry habitats on sandy or gravel substrata with low vegetation cover. Most records are from around railway stations, river harbours, transit sheds, agricultural and industrial areas dealing now or in the past with soya beans, and neighbouring ruderal areas (Jehlík 1998).

The present distribution of *A. artemisiifolia* in the Czech Republic does not correspond to the climate-based prediction that the whole country is suitable for this species to complete its life cycle and set seed (Cunze *et al.* 2013; Chapman *et al.* 2014; Storkey *et al.* 2014). Some predictions highlight the importance of taking local environmental conditions into account (Stratonovitch *et al.* 2012). These authors indicate that this plant's response to climate change is confounded by variation in soil properties, which accords with the reported effects of soil nutrients on *A. artemisiifolia* distribution (Pinke *et al.* 2011, 2013). The role of nutrient availability is indicated by analyses of the factors that affect the distribution of this species in Europe (Pinke *et al.* 2011, 2013) and by greenhouse experiments (Leskovšek *et al.* 2012a, b), but its interaction with temperature has not been experimentally tested.

To obtain a detailed insight into the ecological factors that are likely to co-determine the performance of *A. artemisiifolia* in the field, we investigated the effect of temperature and nutrient availability on the rate of development (RD) of its seedlings. The good survival of seedlings is crucial for the establishment and successful population regeneration of annual species and there are several studies directly linking seedling traits with invasion success (Grotkopp and Rejmánek 2007; Morrison and Mauck 2007; Zheng *et al.* 2009; Skálová *et al.* 2012). The effect of temperature on seedling development has been previously studied (Granier *et al.* 2002; Gramig and Stoltenberg 2007), with one of the few studies on *A. artemisiifolia* (Deen *et al.* 1998a). This approach, which is based on measuring the rate of plant development at different temperatures, provides results that can be used to calculate the following characteristics: the lower developmental threshold—LDT (the temperature below which development ceases); the sum of effective temperatures—SET [the amount of heat needed to complete a developmental stage measured in degree days (DD) above the LDT] and thermal window (the difference between the minimum and maximum temperatures of the range over which development occurs) (Jarošík et al. 2002, 2004).

In this paper, we ask the following questions: What are the effects of temperature and nutrient availability on the RD of *A. artemisiifolia* seedlings? What is the SET and LDT for development and are they modified by nutrient availability? How does nutrient availability affect variations in the thermal window of this species? What are the effects of temperature and nutrient availability on size and biomass allocation in seedlings?

Methods

Experimental design

The seeds of *A. artemisiifolia* were collected at a ruderal site (48°43′35.0″N; 16°58′42.7″E) 1.4 km south-east of the town of Lanžhot in S Moravia, the Czech Republic, in October 2009. We collected ~1000 seeds from at least 50 plants in a population of hundreds of individuals growing in an area of ~300 m^2 at a soil dump. After collection, the seeds were stored in paper bags at room temperature. Before the experiment, the seeds were cold-stratified in wet sand in the dark at 4 °C for 3 months and then germinated at a diurnally fluctuating temperature of 25/10 °C (day/night cycle 12 h/12 h with a corresponding light/dark alternation). Germinated seeds with a radicle length of 5–10 mm were moved to containers filled with pure silica sand. Sixteen germinated seeds were planted and grown at each of the temperatures and nutrient regimes, giving a total of 336 plants (i.e. 7 temperature regimes × 3 nutrient regimes × 16 replicates). Some of the seeds did not germinate or the seedlings died during cultivation and in total 48 individuals (14.2 % of the initial set) were lost.

The seedlings were grown in growth chambers (Vötsch 1014) under identical irradiation [14/10 h light/ dark regime; photosynthetically active radiation of 360 μmol m^{-2} s^{-1}, red radiation (R, λ = 660 nm) of 26 μmol m^{-2} s^{-1} and far-red radiation (FR, λ = 730 nm) of 15 μmol m^{-2} s^{-1}, R/FR = 1.73; radiation measured using an SPh 2020 photometer from Optické dílny Turnov, Czech Republic] and air moisture (80 %) conditions. The seedlings were cultivated at constant temperatures of 10, 14, 18, 22, 26, 30 and 34 °C, which is the full range of temperatures possible using the above growth chambers. The three different nutrient levels used in our experiments were obtained using 10, 50 and 100 % Knop nutrient solution; 100 % Knop solution contained 152 mg N/L. To achieve a stable nutrient supply, conductivity of the solutions was measured daily and nutrient solution or demineralized water was added to keep the conductivity at 370, 1770, 3450 μS cm^{-1}, respectively. In addition,

the nutrient solutions were changed every second day to prevent the growth of algae.

Recording the rate of seedling development

To record the time between the appearance of the first and seventh pair of stem leaves (excluding cotyledons), the plants were checked and measured daily (following Gramig and Stoltenberg 2007). The leaves were assumed to have appeared when their size was equal to or exceeded 1 mm. After appearance of the seventh pair of leaves, height was measured and the plants harvested. The biomass was divided into shoots and roots, dried at 70 °C for 8 h and weighed.

Data analysis

The rate of development was defined as 1/(time in days between appearance of the first and seventh pair of stem leaves) = 1/d. For each nutrient level, the relationship between the RD and temperature (*t*) was fitted by a linear equation in which *a* is the intercept with the *y*-axis and *b* is the slope: RD = *a* + *bt* (Jarošík et al. 2004). The LDT, i.e. base temperature, at which the RD ceases (i.e. RD = 0), was then calculated as LDT = −*a*/*b*. The values were calculated separately for individual nutrient levels. The SET thus corresponds to 1/*b*, which indicates the number of DD above the LDT. Thermal window was expressed as the difference between the LDT, and the temperature at which the maximum RD was recorded.

The effect of temperature and nutrients was tested using analysis of variance (ANOVA). A square-root transformation was used to normalize the distributions of plant heights, total biomasses and root/shoot ratios. Logarithmic transformation was similarly used for the data on the time between appearance of the first and seventh pair of stem leaves.

Results

Both the decrease in temperature and availability of nutrients resulted in a decrease in the RD of the seedlings (Fig. 1) and increase in the time from the appearance of the first and seventh pair of stem leaves (Fig. 1). The slopes of the regression lines arising from the three nutrient levels are significantly different [analysis of covariance (ANCOVA), F = 12.37, P = 0.001]. The highest temperature of 34 °C resulted in 5.1, 5.3 and 5.8 times faster development than at 10 °C for the low, moderate and high nutrient levels, respectively. A decrease in the availability of nutrients also resulted in a decrease in the LDT from 5.4 to 3.0 °C (Table 1) and increase in the SET needed to complete this stage in the development of the plants from 392.5 to 546.9 DD (Table 1). Consequently, the thermal window for development was increased from 28.7 °C when the plants were provided with a full nutrient supply to 31.0 °C when provided with the weakest nutrient supply.

The seedlings grew taller with increase in temperature and decrease in nutrient availability (Table 2, Fig. 2A). Despite their rather quick development and increase in stem height, a marked decrease in biomass was recorded

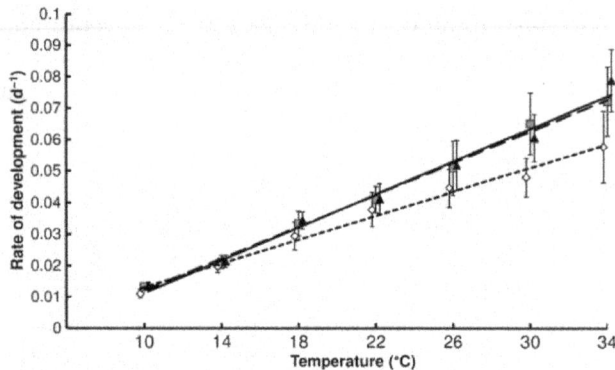

Figure 1. Dependence of the RD of *A. artemisiifolia* seedlings measured as 1/(time from the appearance of the first and seventh pair of stem leaves) on temperature; continuous line and black triangles indicate 100 % Knop solution, dashed line and grey squares 50 % solution, dotted line and white diamonds 10 % solution and bars indicate SD; for the regression equations, coefficient of determination (R^2) and calculated values—LDT and SET, see Table 1.

for seedlings grown at the highest temperature of 34 °C (Table 2, Fig. 2B). With decrease in nutrients and decrease in temperature, the seedlings allocated more biomass to roots (Table 2, Fig. 2C).

Discussion

This experiment revealed a pronounced decrease in developmental rate of *A. artemisiifolia* seedlings when reared at low temperatures and provided with weak solutions of nutrients. This suggests that physiological mechanisms at an early stage in their development contribute to shaping the current distribution of this species in the Czech Republic. The negative response to low temperature is in accordance with its current distribution only in the warmest parts of the country (Jehlík 1998; Pyšek *et al.* 2012*a*). This corresponds to the distribution pattern in Austria (Essl *et al.* 2009), but differs from that in Hungary where the occurrence correlates with low temperatures (Pinke *et al.* 2011, 2013). As the spring and summer temperatures in Hungary are within the thermal window identified by our experiments, this might indicate the importance of other environmental factors such as precipitation and soil characteristics in determining the distribution in this country (Pinke *et al.* 2011). The spring and summer temperatures in most of the Czech Republic are also within the common

Table 1. Dependence of the RD on temperature, t, fitted using linear regression. The table shows regression equations, R^2, LDT, LDT in °C and SET, SET in DD for nutrient levels equal to 10, 50 and 100 % Knop solution.

Nutrients (% Knop solution)	Regression equation	R^2	LDT ($\pm$ SD) (°C)	SET (DD)
10	RD = 0.0018t − 0.006	0.99	3.01 ± 2.13	546.9
50	RD = 0.0024t − 0.012	0.99	4.89 ± 1.29	411.8
100	RD = 0.0025t − 0.014	0.98	5.35 ± 2.06	392.5

Table 2. Effects of temperature and nutrient level on (i) the RD of *A. artemisiifolia* seedlings, expressed as the time between the appearance of the first and seventh pair of stem leaves, (ii) SET (those above the LDT) between the appearance of the first and seventh pair of stem leaves, (iii) time between the appearance of the first and seventh pair of stem leaves, (iv) height, (v) biomass and (vi) root/shoot ratio of harvested plants. Significance of the effects was tested using ANOVA, significant values are in bold. For all the variables tested, the numbers of degrees of freedom for the effect of temperature, nutrients, the interaction and residuals were equal to 1, 1, 1 and 284, respectively.

	Temperature		Nutrients		Temperature × nutrients	
	F	P	F	P	F	P
RD	2191.33	**<0.001**	63.8	**<0.001**	36.5	**<0.001**
SET (DD)	3.5	0.062	431.3	**<0.001**	24.1	**<0.001**
Developmental time (days)	932.1	**<0.001**	13.9	**<0.001**	3.6	0.059
Final seedling height	206.8	**<0.001**	4.8	**0.029**	0.4	0.555
Total biomass	50.6	**<0.001**	0.5	0.468	2.6	0.106
Root/shoot ratio	188.6	**<0.001**	265.6	**<0.001**	13.3	**<0.001**

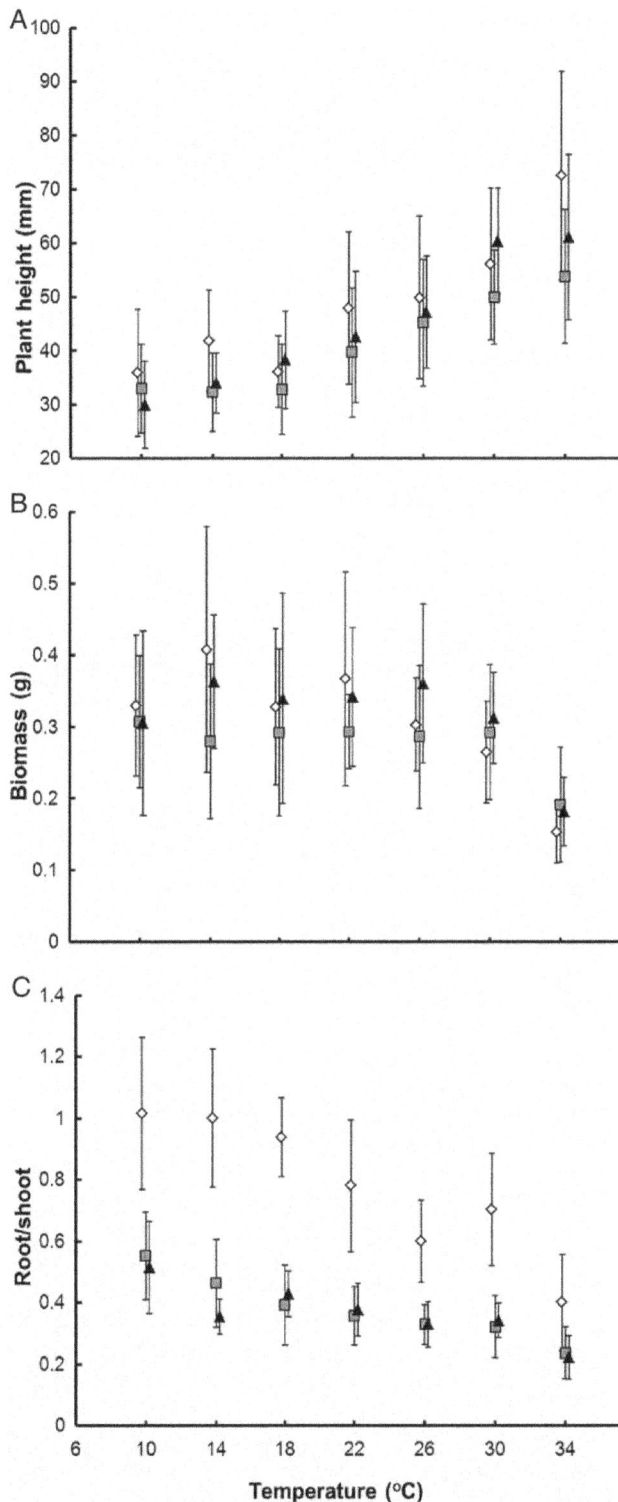

Figure 2. The height (A), biomass (B) and root/shoot ratio (C) of *A. artemisiifolia* seedlings at the seventh pair of stem leaves stage recorded at different temperatures. For significance of these dependences, see Table 2. Black triangles indicate 100 % Knop solution, grey squares 50 % solution, white diamonds 10 % solution and bars indicate SD.

ragweed's thermal window, but at low temperatures the plants' development is delayed. For *A. artemisiifolia*, an opportunistic species confined to open, disturbed and nutrient-rich ruderal sites and arable land (Pyšek *et al.* 2012*a*), the above factors are likely to influence its already rather poor competitive ability (Leskovšek *et al.* 2012*a*) in regions where temperatures and nutrient supply are suboptimal, and adversely affect the early stage of population establishment.

The maximum rate of leaf development of seedlings of *A. artemisiifolia* was recorded in our experiment at 34 °C, which is close to the 31.7 °C reported by Deen *et al.* (1998*b*) and the maximum rates of development are also similar. Here it needs to be noted that the rate of seedling development increased up to the highest temperature we used, and testing the response to temperatures above 34 °C was not possible due to the technical limitations of the growth chambers. Nevertheless, the abrupt decrease in biomass recorded at the highest temperature and taking into account the results of Deen *et al.* (1998*b*) it is reasonably certain that 34 °C is indeed likely to be the maximum temperature at which *A. artemisiifolia* seedlings can develop. Our results also reveal that the optimum temperature for leaf development is much higher than that for seed germination, which is 18.3 °C (Leiblein-Wild *et al.* 2014). On the other hand, the temperature range of germination, 32.5 °C, i.e. from 2 to 34.5 °C, is similar to that for seedling development (Leiblein-Wild *et al.* 2014).

The width of the thermal window for seedling development varied from 28.7 to 31.0 °C depending on the nutrient supply, which differs from the theoretical prediction based on the biochemical kinetics of metabolism of a constant width of 20 °C (Gillooly *et al.* 2002; Charnov and Gillooly 2003). The deviation from this theoretical prediction, as well as the recorded variation in the RD, LDT and SET, is associated with the availability of nutrients; to the best of our knowledge, this is the first study assessing the effect of an environmental factor on thermal constants of a plant. Still, there are some commonalities if our results are compared with those for insects; the increased rate of seedling development at high nutrient supply is analogous to faster development of ladybirds provided with either more and/or better quality food (Hodek *et al.* 2012; Jarošík *et al.* 2014). For plants, the accelerating effect of nutrients on the development of *A. artemisiifolia* corresponds to the faster leaf development previously recorded at higher levels of radiance for this species (Deen *et al.* 1998*b*), or the increase in the rate of leafing and tillering of the tropical grass *Brachiaria brizantha* recorded when provided with nitrogen (N) and sulfur (S) fertilizers (de Bona and Monteiro 2010). Unlike in aphids and ladybirds (Dixon *et al.* 2013; Jarošík *et al.* 2014), the increased nutrient supply did not result in a decrease in the LDT; on the contrary, nutrient-limited seedlings of the common ragweed had lower LDT. However, it needs to be noted that we derived the LDT from a linear model of plant development and possible non-linearity may occur at temperatures <10 °C. The thermal window of *A. artemisiifolia* is also very wide compared with experimentally obtained windows of a large set of both native and invasive species in central Europe (L. Moravcová and H. Skálová, unpubl. data). Whether a flexible thermal window is a trait associated with invasiveness, which resulted from the evolution of highly invasive genotypes of *A. artemisiifolia* due to seed-mediated gene flow promoted by agricultural disturbance during the westward expansion of human populations in the USA (Martin *et al.* 2014), requires further study. The same holds for possible development of genotypes with wider thermal windows or those shifted towards lower temperatures driven by the existence of locally adapted genotypes, similar to those reported to occur in *A. artemisiifolia* for salinity (DiTommaso 2004).

The faster increase in height of *A. artemisiifolia* seedlings recorded at high temperatures may prevent them from being shaded by neighbouring vegetation, and increase the competitive ability of plants in early stages of population establishment. On the other hand, the tall seedlings that developed at high temperatures weighed less and were weak plants, which may have an opposite effect. In addition, the competitiveness at high temperatures is probably further decreased by reduced allocation to the roots, which constrains nutrient acquisition. The negative effect of nutrient limitation on the performance of *A. artemisiifolia* and the interaction of nutrient availability with temperature explains why this species is confined to nutrient-rich disturbed habitats (Chauvel *et al.* 2006; Essl *et al.* 2009; Šikoparija *et al.* 2009).

Conclusions

We found that *A. artemisiifolia* plants grow within a temperature range exceeding the 20 °C thermal window, predicted based on the biochemical kinetics of metabolism. The LDT and SET were influenced by growing condition, which contradicts the thermal constant concept. Our results highlight temperature as the main determinant of common ragweed's distribution and identify nutrient availability as a factor that results in the realized niche being smaller than the fundamental niche; both of which need to be taken into account when predicting the future spread of *A. artemisiifolia*.

Sources of Funding

This research was funded by grants GACR 206/09/0563 and 14-36098G, long-term research development project no. RVO 67985939 from the Czech Academy of Sciences, and by the Praemium Academiae award to P.P.

Contributions by the Authors

Conceived and designed the experiments: H.S., L.M., P.P. and A.F.G.D. Performed the experiments: H.S. and L.M. Analysed the data: P.K. and H.S. Contributed reagents/materials/analysis tools: H.S., L.M., P.P. and P.K. Wrote the paper: H.S., L.M., P.P., P.K. and A.F.G.D.

Acknowledgements

We thank the journal editors and two anonymous reviewers for their valuable comments on the manuscript, the late Vojta Jarošík for consultation on the thermal time concept and the experiment design and Michal Pyšek, Vendula Havlíčková and Šárka Dvořáčková for logistic support.

Literature Cited

Baskin JM, Baskin CC. 1980. Ecophysiology of secondary dormancy in seeds of *Ambrosia artemisiifolia*. *Ecology* **61**:475–480.

Bazzaz FA. 1974. Ecophysiology of *Ambrosia artemisiifolia*: a successional dominant. *Ecology* **55**:112–119.

Blackburn TM, Pyšek P, Bacher S, Carlton JT, Duncan RP, Jarošík V, Wilson JRU, Richardson DM. 2011. A proposed unified framework for biological invasions. *Trends in Ecology and Evolution* **26**: 333–339.

Brandes D, Nitzsche J. 2007. Biology, introduction, dispersal, and distribution of common ragweed (*Ambrosia artemisiifolia* L.) with special regard to Germany. *Nachrichtenblatt des Deutschen Pflanzenschutzdienstes* **58**:286–291.

Chapman DS, Haynes T, Beal S, Essl F, Bullock JM. 2014. Phenology predicts the native and invasive range limits of common ragweed. *Global Change Biology* **20**:192–202.

Charnov EL, Gillooly J. 2003. Thermal time: body size, food quality and the 10°C rule. *Evolutionary Ecology Research* **5**:43–51.

Chauvel B, Dessaint F, Cardinal-Legrand C, Bretagnolle F. 2006. The historical spread of *Ambrosia artemisiifolia* L. in France from herbarium records. *Journal of Biogeography* **33**:665–673.

Cunze S, Leiblein MC, Tackenberg O. 2013. Range expansion of *Ambrosia artemisiifolia* in Europe is promoted by climate change. *ISRN Ecology* **2013**: Article 610126.

de Bona FD, Monteiro FA. 2010. The development and production of leaves and tillers by *Marandu palisadegrass* fertilised with nitrogen and sulphur. *Tropical Grasslands* **44**:192–201.

Deen W, Hunt LA, Swanton CJ. 1998a. Photothermal time describes common ragweed (*Ambrosia artemisiifolia* L.) phenological development and growth. *Weed Science* **46**:561–568.

Deen W, Hunt T, Swanton CJ. 1998b. Influence of temperature, photoperiod, and irradiance on the phenological development of common ragweed (*Ambrosia artemisiifolia*). *Weed Science* **46**:555–560.

DiTommaso A. 2004. Germination behavior of common ragweed (*Ambrosia artemisiifolia*) populations across a range of salinities. *Weed Science* **52**:1002–1009.

Dixon AFG, Honěk A, Jarošík V. 2013. Physiological mechanism governing slow and fast development in predatory ladybirds. *Physiological Entomology* **38**:26–32.

Dullinger S, Kleinbauer I, Peterseil J, Smolik M, Essl F. 2009. Niche based distribution modelling of an invasive alien plant: effects of population status, propagule pressure and invasion history. *Biological Invasions* **11**:2401–2414.

Essl F, Dullinger S, Kleinbauer I. 2009. Changes in the spatio-temporal patterns and habitat preferences of *Ambrosia artemisiifolia* during its invasion of Austria. *Preslia* **81**:119–133.

Fumanal B, Chauvel B, Bretagnolle F. 2007. Estimation of pollen and seed production of common ragweed in France. *Annals of Agricultural and Environmental Medicine* **14**:233–236.

Fumanal B, Gaudot I, Bretagnolle F. 2008. Seed-bank dynamics in the invasive plant, *Ambrosia artemisiifolia* L. *Seed Science Research* **18**:101–114.

Gillooly JF, Charnov EL, West GB, Savage VM, Brown JH. 2002. Effects of size and temperature on developmental time. *Nature* **417**:70–73.

Gramig GG, Stoltenberg DE. 2007. Leaf appearance base temperature and phyllochron for common grass and broadleaf weed species. *Weed Technology* **21**:249–254.

Granier C, Massonnet C, Turc O, Muller B, Chenu K, Tardieu F. 2002. Individual leaf development in *Arabidopsis thaliana*: a stable thermal-time-based programme. *Annals of Botany* **89**:595–604.

Grotkopp E, Rejmánek M. 2007. High seedling relative growth rate and specific leaf area are traits of invasive species: phylogenetically independent contrasts of woody angiosperms. *American Journal of Botany* **94**:526–532.

Hodek I, van Emden HF, Honěk A, eds. 2012. *Ecology and behaviour of the ladybird beetles (Coccinellidae)*. Chichester: Wiley-Blackwell.

Jarošík V, Honěk A, Dixon AFG. 2002. Developmental rate isomorphy in insects and mites. *The American Naturalist* **160**:497–510.

Jarošík V, Kratochvíl L, Honěk A, Dixon AFG. 2004. A general rule for the dependence of developmental rate on temperature in ectothermic animals. *Proceedings of the Royal Society of London B: Biological Sciences* **271**:S219–S221.

Jarošík V, Kumar G, Omkar, Dixon AFG. 2014. Are thermal constants constant? A test using two species of ladybird. *Journal of Thermal Biology* **40**:1–8.

Jehlík V. 1998. *Cizí expanzivní plevele České republiky a Slovenské republiky [Alien expansive weeds of the Czech Republic and Slovak Republic]*. Praha: Academia.

Kartesz JT, Meacham CA. 1999. *Synthesis of the North American flora*. Version 1.0. Chapel Hill: North Carolina Botanical Garden.

Kömives T, Béres I, Reisinger P, Lehoczky E, Berke J, Tamás J, Páldy A, Csornai G, Nándor G, Kardeván P, Mikulás J, Gólya G, Molnar J. 2006. New strategy of the integrated protection against common ragweed (*Ambrosia artemisiifolia* L). *Hungarian Weed Research and Technology* **6**:5–50.

Lambdon PW, Pyšek P, Basnou C, Hejda M, Arianoutsou M, Essl F, Jarošík V, Pergl J, Winter M, Anastasiu P, Andriopoulos P, Bazos I, Brundu G, Celesti-Grapow L, Chassot P, Delipetrou P, Josefsson M, Kark S, Klotz S, Kokkoris Y, Kuehn I, Marchante H, Perglova I, Pino J, Vila M, Zikos A, Roy D, Hulme PE. 2008. Alien

flora of Europe: species diversity, temporal trends, geographical patterns and research needs. *Preslia* **80**:101–149.

Leiblein-Wild MC, Kaviani R, Tackenberg O. 2014. Germination and seedling frost tolerance differ between the native and invasive range in common ragweed. *Oecologia* **174**:739–750.

Leskovšek R, Eler K, Batič F, Simončic A. 2012*a*. The influence of nitrogen, water and competition on the vegetative and reproductive growth of common ragweed (*Ambrosia artemisiifolia* L.). *Plant Ecology* **213**:769–781.

Leskovšek R, Datta A, Simončic A, Knezevic SZ. 2012*b*. Influence of nitrogen and plant density on the growth and seed production of common ragweed (*Ambrosia artemisiifolia* L.). *Journal of Pest Science* **85**:527–539.

Martin MD, Zimmer EA, Olsen MT, Foote AD, Gilbert MTP, Brush GS. 2014. Herbarium specimens reveal a historical shift in phylogeographic structure of common ragweed during native range disturbance. *Molecular Ecology* **23**:1701–1716.

Moravcová L, Pyšek P, Jarošík V, Havlíčková V, Zákravský P. 2010. Reproductive characteristics of neophytes in the Czech Republic: traits of invasive and non-invasive species. *Preslia* **82**:365–390.

Morrison JA, Mauck K. 2007. Experimental field comparison of native and non-native maple seedlings: natural enemies, ecophysiology, growth and survival. *Journal of Ecology* **95**:1036–1049.

Pinke G, Karácsony P, Czúcz B, Botta-Dukát Z. 2011. Environmental and land-use variables determining the abundance of *Ambrosia artemisiifolia* in arable fields in Hungary. *Preslia* **83**:219–235.

Pinke G, Karácsony P, Botta-Dukát Z, Czúcz B. 2013. Relating *Ambrosia artemisiifolia* and other weeds to the management of Hungarian sunflower crops. *Journal of Pest Science* **86**:621–631.

Pyšek P, Chytrý M, Pergl J, Sádlo J, Wild J. 2012*a*. Plant invasions in the Czech Republic: current state, introduction dynamics, invasive species and invaded habitats. *Preslia* **84**:575–630.

Pyšek P, Danihelka J, Sádlo J, Chrtek J, Chytrý M, Jarošík V, Kaplan Z, Krahulec F, Moravcová L, Pergl J, Štajerová K, Tichý L. 2012*b*. Catalogue of alien plants of the Czech Republic (2nd edition): checklist update, taxonomic diversity and invasion patterns. *Preslia* **84**:155–255.

Richardson DM, Pyšek P, Rejmánek M, Barbour MG, Panetta FD, West CJ. 2000. Naturalization and invasion of alien plants: concepts and definitions. *Diversity & Distributions* **6**:93–107.

Richter R, Dullinger S, Essl F, Leitner M, Vogl G. 2013*a*. How to account for habitat suitability in weed management programmes? *Biological Invasions* **15**:657–669.

Richter R, Berger UE, Dullinger S, Essl F, Leitner M, Smith M, Vogl G. 2013*b*. Spread of invasive ragweed: climate change, management and how to reduce allergy costs. *Journal of Applied Ecology* **50**:1422–1430.

Rybníček K, Jäger S. 2001. *Ambrosia* (Ragweed) in Europe. *ACI International* **13**:60–66.

Šikoparija B, Smith M, Skjøth CA, Radišić P, Milkovska S, Šimić S, Brandt J. 2009. The Pannonian plain as a source of *Ambrosia* pollen in the Balkans. *International Journal of Biometeorology* **53**: 263–272.

Skálová H, Havlíčková V, Pyšek P. 2012. Seedling traits, plasticity and local differentiation as strategies of invasive species of *Impatiens* in central Europe. *Annals of Botany* **110**:1429–1438.

Storkey J, Stratonovitch P, Chapman DS, Vidotto F, Semenov MA. 2014. A process-based approach to predicting the effect of cli-

12

Attract them anyway: benefits of large, showy flowers in a highly autogamous, carnivorous plant species

A. Salces-Castellano[1,2,†], M. Paniw[1,†], R. Casimiro-Soriguer[1] and F. Ojeda[1]*

[1] Departamento de Biología and IVAGRO, Universidad de Cádiz, Campus Río San Pedro, E-11510 Puerto Real, Spain
[2] Present address: IPNA-CSIC, C/Astrofísico Francisco Sánchez 3, 38206-La Laguna, Tenerife, Canary Islands, Spain

Associate Editor: Dennis F. Whigham

Abstract. Reproductive biology of carnivorous plants has largely been studied on species that rely on insects as pollinators and prey, creating potential conflicts. Autogamous pollination, although present in some carnivorous species, has received less attention. In angiosperms, autogamous self-fertilization is expected to lead to a reduction in flower size, thereby reducing resource allocation to structures that attract pollinators. A notable exception is the carnivorous pyrophyte *Drosophyllum lusitanicum* (Drosophyllaceae), which has been described as an autogamous selfing species but produces large, yellow flowers. Using a flower removal and a pollination experiment, we assessed, respectively, whether large flowers in this species may serve as an attracting device to prey insects or whether previously reported high selfing rates for this species in peripheral populations may be lower in more central, less isolated populations. We found no differences between flower-removed plants and intact, flowering plants in numbers of prey insects trapped. We also found no indication of reduced potential for autogamous reproduction, in terms of either seed set or seed size. However, our results showed significant increases in seed set of bagged, hand-pollinated flowers and unbagged flowers exposed to insect visitation compared with bagged, non-manipulated flowers that could only self-pollinate autonomously. Considering that the key life-history strategy of this pyrophytic species is to maintain a viable seed bank, any increase in seed set through insect pollinator activity would increase plant fitness. This in turn would explain the maintenance of large, conspicuous flowers in a highly autogamous, carnivorous plant.

Keywords: Autogamous selfing; *Drosophyllum lusitanicum*; floral display; pollination biology; prey capture; pyrophyte; seed set.

Introduction

Carnivorous plants have long captivated naturalists and scientists worldwide (Chase *et al.* 2009; Król *et al.* 2012). Charles Darwin himself was most fascinated by them and was the first to demonstrate plant carnivory experimentally (Darwin 1875). Carnivory has evolved several times independently in the angiosperms and ~600 species of carnivorous plants can be found today across the globe, most prominently in tropical and temperate regions (Heubl *et al.* 2006; Ellison and Gotelli 2009). They are largely restricted to infertile, wet, open habitats (Givnish *et al.* 1984) where they have adapted to extremely low nutrient levels by evolving elaborately modified leaves

* Corresponding author's e-mail address: fernando.ojeda@uca.es
† These authors have contributed equally.

that trap small animals, mainly insects, as prey (Ellison and Gotelli 2001, 2009; Gibson and Waller 2009) and absorb the necessary mineral nutrients from them, particularly nitrogen and phosphorus (Adamec 1997).

Since most carnivorous plants are also entomophilous (i.e. they rely on pollinating insects to facilitate sexual reproduction), a pollinator–prey conflict might occur if they trapped potentially efficient pollinators (Zamora 1999; Ellison and Gotelli 2001). However, there are mechanisms in carnivorous plants to avoid or minimize this conflict, such as separation (spatial or temporal) of flowers from leaf traps to avoid pollinators being trapped as prey, or the occurrence of autogamous self-pollination to become somewhat independent of the role of insect vectors for reproduction (Ellison and Gotelli 2001; Jürgens et al. 2012). Autogamous self-pollination is actually common in some species from different carnivorous genera (see references in Jürgens et al. 2012).

Drosophyllum lusitanicum (Drosophyllaceae), the only extant species of the family Drosophyllaceae (Heubl et al. 2006), is an example of autogamous self-pollination in carnivorous plant species (Ortega-Olivencia et al. 1995, 1998). This species (*Drosophyllum*, hereafter) is endemic to the western Iberian Peninsula and northern Morocco (Garrido et al. 2003; Paniw et al. 2015), where it is restricted to acidic, nutrient-poor Mediterranean heathlands (Müller and Deil 2001; Adlassnig et al. 2006) and tightly associated to post-fire habitats (Correia and Freitas 2002; Paniw et al. 2015). *Drosophyllum* is a short-lived subshrub up to 45 cm tall with circinate, linear leaves grouped in dense rosettes and covered with stalked mucilage-producing glands (Paiva 1997). It produces large, sulfur-yellow, hermaphrodite flowers, radiate and pentamerous, borne in stalked, cymose inflorescences (Paiva 1997; Correia and Freitas 2002; Fig. 1). Flowers are homogamous, i.e. possess a spatial and temporal closeness between dehiscing anthers and receptive stigmas, with high selfing capability even in pre-anthesis (Ortega-Olivencia et al. 1995, 1998).

It is well established that autogamous selfing in angiosperms is favoured under pollinator limitation (Schemske and Lande 1985; Morgan and Wilson 2005), and it is usually accompanied by morphological changes in floral traits such as the occurrence of homogamy and a dramatic reduction in corolla size (Goodwillie et al. 2010; Sicard and Lenhard 2011). This reduction in flower size and other floral traits (e.g. showiness) is explained as a way to minimize resource allocation to floral display when pollinator attraction is no longer necessary (e.g. Andersson 2005; Celedón-Neghme et al. 2007). However, one of the noticeable features of the autogamous *Drosophyllum* is the production of large, showy flowers on pedunculed inflorescences (Fig. 1). Therefore, considering the

high allocation costs of flower production (Galen 1999; Andersson 2005), what are the benefits of large, conspicuous flowers in a carnivorous plant species presumably independent of the role of pollinating insects for reproduction (Ortega-Olivencia et al. 1995)?

Here, we present two field experiments on the floral and reproductive biology of *Drosophyllum* aimed to determine fitness benefits from the production of large, conspicuous flowers. First, assuming independence of pollinating insects for reproduction (Ortega-Olivencia et al. 1995), we explored whether the large, bright yellow corollas in this carnivorous species act as attracting devices for enhancing prey capture onto the sticky leaf traps, thereby supporting plant growth. Although there is virtually no overlap between prey and flower-visiting insect faunas (Bertol et al. 2015), it is well established that the bright yellow colour is attractive to many insect species, particularly flies (e.g. Neuenschwander 1982; Yee 2015), which are the most common prey in *Drosophyllum* (Bertol et al. 2015). Specifically, we hypothesized that flowering *Drosophyllum* plants whose flowers are removed would trap fewer prey insects than co-occurring, intact flowering plants, which would indicate an increase in plant fitness through insect capture resulting from maintenance of large, yellow flowers.

Second, we conducted a controlled pollination experiment to investigate the actual contribution of pollinators to fecundity (i.e. seed production) of this species. Unlike previous pollination experiments on this species (Ortega-Olivencia et al. 1995, 1998), which have been performed in geographically isolated, small populations, our experimental populations were located in the northern side of the Strait of Gibraltar, where populations are larger and more abundant (Garrido et al. 2003; Paniw et al. 2015). Since marginal populations of normally outcrossing plant species frequently show a considerable increase in the selfing rate (Lloyd 1980; Pujol et al. 2009), the highly autogamous self-fertilization of *Drosophyllum* reported previously might be contingent on geographical isolation. We predicted that attraction of pollinating insects by *Drosophyllum* flowers would increase fitness through an increase in fecundity in this carnivorous species, thus accounting for its large, conspicuous flowers.

Methods

Ecology of *Drosophyllum*

Drosophyllum is a disturbance-adapted, carnivorous species, colonizing (from a persistent seed bank) recently burned heathlands or heathland patches where small-scale disturbances create open space (Garrido et al. 2003; Paniw et al. 2015). Within 4–6 years after fire, regenerating heathland shrubs outcompete above-ground

Figure 1. Visual description of *Drosophyllum*. (A) Young reproductive individual with a single rosette of leaves and a stalked inflorescences with two open flowers. (B) Lateral view of the flower showing the five large, bright yellow petals (scale bar = 10 mm). (C) Frontal view of the flower, showing the homogamous lack of separation between anthers and stigmas (scale bar = 10 mm). (D) Schematic description of the plant's life cycle.

Drosophyllum individuals, making the formation of a seed bank—in which populations may persist for several decades until another fire—a critical life-history strategy (Paniw *et al.* 2015; M. Paniw, P. Quintana-Ascencio, F. Ojeda and R. Salguero-Gómez, unpublished). In habitats where small-scale disturbances, e.g. browsing, create and maintain open space, individuals may reach up to 10 years of age (Juniper *et al.* 1989). Individuals grow in rosettes, and number of rosettes is a good proxy for age. Plants 1–2 rosettes in size initially reproduce in the second year after emergence and the number of rosettes per plant increases each growing season (Ortega-Olivencia *et al.* 1995; Garrido *et al.* 2003; Fig. 1D). Demographic censuses of

populations across southern Spain determined that each rosette produces one floral scape with an average ($\pm$ SD) of 3.5 $\pm$ 2.1 flowers (M. Paniw, P. Quintana-Ascencio, F. Ojeda and R. Salguero-Gómez, unpublished). Bright sulfur-yellow flowers on each scape open gradually and last 1 day in full anthesis, so that no more than two flowers per rosette are in anthesis at the same time (Fig. 1). Flowers are large (Correia and Freitas 2002), with an average petal length of 2.84 $\pm$ 0.21 cm and petal width of 1.89 $\pm$ 0.17 cm (A. Salces-Castellano, unpubl. data), and show high autogamy rates (Ortega-Olivencia *et al.* 1995, 1998). High autogamy is also supported by the high inbreeding coefficients found in *Drosophyllum* populations

(Paniw *et al.* 2014). Each flower produces a dehiscent capsule with an average of 9.8 ± 2.4 seeds (M. Paniw, P. Quintana-Ascencio, F. Ojeda and R. Salguero-Gómez, unpublished).

Study region and sites

Two field experiments were conducted in five natural *Drosophyllum* populations, located at five sites within the southern Aljibe Mountains, at the European side of the Strait of Gibraltar (Table 1; Fig. 2). From all its distribution range, this is where *Drosophyllum* is more abundant and populations are largest (Garrido *et al.* 2003; Paniw *et al.* 2015). This region is characterized by a mild Mediterranean climate (~18 °C mean annual temperature and ~1200 mm annual rainfall) and a rough topography dominated by Oligo-Miocene sandstone mountains and hills, which produce acidic, nutrient-poor soils in ridges and upper slopes (Ojeda *et al.* 2000). These infertile soils are covered by Mediterranean heathlands, dominated by dwarf shrubs like *Erica australis*, *Pterospartum tridentatum*, *Quercus lusitanica*, *Calluna vulgaris* and *Halimium lasianthum*, and are the primary habitat of *Drosophyllum* (Müller and Deil 2001; Paniw *et al.* 2015). Although this species is highly pyrophytic (i.e. associated with the recurrent presence of fire) and therefore threatened by large-scale anthropogenic activities such as afforestation (Andrés and Ojeda 2002) and fire suppression (Correia and Freitas 2002), it profits from small-scale vegetation clearances, where populations can still thrive (Garrido *et al.* 2003; Paniw *et al.* 2015).

We chose the study sites to represent the most common habitats of *Drosophyllum* populations (Paniw *et al.* 2015). Monte Murta is an open, rocky sandstone ridge with sparse heathland vegetation, which had been mechanically removed about 30 years ago for pine afforestation. In 2014, the *Drosophyllum* population consisted

of ~5000 individuals, where young flowering plants, consisting of 1–2 rosettes, and old flowering plants (>2 rosettes) co-occurred. Sierra Carbonera is a regenerating heathland patch from a fire suffered in early autumn 2011. The *Drosophyllum* population here was also large (~3000 individuals) and consisted mainly of young flowering plants (2–3 years old), plus juveniles and a few seedlings. Montera del Torero is an old firebreak line across a heathland created by mechanical clearance of the vegetation. The *Drosophyllum* population at this site consisted of ~3700 individuals and has persisted for >30 years, being dominated by old (>5 years) flowering plants. Lastly, two populations with different relative abundance of old reproductive individuals were encountered in Monte Retin. The population in Monte Retin North has persisted for >20 years in an open heathland on a rocky sandstone ridge. It consisted of ~1500 individuals where old and young flowering individuals co-occurred. The population in Monte Retin South is found on a regenerating heathland patch from a fire suffered in early autumn 2010. This population, which has been heavily disturbed by cattle grazing and trampling, consisted of ~500 individuals, with an even distribution of young and old reproductive individuals.

Flower contribution to prey attraction

To test whether flowers in *Drosophyllum* functioned to attract prey insects, we carried out a field experiment at three of the five study sites, Monte Murta, Monte Retin North and Monte Retin South (Fig. 2) in April 2014, during peak flowering. At each site, we located 'isolated' flowering plants growing in open microhabitats (>1 m from the nearest conspecific and >30 cm from the nearest interspecific neighbour), in order to avoid potential influences of conspecific flowering neighbours on prey capture. We randomly marked 14 plants and recorded the number

Table 1. Description of sites used in the flower removal and pollination experiments quantifying the role of *Drosophyllum* flowers in prey capture and pollinator attraction, respectively. *N*, total number of *Drosophyllum* individuals found in 2014.

Site	Location	Experiment	Site characteristics	Population characteristics
Monte Murta	36°19′16″N 5°33′03″W	Flower removal	Open, rocky sandstone ridge	N = 5000; mixed-aged population
Monte Retin North	36°11′53″N 5°49′25″W	Flower removal	Open heathland patch	N = 1500; mixed-aged population
Monte Retin South	36°10′23″N 5°50′53″W	Flower removal	Post-fire regenerating heathland (fire 2010); browsed and trampled by cattle	N = 500; mixed-aged population
Sierra Carbonera	36°12′35″N 5°21′37″W	Pollination	Post-fire regenerating heathland (fire 2011)	N = 3000; mainly young reproductive individuals
Montera del Torero	36°13′35″N 5°35′08″W	Pollination	Mechanically built firebreak	N = 3700; mainly old reproductive individuals

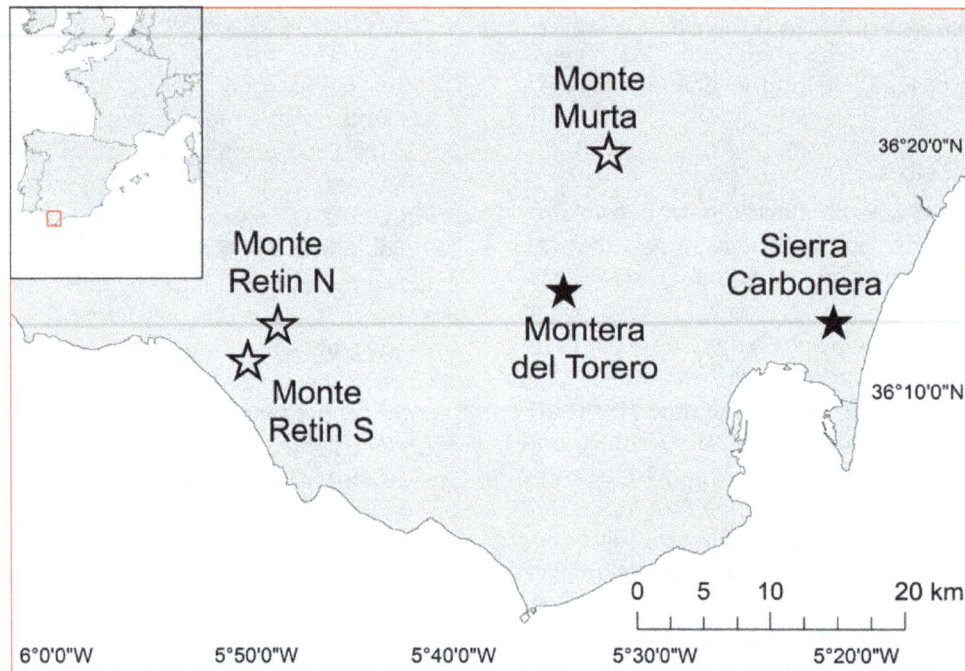

Figure 2. Study area and location of the sites where the flower contribution to prey attraction (open star) and pollination experiments (filled star) were performed. See Table 1 for detailed description of the *Drosophyllum* populations at each site.

of rosettes and leaves per rosette of each plant. All prey insects were then carefully hand-removed with tweezers from each plant. Next, we randomly selected 7 plants out of those 14 and removed all their flowers by cutting off the inflorescence stalks with scissors. After 1 week, we returned to each of the three populations and recorded the number of prey insects attached to the leaves of the 14 plants.

We analysed the differences in insect capture between flower-removed plants (treatment) and intact ones (control) for each site separately by fitting a generalized linear model with a Poisson error distribution on the total number of insects, using the 'flower-cut' treatment as fixed effect and total number of leaves per plant as the offset. Using an offset allowed us to treat the response (number of insects) as proportions (insects per leaf) but allowing the models to be fit as count data in a generalized linear mixed model framework. The analyses were performed separately for each site because we did not have enough spatial replicates to include site as a random effect in our models (Bolker *et al.* 2009).

Pollination experiment

We carried out an experiment at two of the five study sites, Sierra Carbonera and Montera del Torero (Fig. 2), to investigate the contribution of pollinators to *Drosophyllum* fecundity (i.e. seed production). In mid-April 2014, at the beginning of the flowering season, we labelled 56 and 43 plants in Sierra Carbonera and Montera del Torero, respectively. On each plant, flowers were randomly

assigned to one of four treatments: hand cross-pollination (HCP), hand self-pollination (HSP), spontaneous self-pollination (SSP) and control or open pollination (OP). In the first three treatments, flowers were covered with nylon-mesh bags (0.15-mm mesh) before anthesis to exclude potential insect visitors. For the two hand-pollination treatments, HCP and HSP, we collected ripe anthers from plants separated >300 m (HCP) or from the same flower (HSP) and brushed the stigmas with them, taking care of bagging them back after this artificial pollination. Flowers in the SSP treatment were not hand-pollinated and remained bagged in order to account for spontaneous autogamy. Finally, flowers in the OP treatment (control) were left exposed to natural pollinator activity. In most plants, there was more than one flower for each treatment. We also collected a single petal from an extra flower per plant to measure petal length as a surrogate for flower size.

In July 2014, after fruit (capsule) ripening and before seed dispersal (dehiscence), we collected the fruits of the four treatments on each individual plant from the two sites. They were stored individually in labelled paper bags and taken to the laboratory, where we calculated fruit set (percentage of flowers within each treatment developing into fruits) and seed set (percentage of ovules per flower maturing into seeds) per treatment. Additionally, three randomly chosen seeds per fruit were weighed on an electronic balance to the nearest 0.1 mg and their length (as a surrogate for size) measured using an image analyser (Leica Application Suite v4.4.0, LAS v4.4, Leica Microsystems).

We tested for differences in fruit set, seed set, seed weight and seed size among pollination treatments by means of a mixed effect models with a binomial error distribution for the response variables fruit set and seed set and normal error distribution for the response variables seed weight and size. We considered treatment (OP, HCP, SCP and SSP) as fixed effect and plant individual as a random effect in all models. We fitted the models for each of the two sites separately.

All analyses were performed with R software (R Development Core Team 2015). We used the R package lme4 (Bates *et al.* 2013) to fit the mixed effect models. In both experiments described above, we used likelihood ratio tests to determine significant differences between treatments (Vuong 1989). These tests compare the log-likelihoods of increasingly complex, or nested, models to ones of simpler models (starting with intercept-only models) and determine the significance of the deviance between the log-likelihoods using a χ^2 test. When significant differences between treatment levels were found, a *post hoc* Tukey's honestly significant difference (HSD) test was applied to the linear predictors using the R package multcomp (Hothorn *et al.* 2008) to detect significant pairwise differences between treatments.

Results

Flower contribution to prey attraction

Overall, insect capture levels differed between the three sites, being considerably higher in Monte Retin South (Fig. 3). However, we detected no significant differences in insect capture rates between 'flower-removed' plants and control plants across the three sites (Table 2).

Pollination experiment

Flowers had an overall smaller size (i.e. petal length) in *Drosophyllum* plants from Montera del Torero (average petal length $\pm$ SD: 2.64 $\pm$ 0.89 cm) than in those from Sierra Carbonera (2.98 $\pm$ 0.59 cm; Welch's *t*-test: $t_{64.57} =$ 6.46, $P < 0.0001$).

Fruit set was very high in *Drosophyllum*, with no differences across the four treatments in the two sites (Table 3) and almost 100 % flowers developing into fruits (Table 4). In contrast, we detected significant differences in seed set among treatments in the two study sites (Table 3). These significant differences were due to the OP treatment, which produced significantly higher seed set than the other three treatments in Montera del Torero (but not in Sierra Carbonera; Table 4; Fig. 4), and particularly the SSP treatment, which produced significantly lower seed set values than the other three treatments at both sites

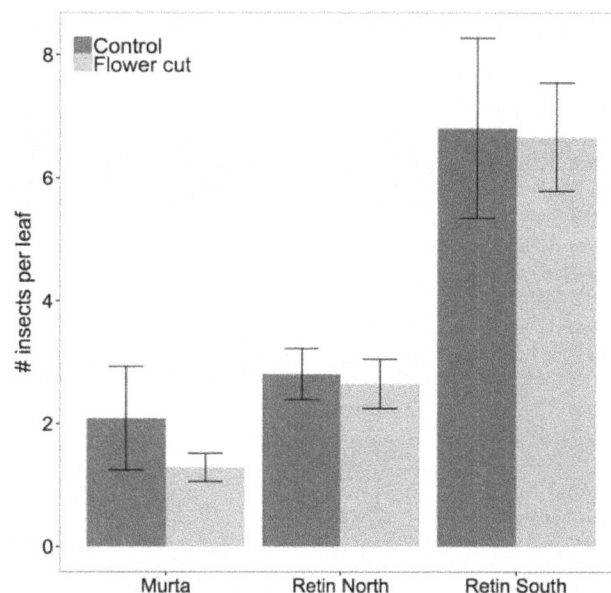

Figure 3. Average number of insects per leaf ($\pm$ SE) at three sites (Monte Murta, Monte Retin North and Monte Retin South) caught by seven intact flowering plants (control; dark grey bar) and seven plants whose flowers were removed (flower cut; light grey bar).

Table 2. Results of the likelihood ratio tests for all considered models testing the role of *Drosophyllum* flowers in attracting insects as prey. The response variable (number of insects/leaf) was measured in a field experiment performed at three sites. For each response, a likelihood ratio test compares nested models assuming a chi-square distribution, χ^2, with the critical value given by the model deviance, D, and the degrees of freedom, df, corresponding to the difference in parameters between the models compared.

Site	Model	df	χ^2	D	P
Murta	Intercept		206.9		
	Flower cut	1	204.7	2.2	0.14
Retin North	Intercept		97.6		
	Flower cut	1	95.8	1.9	0.17
Retin South	Intercept		544.7		
	Flower cut	1	541.8	2.7	0.11

as determined by the HSD test (Table 4; Fig. 4). Seeds were larger and heavier in Sierra Carbonera than in Montera del Torero (Table 4). However, while seeds from the OP treatment in Montera del Torero produced slightly but significantly smaller seeds, no differences in seed size nor weight were detected among treatments in Sierra Carbonera (Tables 3 and 4).

Discussion

Although there are no closely related extant species to *Drosophyllum* for comparison (Heubl *et al.* 2006), its

Table 3. Results of the likelihood ratio tests for all considered models testing the role of *Drosophyllum* flowers in attracting insects as pollinators. The response variables (fruit set, seed set, seed size and seed weight) were measured in a field experiment performed at two sites (Sierra Carbonera and Montera del Torero). For each response, a likelihood ratio test compares nested models assuming a chi-square distribution, χ^2, with the critical value given by the model deviance, D, and the degrees of freedom, df, corresponding to the difference in parameters between the models compared. Significant differences between models are in bold.

Site	Model	df	χ^2	D	P
Response variable: fruit set					
Sierra Carbonera	Intercept		24.6		
	Pollination	3	22.6	2	0.58
Montera del	Intercept		13.2		
Torero	Pollination	3	11.8	1.4	0.71
Response variable: seed set					
Sierra Carbonera	Intercept		2817.8		
	Pollination	**3**	**2706.6**	**140.4**	**<0.01**
Montera del	Intercept		2156.7		
Torero	**Pollination**	**3**	**1958.4**	**198.3**	**<0.01**
Response variable: seed size					
Sierra Carbonera	Intercept		279.3		
	Pollination	3	278.5	0.7	0.9
Montera del	Intercept		291.5		
Torero	**Pollination**	**3**	**282.3**	**9.1**	**0.03**
Response variable: seed weight					
Sierra Carbonera	Intercept		306.4		
	Pollination	3	305.4	1.0	0.8
Montera del	Intercept		128.7		
Torero	**Pollination**	**3**	**110.6**	**18.1**	**<0.01**

large, bright yellow flowers seem to contradict the paradigm of dramatic flower size reduction in highly autogamous angiosperms (Goodwillie *et al.* 2010; Sicard and Lenhard 2011). Considering the presumably high allocation costs of flower production (e.g. Galen 1999; Andersson 2005), we have explored the advantages or benefits that large, conspicuous flowers confer on this highly autogamous, carnivorous plant species.

Since small Diptera (flies) are the main prey insects in *Drosophyllum* (Bertol *et al.* 2015), and the yellow colour is particularly attractive to flies (Neuenschwander 1982; Yee 2015), we tested the hypothesis that large, showy flowers might not be directly related to reproduction, but would instead support plant growth by enhancing prey capture. An increase in prey capture might cause an increase in seed production, as it has been reported

in *Drosera* species (Thum 1988), and would therefore have indirect benefits on the reproductive output. However, insect capture rates between intact blooming plants and those plants whose flowers were removed did not differ in any of the three populations (Fig. 3), so we rejected the role of large yellow flowers as significant contributors to prey attraction in *Drosophyllum*.

Considering that the *Drosophyllum* population at Montera del Torero was dominated by old reproductive plants while most reproductive individuals in Sierra Carbonera were young (Table 1), the differences in flower size between the populations can be explained as an allometric effect of plant age. Branching (i.e. number of rosettes) in this species increases with age (Ortega-Olivencia *et al.* 1995; Garrido *et al.* 2003), and flower (or inflorescence) size is known to decrease with branching (Midgley and Bond 1989).

Regarding the controlled pollination experiments, fruit set was very high, with nearly 100 % of the flowers developing into fruit in the four treatments at the two sites (Table 2). Therefore, our results concur with those of Ortega-Olivencia *et al.* (1995, 1998), suggesting that *Drosophyllum* is a highly autogamous species regardless of geographic isolation and population size (Garrido *et al.* 2003; Paniw *et al.* 2015). However, when looking at seed production, some interesting patterns emerged. First, seeds were overall smaller in size and weight in plants from Montera del Torero than in those from Sierra Carbonera (Table 2). Again, this can be attributed to an allometric effect derived from plant age (see above), as there is a strong direct relationship between petal size and seed size in angiosperms (Primack 1987). The slightly but significantly smaller and lighter seeds from the OP treatment in Montera del Torero (Table 2) might be due to the existence of a trade-off between seed number per fruit and seed size/weight (e.g. Baker *et al.* 1994).

Second, while seed set values after the two hand-pollination treatments (HCP and HSP) were remarkably high in Sierra Carbonera, significantly higher than after control, OP, they were significantly lower than after OP in Montera del Torero (Fig. 4). These differences could also be explained by the overall large differences in plant age between reproductive plants of the two populations (Table 1). Since most reproductive plants from Montera del Torero were old, their siring ability might be low, as pollen viability in plants decreases with ageing (Aizen and Rovere 1995; Marshall *et al.* 2010). As only a single anther brush was applied to stigmas of flowers in both HCP and HSP hand-pollination treatments, this could have been sufficient in Sierra Carbonera, where all reproductive plants were young, but not in Montera del Torero. However, we cannot discard differences in weather conditions between populations during the pollination experiments

Table 4. Fecundity variables (fruit set, seed set, seed weight and seed length; mean $\pm$ SD) of *D. lusitanicum* per treatment in the two sites. Pairwise significant differences ($P < 0.05$; Tukey's HSD tests) between treatments are indicated by different superscript letters. HCP, hand cross-pollination; HSP, hand self-pollination; SSP, spontaneous self-pollination; OP, control, open pollination.

Treatment	No. of flowers	Fruit set (%)	Seed set (%)	Seed weight (mg)	Seed length (mm)
Sierra Carbonera					
HCP	67	98.5 ($\pm$12.2)	77.7 ($\pm$18.9)[A]	4.36 ($\pm$0.35)	2.48 ($\pm$0.13)
HSP	36	100 ($\pm$0.0)	77.4 ($\pm$22.6)[A]	4.40 ($\pm$0.31)	2.48 ($\pm$0.15)
SSP	167	99.4 ($\pm$7.7)	61.0 ($\pm$30.7)[B]	4.35 ($\pm$0.45)	2.50 ($\pm$0.16)
OP	76	100 ($\pm$0.0)	70.6 ($\pm$29.7)[C]	4.39 ($\pm$0.41)	2.49 ($\pm$0.19)
Montera del Torero					
HCP	43	100 ($\pm$0.0)	60.0 ($\pm$29.1)[a]	3.29 ($\pm$0.32)[a]	2.15 ($\pm$0.13)[a]
HSP	24	100 ($\pm$0.0)	54.6 ($\pm$28.2)[a]	3.28 ($\pm$0.23)[a]	2.15 ($\pm$0.12)[a]
SSP	135	99.3 ($\pm$8.6)	47.0 ($\pm$31.5)[b]	3.38 ($\pm$0.37)[b]	2.15 ($\pm$0.17)[a]
OP	65	100 ($\pm$0.0)	73.0 ($\pm$25.8)[c]	3.16 ($\pm$0.31)[a]	2.10 ($\pm$0.13)[b]

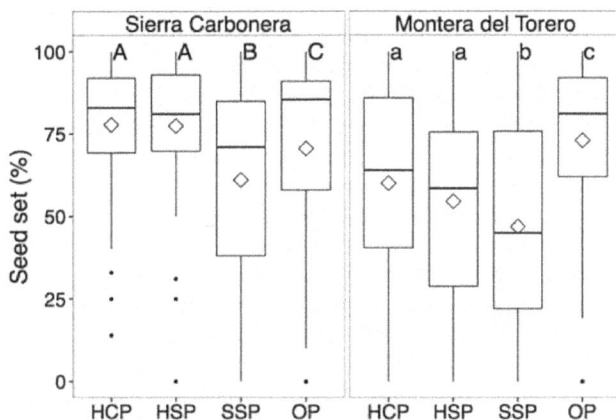

Figure 4. Boxplots of seed set of *D. lusitanicum* after HCP, HSP, SSP and control, OP across two experimental sites (Sierra Carbonera and Montera del Torero). Different letters represent significant pairwise differences (Tukey's HSD, $P < 0.05$) of group means between the four pollination treatments at each site.

that might have produced different bagging effects. All the same, the lack of differences in seed set between both HCP and HSP treatments in the two populations confirms that no mechanism of self-incompatibility is operating in this species (Ortega-Olivencia *et al.* 1998).

But the most remarkable result found in this study has been the significantly lower seed set values in the SSP treatment at both sites (Table 2; Fig. 4). This means that, even though *Drosophyllum* flowers are readily able to self-pollinate spontaneously, as Ortega-Olivencia *et al.* (1995) had already reported, insect visitation significantly increases seed production by 15–25 % in this species, either by cross-assisted or by insect-assisted self-pollination (facilitated selfing sensu Lloyd 1992). Considering the relatively high rates of seed set after SSP (Ortega-Olivencia *et al.* 1995; this study), may a

15–25 % increase in seed set through insect-assisted pollination offset the costs associated with maintaining large, showy flowers in this highly autogamous species? Its life history and population dynamics suggest an affirmative answer. Adult individuals of this early-successional pyrophyte cannot persist in mature, dense vegetation stands (Paniw *et al.* 2015), whose germination and growth are largely confined to a short post-fire window (Correia and Freitas 2002; M. Paniw, P. Quintana-Ascencio, F. Ojeda and R. Salguero-Gómez, unpublished). In this short temporal window, producing seeds to replenish the seed bank is critical for *Drosophyllum*, as it happens in other pyrophytes (Quintana-Ascencio *et al.* 2003; Menges and Quintana-Ascencio 2004). Therefore, any increase in seed set over autonomous selfing caused by insect visitation—either by facilitated selfing (Lloyd 1992) or by favouring some outcrossing—would increase plant fitness. This, in turn, would account for the maintenance of large, conspicuous flowers in this highly autogamous plant species.

Conclusions

Although *Drosophyllum* flowers are certainly homogamous (Fig. 1C; Ortega-Olivencia *et al.* 1995), their relatively large, bright yellow corollas challenge the paradigm of autogamous flowers being characterized by a dramatic reduction in corolla size and showiness (Goodwillie *et al.* 2010; Sicard and Lenhard 2011). We rejected the possible role of these flowers as attracting devices for enhancing insect prey capture in this carnivorous species. On the other hand, since the key life-history strategy of this early colonizing pyrophyte is to produce a large, persistent seed bank to maximize post-fire germination (Correia and Freitas 2002; M. Paniw, P. Quintana-Ascencio, F. Ojeda and R. Salguero-Gómez, unpublished), any investment into increasing seed production would have a positive fitness effect. This would thus account for the maintenance of large, showy flowers in a highly autogamous plant species.

Sources of Funding

This study was supported by the Spanish Ministerio de Economía y Competitividad (project BREATHAL; Geographical barrier, habitat fragmentation and vulnerability of endemics: Biodiversity patterns of the Mediterranean heathland across the Strait of Gibraltar, CGL2011-28759).

Contributions by the Authors

F.O. conceived the study; F.O., M.P. and R.C.-S. designed the experiments and M.P. A.S.-C. and R.C.-S. carried them out; M.P. and A.S.-C. analysed the data; all authors contributed to writing the manuscript.

Acknowledgements

We thank the Spanish military Campo de Adiestramiento de la Armada in Sierra del Retín (Barbate, Cádiz) for allowing access to the Monte Retin sites and facilitating the study. We also thank the Director and Staff of Los Alcornocales Natural Park for authorizing and supporting the study. The Andalusian Consejería de Medio Ambiente provided the necessary permits to work with *Drosophyllum lusitanicum*, an endemic, red-listed species. Dennis Whigham and two anonymous reviewers made helpful comments on an earlier version of the manuscript.

Literature Cited

Adamec L. 1997. Mineral nutrition of carnivorous plants: a review. *Botanical Review* **63**:273–299.

Adlassnig W, Peroutka M, Eder G, Pois W, Lichtscheidl IK. 2006. Ecophysiological observations on *Drosophyllum lusitanicum*. *Ecological Research* **21**:255–262.

Aizen MA, Rovere AE. 1995. Does pollen viability decrease with aging? A cross-population examination in *Austrocedrus chilensis* (Cupressaceae). *International Journal of Plant Sciences* **156**:227–231.

Andersson S. 2005. Floral costs in *Nigella sativa* (Ranunculaceae): compensatory responses to perianth removal. *American Journal of Botany* **92**:279–283.

Andrés C, Ojeda F. 2002. Effects of afforestation with pines on woody plant diversity of Mediterranean heathlands in southern Spain. *Biodiversity and Conservation* **11**:1511–1520.

Baker K, Richards AJ, Tremayne M. 1994. Fitness constraints on flower number, seed number and seed size in the dimorphic species *Primula farinosa* L. and *Armeria maritima* (Miller) Willd. *New Phytologist* **128**:563–570.

Bates D, Maechler M, Bolker B, Walker S. 2013. *lme4: linear mixed-effects models using Eigen and S4. R Package Version 1.* http://lme4.r-forge.r-project.org/ (22 October 2015).

Bertol N, Paniw M, Ojeda F. 2015. Effective prey attraction in the rare *Drosophyllum lusitanicum*, a flypaper-trap carnivorous plant. *American Journal of Botany* **102**:689–694.

Bolker BM, Brooks ME, Clark CJ, Geange SW, Poulsen JR, Stevens MHH, White JSS. 2009. Generalized linear mixed models: a practical guide for ecology and evolution. *Trends in Ecology and Evolution* **24**:127–135.

Celedón-Neghme C, Gonzáles WL, Gianoli E. 2007. Cost and benefits of attractive floral traits in the annual species *Madia sativa* (Asteraceae). *Evolutionary Ecology* **21**:247–257.

Chase MW, Christenhusz MJM, Sanders D, Fay MF. 2009. Murderous plants: Victorian Gothic, Darwin and modern insights into vegetable carnivory. *Botanical Journal of the Linnean Society* **161**:329–356.

Correia E, Freitas H. 2002. *Drosophyllum lusitanicum*, an endangered West Mediterranean endemic carnivorous plant: threats and its ability to control available resources. *Botanical Journal of the Linnean Society* **140**:383–390.

Darwin C. 1875. *Insectivorous plants*. London, UK: John Murray.

Ellison AM, Gotelli NJ. 2001. Evolutionary ecology of carnivorous plants. *Trends in Ecology and Evolution* **16**:623–629.

Ellison AM, Gotelli NJ. 2009. Energetics and the evolution of carnivorous plants—Darwin's 'most wonderful plants in the world'. *Journal of Experimental Botany* **60**:19–42.

Galen C. 1999. Why do flowers vary? The functional ecology of variation in flower size and form within natural plant populations. *Bioscience* **49**:631–640.

Garrido B, Hampe A, Marañón T, Arroyo J. 2003. Regional differences in land use affect population performance of the threatened insectivorous plant *Drosophyllum lusitanicum* (Droseraceae). *Diversity and Distributions* **9**:335–350.

Gibson TC, Waller DM. 2009. Evolving Darwin's 'most wonderful' plant: ecological steps to a snap-trap. *New Phytologist* **183**:575–587.

Givnish TJ, Burkhardt EL, Happel RE, Weintraub JD. 1984. Carnivory in the bromeliad *Brocchinia reducta*, with a cost/benefit model for the general restriction of carnivorous plants to sunny, moist, nutrient-poor habitats. *The American Society of Naturalists* **124**:479–497.

Goodwillie C, Sargent RD, Eckert CG, Elle E, Geber MA, Johnston MO, Kalisz S, Moeller DA, Ree RH, Vallejo-Marin M, Winn AA. 2010. Correlated evolution of mating system and floral display traits in flowering plants and its implications for the distribution of mating system variation. *New Phytologist* **185**:311–321.

Heubl G, Bringmann G, Meimberg H. 2006. Molecular phylogeny and character evolution of carnivorous plant families in Caryophyllales—revisited. *Plant Biology* **8**:821–830.

Hothorn T, Bretz F, Westfall P. 2008. Simultaneous inference in general parametric models. *Biometrical Journal* **50**:346–363.

Juniper BE, Robins RJ, Joel DM. 1989. *The carnivorous plants*. San Diego, CA: Academic Press.

Jürgens A, Sciligo A, Witt T, El-sayed AM, Suckling DM. 2012. Pollinator-prey conflict in carnivorous plants. *Biological Reviews* **87**:602–615.

Król E, Płachno BJ, Adamec L, Stolarz M, Dziubińska H, Trębacz K. 2012. Quite a few reasons for calling carnivores 'the most wonderful plants in the world'. *Annals of Botany* **109**:47–64.

Lloyd DG. 1980. Demographic factors and mating patterns in angiosperms. In: Solbrig OT, ed. *Demography and evolution in plant populations*. Oxford, UK: Blackwell, 67–88.

Lloyd DG. 1992. Self- and cross-fertilization in plants. II. The selection of self-fertilization. *International Journal of Plant Science* **153**: 370–380.

Marshall DL, Avritt JJ, Maliakal-Witt S, Medeiros JS, Shaner MGM. 2010. The impact of plant and flower age on mating patterns. *Annals of Botany* **105**:7–22.

Menges ES, Quintana-Ascencio PF. 2004. Population viability with fire in *Eryngium cuneifolium*: deciphering a decade of demographic data. *Ecological Monographs* **74**:79–99.

Midgley J, Bond W. 1989. Leaf size and inflorescence size may be allometrically related traits. *Oecologia* **78**:427–429.

Morgan MT, Wilson WG. 2005. Self-fertilization and the escape from pollen limitation in variable pollination environments. *Evolution* **59**:1143–1148.

Müller J, Deil U. 2001. Ecology and structure of *Drosophyllum lusitanicum* (L.) links populations in the south-west of the Iberian Peninsula. *Acta Botánica Malacitana* **26**:47–68.

Neuenschwander P. 1982. Beneficial insects caught by yellow traps used in mass-trapping of the olive fly, *Dacus oleae*. *Entomologia Experimentalis et Applicata* **32**:286–296.

Ojeda F, Marañón T, Arroyo J. 2000. Plant diversity patterns in the Aljibe Mountains (S. Spain): a comprehensive account. *Biodiversity and Conservation* **9**:1323–1343.

Ortega-Olivencia A, Carrasco Claver JP, Devesa Alcaraz JA. 1995. Floral and reproductive biology of *Drosophyllum lusitanicum* (L.) Link (Droseraceae). *Botanical Journal of the Linnean Society* **118**:331–351.

Ortega-Olivencia A, López Paredes JA, Rodríguez-Riaño T, Devesa JA. 1998. Modes of self-pollination and absence of cryptic self-incompatibility in *Drosophyllum lusitanicum* (Droseraceae). *Botanica Acta* **111**:474–480.

Paiva J. 1997. *Drosophyllum* link. In: Castroviejo S, Aedo C, Laínz M, Muñoz Garmendia F, Nieto Feliner G, Paiva J, Benedí C, eds. *Flora iberica*, 5th edn. Madrid, Spain: CSIC, Real Jardín Botánico, 78–80.

Paniw M, Gil-López MJ, Segarra-Moragues JG. 2014. Isolation and characterization of microsatellite loci in the carnivorous subshrub *Drosophyllum lusitanicum* (Drosophyllaceae). *Biochemical Systematics and Ecology* **57**:416–419.

Paniw M, Salguero-Gómez R, Ojeda F. 2015. Local-scale disturbances can benefit an endangered, fire-adapted plant species in Western Mediterranean heathlands in the absence of fire. *Biological Conservation* **187**:74–81.

Primack RB. 1987. Relationships among flowers, fruits, and seeds. *Annual Review of Ecology and Systematics* **18**:409–430.

Pujol B, Zhou S-R, Vilas JS, Pannell JR. 2009. Reduced inbreeding depression after species range expansion. *Proceedings of the National Academy of Sciences of the USA* **106**:15379–15383.

Quintana-Ascencio PF, Menges ES, Weekley CW. 2003. A fire-explicit population viability analysis of *Hypericum cumulicola* in Florida rosemary scrub. *Conservation Biology* **17**:433–449.

R Development Core Team. 2015. *R: a language and environment for statistical computing*. Vienna, Austria: R Foundation for Statistical Computing. http://www.R-project.org/ (29 August 2015).

Schemske DW, Lande R. 1985. The evolution of self-fertilization and inbreeding depression in plants. II. Empirical observations. *Evolution* **39**:41–52.

Sicard A, Lenhard M. 2011. The selfing syndrome: a model for studying the genetic and evolutionary basis of morphological adaptation in plants. *Annals of Botany* **107**:1433–1443.

Thum M. 1988. The significance of carnivory for the fitness of *Drosera* in its natural habitat. *Oecologia* **75**:472–480.

Vuong HQ. 1989. Likelihood ratio tests for model selection and nonnested hypotheses. *Econometrica* **57**:307–333.

Yee WL. 2015. Commercial Yellow Sticky Strips more attractive than yellow boards to western cherry fruit fly (Dipt., Tephritidae). *Journal of Applied Entomology* **139**:289–301.

Zamora R. 1999. Conditional outcomes of interactions: the pollinator-prey conflict of an insectivorous plant. *Ecology* **80**:786–795.

13

Impact of an invasive nitrogen-fixing tree on arbuscular mycorrhizal fungi and the development of native species

Alejandra Guisande-Collazo*, Luís González and Pablo Souza-Alonso
Department of Plant Biology and Soil Science, University of Vigo, 36310 Vigo, Spain

Associate Editor: James F. Cahill

Abstract. Arbuscular mycorrhizal fungi (AMF) are obligate soil biotrophs that establish intimate relationships with 80 % of terrestrial plant families. Arbuscular mycorrhizal fungi obtain carbon from host plants and contribute to the acquisition of mineral nutrients, mainly phosphorus. The presence of invasive plants has been identified as a soil disturbance factor, often conditioning the structure and function of soil microorganisms. Despite the investigation of many aspects related to the invasion of *Acacia dealbata*, the effect produced on the structure of AMF communities has never been assessed. We hypothesize that *A. dealbata* modifies the structure of AMF community, influencing the establishment and growth of plants that are dependent on these mutualisms. To validate our hypothesis, we carried out denaturing gradient gel electrophoresis (DGGE) analysis and also grew plants of *Plantago lanceolata* in pots using roots of native shrublands or from *A. dealbata*, as inoculum of AMF. Cluster analyses from DGGE indicated an alteration in the structure of AMF communities in invaded soils. After 15 weeks, we found that plants grown in pots containing native roots presented higher stem and root growth and also produced higher biomass in comparison with plants grown with *A. dealbata* inoculum. Furthermore, plants that presented the highest biomass and growth exhibited the maximum mycorrhizal colonization and phosphorus content. Moreover, fluorescence measurements indicated that plants grown with *A. dealbata* inoculum even presented higher photosynthetic damage. Our results indicate that the presence of the invader *A. dealbata* modify the composition of the arbuscular fungal community, conditioning the establishment of native plants.

Keywords: *Acacia dealbata*; DGGE; microbial community structure; plant invasion; *Plantago lanceolata*; root inoculum; soil sterilization.

Introduction

Plants cope with environmental stresses daily, without the possibility of escape, being forced to relate and extend intimate relationships with their immediate neighbours. Among them, arbuscular mycorrhizal fungi (AMF) are one of the most important symbiotic associations in nature (Willis *et al.* 2013). These plant–fungus associations form specialized interfaces, where the exchange of materials occurs between living cells (Pfeffer *et al.* 2001).

Belonging to *Glomeromycota* phylum, these fungi are usually obligated mutualists that establish intimate relationships with 80 % of terrestrial plant families (Smith and Read 1997; Brundrett 2004; Harrison 2005; Willis *et al.* 2013). The relationship can also be parasitic, depending on the combination of fungal–plant species

* Corresponding author's e-mail address: aleguisande@uvigo.es

and environmental conditions (Klironomos et al. 2000; Klironomos 2003). In their mutualistic associations with plants, they obtain carbon from the host plant, contributing simultaneously with the acquisition of mineral nutrients, mainly phosphorus (P) (Harrison 2005; Willis et al. 2013). Arbuscular mycorrhizal fungi are considered non-host plant specific (Giovannetti and Hepper 1985; Zhang et al. 2010), but there are studies that have shown preference between fungus and plant species (Vandenkoornhuyse et al. 2002; Croll et al. 2008). Therefore, there are plant species that completely depend on AMF associations to survive (Van der Heijden et al. 2008). Arbuscular mycorrhizal fungi – plant interactions, at individual or community levels, are conditioned by several factors. Abiotic conditions, such as nutrients or microenvironmental soil conditions, entail small-scale patchiness in the abundance of AMF (Mummey and Rillig 2008), and biotic factors such as competence or predation can contribute to the modification of plant – fungi associations (Lin et al. 2015). Among them, the presence of invasive alien plants (IAPs) has been identified as a factor altering soil communities, despite interactions between fungi and host plants (Zhang et al. 2010). In their relationships with AMF, successful IAPs are usually related with three criteria: nonmycorrhizal or facultative symbionts, obligated symbionts but flexible in its associations or transported together with their symbionts (Richardson et al. 2000; Pringle et al. 2009). There are many examples in the literature, which show the impact of IAPs on the structure of AMF (Richardson et al. 2000; Klironomos 2003; Hawkes et al. 2006; Broz et al. 2007; Vogelsang and Bever 2009; Yang et al. 2014).

It is well known that the presence of IAPs is usually accompanied by a reduction of native biodiversity, as they are capable of replacing native species (Richardson et al. 2000; Reinhart et al. 2005; Broz et al. 2007; Brewer 2008; Hoyos et al. 2010; de Abreu and Durigan 2011). Modifications produced can alter biotic or abiotic components of the soil environment, influencing the growth of plant species that depend on soil microorganisms (Bever et al. 1997). Nevertheless, due to the visual impact that the invasive species produces in the aboveground, in many cases, the belowground effects remain unexplored. There are several mechanisms that IAPs may be using to outcompete native plants and one of the most important is the direct modification of the structure and function of soil microbial communities (Richardson et al. 2000; Hawkes et al. 2006; Zhang et al. 2010; Tanner and Gange 2013; Souza-Alonso et al. 2014). Acacia dealbata is a leguminous tree native to Australia that has become a dangerous invader throughout the world (Richardson and Rejmánek 2011). The plant causes damage at several different levels on the invaded ecosystems, and this can include a severe decrease in native biodiversity (Fuentes-Ramírez et al. 2010; Lorenzo et al. 2010, 2012), seed bank composition (González-Muñoz et al. 2012), modification of decomposition processes (Castro-Díez et al. 2012), soil biochemical composition (Lorenzo et al. 2010), changes in soil microbial communities (Lorenzo et al. 2010) or changes in soil microbial function (Souza-Alonso et al. 2014). It has also been suggested that the alteration, in both soil chemistry and microbial community, is highly related to the type of ecosystem and with the age of invasion (Souza-Alonso et al. 2014, 2015). Despite the many aspects that have been investigated in relation with A. dealbata invasion, the effect produced by the entrance of the invader on the structure of native AMF communities has never been assessed. It was suggested that, at least, this species does not seem to be benefited by specific associations with AMF (Crisóstomo 2012). Therefore, we hypothesize that A. dealbata could change the structure of AMF community and these changes influence the establishment and growth of plants that are dependent on AMF, as in the case of Plantago lanceolata (fam. Plantaginaceae). Due to its AMF dependence, P. lanceolata is commonly used as a model species in mycorrhizal studies (Gange and West 1994; De Deyn et al. 2009; Cotuna et al. 2013; Lorenzo et al. 2013). This species is commonly found in the studied region associated with several ecosystems, including Atlantic shrublands.

Therefore, we have two main objectives: firstly, to compare AMF diversity in soils invaded by A. dealbata with non-invaded soils by using denaturing gradient gel electrophoresis (DGGE) technique, and, secondly, to evaluate the effect of changes in AMF composition on the growth of P. lanceolata, under controlled conditions in a greenhouse experiment.

Methods

Preliminary soil sampling

With the aim of evaluating the global effect of A. dealbata on AMF communities and justifying an extended greenhouse assay, we carried out a preliminary study of AMF structure in the invaded communities. With this objective, we sampled three separate shrublands (S_1, 42.266397, −8.208299; S_2, 42.305789, −8.171543; and S_3, 42.306568, −8.172608) invaded by A. dealbata in O Ribeiro Region (Galicia, NW Spain). Atlantic shrublands are very common in this region and they are mainly dominated by Ulex europaeus, Pterospartum tridentatum, Erica cinerea, E. umbellata, Calluna vulgaris, Anchusa sp. and Lotus spp. In each shrubland, we clearly identified an invaded area (totally covered by A. dealbata) and a native area (without A. dealbata presence). Native areas were located contiguously to the invaded areas, assuring that soil characteristics were the same. At each sampling point, surface litter was removed and 30 soil samples were collected from the rhizosphere of at least 10 mature plants of A. dealbata (in the invaded zone) and from at least 10 mature plants of the native shrubland (a mix of the dominant species mentioned above) and pooled. Soil was immediately taken to the laboratory, freshly sieved (0.2 mm) and frozen at −20 °C until DGGE analyses were carried out.

DNA extraction and DGGE analyses

Soil DNA extractions were performed using an UltraClean Soil DNA Isolation Kit (MO BIO Laboratories, Inc., Carlsbad, CA). An aliquot of 0.2 g soil of six replicates was used per extraction and stored at −20 °C. DNA extracted from soil samples was amplified using the primers AM1 (Helgason et al. 1998) and NS31-GC (Kowalchuk et al. 1997). All reactions were carried out in a final volume of 25 μL containing 1× polymerase chain reaction (PCR) buffer, 2.5 U Taq DNA polymerase (VWR), 0.25 mM dNTP, 0.5 μM AM1, 0.003 g bovine serum albumin, 96 % electrophoresis degree (SIGMA) and 1 μL of extracted genomic DNA. Polymerase chain reaction conditions are described in Hassan et al. (2013): one cycle at 94 °C for 3 min, followed by 30× (94 °C, 45 s; 58 °C, 45 s; 72 °C, 45 s) and a final extension step at 72 °C for 10 min. Polymerase chain reaction products, in aliquots of 5 μL, were analysed in 1 % agarose gel electrophoresis stained with GelRed™ and then subjected to DGGE. The PCRs were performed using a T100 Thermal Cycler (Bio-Rad, Hercules, CA, USA).

Denaturing gradient gel electrophoresis was performed with a DGGE-2401 system from CBS Scientific (San Diego, CA, USA). Twenty microlitres of each PCR product of the soil samples were analysed. Denaturing gradient gel electrophoresis analyses were conducted in 1× TAE buffer at a constant temperature of 60 °C at 20 V for 15 min, followed by 16 h at 70 V. Gels contained 8 % (w/v) acrylamide for fungi PCR products.

The linear gradient used was from 26 to 67 %, while 100 % denaturing acrylamide was defined as containing 7 M urea and 40 % (v/v) formamide. Gels were stained with 1× GelStar for 20 min and destained in distilled water for 30 min, after which they were visualized in a UV-transilluminator.

Experimental design and sampling

After the differences found due to A. dealbata presence in the structure of AMF communities inferred from DGGE results, the consequences of structural changes in the growth of P. lanceolata were evaluated. We compared the effect of AMF from native sites (shrublands) with AMF from invaded sites using roots as inoculum (Klironomos and Hart 2002; Gu et al. 2011; Hassan et al.

2013). Mycorrhizal root fragments or active hyphal networks are both viable infection units, especially in thriving habitats (Smith and Read 2008).

To create our inoculum, soil and roots were collected in the first week of March 2013, from the same places described in the Preliminary soil sampling. In each area, two different plant materials were selected, forming the inoculum of our two treatments: from the invasive A. dealbata and from the dominant species of native shrubland. At the same time, soil was collected to fill pots in which P. lanceolata would be sown. At each sampling area, roots from at least 25 different plants of A. dealbata were collected using hand-scissors. Similarly, 25 different plants of shrubland species were removed with the use of a shovel, and roots carefully cut and placed in plastic bags. At the same time, 30 samples of soil were collected randomly in each zone within the first 15 cm (±1 cm) with a hand shovel and pooled together. Soil and plant material were immediately taken to the laboratory for further processing. Once in the laboratory, the soil was sieved (2 mm) and sterilized by autoclaving for three consecutive days (Nazeri et al. 2013). Roots from acacia and shrubland were chopped into small pieces (±1 cm) to facilitate incorporation and homogenize their distribution in the pots. After that, roots were separated into two different fractions; one part was individually sterilized together with soils and the other remained untreated. The roots were mixed with sterilized soil in plastic pots (375 cm³) previously sterilized (ethanol, 80 % and UV-light, 30 min) and then filled in a laminar chamber with a mixture of sterile soil : perlite : roots in a ratio of 300 : 10 : 1; perlite was added to ameliorate water retention and facilitate root development. A total of four treatments were created: (i) sterilized acacia roots (SA), (ii) non-sterilized acacia roots (NSA), (iii) sterilized shrub roots (SS) and (iv) non-sterilized shrub roots (NSS) and control without roots. The inclusion of two treatments with sterilized roots is due to the putative effect of root decomposition (Acacia vs. shrub) providing a different source of nutrients. The pots were arranged in a completely randomized design with nine replicates. The experimental set-up was carried out in a greenhouse at the University of Vigo (NW Spain).

In each pot, five seedlings of P. lanceolata were sown. This species was selected by its use as a model species in mycorrhizal studies (Gange and West 1994; Cotuna et al. 2013; Lorenzo et al. 2013). Seeds were previously sterilized in a sodium hypochlorite solution (1 %) for 5 min and thoroughly rinsed in distilled water. After that, seeds were germinated in a growth chamber at 24/21 °C and 16/8 h light/darkness conditions. After 3 days, germinated seeds were selected and carefully sown in a laminar flow chamber to minimize fungal contamination. Pots were maintained for 15 weeks (from May to July 2014) under greenhouse conditions.

Fluorescence measurements

After 15 weeks and before *P. lanceolata* individuals were harvested, fluorescence parameters were evaluated. As we stated above, *P. lanceolata* is dependent on AMF, and therefore, we suspect that the absence of fungal relationship can be translated into a loss of photosynthetic efficiency. Chlorophyll *a* emission was monitored with a fluorescence imaging system (Imaging-PAM M-Series, Walz, Effeltrich, Germany). Five plants per treatment were kept in darkness for 5 min to allow all reaction centres to open and to minimize fluorescence associated with the energization of the thylakoid membrane (Krause *et al.* 1982). After this, the plants were successively illuminated at an intensity of 0.5 mol m^{-2} s^{-1} for a measurement of F_0 (the minimum fluorescence of dark-adapted leaves), with a saturating pulse of intensity 2700 mol m^{-2} s^{-1} for measurement of F_m (the maximum fluorescence of dark-adapted leaves). With this procedure, we measured fluorescence parameters as PSII efficiency (Ψ_{II}), regulated and non-regulated dissipated energy rate (Ψ_{NPQ} and Ψ_{NO}, respectively), non-photochemical quenching rate (qN) and electronic transporting rate (ETR).

Plant harvest

After the fluorescence measurement, the plants were gently removed from the pots and the roots of each plant were gently washed to remove any soil adhered, and then roots, shoots and hypocotyls were measured in length. Subsequently, the fresh weight of aerial and belowground material was measured and then plant material was dried in an oven at 70 °C for a minimum of 72 h to collect dry weights. Due to the key role of AMF in P uptake, we also measured the content of P in leaves and roots of *P. lanceolata* using ICP-OES (Perkin Elmer Optima 4300 DV).

Additionally, the root material from three plants of each pot was used for the measurements of root colonization. Roots from these plants were washed under tap water to remove soil particles. After that, roots were cut into small fragments (± 1 cm), and these fragments were stained following the method of Walker (2005), slightly modified. The root fragments were briefly introduced into amber glass vials containing Coomassie Blue, the dye used for root staining, and covered with a cotton mesh (1 mm diameter) in order to prevent root loss during staining stages. The AMF colonization was assessed using a modified grid-line intersection method according to Lorenzo *et al.* (2013). To avoid misinterpretation, percentages of control, SA and SS were averaged and compared with non-sterile treatments since a slight percentage of colonization in sterilized treatments was found.

Statistical analyses

To valorize the effect of *A. dealbata* on soil fungal structure, GelCompar II (Applied Maths, Belgium) was used in the cluster analysis of AMF based on the DGGE results. The unweighted pair-group method with arithmetic mean algorithm and the Pearson product–moment correlation coefficient were used for the analysis. Richness, defined as the number of species, was calculated as the total number of bands per sample. To calculate abundance and diversity, defined as the number of different species and their relative frequency, gel bands were classified according to their intensity in six categories. Diversity was calculated using a modification of the Shannon index, $H' = -\Sigma[(n_i/N) \, Ln(n_i/N)]$, where n_i had one of four possible values (1–6), depending on band intensity. However, we assume that ecological parameters as defined here require a cautious interpretation since bands from DGGE cannot be unmistakably translated into AMF species (van Elsas and Boersma 2011).

The effect of the independent variables of the model (sterilization or not and species) on stem, hypocotyl and root length, aboveground and belowground biomass, P content, mycorrhizal colonization and fluorescence was evaluated using two-way analysis of variance (ANOVA). When interactions between independent variables were found, the effects were investigated through pairwise comparisons using Tukey's honest significant difference or Dunnett's T3 as the *post hoc* test. All data were previously subjected to Kolmogorov–Smirnov test for normality and Levene's test to check homoscedasticity of the variances. The statistical analyses were carried out using the SPSS v.19 (Chicago, IL) software for Windows.

Results

Soil characteristics and microbial community analysis

Data from PCR-DGGE showed us alterations in the structure of soil AMF community. The cluster analyses revealed differences between the invaded and native zones in each shrubland (S1, S2 and S3), as indicated in the tree diagram (Fig. 1). In each native area studied, AMF community of invaded soils clustered together and separated from non-invaded samples. In contrast, AMF diversity, richness and density did not present significant differences between invaded and non-invaded areas (data not shown).

Fluorescence measurements

Parameters related to fluorescence were significantly affected in *P. lanceolata* due to the treatments during the timing of the assay (Table 1). In general, we observed that plants with the non-sterile shrub inoculum presented higher photosynthetic activity, indicated as a

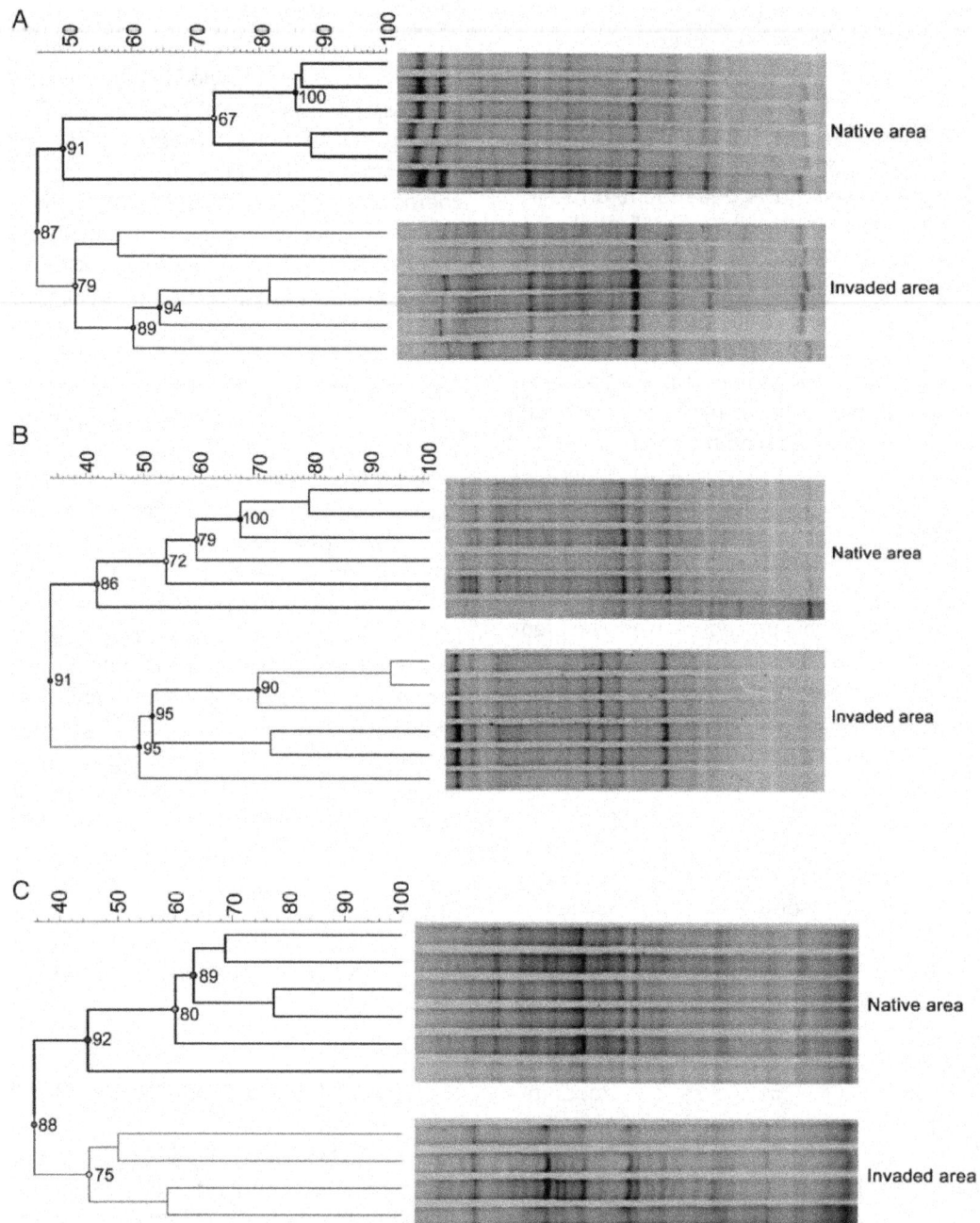

Figure 1. Dendrograms of AMF community structure based on PCR-DGGE bands, using the unweighted pair-group method with arithmetic mean algorithm and the Pearson product–moment correlation coefficient. S1 (A), S2 (B) and S3 (C) are the soils used in the assay.

significant increase in the effective quantum yield at PSII (Ψ_{II}), or a reduction in the qN. While the ETR was reduced or virtually inhibited in other treatments, values of NSS were significantly increased.

Plant growth and P content

We found a general effect of species and treatment in the two-way ANOVA in the growth and development of *P. lanceolata* (Table 2). Additionally, significant values of

interaction in the model between the independent variables *species × treatment* were observed. Therefore, we can anticipate that differences found in the variables considered are not the same within species (*A. dealbata* or native species inoculum) than within treatments (sterilized or non-sterilized).

The two-way ANOVA showed that biometric parameters were affected by species and treatment. There were significant differences between treatments in

Table 1. Values of the effective quantum yield at PSII (Ψ_{II}), the quantum yield of light-induced heat dissipation (non-photochemical quenching, Ψ_{NPQ}) and the quantum yield of all processes alternative to light-induced heat dissipation and photochemical use (Ψ_{NO}). The ETR and the qN in *P. lanceolata* plants. Different letters indicate significant differences in Tukey's or Dunnett's T3 *post hoc* test at $P \leq 0.05$ level.

	Control	NSA	SA	NSS	SS
Ψ_{II}	0.004 ($\pm$0.01)**b**	0.012 ($\pm$0.02)**b**	0.005 ($\pm$0.01)**b**	0.14 ($\pm$0.05)**a**	0.022 ($\pm$0.02)**b**
Ψ_{NPQ}	0.62 ($\pm$0.08)	0.66 ($\pm$0.06)	0.65 ($\pm$0.05)	0.61 ($\pm$0.05)	0.61 ($\pm$0.02)
Ψ_{NO}	0.38 ($\pm$0.09)	0.33 ($\pm$0.06)	0.35 ($\pm$0.06)	0.26 ($\pm$0.01)	0.37 ($\pm$0.03)
ETR	<0.001 ($\pm$0.00)**b**	<0.001 ($\pm$0.00)**b**	2.02 ($\pm$2.58)**b**	28.77 ($\pm$8.41)**a**	4.45 ($\pm$4.01)**b**
qN	11.26 ($\pm$0.28)**ab**	10.92 ($\pm$0.5)**ab**	11.40 ($\pm$0.29)**a**	10.61 ($\pm$0.27)**b**	11.19 ($\pm$0.19)**ab**

length and biomass of *P. lanceolata*. In general, we found an overall increase in plant growth in pots containing non-sterile inoculum of native shrubs. More specifically, we found an evident increase in the length of *Plantago* leaves ($P < 0.001$) when plants were grown with non-sterile shrub inoculum (Fig. 2).

As occurred with growth, the two-way ANOVA indicated that leaf and root biomass were affected by species, treatment and their interaction. Overall, the effect of NSS was particularly evident in biomass production (Fig. 3). When *P. lanceolata* grew with native shrub inoculum, we found a significant increase in the production of aerial biomass in comparison with control (341 %), NSA (500 %), SA (492 %) and SS (517 %; $P < 0.001$, in all cases). Complementarily, the production of root biomass in NSS was also significantly increased in comparison with control (110 %), NSA (263 %), SA (214 %) and SS treatment (184 %; $P < 0.001$ in all cases). A general increase in the P content related to NSS treatment in both the aerial and root content was also evident (Fig. 4).

Mycorrhizal colonization

Significant differences ($P < 0.001$) were found between treatments in mycorrhizal colonization of *P. lanceolata* roots (Fig. 5). Plants treated with NSS inoculum had the highest colonization rate, 75 % of root surface colonized compared with the 13 % of colonization in NSA treatment. We found a slight percentage of colonization in treatments with sterile inoculum. Nevertheless, there were no differences in colonization rates between treatments with sterile roots and control.

Discussion

As we hypothesized, cluster analysis from DGGE revealed that the structure of AMF community in non-invaded areas was different than community from areas with *A. dealbata* presence. In some cases, invasive plants have the capacity to disrupt mutualisms between AMF and native species (Klironomos 2003; Meinhardt and

Gehring 2012), even decreasing the competitiveness of native plants (Yang *et al.* 2014). In this sense, it is interesting to note that previous studies indicated that *A. dealbata* does not seem to be benefited by specific associations with AMF (Crisóstomo 2012) and so the change in the AMF structure could be produced indirectly. Apparently, the low dependence on AM symbionts could be related with their highly invasive success (Pringle *et al.* 2009). Arbuscular mycorrhizal fungi are obligate biotrophs that need host plants in order to live (Smith and Read 1997), and the change in the dominant plant species in the ecosystem could unbalance the established equilibrium favouring some fungal species. Nevertheless, the reduction of plant community diversity that takes place in areas invaded by *A. dealbata* (Lorenzo *et al.* 2012) should decrease AMF community richness (Van der Heijden *et al.* 1998), which does not occur. We suggest that the change in plant composition and dominance in the structure of the aboveground ecosystem could be the main force that drives the change in AMF structure. The reduction in plant diversity under *A. dealbata* canopy (Lorenzo *et al.* 2012) decreases the possibilities of finding compatible mutualisms. Therefore, selective forces towards more competitive and adaptable fungal species are probably active, contributing to the explanation of the shift in the AMF community structure. In this sense, impact is the term usually employed to describe the effects produced after the entrance of the invasive plant (Hejda *et al.* 2009). Following interpretation of data from DGGE, we consider that the term 'restructuration' or 'addressing' can be more appropriate than 'impact' since the diversity and richness of AMF species were not affected in the invaded soils.

Native plants can vary widely in response to the AMF change produced by IAPs. It was observed that the inhibition of native mycorrhizas produced by the nonmycorrhizal IAP *Alliaria petiolata* modifies the scenario and reduces the competitive ability of native plants (Stinson *et al.* 2006). In the case of *A. dealbata*, the structural change produced in AMF in the invaded soils could not

Table 2. Two-way ANOVA results, including independent variables (treatment and species) and interaction (T × S) for the leaf, hypocotyl and root length; dry weight of leaves and roots (aboveground and belowground biomass), P content in leaves and roots and percentage of AMF colonization in roots.

	df	Leaf length		Hypocotyl length		Root length		DW (leaves)		DW (roots)		P (leaves)		P (roots)		Colonization	
		F	p	F	p	F	p	F	p	F	p	F	p	F	p	F	p
Treatment	2	169.39	<0.001	2.66	0.154	0.56	0.484	1778.28	<0.001	13.27	<0.05	1250.57	<0.001	147.31	<0.001	22.42	<0.01
Species	2	193.28	<0.001	4.16	0.088	1.35	0.289	1312.97	<0.001	13.92	<0.05	1144.25	<0.001	58.55	<0.001	18.47	<0.01
T × S	4	219.10	<0.001	0.25	0.638	22.86	<0.01	1668.54	<0.001	23.99	<0.01	1293.86	<0.001	74.27	<0.001	21.06	<0.01

be related with the main hypothesis used to explain its invasive success (fast growth, sprouting, allelopathy), but with the low dependence on AM symbionts. Studied ecosystems are generally dominated by plants from different functional groups, as we found in agricultural fields, shrublands, grasslands or even forest dominated by non-nitrogen-fixing species (Lorenzo et al. 2012; Hernández et al. 2014; Souza-Alonso et al. 2014, 2015). Therefore, observed changes in AMF structure are produced by the obvious replacement in the identity of plant species dominance in areas invaded by A. dealbata. Nevertheless, it was also suggested that substitution in the main functional group can also be relevant in the response of the AMF community (Hoeksema et al. 2010; Lin et al. 2015).

Our second objective was to evaluate the change shown on AMF structure induced by A. dealbata on the growth and development of P. lanceolata. This species is highly colonized and dependent on AMF for development (Gange and West 1994; Ayres et al. 2006; De Deyn et al. 2009; Cotuna et al. 2013). Additionally, in terms of symbiotic relationships, P. lanceolata seems to be favoured by the coexistence with plant-related species (Bever 2002). Therefore, changes in the composition or structure of the established AMF communities probably entail consequences on plant development and growth.

Results from fluorescence measurements, mainly the significant increase in Ψ_{II} and ETR in plants growing with native roots inoculum, indicate that the new structure of AMF present in the invaded area produced a strong decline in photosynthetic efficiency, suggesting a possible damage to the photosynthetic machinery (Martínez-Peñalver et al. 2012) of P. lanceolata. This fact is supported by the external appearance of plants grown without native inoculum, presenting signs of chlorosis and nechrosis with evident tissue damage. A reduction in the amount of light energy–indicated by the increase in Ψ_{II}—reaching the photosynthetic apparatus usually entails a significant decrease in the production of carbon-derived compounds (Genty et al. 1989; Rolfe and Scholes 1995; Li et al. 2008). Therefore, the damage in the photosynthetic apparatus contributes to the evident decrease in P. lanceolata growth.

Plants grown in pots with NSS inoculum presented maximum growth, biomass, P content and colonization. This occurs because AMF establish mutualistic associations with host plants, which raises the nutrient uptake, mainly P (Jeffries et al. 2003; Hassan et al. 2013). Additionally, it is important to consider that the difference found in AMF structure in invaded soils does not necessarily correspond to AMF species that effectively infect P. lanceolata roots. Instead of the general trend that indicates that AMF do not present high host specificity and

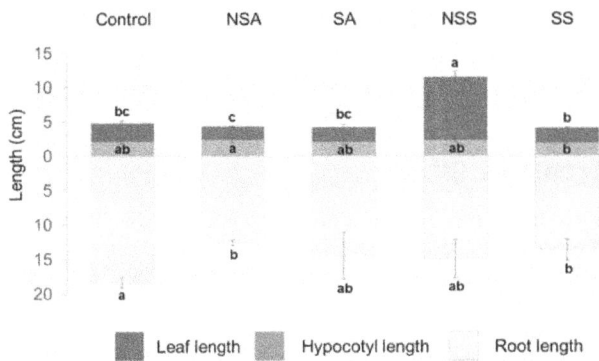

Figure 2. Length of root, leaf and hypocotyl in all treatments. NSA, non-sterilized acacia roots; SA, sterilized acacia roots; NSS, non-sterilized shrub roots and SS, sterilized shrub roots. Different letters indicate significant differences at $P \leq 0.05$ level.

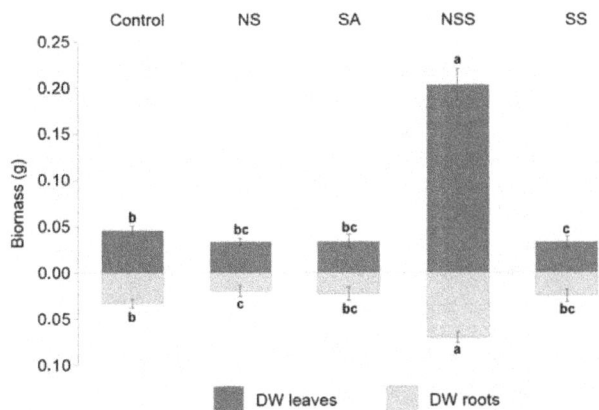

Figure 4. Phosphorus content of roots and leaves in all treatments. NSA, non-sterilized acacia roots; SA, sterilized acacia roots; NSS, non-sterilized shrub roots and SS, sterilized shrub roots. Different letters indicate significant differences at $P \leq 0.05$ level.

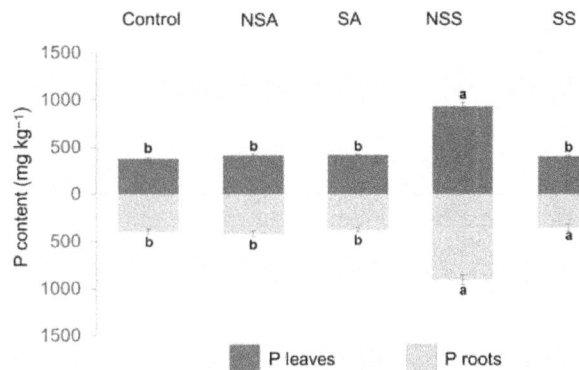

Figure 3. Dry weight (DW) of roots and leaves in all treatments. NSA, non-sterilized acacia roots; SA, sterilized acacia roots; NSS, non-sterilized shrub roots and SS, sterilized shrub roots. Different letters indicate significant differences at $P \leq 0.05$ level.

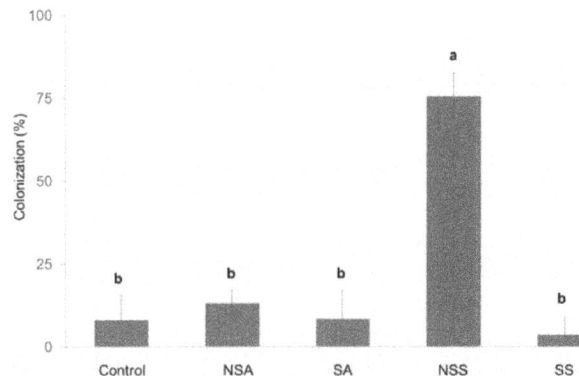

Figure 5. Percentage of colonization in *P. lanceolata* roots. NSA, non-sterilized acacia roots; SA, sterilized acacia roots; NSS, non-sterilized shrub roots and SS, sterilized shrub roots. Different letters indicate significant differences at $P \leq 0.05$ level.

plants are able to relate to almost every AMF (Giovannetti and Hepper 1985; Zhang *et al.* 2010), our results, surprisingly, do not go along the same line. We found different fungal composition, whereas density, diversity and richness values were not significantly altered. Regardless of the change in the 'AMF species identity'—despite not being specifically addressed—we assume that the presence of propagules, which grants opportunities to establish plant–fungal relationships, was similar between invaded and non-invaded areas. Therefore, we should expect that the number of infections were similar, but this was not the case. Consequently, we cautiously suggest that, in some cases, *P. lanceolata* presents some level of specificity in selecting their partners, or vice versa, probably influenced by nutrient requirements, soil environmental conditions or the absence of local adaptation. Another plausible explanation could be the different

form of AMF propagules identified in DGGE analyses. The composition of AMF community is important, but the form in which the AMF species is present—in the form of hyphae or spores—could also be relevant.

Differences found in root colonization can be related to the soil environment produced by *A. dealbata* in the field. Root–AMF association is a chemically modulated process and AMF can sense components of the rhizosphere (Harrison 2005 and references therein). The extent of the rhizomatous system of *A. dealbata* produces severe physico-chemical changes in soils under its canopy (Lorenzo *et al.* 2010; Souza-Alonso *et al.* 2014), producing an unfavourable ambient in which the association of AMF with plant roots can be challenging. These difficulties can be produced mainly at two levels: diminishing spore germination or limiting the growth of the hyphal tube in the search of a host root. In this sense, the chemoactive compounds that this species releases (Reigosa *et al.* 1999; Lorenzo *et al.* 2010, 2011; Aguilera *et al.* 2015) do not seem to affect native AMF colonization of *P. lanceolata* in field conditions (Lorenzo *et al.* 2013).

Structural changes produced in the AMF community in invaded soils could also have further consequences. Native plants and mycorrhizal fungal communities show interdependence, and so reassembly of one community may be limited by the reassembly of the other (Lankau *et al.* 2014). It has previously been indicated that the influence of invasive species on soil characteristics remains even after their removal, an effect known as 'legacy effect'. Residual effects are usually related to the age (short- or long-term invasion) and degree of invasion (low or high level of invasion) and takes place at several levels: soil nutrient changes (Marchante *et al.* 2008), organic matter content (Novoa *et al.* 2013) and even those that alter the AMF community (Lankau *et al.* 2014; Shannon *et al.* 2014). Therefore, it should be noted that the modification in the AMF structure could condition individual plant establishment after *A. dealbata* management, complicating ecosystem restoration processes.

Conclusions

Our results indicate that *A. dealbata* effectively changes the structure of the AMF community in the invaded shrublands with negative consequences. The change in the identity of AMF species constrained the growth of plants that depends on AMF, such as *P. lanceolata*. Our work highlights the importance of maintaining soil communities, particularly in regards to the entrance of invasive dominant species.

Sources of Funding

This work was supported by The Agroalimentary Research Center (CIA) of the Regional Government of Galicia, also A.G.-C. was awarded by the PhD Student grant of University of Vigo (00VI 131H 641.02). This is a contribution from the Alien Species Network (Ref. R2014/036 – Xunta de Galicia, Regional Government of Galicia).

Contributions by the Authors

P.S.-A., L.G. and A.G.-C. conceived and designed the idea. A.G.-C. and P.S.-A collected the data. A.G.-C. and P.S.-A. ran the statistics. A.G.-C., P.S.-A. and L.G. discussed the results and wrote the manuscript.

Acknowledgements

We would like to give special thanks to Antonio López Nogueira for his helpful support during field work and our colleague Óscar Martínez for his technical support. We also thank the reviewers who significantly improved the final version of the manuscript with their comments.

Literature Cited

Aguilera N, Becerra J, Villaseñor-Parada C, Lorenzo P, González L, Hernández V. 2015. Effects and identification of chemical compounds released from the invasive *Acacia dealbata* Link. *Chemistry and Ecology* **31**:479–493.

Ayres RL, Gange AC, Aplin DM. 2006. Interactions between arbuscular mycorrhizal fungi and intraspecific competition affect size, and size inequality, of *Plantago lanceolata* L. *Journal of Ecology* **94**:285–294.

Bever JD. 2002. Negative feedback within a mutualism: host–specific growth of mycorrhizal fungi reduces plant benefit. *Proceedings of the Royal Society B Biological Sciences* **269**:2595–2601.

Bever JD, Westover KM, Antonovics J. 1997. Incorporating the soil community into plant population dynamics: the utility of the feedback approach. *Journal of Ecology* **85**:561–573.

Brewer S. 2008. Declines in plant species richness and endemic plant species in longleaf pine savannas invaded by *Imperata cylindrica*. *Biological Invasions* **10**:1257–1264.

Broz AK, Manter DK, Vivanco JM. 2007. Soil fungal abundance and diversity: another victim of the invasive plant *Centaurea maculosa*. *The ISME Journal* **1**:763–765.

Brundrett M. 2004. Diversity and classification of mycorrhizal associations. *Biological Reviews* **79**:473–495.

Castro-Díez P, Fierro-Brunnenmeister N, González-Muñoz N, Gallardo A. 2012. Effects of exotic and native tree leaf litter on soil properties of two contrasting sites in the Iberian Peninsula. *Plant and Soil* **350**:179–191.

Cotuna O, Sărățeanu V, Durău CC. 2013. Influence of arbuscular mycorrhizae (AM) colonization on plant growth: *Plantago lanceolata* L. case study. *Journal of Food, Agriculture & Environment* **11**:2005–2008.

Crisóstomo JA. 2012. *Belowground mutualisms and plant genetic diversity: insights into the invasion process of Acacia dealbata and Acacia saligna*. PhD Thesis, University of Coimbra, Portugal.

Croll D, Wille L, Gamper HA, Mathimaran N, Lammers PJ, Corradi N, Sanders IR. 2008. Genetic diversity and host plant preferences revealed by simple sequence repeat and mitochondrial markers in a population of the arbuscular mycorrhizal fungus *Glomus intraradices*. *New Phytologist* **178**:672–687.

de Abreu RCR, Durigan G. 2011. Changes in the plant community of a Brazilian grassland savannah after 22 years of invasion by *Pinus elliottii* Engelm. *Plant Ecology and Diversity* **4**:269–278.

de Deyn GB, Biere A, Van der Putten WH, Wagenaar R, Klironomos JN. 2009. Chemical defense, mycorrhizal colonization and growth responses in *Plantago lanceolata* L. *Oecologia* **160**:433–442.

Fuentes-Ramírez A, Pauchard A, Marticorena A, Sánchez P. 2010. Relationship between the invasion of *Acacia dealbata* Link (Fabaceae: Mimosoideae) and plant species richness in South-Central Chile. *Gayana Botanica* **67**:188–197.

Gange AC, West HM. 1994. Interactions between arbuscular mycorrhizal fungi and foliar-feeding insects in *Plantago lanceolata* L. *New Phytologist* **128**:79–87.

Genty B, Briantais J-M, Baker NR. 1989. The relationship between the quantum yield of photosynthetic electron transport and quenching of chlorophyll fluorescence. *Biochimica et Biophysica Acta (BBA), General Subjects* **990**:87–92.

Giovannetti M, Hepper CM. 1985. Vesicular-arbuscular mycorrhizal infection in *Hedysarum coronarium* and *Onobrychis viciaefolia*: host-endophyte specificity. *Soil Biology and Biochemistry* **17**: 899–900.

González-Muñoz N, Costa-Tenorio M, Espigares T. 2012. Invasion of alien *Acacia dealbata* on Spanish *Quercus robur* forests: impact on soils and vegetation. *Forest Ecology Management* **269**:214–221.

Gu M, Chen A, Dai X, Liu W, Xu G. 2011. How does phosphate status influence the development of the arbuscular mycorrhizal symbiosis? *Plant Signaling & Behavior* **6**:1300–1304.

Harrison JM. 2005. Signaling in the arbuscular mycorrhizal symbiosis. *Annual Review of Microbiology* **59**:19–42.

Hassan SED, Liu A, Bittman S, Forge TA, Hunt DE, Hijri M, St-Arnaud M. 2013. Impact of 12-year field treatments with organic and inorganic fertilizers on crop productivity and mycorrhizal community structure. *Biology and Fertility of Soils* **49**:1109–1121.

Hawkes CV, Belnap J, D'antonio C, Firestone MK. 2006. Arbuscular mycorrhizal assemblages in native plant roots change in the presence of invasive exotic grasses. *Plant and Soil* **281**:369–380.

Hejda M, Pyšek P, Jarošík V. 2009. Impact of invasive plants on the species richness, diversity and composition of invaded communities. *Journal of Ecology* **97**:393–403.

Helgason T, Daniell TJ, Husband R, Fitter AH, Young JPW. 1998. Ploughing up the wood-wide web? *Nature* **394**:431.

Hernández L, Martínez-Fernández J, Cañellas I, De la Cueva AV. 2014. Assessing spatio-temporal rates, patterns and determinants of biological invasions in forest ecosystems. The case of *Acacia* species in NW Spain. *Forest Ecology and Management* **329**:206–213.

Hoeksema JD, Chaudhary VB, Gehring CA, Johnson NC, Karst J, Koide RT, Pringle A, Zabinski C, Bever JD, Moore JC, Wilson GWT, Klironomos JN, Umbanhower J. 2010. A meta-analysis of context-dependency in plant response to inoculation with mycorrhizal fungi. *Ecology Letters* **13**:394–407.

Hoyos LE, Gavier-Pizarro GI, Kuemmerle T, Bucher EH, Radeloff VC, Tecco PA. 2010. Invasion of glossy privet (*Ligustrum lucidum*) and native forest loss in the Sierras Chicas of Córdoba, Argentina. *Biological Invasions* **12**:3261–3275.

Jeffries P, Gianinazzi S, Perotto S, Turnau K, Barea J-M. 2003. The contribution of arbuscular mycorrhizal fungi in sustainable maintenance of plant health and soil fertility. *Biology and Fertility of Soils* **37**:1–16.

Klironomos JN. 2003. Variation in plant response to native and exotic arbuscular mycorrhizal fungi. *Ecology* **84**:2292–2301.

Klironomos JN, Hart MM. 2002. Colonization of roots by arbuscular mycorrhizal fungi using different sources of inoculum. *Mycorrhiza* **12**:181–184.

Klironomos JN, McCune J, Hart M, Neville J. 2000. The influence of arbuscular mycorrhizae on the relationship between plant diversity and productivity. *Ecology Letters* **3**:137–141.

Kowalchuk GA, Gerards S, Woldendorp JW. 1997. Detection and characterization of fungal infections of *Ammophila arenaria* (Marram Grass) roots by denaturing gradient gel electrophoresis of specifically amplified 18S rDNA. *Applied and Environmental Microbiology* **63**:3858–3865.

Krause GH, Vernotte C, Briantais J-M. 1982. Photoinduced quenching of chlorophyll fluorescence in intact chloroplasts and algae.

Resolution into two components. *Biochimica Biophysica Acta (BBA), Bioenergetics* **679**:116–124.

Lankau RA, Bauer JT, Anderson MR, Anderson RC. 2014. Long-term legacies and partial recovery of mycorrhizal communities after invasive plant removal. *Biological Invasions* **16**: 1979–1990.

Li Q-M, Liu B-B, Wu Y, Zou Z-R. 2008. Interactive effects of drought stresses and elevated CO_2 concentration on photochemistry efficiency of cucumber seedlings. *Journal of Integrative Plant Biology* **50**:1307–1317.

Lin G, Mccormack ML, Guo D. 2015. Arbuscular mycorrhizal fungal effects on plant competition and community structure. *Journal of Ecology* **103**:1224–1232.

Lorenzo P, Rodríguez-Echeverría S, González L, Freitas H. 2010. Effect of invasive *Acacia dealbata* link on soil microorganisms as determined by PCR-DGGE. *Applied Soil Ecology* **44**:245–251.

Lorenzo P, Palomera-Pérez A, Reigosa MJ, González L. 2011. Allelopathic interference of invasive *Acacia dealbata* link on the physiological parameters of native understory species. *Plant Ecology* **212**:403–412.

Lorenzo P, Pazos-Malvido E, Rubido-Bará M, Reigosa JM, González L. 2012. Invasion by the leguminous tree *Acacia dealbata* (Mimosaceae) reduces the native understorey plant species in different communities. *Australian Journal of Botany* **60**:669–675.

Lorenzo P, Rodríguez-Echeverría S, Freitas H. 2013. No allelopathic effect of the invader *Acacia dealbata* on the potential infectivity of arbuscular mycorrhizal fungi from native soils. *European Journal of Soil Biology* **58**:42–44.

Marchante E, Kjøller A, Struwe S, Freitas H. 2008. Short-and long-term impacts of *Acacia longifolia* invasion on the belowground processes of a Mediterranean coastal dune ecosystem. *Applied Soil Ecology* **40**:210–217.

Martínez-Peñalver A, Graña E, Reigosa MJ, Sánchez-Moreiras AM. 2012. Early photosynthetic response of *Arabidopsis thaliana* to temperature and salt stress conditions. *Russian Journal of Plant Physiology* **59**:640–647.

Meinhardt KA, Gehring CA. 2012. Disrupting mycorrhizal mutualisms: a potential mechanism by which exotic tamarisk outcompetes native cottonwoods. *Ecological Applications* **22**:532–549.

Mummey DL, Rillig MC. 2008. Spatial characterization of arbuscular mycorrhizal fungal molecular diversity at the submetre scale in a temperate grassland. *FEMS Microbiology Ecology* **64**: 260–270.

Nazeri NK, Lambers H, Tibbett M, Ryan MH. 2013. Do arbuscular mycorrhizas or heterotrophic soil microbes contribute toward plant acquisition of a pulse of mineral phosphate? *Plant Soil* **373**:699–710.

Novoa A, González L, Maravcová L, Pyšek P. 2013. Constraints to native plant species establishment in coastal dune communities invaded by *Carpobrotus edulis*: implications for restoration. *Biological Conservation* **164**:1–9.

Pfeffer PE, Bago B, Shachar-hill Y. 2001. Exploring mycorrhizal function with NMR spectroscopy. *New Phytologist* **150**:543–553.

Pringle A, Bever JD, Gardes M, Parrent JL, Rillig MC, Klironomos JN. 2009. Mycorrhizal symbioses and plant invasions. *Annual Review of Ecology, Evolution, and Systematics* **40**:699–715.

Reigosa MJ, Sánchez-Moreiras A, González L. 1999. Ecophysiological approach in Allelopathy. *Critical Reviews in Plant Sciences* **18**: 577–608.

Reinhart KO, Greene E, Callaway RM. 2005. Effects of *Acer platanoides* invasion on understory plant communities and tree regeneration in the northern Rocky Mountains. *Ecography* **28**:573–582.

Richardson DM, Rejmánek M. 2011. Trees and shrubs as invasive alien species—a global review. *Diversity and Distributions* **17**: 788–809.

Richardson DM, Allsopp N, D'antonio CM, Milton SJ, Rejmánek M. 2000. Plant invasions—the role of mutualisms. *Biological Reviews* **75**:65–93.

Rolfe SA, Scholes JD. 1995. Quantitative imaging of chlorophyll fluorescence. *New Phytologist* **131**:69–79.

Shannon SM, Bauer JT, Anderson WE, Reynolds HL. 2014. Plant-soil feedbacks between invasive shrubs and native forest understory species lead to shifts in the abundance of mycorrhizal fungi. *Plant and Soil* **382**:317–328.

Smith SE, Read DJ. 1997. *Mycorrhizal symbiosis*, 2nd edn. London: Academic Press.

Smith SE, Read DJ. 2008. *Mycorrhizal symbiosis*, 3rd edn. London: Academic Press.

Souza-Alonso P, Novoa A, González L. 2014. Soil biochemical alterations and microbial community responses under *Acacia dealbata* link invasion. *Soil Biology and Biochemistry* **79**:100–108.

Souza-Alonso P, Guisande-Collazo A, González L. 2015. Gradualism in *Acacia dealbata* link invasion: impact on soil chemistry and microbial community over a chronological sequence. *Soil Biology and Biochemistry* **80**:315–323.

Stinson KA, Campbell SA, Powell JR, Wolfe BE, Callaway RM, Thelen GC, Hallett SG, Prati D, Klironomos JN. 2006. Invasive plant suppresses the growth of native tree seedlings by disrupting belowground mutualisms. *PLoS Biology* **4**:e140.

Tanner RA, Gange AC. 2013. The impact of two non-native plant species on native flora performance: potential implications for habitat restoration. *Plant Ecology* **214**:423–432.

Vandenkoornhuyse P, Husband R, Daniell TJ, Watson IJ, Duck JM, Fitter AH, Young JPW. 2002. Arbuscular mycorrhizal community composition associated with two plant species in a grassland ecosystem. *Molecular Ecology* **11**:1555–1564.

Van der Heijden MGA, Klironomos JN, Ursic M, Moutoglis P, Streitwolf-Engel R, Boller T, Wiemken A, Sanders IR. 1998. Mycorrhizal fungal diversity determines plant biodiversity, ecosystem variability and productivity. *Nature* **396**:69–72.

Van der Heijden MGA, Bardgett RD, Van Straalen NM. 2008. The unseen majority: soil microbes as drivers of plant diversity and productivity in terrestrial ecosystems. *Ecology Letters* **11**:296–310.

Van Elsas JD, Boersma FGH. 2011. A review of molecular methods to study the microbiota of soil and the mycosphere. *European Journal of Soil Biology* **47**:77–87.

Vogelsang KM, Bever JD. 2009. Mycorrhizal densities decline in association with nonnative plants and contribute to plant invasion. *Ecology* **90**:399–407.

Walker C. 2005. A simple blue staining technique for arbuscular mycorrhizal and other root-inhabiting fungi. *Inoculum* **56**:68–69.

Willis A, Rodrigues BF, Harris PJC. 2013. The ecology of arbuscular mycorrhizal fungi. *Critical Reviews in Plant Sciences* **32**:1–20.

Yang R, Zhou G, Zan S, Guo F, Su N, Li J. 2014. Arbuscular mycorrhizal fungi facilitate the invasion of *Solidago canadensis* L. in southeastern China. *Acta Oecologica* **61**:71–77.

Zhang Q, Yang R, Tang J, Yang H, Hu S, Chen X. 2010. Positive feedback between mycorrhizal fungi and plants influences plant invasion success and resistance to invasion. *PLoS ONE* **5**:1–10.

14

Conflicting genomic signals affect phylogenetic inference in four species of North American pines

Tomasz E. Koralewski[1]*, Mariana Mateos[2] and Konstantin V. Krutovsky[1,3,4,5]

[1] Department of Ecosystem Science and Management, Texas A&M University, 2138 TAMU, College Station, TX 77843-2138, USA
[2] Department of Wildlife and Fisheries Sciences, Texas A&M University, 2258 TAMU, College Station, TX 77843-2258, USA
[3] Department of Forest Genetics and Forest Tree Breeding, Büsgen-Institute, Georg-August University of Göttingen, Büsgenweg 2, D-37077 Göttingen, Germany
[4] N.I. Vavilov Institute of General Genetics, Russian Academy of Sciences, Moscow 119333, Russia
[5] Genome Research and Education Center, Siberian Federal University, 50a/2 Akademgorodok, Krasnoyarsk 660036, Russia

Associate Editor: Chelsea D. Specht

Abstract. Adaptive evolutionary processes in plants may be accompanied by episodes of introgression, parallel evolution and incomplete lineage sorting that pose challenges in untangling species evolutionary history. Genus *Pinus* (pines) is one of the most abundant and most studied groups among gymnosperms, and a good example of a lineage where these phenomena have been observed. Pines are among the most ecologically and economically important plant species. Some, such as the pines of the southeastern USA (southern pines in subsection *Australes*), are subjects of intensive breeding programmes. Despite numerous published studies, the evolutionary history of *Australes* remains ambiguous and often controversial. We studied the phylogeny of four major southern pine species: shortleaf (*Pinus echinata*), slash (*P. elliottii*), longleaf (*P. palustris*) and loblolly (*P. taeda*), using sequences from 11 nuclear loci and maximum likelihood and Bayesian methods. Our analysis encountered resolution difficulties similar to earlier published studies. Although incomplete lineage sorting and introgression are two phenomena presumptively underlying our results, the phylogenetic inferences seem to be also influenced by the genes examined, with certain topologies supported by sets of genes sharing common putative functionalities. For example, genes involved in wood formation supported the clade *echinata–taeda*, genes linked to plant defence supported the clade *echinata–elliottii* and genes linked to water management properties supported the clade *echinata–palustris*. The support for these clades was very high and consistent across methods. We discuss the potential factors that could underlie these observations, including incomplete lineage sorting, hybridization and parallel or adaptive evolution. Our results likely reflect the relatively short evolutionary history of the subsection that is thought to have begun during the middle Miocene and has been influenced by climate fluctuations.

Keywords: *Australes*; drought tolerance; parallel evolution; phylogeny; plant defence; southern pines; wood formation.

Introduction

Plant genome structure and evolution are subjects of intensive investigations with recent milestone advances spanning whole-genome sequencing of some of the largest plant genomes, such as those of pines and spruces (Birol *et al.* 2013; Nystedt *et al.* 2013; Krutovsky *et al.* 2014; Neale *et al.* 2014; Zimin *et al.* 2014). These advancements bring opportunities to ask more focussed

* Corresponding author's e-mail address: tkoral@tamu.edu

questions about the dynamic processes that have con-tributed to plant genome evolution and influenced phylo-genetic relationships among plant species. Duplications, from single gene to whole genome, are an intrinsic pro-cess that appears to be the major force driving genome evolution in plants (Freeling 2009; Li *et al.* 2015). Other processes, such as introgression, incomplete lineage sort-ing and parallel evolution, primarily influenced by extrin-sic factors, additionally contribute to plant genome complexity and evolution (Martinsen *et al.* 2001; Wood *et al.* 2005; Willyard *et al.* 2009). The latter processes may also shape phenotypic and genotypic similarities among species becoming a challenge in phylogenetic studies.

Pines (genus *Pinus*, family Pinaceae) comprise a group of coniferous tree species occurring almost exclusively in the Northern Hemisphere, with populations often domin-ant across areas of the North Temperate Zone (Critchfield and Little 1966). They are keystone species of the vast boreal forest ecosystem. They provide pulp and timber products as primary commodities (Lowe *et al.* 1999; Borders and Bailey 2001; Alexander *et al.* 2002; Wear and Greis 2002; Croitoru 2007), but some additional pine forest products may include, for example, pine nuts, berries, herbs and mushrooms (Alexander *et al.* 2002; Bürgi *et al.* 2013; Nagasaka 2013). They provide a pivotal habitat for wildlife species (Brockerhoff *et al.* 2008), and greatly contribute to carbon storage and other ecosystem services (Goodale *et al.* 2002; McKinley *et al.* 2011). Due to their aesthetic value, they add recre-ational and ornamental dimensions in urban and subur-ban areas (Tyrväinen and Väänänen 1998; Knoth *et al.* 2002). These qualities extend to southeastern forest eco-systems of the USA, where pines may grow also in mixed forests with hardwoods (Sharitz *et al.* 1992), and where they additionally play an important role in ecosystem recovery from natural disturbances. For example, longleaf pine has developed complex adaptations to fire that allow fast stand regeneration (Outcalt 2000), and loblolly pine helps minimize soil erosion and provides watershed protection due to its fast growth (Schultz 1997).

Four major pines of the southeastern USA were investi-gated in this study: shortleaf (*Pinus echinata*), slash (*P. elliottii*), longleaf (*P. palustris*) and loblolly (*P. taeda*) (four of the 'southern pines'; subsection *Australes*, section *Trifoliae*, genus *Pinus*). They are widely cultivated, greatly dominating the southern US forest inventory (Sternitzke and Nelson 1970), and are, therefore, a subject of breeding programmes in the region (Wear and Greis 2002; McKeand *et al.* 2003; Fox *et al.* 2007). The traditional classification (Little and Critchfield 1969) considered 11 species in this subsection with habitat stretching cumulatively from the southeastern USA, through Mexico, to the Caribbean and Central America: slash (*P. elliottii*), spruce (*P. glabra*), long-leaf (*P. palustris*), pond (*P. serotina*) and loblolly (*P. taeda*) pines in the southeastern USA; shortleaf (*P. echinata*), Table Mountain (*P. pungens*) and pitch (*P. rigida*) pines in the eastern USA; Cuban pine (*P. cubensis*) in Cuba; West Indian pine (*P. occidentalis*) in the West Indies; and Carib-bean pine (*P. caribaea*) in both the West Indies and adjacent Central America. Attempts to refine the phyl-ogeny of *Australes* are ongoing, but typically have been approached in the broader context of the Pinaceae family or along with other subsections.

Several previous investigations have provided insights to the phylogenetic relationships of the four species on which we focus in our study (i.e. *P. echinata*, *P. elliottii*, *P. palustris* and *P. taeda*; Fig. 1). Adams and Jackson (1997) found a very close relationship between *P. palustris* and *P. taeda* based on 21 morphological characters, but their relation-ship to *P. echinata* and *P. elliottii* remained unresolved, and no statistical support was provided. Later, using RAPD mar-kers and the neighbour-joining method, Dvorak *et al.* (2000) suggested a close relationship among *P. echinata*, *P. palustris* and *P. taeda*, in which *P. echinata* and *P. taeda* were sister lineages, and *P. elliottii* was sister to *P. caribaea*. Using a supertree approach and previously published phy-logenies based on both morphological and molecular data, Grotkopp *et al.* (2004) also inferred a close relationship among *P. echinata*, *P. palustris* and *P. taeda*, but found that *P. palustris* was sister to *P. taeda*; *P. elliottii* was, again, sister to *P. caribaea*. Gernandt *et al.* (2005) used

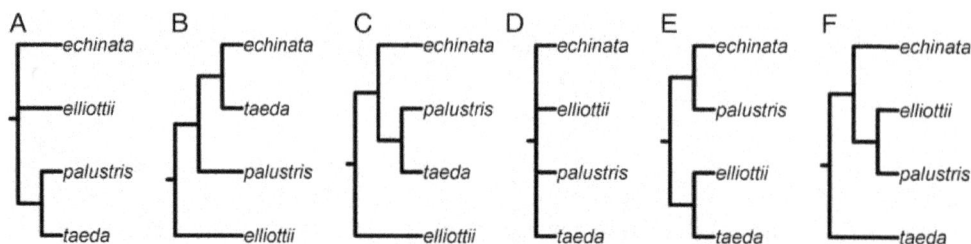

Figure 1. Cladograms for the four *Australes* species—*P. echinata, P. elliottii, P. palustris* and *P. taeda*—from published studies. (A) Adams and Jackson (1997). (B) Dvorak *et al.* (2000). (C) Grotkopp *et al.* (2004). (D) Gernandt *et al.* (2005). (E) Eckert and Hall (2006). (F) Hernández-León *et al.* (2013). The original studies presented the four species in a broader context along with other pines.

two chloroplast genes in 101 pine species, but relationships among the four *Australes* species remained unresolved. Eckert and Hall (2006) used four chloroplast genes in 83 pine species. In their study, *P. echinata* and *P. palustris* were placed in one clade with low support, and *P. elliottii* and *P. taeda* were placed in another. Most recently, based on five chloroplast DNA (cpDNA) markers, Hernández-León *et al.* (2013) placed *P. echinata*, *P. elliottii* and *P. palustris* within the same clade, where *P. elliottii* and *P. palustris* were sisters, and *P. taeda* was in a separate clade, albeit with low bootstrap support (BS). Consequently, the monophyly of the four species was placed under question.

Evidence from cpDNA used in recent studies additionally questioned monophyly of the subsection *Australes* as defined by Little and Critchfield (1969), and suggested that the 11 *Australes* species may be scattered throughout a larger clade of over twice as many species (Gernandt *et al.* 2005; Eckert and Hall 2006; Hernández-León *et al.* 2013). Chloroplast genomes, however, may follow different evolutionary trajectories than nuclear genomes, with potentially confounding effects in phylogenetic studies (Rieseberg and Soltis 1991). In pines, chloroplast genomes are strictly paternally inherited, thus having smaller effective population sizes than nuclear genomes. Additionally, cpDNA experiences lower substitution rates (Willyard *et al.* 2007). Consequently, these factors may lead to discordant patterns of polymorphism between the two genomes (Soltis *et al.* 1992; Hong *et al.* 1993*b*). Moreover, cpDNA differentiation among populations of a species can be high (Hong *et al.* 1993*a*), and the cpDNA sequence variation in pines may be unevenly distributed throughout the genome requiring a representative locus sampling (Whittall *et al.* 2010). Finally, foreign chloroplast genomes introduced through hybridization may undergo a rapid fixation—chloroplast capture (Rieseberg and Soltis 1991; Matos and Schaal 2000; Liston *et al.* 2007).

Apart from the specificity of cpDNA evolution, additional problems with taxonomic classification of *Australes* may stem from incomplete lineage sorting, a phenomenon reported in pines (Syring *et al.* 2007; Willyard *et al.* 2009), especially considering the relatively recent evolutionary history of the subsection. Radiation within *Australes* is thought to have begun only 7–15 million years ago (MYA) (Willyard *et al.* 2007; Hernández-León *et al.* 2013). Large overlap in present-day areas of the four species studied here has fostered occasional or historic hybridization between some of them that may have additionally contributed to insufficient resolution in phylogenetic studies. Natural hybridization has been observed between *P. echinata* and *P. taeda* (Zobel 1953; Mergen *et al.* 1965; Smouse and Saylor 1973; Edwards-Burke *et al.* 1997), between *P. palustris* and *P. taeda* (Chapman 1922) and between *P. elliottii* and *P. palustris* (Mergen 1958). Almost all possible

hybrids for these species can be artificially generated in controlled crosses, with the exception of *P. echinata* and *P. palustris* that are not considered interfertile despite a reported hybrid case (Snyder and Squillace 1966; Campbell *et al.* 1969). Additionally, similar environmental constraints could lead to a parallel support for beneficial alleles and/or removal of detrimental ones. Nevertheless, sympatry could have been repeatedly disturbed in the past; for example, during the Pleistocene, *P. taeda* was likely constricted to two refugia, *P. elliottii* and *P. palustris* could have been separated, each one occupying one of the two refugia, while *P. echinata*'s range was probably continuous due to its cold-hardiness (Wells *et al.* 1991; Schmidtling and Hipkins 1998; Schmidtling 2003). Each of these factors, or their combination, could confound phylogenetic inference.

The challenge with inferring relationships among *Australes* indicates that more work is needed. Nuclear markers are recommended to use when introgression, cytoplasmic in particular, or lineage sorting phenomena are present (Soltis *et al.* 1992). Consequently, we used 11 nuclear protein coding genes. Our objective was to refine and potentially clarify the phylogenetic relationships among *P. echinata*, *P. elliottii*, *P. palustris* and *P. taeda*. *Pinus pinaster* was used as an outgroup. Despite the larger dataset and application of advanced methods, untangling their phylogeny was not straightforward. We discuss potential factors that likely contributed to this problem and could explain the difficulties in inferring phylogenies using multiple genes observed in previous studies.

Methods

Source of data

Data from 32 genes recently sequenced and annotated in four pines from subsection *Australes*, namely *P. taeda* (Brown *et al.* 2004; González-Martínez *et al.* 2006), *P. echinata*, *P. elliottii* and *P. palustris* (Koralewski *et al.* 2014), were used to query the NCBI GenBank database (Benson *et al.* 2013) for orthologs in related pine (genus *Pinus*) taxa that could be used as appropriate outgroup species. Eleven putative orthologous genes were identified in *P. pinaster* (Table 1), which was used as an outgroup. Given that the ingroup species belong to a single subsection, it is possible that a species from a more closely related subsection could be a more optimal outgroup for phylogenetic analyses; however, *P. pinaster* is the best studied species with respect to the genes investigated in the southern pines, allowing us to utilize more sequence data in the analyses. We assumed orthology of the genes based on exon–intron structure and very high sequence similarity. For each putative ortholog, the E-value was 3E-46 or less and identity score was 95 % or more in at least one comparison of an ingroup species

Table 1. Genes used in the study and their NCBI accession numbers. Genomic DNA sequences for the ingroup (*P. echinata*, *P. elliottii*, *P. palustris* and *P. taeda*) were newly generated and presented in Koralewski *et al.* (2014), and both genomic DNA and mRNA sequences for the outgroup (*P. pinaster*) were already available in NCBI.

Gene name	Gene abbreviation	Pinus echinata	Pinus elliottii	Pinus palustris	Pinus taeda	Pinus pinaster
4-coumarate:CoA ligase	4cl	KF158811	KF158813	KF158814	KF158816	HM482497
arabinogalactan 4	agp-4	KF158819	KF158821	KF158822	KF158824	AM501931
trans-cinnamate 4-hydroxylase 2	c4h-2	KF158875	KF158877	KF158878	KF158880	JN013973
cinnamyl alcohol dehydrogenase	cad	KF158882	KF158883	KF158884	KF158885	FN824799
cellulose synthase	cesA3	KF158898	KF158900	KF158901	KF158903	FN257074
caffeate O-methyltransferase	comt-2	KF158906	KF158908	KF158909	KF158911	HE574557
dehydrin 2	dhn-2	KF158924	KF158926	KF158927	KF158929	HE796687
early response to drought 3	erd3	KF158931	KF158932	KF158933	KF158934	EU020011
glycine hydroxymethyltransferase	glyhmt	KF158935	KF158936	KF158937	KF158938	HE574564
ABII protein phosphatase 2C-like	pp2c	KF158952	KF158953	KF158954	KF158955	EU020014
chloroplast Cu/Zn superoxide dismutase	sod-chl	KF158978	KF158979	KF158980	KF158981	AF434186

with *P. pinaster*; additionally, in comparisons where the *E*-value was higher than E-100, the identity score was 99 %. However, well-annotated, mapped and well-assembled whole-genome data are ultimately needed to validate orthology. Such data are still unavailable, although the first incomplete draft genome assembly was published recently for loblolly pine (Neale *et al.* 2014; Zimin *et al.* 2014).

Multiple alignment

The FASTA sequences for individual genes were aligned using BioEdit (ver. 7.0.9.0) (Hall 1999) and merged into one dataset in SeaView (ver. 4.0) (Galtier *et al.* 1996). Conversion from FASTA to NEXUS was done in SeaView, and from FASTA to PHYLIP manually in the text editor Notepad++ (ver. 5.9.3) (Ho 2011). Coding sites were assigned based on annotation data in NCBI GenBank for each individual gene separately using DnaSP (ver. 5.10.01) (Librado and Rozas 2009). The merged NEXUS file was further manually annotated, and one of five categories was assigned to each site: a codon position (1, 2 or 3), intron or 3′UTR. All sites with missing data were removed from the analysis **[see Supporting Information—Dataset S1]**.

Partitioning

Partitioning schemes and models of molecular evolution were evaluated using PartitionFinder (ver. 1.1.1) (Lanfear *et al.* 2012). Model selection was restricted to the set of models implemented in the software in which we intended to run the phylogenetic analysis (parameter 'model =' in PartitionFinder). At most 42 models were evaluated simultaneously based on Akaike information criterion (AIC) and Bayesian information criterion (BIC). The parameters describing gamma distribution of rates among sites (Γ or G) and a proportion of invariable sites (I) were considered among the models exclusively, i.e. none of the evaluated models accounted for both Γ and I jointly, because of reported problems with independent optimization of these two parameters (see discussions and relevant references in Stamatakis 2008: pp. 20–21 and in Stamatakis 2014: p. 49). Additionally, in the group of models evaluated for GARLI (Zwickl 2006) (see below), one or more of the K81 models (K81, K81 + I and K81 + G, depending on the subset of data; a few data subsets were affected) caused convergence problems and were not considered. Depending on the intended phylogenetic analysis, alternative partitioning schemes were then implemented **[see Supporting Information—Table S1]**: (i) by-gene-site: best partitioning schemes for 39 sets corresponding to different genes and site categories within genes, or (ii) by-gene: best models identified for each of the 11 genes in the set (no partitioning within a gene). Additionally, in the cases where each gene was analysed separately, models were identified for every gene independently with sites assigned to one of the five site categories (by-site).

Phylogenetic analysis

The combined dataset of 11 genes was subjected to two gene tree methods with partitioning by-gene-site: maximum likelihood (ML; GARLI, ver. 2.01) (Zwickl 2006) and

Bayesian inference (BI; MrBayes, ver. 3.2.2) (Ronquist *et al.* 2012). Species trees were reconstructed using the Bayesian method BEST (ver. 2.3) (Liu 2008) with partitioning by-gene. To account for potential polytomies, and thus potential for the star tree paradox (Suzuki *et al.* 2002; Lewis *et al.* 2005), we ran Phycas (ver. 1.2.0) (Lewis *et al.* 2010), also a Bayesian method, with partitioning by-gene-site. We then analysed each gene separately using GARLI and MrBayes (partitioning by-site). The MrBayes outputs were further investigated using Bayesian concordance analysis (Ané *et al.* 2007), as implemented in BUCKy (ver. 1.4.0, mbsum ver. 1.4.2) (Larget *et al.* 2010), for seven values of α: 0.01, 0.5, 1, 2, 5, 10 and 10 000. The value of α corresponds to the probability that loci share the same tree (the lower α the higher the probability, and vice versa), thus affecting the clade support (the higher the probability the higher support). Formally, BUCKy is not considered a species-tree method; however, the resulting primary concordance tree can be generally considered comparable with species trees (for an in-depth discussion about the species methods, see Mateos *et al.* 2012).

All analyses were run in Windows 7 **[see Supporting Information—Table S1]** except for BUCKy, which was compiled and run under Cygwin. The tree figures were visualized in FigTree (ver. 1.4) (Rambaut 2013) and further edited manually to increase their readability and compactness.

Clade support and convergence

In order to determine the clade support, BS was calculated in GARLI using 1000 replicates **[see Supporting Information—Table S1]**. SumTrees (ver. 3.3.1) (Sukumaran and Holder 2010) was used to generate majority consensus trees. Posterior probability (PP) estimates were used for the BI methods. To verify that the Markov chain Monte Carlo analysis converged on a stationary distribution and that the sampling of the distribution was adequate, the following criteria were applied for MrBayes and BEST: (i) stable PP values, (ii) small and stable average standard deviation of the split frequencies of independent runs, (iii) potential

scale reduction factor close to 1 and (iv) an effective sample size of at least 200 for the posterior probabilities. The conditions (i) and (iv) were evaluated also for Phycas, and additionally the split PP plot and split sojourn plot were examined. Samples prior to reaching stationarity were discarded as 'burnin'. The conditions (i) and (iv) were evaluated in Tracer (ver. 1.5.0) (Rambaut and Drummond 2009). Average standard deviation of mean sample-wide concordance factor (CF) was examined for BUCKy, and the CF was used to determine clade support.

Results

Combined evidence from all genes

Three methods, GARLI, MrBayes and Phycas, recovered the clade *echinata–elliottii* for the concatenated matrix of 11 genes. Support from the ML method was much lower (highest BS = 65 %, AIC partitioning) than that from the BI methods (lowest PP = 0.98 in MrBayes, BIC partitioning). The BI methods additionally supported the clade *echinata–elliottii–palustris*, but support varied considerably (from PP < 0.50 in Phycas, AIC, to PP = 0.94 in MrBayes, BIC). Significant differences between the ML and BI methods in support for the clade *echinata–elliottii*, and in the case of the clade *echinata–elliottii–palustris*, also between Phycas and MrBayes (Fig. 2) are characteristic signatures of true or approximate star phylogenies (Suzuki *et al.* 2002; Lewis *et al.* 2005).

BEST supported the clade *echinata–taeda*, absent in the gene-tree methods, with PP = 0.53 (AIC and BIC; Fig. 3). This clade was present also in the primary concordance tree in BUCKy (CF ranging from 0.35 to 0.44, depending on α and model selection criterion). Another clade present in the BUCKy's primary concordance tree was *elliottii–palustris* (CF ranging from 0.24 to 0.44, depending on α and model selection criterion).

Individual gene approach

Given the results from the analysis of the combined dataset resulting in mostly unresolved relationships among the studied species, and the conflicts found between

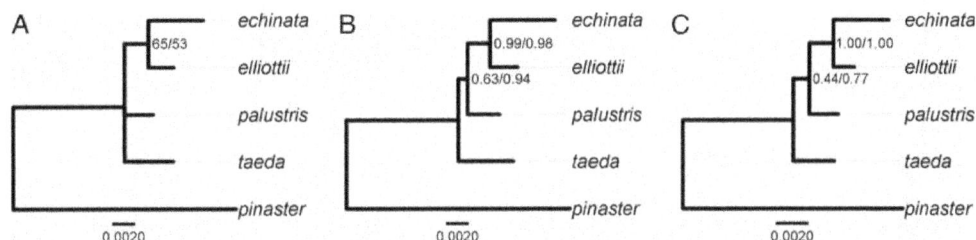

Figure 2. Joint analysis of 11 genes in *P. echinata*, *P. elliottii*, *P. palustris* and *P. taeda* using GARLI (A), MrBayes (B) and Phycas (C). Numbers at nodes correspond to BS for GARLI (%; AIC/BIC), and PP for MrBayes and Phycas (AIC/BIC). Branch lengths are shown for the BIC partitioning schemes.

the gene-tree and species-tree methods, we looked at each gene separately using ML (GARLI) and BI (MrBayes). The two approaches produced consistent results for each gene, although the difference in clade support varied among the genes [see Supporting Information—Fig. S1]. We, therefore, clustered genes into three groups following similarity in topologies. The genes *4cl*, *c4h-2* and *cesA3* (Group A) supported the clade *echinata–taeda*, the genes *agp-4* and *sod-chl* (Group B) supported the clade *echinata–elliottii* and the genes *dhn-2* and *erd3* (Group C) supported the clade *echinata–palustris*. Group C genes disagreed in the placement of the species *P. elliottii* and *P. taeda*; however, we decided to focus on the similarity. Additional analysis in the R environment (ver. 2.15.2) (R Core Team 2012) supported these groupings—principal coordinate analysis (function pcoa, package ape ver. 3.0-11) (Paradis *et al.* 2004) run on Robinson–Foulds distance matrix for the individual gene trees (function RF.dist, package phangorn ver. 1.99-7) (Schliep 2011) placed the three groups of genes in distinct clusters [see Supporting Information—Fig. S2]. Each set was partitioned by-gene-site, and analysed using GARLI and MrBayes. In general, the BS for the clades jointly supported by genes within each group increased, the PP reached (or stayed at) 1.00 and the level of support for these clades became consistent between GARLI and

MrBayes (Fig. 4) [see Supporting Information—Fig. S3]. In the case of Group B, the clade *palustris–taeda*, previously present only in the *agp-4* gene tree, was also recovered, although with low support. In Group C, *P. elliottii* and *P. taeda* were placed as sister taxa in a clade with low BS and high PP, which reflects the conflict between the two individual gene trees.

Discussion

The discordance among various approaches applied in this study mirrors the conflicting results from previously published work (see Introduction). Comparison of the topologies supported by Group A vs. Group B vs. Group C genes shows highly consistent results within each group across different methods, yet the conclusions contradict each other among the groups. The clade *echinata–taeda*, supported by Group A genes, was previously recovered by Dvorak *et al.* (2000) using RAPD markers and the neighbour-joining method with BS = 0.90. The species in the clade *palustris–taeda*, supported by the gene *agp-4* (Group B), were placed in one clade by Adams and Jackson (1997) based on parsimony analysis of morphological characters, and by Grotkopp *et al.* (2004) through a supertree approach (57 % of individual trees agreed with the supertree at the node). Both clades *echinata–palustris* and *elliottii–taeda* supported by the Group C genes corresponded to grouping in Eckert and Hall (2006), who used chloroplast data.

Given life history of the subsection (see Introduction), long-distance gene flow through pollen, hybridization events that still occur today and large (but variable in the past) population sizes of the investigated species, incomplete lineage sorting could be the simplest and most straightforward explanation of the pattern observed in our study. It is a known phenomenon in pines (Syring *et al.* 2007; Willyard *et al.* 2009), and it is likely to confound phylogenetic inference in a clade as recent as *Australes*. Interestingly, however, the genes forming the three groups

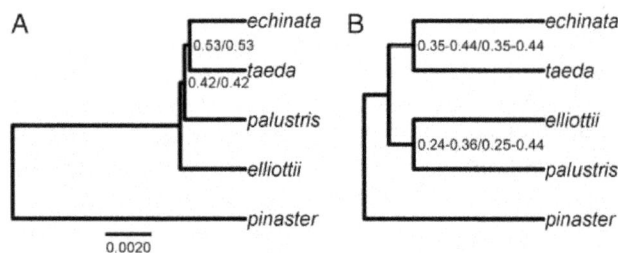

Figure 3. Joint analysis of 11 genes in *P. echinata*, *P. elliottii*, *P. palustris* and *P. taeda* using BEST (A) and BUCKy (cladogram; B). Numbers at nodes correspond to PP for BEST (AIC/BIC) and CFs for BUCKy (AIC/BIC; range of CFs for α values from 10 000 to 0.01).

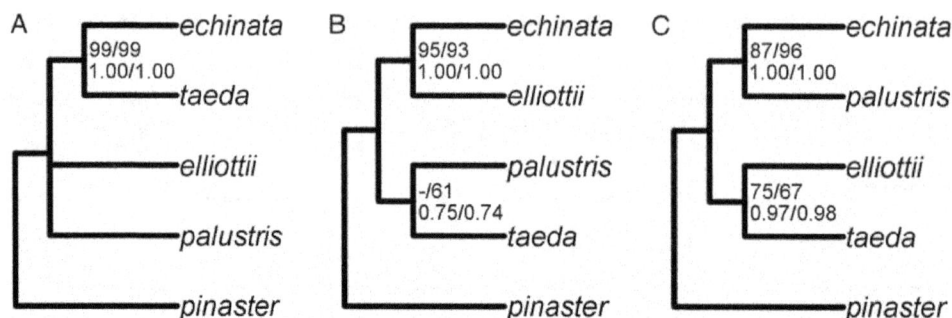

Figure 4. Analysis of three groups of genes: Group A (*4cl*, *c4h-2* and *cesA3*; A), Group B (*agp-4* and *sod-chl*; B) and Group C (*dhn-2* and *erd3*; C) in *P. echinata*, *P. elliottii*, *P. palustris* and *P. taeda* using GARLI and MrBayes. Cladograms are shown. Numbers at nodes correspond to clade support: GARLI (AIC/BIC; top row) and MrBayes (AIC/BIC; bottom row). The BS for the clade *palustris–taeda* in Group B was <0.50.

Table 2. Groups of genes based on phylogenetic analysis of *P. echinata*, *P. elliottii*, *P. palustris* and *P. taeda*, their selected putative functions and numbers of nonsynonymous (N) and silent (S) nucleotide substitutions in each group.

Group	Genes	Putative function	N	S
A	*4cl*, *c4h-2* and *cesA3*	Wood properties	2	10
B	*agp-4* and *sod-chl*	Plant defence, water management	4	10
C	*dhn-2* and *erd3*	Water stress recognition and response	2	11

are also connected at the putative functional level (Table 2). The Group A genes all directly affect wood properties: *4cl* and *c4h-2* are both part of the lignin biosynthesis pathway (Whetten and Sederoff 1995; Boerjan *et al.* 2003), and *cesA3* is involved in cellulose synthesis (Pot *et al.* 2005). Another gene in our dataset, *comt-2*, is also involved in lignin biosynthesis (Whetten and Sederoff 1995; Boerjan *et al.* 2003) but did not recover the *echinata–taeda* clade. Unlike the other three genes in Group A, it supported the clade *elliottii–palustris*. An additional joint analysis of all four genes, using the same methodology, showed that the support for the previously recovered clade *echinata–taeda* almost did not change, while the clade *elliottii–palustris* received a low-to-moderate support **[see Supporting Information—Fig. S4]**. The latter two species were previously placed in one clade (BS ≤ 0.50) by Hernández-León *et al.* (2013) based on an ML method applied to plastid data. Conversely, the Group B and C genes are not involved in the lignin biosynthesis pathway directly. The Group B genes (*agp-4* and *sod-chl*) appear to contribute to plant defence, water management and other functions. Proteins from the arabinogalactan family play signalling and protective roles, and participate in xylem development, cell growth and expansion, programmed cell death and other processes (Loopstra and Sederoff 1995; Loopstra *et al.* 2000; Zhang *et al.* 2000, 2003; Showalter 2001; Seifert and Roberts 2007). *Sod-chl* has antioxidative properties (Huttunen and Heiska 1988; Bowler *et al.* 1992). It was also drought responsive (Costa *et al.* 1998) and hence was considered a drought-tolerance candidate gene in loblolly pine (González-Martínez *et al.* 2006). *Dhn-2* and *erd3* (Group C) are also associated with water stress recognition and response. *Dhn-2*, a *dehydrin*, is responsive to dehydration stress and also plays other protective functions (González-Martínez *et al.* 2006; Eveno *et al.* 2008). The role of *erd3* is less known, but it has been found active in an early stage of dehydration stress (González-Martínez *et al.* 2006; Eveno *et al.* 2008).

Regardless of the extent of the potential effects of shared ancestral polymorphisms, two other hypotheses of parallel evolution within the functional gene groups could be pursued in follow-up studies. The functionally bound clades could have resulted from a transfer of adaptive alleles via introgressive hybridization from an adapted pine donor, followed by positive selection and subsequent purifying or balancing selection acting in both donor and acceptor populations. Alternatively, speciation could have begun while the locus was already under purifying or balancing selection, which continued working simultaneously in both populations in parallel, given common habitat locations and environmental pressures. Sympatry, shifting locale of the optimum habitat, population size changes and the recent evolutionary origin of the southern pines could have facilitated hybridization, while crucial roles of the genes belonging to the three groups could have resulted in preservation of adaptive variants, especially under similar environmental selection pressures. A tight physical linkage among the loci within each group can be excluded as an alternative possible explanation for the observed phenomenon. Most of the genes studied here are mapped to different linkage groups in *P. taeda*, or far from each other, although no linkage data were found for *dhn-2* and *erd3* **[see Supporting Information—Table S2]**. These two hypotheses require more assumptions to explain the observed pattern when compared with the hypothesis of incomplete lineage sorting, and therefore, the latter should be preferred, although some form of interplay of all three is certainly possible.

In order to reject the hypothesis of incomplete lineage sorting, and to pursue an alternative, limitations of our study need to be overcome. Multiple individuals sampled for each species would allow for thorough identification of shared ancestral polymorphisms and variation fixed at the species level. Including more genes per functional group would allow to examine whether the observed pattern holds also for other members of a given pathway or process, or if it is purely stochastic. Samples from all members of the clade *Australes* would likely help improve overall robustness of the phylogenetic inference. Additionally, to test the monophyly of the clade *Australes* in the traditional sense (Little and Critchfield 1969), which has been questioned by cpDNA-based studies (Gernandt *et al.* 2005; Eckert and Hall 2006; Hernández-León *et al.* 2013), species traditionally classified as *Attenuatae*, *Leiophyllae* and *Oocarpae* should be included in the nuclear marker-based study. Alternative outgroup species could also be considered, especially those from subsections *Contortae* and *Ponderosae* that are more closely related to the ingroup than *P. pinaster*. Solving these caveats, however, requires additional resources and cannot be done purely analytically. Given the limitations of the data, our primary intention was to apply a spectrum of methods to a gene sample that would maximize genome coverage.

Recent estimates of radiation within *Australes* suggested that it could have begun as recently as 7 MYA (*P. taeda – P. radiata* split), although the split between the ancestors of *Ponderosae* and *Australes* might have happened as early as 15 MYA (Willyard *et al.* 2007). This timeframe overlaps with the mid- to late-Miocene (about 14–15 MYA), starting at or directly following the middle Miocene climate transition (MMCT), a period of cooling and ice-sheet expansion that took place about 14 MYA (Shevenell *et al.* 2004). Pines experienced habitat locale shifts both before and after the MMCT, for example, during the Eocene (56–34 MYA), interpreted as the major stimulus for pine divergence at the time, and during the Pleistocene (2.6–0.01 MYA) (Millar 1993). The MMCT likely affected population sizes, species range and distribution of the allelic variation, and probably had the momentum to trigger radiation within *Australes*. The potential for hybridization events resulting in introgression was likely greater back in time, especially when the ancestral species were far less diverged and stressed by recurrent changes in environmental pressures and by range shifts. Range expansions could have then brought multiple genetic effects (Excoffier *et al.* 2009) including increase in frequency of rare (and also newly introduced) variants. This process would have happened much faster if the newly acquired alleles were advantageous.

The adaptations shared among the southern pines and shaped by the vibrant historic climate are particularly interesting in the light of the ongoing and forthcoming climate changes, amidst the discussion on assisted migration (Vitt *et al.* 2010; Krutovsky *et al.* 2012; Koralewski *et al.* 2015). The historical events might have led to increased standing genetic variation in these species, directly influencing their level of adaptability and making them somewhat 'climate-change ready'. Additional inquiries directed towards the loci studied here could improve breeding strategies in the face of climate change (Krutovsky *et al.* 2013).

Conclusions

Incomplete lineage sorting, introgression and parallel evolution can explain inconsistencies observed in the phylogenetic analysis of the four southern pines. However, more data are needed to discriminate among these hypotheses. The conflicting signals were vigorously tested, but evidence in the current data was not robust enough to support potent claims, and thus, the simplest hypothesis of incomplete lineage sorting may be preferred, while the alternatives may be pursued in future studies. To overcome limitations of our study, additional sampling should include multiple individuals per species, additional species that form one clade with the four pines investigated here, less distant outgroup species and additional functionally related genes. Our work provided new insights into the *Australes* phylogeny, but their evolutionary history remains elusive.

Sources of Funding

The project was supported by the United States Department of Agriculture (USDA) Cooperative State Research, Education, and Extension Service (CSREES) and Texas Agricultural Experiment Station (TAES) McIntire-Stennis Project (TEX09122-0210381). The Pine Integrated Network: Education, Mitigation, and Adaptation Project (PINEMAP), a Coordinated Agricultural Project funded by the USDA National Institute of Food and Agriculture, Award #2011-68002-30185, provided support during preparation of this manuscript.

Contributions by the Authors

T.E.K., M.M. and K.V.K. conceived and designed the study. T.E.K. arranged the data and ran the software. T.E.K., M.M. and K.V.K. interpreted results. T.E.K. drafted the manuscript. T.E.K., M.M. and K.V.K. edited, read and approved the final version of the manuscript.

Acknowledgements

We are grateful to Dr Alan E. Pepper, Dr Clare A. Gill (Texas A&M University) and Dr Ruzong Fan (National Institute of Child Health and Human Development, National Institutes of Health) for valuable discussions, suggestions and comments. We thank Dr Earl M. 'Fred' Raley (Texas A&M Forest Service) for comments on the final version of the manuscript. T.E.K. would like to thank Dr Thomas D. Byram (Texas A&M Forest Service, Texas A&M University) for his support during this study. We thank anonymous reviewers, the Associate Editor, Dr Chelsea D. Specht (University of California – Berkeley), and the Chief Editor, Dr J. Hall Cushman (Sonoma State University), for suggestions that helped us improve the manuscript.

Supporting Information

The following additional information is available in the online version of this article —

Dataset S1. Multiple alignment in NEXUS format.

Figure S1. Separate analysis of each gene for the four *Australes* species: *P. echinata*, *P. elliottii*, *P. palustris* and *P. taeda*. Cladograms are shown. From top to bottom: *4cl*, *c4h-2* and *cesA3* (A); *agp-4* and *sod-chl* (B); *dhn-2* and *erd3* (C); *cad*, *glyhmt*, *pp2c* and *comt-2* (D). Numbers at nodes correspond to clade support: GARLI (AIC/BIC; top row) and MrBayes (AIC/BIC; bottom row).

Figure S2. Principal coordinate analysis run on Robin-son–Foulds distance matrix for the individual gene trees. Group A gene names (*4cl*, *c4h-2* and *cesA3*) are in green, Group B gene names (*agp-4* and *sod-chl*) are in orange and Group C gene names (*dhn-2* and *erd3*) are in blue.

Figure S3. Joint analysis of the genes from Group A (*4cl*, *c4h-2* and *cesA3*; A), Group B (*agp-4* and *sod-chl*; B) and Group C (*dhn-2* and *erd3*; C) in *P. echinata*, *P. elliottii*, *P. palustris* and *P. taeda* using GARLI and MrBayes. From top to bottom: GARLI (AIC), GARLI (BIC), MrBayes (AIC) and MrBayes (BIC). Numbers at nodes correspond to clade support.

Figure S4. Joint analysis of the genes from Group A (*4cl*, *c4h-2* and *cesA3*) and *comt-2* in *P. echinata*, *P. elliottii*, *P. palustris* and *P. taeda* using GARLI and MrBayes. From top to bottom: GARLI (AIC), GARLI (BIC), MrBayes (AIC) and MrBayes (BIC). Numbers at nodes correspond to clade support.

Table S1. Partitioning schemes and software settings.

Table S2. Linkage information for the studied genes.

Literature Cited

Adams DC, Jackson JF. 1997. A phylogenetic analysis of the southern pines (*Pinus* subsect. *Australes* Loudon): biogeographical and ecological implications. *Proceedings of the Biological Society of Washington* 110:681–692.

Alexander SJ, Pilz D, Weber NS, Brown E, Rockwell VA. 2002. Mushrooms, trees, and money: value estimates of commercial mushrooms and timber in the Pacific Northwest. *Environmental Management* 30:129–141.

Ané C, Larget B, Baum DA, Smith SD, Rokas A. 2007. Bayesian estimation of concordance among gene trees. *Molecular Biology and Evolution* 24:412–426.

Benson DA, Cavanaugh M, Clark K, Karsch-Mizrachi I, Lipman DJ, Ostell J, Sayers EW. 2013. GenBank. *Nucleic Acids Research* 41:D36–D42.

Birol I, Raymond A, Jackman SD, Pleasance S, Coope R, Taylor GA, Saint Yuen MM, Keeling CI, Brand D, Vandervalk BP, Kirk H, Pandoh P, Moore RA, Zhao YJ, Mungall AJ, Jaquish B, Yanchuk A, Ritland C, Boyle B, Bousquet J, Ritland K, MacKay J, Bohlmann J, Jones SJM. 2013. Assembling the 20 Gb white spruce (*Picea glauca*) genome from whole-genome shotgun sequencing data. *Bioinformatics* 29:1492–1497.

Boerjan W, Ralph J, Baucher M. 2003. Lignin biosynthesis. *Annual Review of Plant Biology* 54:519–546.

Borders BE, Bailey RL. 2001. Loblolly pine—pushing the limits of growth. *Southern Journal of Applied Forestry* 25:69–74.

Bowler C, Van Montagu M, Inzé D. 1992. Superoxide dismutase and stress tolerance. *Annual Review of Plant Physiology and Plant Molecular Biology* 43:83–116.

Brockerhoff EG, Jactel H, Parrotta JA, Quine CP, Sayer J. 2008. Plantation forests and biodiversity: oxymoron or opportunity? *Biodiversity and Conservation* 17:925–951.

Brown GR, Gill GP, Kuntz RJ, Langley CH, Neale DB. 2004. Nucleotide diversity and linkage disequilibrium in loblolly pine. *Proceedings of the National Academy of Sciences of the United States of America* 101:15255–15260.

Bürgi M, Gimmi U, Stuber M. 2013. Assessing traditional knowledge on forest uses to understand forest ecosystem dynamics. *Forest Ecology and Management* 289:115–122.

Campbell TE, Hamaker JM, Schmitt DM. 1969. Longleaf pine × shortleaf pine—a new hybrid. *Bulletin of the Torrey Botanical Club* 96:519–524.

Chapman HH. 1922. A new hybrid pine (*Pinus palustris* × *Pinus taeda*). *Journal of Forestry* 20:729–734.

Costa P, Bahrman N, Frigerio J-M, Kremer A, Plomion C. 1998. Water-deficit-responsive proteins in maritime pine. *Plant Molecular Biology* 38:587–596.

Critchfield WB, Little EL. 1966. *Geographic distribution of the pines of the world*. Miscellaneous Publication 991. Washington, DC: USDA Forest Service, U.S. Government Printing Office.

Croitoru L. 2007. How much are Mediterranean forests worth? *Forest Policy and Economics* 9:536–545.

Dvorak WS, Jordon AP, Hodge GP, Romero JL. 2000. Assessing evolutionary relationships of pines in the *Oocarpae* and *Australes* subsections using RAPD markers. *New Forests* 20:163–192.

Eckert AJ, Hall B. 2006. Phylogeny, historical biogeography, and patterns of diversification for *Pinus* (Pinaceae): phylogenetic tests of fossil-based hypotheses. *Molecular Phylogenetics and Evolution* 40:166–182.

Edwards-Burke MA, Hamrick JL, Price RA. 1997. Frequency and direction of hybridization in sympatric populations of *Pinus taeda* and *P. echinata* (Pinaceae). *American Journal of Botany* 84:879–886.

Eveno E, Collada C, Guevara MA, Léger V, Soto A, Díaz L, Léger P, González-Martínez SC, Cervera MT, Plomion C, Garnier-Géré PH. 2008. Contrasting patterns of selection at *Pinus pinaster* Ait. drought stress candidate genes as revealed by genetic differentiation analyses. *Molecular Biology and Evolution* 25:417–437.

Excoffier L, Foll M, Petit RJ. 2009. Genetic consequences of range expansions. *Annual Review of Ecology Evolution and Systematics* 40:481–501.

Fox TR, Jokela EJ, Allen HL. 2007. The development of pine plantation silviculture in the southern United States. *Journal of Forestry* 105:337–347.

Freeling M. 2009. Bias in plant gene content following different sorts of duplication: tandem, whole-genome, segmental, or by transposition. *Annual Review of Plant Biology* 60:433–453.

Galtier N, Gouy M, Gautier C. 1996. SEAVIEW and PHYLO_WIN: two graphic tools for sequence alignment and molecular phylogeny. *Computer Applications in the Biosciences* 12:543–548.

Gernandt DS, López GG, García SO, Liston A. 2005. Phylogeny and classification of *Pinus*. *Taxon* **54**:29–42.

González-Martínez SC, Ersoz E, Brown GR, Wheeler NC, Neale DB. 2006. DNA sequence variation and selection of tag single-nucleotide polymorphisms at candidate genes for drought-stress response in *Pinus taeda* L. *Genetics* **172**:1915–1926.

Goodale CL, Apps MJ, Birdsey RA, Field CB, Heath LS, Houghton RA, Jenkins JC, Kohlmaier GH, Kurz W, Liu SR, Nabuurs G-J, Nilsson S, Shvidenko AZ. 2002. Forest carbon sinks in the Northern Hemisphere. *Ecological Applications* **12**:891–899.

Grotkopp E, Rejmánek M, Sanderson MJ, Rost TL. 2004. Evolution of genome size in pines (*Pinus*) and its life-history correlates: Supertree analyses. *Evolution* **58**:1705–1729.

Hall TA. 1999. BioEdit: a user-friendly biological sequence alignment editor and analysis program for Windows 95/98/NT. *Nucleic Acids Symposium Series* **41**:95–98.

Hernández-León S, Gernandt DS, Pérez de la Rosa JA, Jardón-Barbolla L. 2013. Phylogenetic relationships and species delimitation in *Pinus* section *Trifoliae* inferrred from plastid DNA. *PLoS ONE* **8**:e70501.

Ho D. 2011. *Notepad++ v5.9.3*. http://notepad-plus-plus.org (11 October 2011).

Hong Y-P, Hipkins VD, Strauss SH. 1993*a*. Chloroplast DNA diversity among trees, populations and species in the California closed-cone pines (*Pinus radiata, Pinus muricata* and *Pinus attenuata*). *Genetics* **135**:1187–1196.

Hong Y-P, Krupkin AB, Strauss SH. 1993*b*. Chloroplast DNA transgresses species boundaries and evolves at variable rates in the California closed-cone pines (*Pinus radiata, P. muricata,* and *P. attenuata*). *Molecular Phylogenetics and Evolution* **2**:322–329.

Huttunen S, Heiska E. 1988. Superoxide dismutase (SOD) activity in Scots pine (*Pinus sylvestris* L.) and Norway spruce (*Picea abies* L. Karst.) needles in northern Finland. *European Journal of Forest Pathology* **18**:343–350.

Knoth J, Frampton J, Moody R. 2002. Genetic improvement of Virginia pine planting stock for Christmas tree production in South Carolina. *HortTechnology* **12**:675–678.

Koralewski TE, Brooks JE, Krutovsky KV. 2014. Molecular evolution of drought tolerance and wood strength related candidate genes in loblolly pine (*Pinus taeda* L.). *Silvae Genetica* **63**:59–66.

Koralewski TE, Wang H-H, Grant WE, Byram TD. 2015. Plants on the move: assisted migration of forest trees in the face of climate change. *Forest Ecology and Management* **344**:30–37.

Krutovsky KV, Burczyk J, Chybicki I, Finkeldey R, Pyhäjärvi T, Robledo-Arnuncio JJ. 2012. Gene flow, spatial structure, local adaptation, and assisted migration in trees. In: Schnell RJ, Priyadarshan PM, eds. *Genomics of tree crops*. New York: Springer Science, Inc., 71–116.

Krutovsky K, Byram T, Whetten R, Wheeler N, Neale D, Lu M, Koralewski T, Loopstra C. 2013. PINEMAP + PineRefSeq = future forests. In: Sommer E, ed. *PINEMAP (Pine Integrated Network: Education, Mitigation, and Adaptation Project) Year 2 Annual Report | March 2012-February 2013 "Mapping the future of southern pine management in a changing world"*, 26–27.

Krutovsky KV, Oreshkova NV, Putintseva Yu. A, Ibe AA, Deych KO, Shilkina EA. 2014. Preliminary results of *de novo* whole genome sequencing of the Siberian larch (*Larix sibirica* Ledeb.) and the Siberian stone pine (*Pinus sibirica* Du Tour). *Siberian Journal of Forest Science* **1**:79–83.

Lanfear R, Calcott B, Ho SYW, Guindon S. 2012. PartitionFinder: combined selection of partitioning schemes and substitution models for phylogenetic analyses. *Molecular Biology and Evolution* **29**:1695–1701.

Larget BR, Kotha SK, Dewey CN, Ané C. 2010. BUCKy: gene tree/species tree reconciliation with Bayesian concordance analysis. *Bioinformatics* **26**:2910–2911.

Lewis PO, Holder MT, Holsinger KE. 2005. Polytomies and Bayesian phylogenetic inference. *Systematic Biology* **54**:241–253.

Lewis PO, Holder MT, Swofford DL. 2010. *Phycas ver. 1.2.0.* http://www.phycas.org (17 September 2013).

Li Z, Baniaga AE, Sessa EB, Scascitelli M, Graham SW, Rieseberg LH, Barker MS. 2015. Early genome duplications in conifers and other seed plants. *Science Advances* **1**:e1501084.

Librado P, Rozas J. 2009. DnaSP v5: a software for comprehensive analysis of DNA polymorphism data. *Bioinformatics* **25**:1451–1452.

Liston A, Parker-Defeniks M, Syring JV, Willyard A, Cronn R. 2007. Interspecific phylogenetic analysis enhances intraspecific phylogeographical inference: a case study in *Pinus lambertiana*. *Molecular Ecology* **16**:3926–3937.

Little EL, Critchfield WB. 1969. *Subdivisions of the genus Pinus (pines)*. Miscellaneous Publication No. 1144. Washington, DC: USDA Forest Service, U.S. Government Printing Office.

Liu L. 2008. BEST: Bayesian estimation of species trees under the coalescent model. *Bioinformatics* **24**:2542–2543.

Loopstra CA, Sederoff RR. 1995. Xylem-specific gene expression in loblolly pine. *Plant Molecular Biology* **27**:277–291.

Loopstra CA, Puryear JD, No E-G. 2000. Purification and cloning of an arabinogalactan-protein from xylem of loblolly pine. *Planta* **210**:686–689.

Lowe WJ, Byram TD, Bridgwater FE. 1999. Selecting loblolly pine parents for seed orchards to minimize the cost of producing pulp. *Forest Science* **45**:213–216.

Martinsen GD, Whitham TG, Turek RJ, Keim P. 2001. Hybrid populations selectively filter gene introgression between species. *Evolution* **55**:1325–1335.

Mateos M, Hurtado LA, Santamaria CA, Leignel V, Guinot D. 2012. Molecular systematics of the deep-sea hydrothermal vent endemic brachyuran family Bythograeidae: a comparison of three Bayesian species tree methods. *PLoS ONE* **7**:e32066.

Matos JA, Schaal BA. 2000. Chloroplast evolution in the *Pinus montezumae* complex: a coalescent approach to hybridization. *Evolution* **54**:1218–1233.

McKeand S, Mullin T, Byram T, White T. 2003. Deployment of genetically improved loblolly and slash pines in the South. *Journal of Forestry* **101**:32–37.

McKinley DC, Ryan MG, Birdsey RA, Giardina CP, Harmon ME, Heath LS, Houghton RA, Jackson RB, Morrison JF, Murray BC, Pataki DE, Skog KE. 2011. A synthesis of current knowledge on forests and carbon storage in the United States. *Ecological Applications* **21**:1902–1924.

Mergen F. 1958. Genetic variation in needle characteristics of slash pine and in some of its hybrids. *Silvae Genetica* **7**:1–9.

Mergen F, Stairs GR, Snyder EB. 1965. Natural and controlled loblolly × shortleaf pine hybrids in Mississippi. *Forest Science* **11**:306–314.

Millar CI. 1993. Impact of the Eocene on the evolution of *Pinus* L. *Annals of the Missouri Botanical Garden* **80**:471–498.

Nagasaka K. 2013. Comparative economic value estimation of matsutake mushroom and timber production in Swedish Scots pine forest. Master Thesis no. 218, Alnarp, Southern Swedish Forest Research Centre, Swedish University of Agricultural Sciences, Sweden.

Neale DB, Wegrzyn JL, Stevens KA, Zimin AV, Puiu D, Crepeau MW, Cardeno C, Koriabine M, Holtz-Morris AE, Liechty JD, Martínez-García PJ, Vasquez-Gross HA, Lin BY, Zieve JJ, Dougherty WM, Fuentes-Soriano S, Wu L-S, Gilbert D, Marçais G, Roberts M, Holt C, Yandell M, Davis JM, Smith KE, Dean JFD, Lorenz WW, Whetten RW, Sederoff R, Wheeler N, McGuire PE, Main D, Loopstra CA, Mockaitis K, deJong PJ, Yorke JA, Salzberg SL, Langley CH. 2014. Decoding the massive genome of loblolly pine using haploid DNA and novel assembly strategies. *Genome Biology* **15**:R59.

Nystedt B, Street NR, Wetterbom A, Zuccolo A, Lin Y-C, Scofield DG, Vezzi F, Delhomme N, Giacomello S, Alexeyenko A, Vicedomini R, Sahlin K, Sherwood E, Elfstrand M, Gramzow L, Holmberg K, Hällman J, Keech O, Klasson L, Koriabine M, Kucukoglu M, Käller M, Luthman J, Lysholm F, Niittylä T, Olson Å, Rilakovic N, Ritland C, Rosselló JA, Sena J, Svensson T, Talavera-López C, Theißen G, Tuominen H, Vanneste K, Wu Z-Q, Zhang B, Zerbe P, Arvestad L, Bhalerao R, Bohlmann J, Bousquet J, Gil RG, Hvidsten TR, de Jong P, MacKay J, Morgante M, Ritland K, Sundberg B, Thompson SL, Van de Peer Y, Andersson B, Nilsson O, Ingvarsson PK, Lundeberg J, Jansson S. 2013. The Norway spruce genome sequence and conifer genome evolution. *Nature* **497**:579–584.

Outcalt KW. 2000. The longleaf pine ecosystem of the South. *Native Plants Journal* **1**:42–53.

Paradis E, Claude J, Strimmer K. 2004. APE: analyses of phylogenetics and evolution in R language. *Bioinformatics* **20**:289–290.

Pot D, McMillan L, Echt C, Le Provost G, Garnier-Géré P, Cato S, Plomion C. 2005. Nucleotide variation in genes involved in wood formation in two pine species. *New Phytologist* **167**:101–112.

Rambaut A. 2013. *Figtree v. 1.4*. http://tree.bio.ed.ac.uk/software/figtree (16 January 2014).

Rambaut A, Drummond AJ. 2009. *Tracer v. 1.5*. http://beast.bio.ed.ac.uk/Tracer (26 April 2013).

R Core Team. 2012. *R: a language and environment for statistical computing*. Vienna, Austria: R Foundation for Statistical Computing. http://www.R-project.org (31 October 2012).

Rieseberg LH, Soltis DE. 1991. Phylogenetic consequences of cytoplasmic gene flow in plants. *Evolutionary Trends in Plants* **5**:65–84.

Ronquist F, Teslenko M, Van der Mark P, Ayres DL, Darling A, Höhna S, Larget B, Liu L, Suchard MA, Huelsenbeck JP. 2012. MrBayes 3.2: efficient Bayesian phylogenetic inference and model choice across a large model space. *Systematic Biology* **61**:539–542.

Schliep KP. 2011. phangorn: phylogenetic analysis in R. *Bioinformatics* **27**:592–593.

Schmidtling RC. 2003. The southern pines during the Pleistocene. *ISHS Acta Horticulturae* **615**:203–209.

Schmidtling RC, Hipkins V. 1998. Genetic diversity in longleaf pine (*Pinus palustris*): influence of historical and prehistorical events. *Canadian Journal of Forest Research* **28**:1135–1145.

Schultz RP. 1997. *Loblolly pine. The ecology and culture of loblolly pine (Pinus taeda L.)*. Agricultural Handbook 713. Washington, DC: USDA Forest Service, U.S. Government Printing Office.

Seifert GJ, Roberts K. 2007. The biology of arabinogalactan proteins. *Annual Review of Plant Biology* **58**:137–161.

Sharitz RR, Boring LR, Van Lear DH, Pinder JE III. 1992. Integrating ecological concepts with natural resource management of southern forests. *Ecological Applications* **2**:226–237.

Shevenell AE, Kennett JP, Lea DW. 2004. Middle Miocene Southern Ocean cooling and Antarctic cryosphere expansion. *Science* **305**:1766–1770.

Showalter AM. 2001. Arabinogalactan-proteins: structure, expression and function. *Cellular and Molecular Life Sciences* **58**:1399–1417.

Smouse PE, Saylor LC. 1973. Studies of the *Pinus rigida-serotina* complex II. Natural hybridization among the *Pinus rigida-serotina* complex, *P. taeda* and *P. echinata*. *Annals of the Missouri Botanical Garden* **60**:192–203.

Snyder EB, Squillace AE. 1966. Cone and seed yields from controlled breeding of southern pines. New Orelans, LA: Southern Forest Experiment Station, USDA Forest Service (USDA For. Serv. Res. Pap. SO-22).

Soltis DE, Soltis PS, Milligan BG. 1992. Intraspecific chloroplast DNA variation: systematic and phylogenetic implications. In: Soltis PS, Soltis DE, Doyle JJ, eds. *Molecular systematics of plants*. New York, NY: Chapman & Hall, 117–150.

Stamatakis A. 2008. *The RAxML 7.0.4 manual*. http://sco.h-its.org/exelixis/resource/download/oldPage/RAxML-Manual.7.0.4.pdf (16 January 2014).

Stamatakis A. 2014. *The RAxML v8.0.X manual*. http://sco.h-its.org/exelixis/resource/download/NewManual.pdf (16 January 2014).

Sternitzke HS, Nelson TC. 1970. The southern pines of the United States. *Economic Botany* **24**:142–150.

Sukumaran J, Holder MT. 2010. DendroPy: a Python library for phylogenetic computing. *Bioinformatics* **26**:1569–1571.

Suzuki Y, Glazko GV, Nei M. 2002. Overcredibility of molecular phylogenies obtained by Bayesian phylogenetics. *Proceedings of the National Academy of Sciences of the United States of America* **99**:16138–16143.

Syring J, Farrell K, Businsky R, Cronn R, Liston A. 2007. Widespread genealogical nonmonophyly in species of *Pinus* subgenus *Strobus*. *Systematic Biology* **56**:163–181.

Tyrväinen L, Väänänen H. 1998. The economic value of urban forest amenities: an application of the contingent valuation method. *Landscape and Urban Planning* **43**:105–118.

Vitt P, Havens K, Kramer AT, Sollenberger D, Yates E. 2010. Assisted migration of plants: changes in latitudes, changes in attitudes. *Biological Conservation* **143**:18–27.

Wear DN, Greis JG. 2002. Southern forest resource assessment: summary of findings. *Journal of Forestry* **100**:6–14.

Wells OO, Switzer GL, Schmidtling RC. 1991. Geographic variation in Mississippi loblolly pine and sweetgum. *Silvae Genetica* **40**:105–119.

Whetten R, Sederoff R. 1995. Lignin biosynthesis. *Plant Cell* **7**:1001–1013.

Whittall JB, Syring J, Parks M, Buenrostro J, Dick C, Liston A, Cronn R. 2010. Finding a (pine) needle in a haystack: chloroplast genome sequence divergence in rare and widespread pines. *Molecular Ecology* **19**:100–114.

Willyard A, Syring J, Gernandt DS, Liston A, Cronn R. 2007. Fossil calibration of molecular divergence infers a moderate mutation rate and recent radiations for *Pinus*. *Molecular Biology and Evolution* **24**:90–101.

Willyard A, Cronn R, Liston A. 2009. Reticulate evolution and incomplete lineage sorting among the ponderosa pines. *Molecular Phylogenetics and Evolution* **52**:498–511.

Wood TE, Burke JM, Rieseberg LH. 2005. Parallel genotypic adaptation: when evolution repeats itself. *Genetica* **123**: 157–170.

Zhang Y, Sederoff RR, Allona I. 2000. Differential expression of genes encoding cell wall proteins in vascular tissues from vertical and bent loblolly pine trees. *Tree Physiology* **20**:457–466.

Zhang Y, Brown G, Whetten R, Loopstra CA, Neale D, Kieliszewski MJ, Sederoff RR. 2003. An arabinogalactan protein associated with secondary cell wall formation in differentiating xylem of loblolly pine. *Plant Molecular Biology* **52**:91–102.

Zimin A, Stevens KA, Crepeau MW, Holtz-Morris A, Koriabine M, Marçais G, Puiu D, Roberts M, Wegrzyn JL, de Jong PJ, Neale DB, Salzberg SL, Yorke JA, Langley CH. 2014. Sequencing and assembly of the 22-Gb loblolly pine genome. *Genetics* **196**:875–890.

Zobel BJ. 1953. Are there natural loblolly-shortleaf pine hybrids? *Journal of Forestry* **51**:494–495.

Zwickl DJ. 2006. Genetic algorithm approaches for the phylogenetic analysis of large biological sequence datasets under the maximum likelihood criterion. PhD Dissertation, The University of Texas at Austin, USA.

15

Earlier snowmelt and warming lead to earlier but not necessarily more plant growth

Carolyn Livensperger[1]*, Heidi Steltzer[2], Anthony Darrouzet-Nardi[3], Patrick F. Sullivan[4], Matthew Wallenstein[1] and Michael N. Weintraub[5]

[1] Department of Ecosystem Science and Sustainability, Colorado State University, Fort Collins, CO 80523, USA
[2] Biology Department, Fort Lewis College, Durango, CO 81301, USA
[3] Department of Biological Sciences, University of Texas at El Paso, El Paso, TX 79968, USA
[4] Environment and Natural Resources Institute, University of Alaska Anchorage, Anchorage, AK 99508, USA
[5] Department of Environmental Sciences, University of Toledo, Toledo, OH 43606, USA

Associate Editor: Colin M. Orians

Abstract. Climate change over the past ~50 years has resulted in earlier occurrence of plant life-cycle events for many species. Across temperate, boreal and polar latitudes, earlier seasonal warming is considered the key mechanism leading to earlier leaf expansion and growth. Yet, in seasonally snow-covered ecosystems, the timing of spring plant growth may also be cued by snowmelt, which may occur earlier in a warmer climate. Multiple environmental cues protect plants from growing too early, but to understand how climate change will alter the timing and magnitude of plant growth, experiments need to independently manipulate temperature and snowmelt. Here, we demonstrate that altered seasonality through experimental warming and earlier snowmelt led to earlier plant growth, but the aboveground production response varied among plant functional groups. Earlier snowmelt without warming led to early leaf emergence, but often slowed the rate of leaf expansion and had limited effects on aboveground production. Experimental warming alone had small and inconsistent effects on aboveground phenology, while the effect of the combined treatment resembled that of early snowmelt alone. Experimental warming led to greater aboveground production among the graminoids, limited changes among deciduous shrubs and decreased production in one of the dominant evergreen shrubs. As a result, we predict that early onset of the growing season may favour early growing plant species, even those that do not shift the timing of leaf expansion.

Keywords: Arctic tussock tundra; climate change; plant phenology; plant production; seasonality.

Introduction

Seasonality in temperate to polar ecosystems is shifting through earlier seasonal warming and changes in precipitation regimes that lead to earlier snowmelt (Schwartz *et al.* 2006; Hayhoe *et al.* 2007; Ernakovich *et al.* 2014). Plant communities are responding through changes in timing of life-cycle events such as leaf expansion and flowering (i.e. phenology) (Fitter and Fitter 2002; Thompson and Clark 2008), shifts in species relative abundance (Harte and Shaw 1995; Willis *et al.* 2008), species' range shifts (Walther *et al.* 2005; Gottfried *et al.* 2012) and greater aboveground plant production (Knapp and Smith 2001; Wang *et al.* 2012). These observations of altered seasonality and plant community changes

* Corresponding author's e-mail address: clivensp@rams.colostate.edu

correspond to a period of increasing global temperatures, but experiments are still needed to determine mechanisms and develop predictive models as climate change continues (Pau et al. 2011; Richardson et al. 2012).

Temperature and photoperiod are known plant phenological cues that determine the timing of spring events, such as bud burst, leaf emergence and canopy development, and flowering (Cleland et al. 2007; Körner and Basler 2010; Polgar and Primack 2011). Experimental warming studies using different techniques, such as active warming through overhead infrared heaters or passive warming with open-top chambers (OTCs), demonstrate that many species begin growth and flowering earlier in warmed versus control plots (Cleland et al. 2006; Sherry et al. 2007; Reyes-Fox et al. 2014). However, responses can vary with some species not shifting or delaying the timing of spring events under warmed conditions (Hollister et al. 2005a; Reyes-Fox et al. 2014; Marchin et al. 2015). Similarly, long-term observations of phenological response to climate warming over time show an overall advance in timing of spring of an estimated 5–6 days $°C^{-1}$ (Wolkovich et al. 2012), but with interspecific variation (Fitter and Fitter 2002; Menzel et al. 2006). The variation in response suggests that phenology is cued by other environmental variables (e.g. photoperiod) for species within and across diverse plant communities from tundra, grassland and forest biomes.

The influence of snow cover on plant phenology is less well understood, in part because temperature change due to climate change and experimental warming influence when an area becomes snow-free. Development of early emerging species may be closely synchronized with timing of snowmelt in Arctic and alpine ecosystems (Galen and Stanton 1995; Høye et al. 2007), and indeed, later snowmelt due to increased snow depth has been shown to delay bud break of deciduous shrubs (Borner et al. 2008; Sweet et al. 2014). However, few experiments have examined the isolated effects of early snowmelt and summer warming in the Arctic or alpine, and often these effects are confounded either by the use of warming treatments such as OTCs to accelerate snowmelt or by the snow removal that reduces water inputs (Wipf and Rixen 2010).

Multifactor global change experiments have shown that plant production is sensitive to manipulations of abiotic factors, including air and soil warming, nutrients, CO_2 and precipitation (Fay et al. 2003; Zavaleta et al. 2003; Dukes et al. 2005; Dawes et al. 2011). Response to these factors is complex, with variation across plant communities due to differences in limiting factors (Smith et al. 2015), and variation within communities due to differences in functional group responses (Zavaleta et al. 2003; Wahren et al. 2005; Muldavin et al. 2008). In the

Arctic, production is strongly limited by nutrient availability, which in turn is sensitive to temperature and ongoing changes in the timing of seasonal climatic events such as snowmelt, soil thaw, the onset of freezing and snowfall (Billings and Mooney 1968; Weintraub and Schimel 2005). In recent years, both observational and experimental studies have linked increased production, specifically that of deciduous shrubs and graminoids, to warmer temperatures (Tape et al. 2006; Walker et al. 2006; Forbes et al. 2010; Elmendorf et al. 2012b; Sistla et al. 2013). However, a number of experiments that have manipulated summer temperature in both Arctic and alpine regions did not find a consistent increase in community-level aboveground net primary production (ANPP); rather, individual species or functional groups varied in their response (Chapin et al. 1995; Harte and Shaw 1995; Hollister et al. 2005b). Evergreen shrubs have responded to warming with positive, negative or no changes in production (Hollister et al. 2005b; Walker et al. 2006; Campioli et al. 2012), and they may be less likely to show short-term growth responses due to their conservative growth strategy (Chapin and Shaver 1996; Starr et al. 2008).

Changes in plant production may also be expected to vary in relation to changes in the timing of growth; for example, earlier leaf expansion may lead to greater productivity (Richardson et al. 2010). There is evidence that phenological 'tracking' of climate change across biomes can result in positive growth responses, through increased abundance, production or flowering effort (Cleland et al. 2012). However, physiological constraints and interactions of the affected species may prevent some plants from taking advantage of an earlier start to the growing season (Schwartz 1998; Richardson et al. 2010; Polgar and Primack 2011). One such constraint could be negative impacts of exposure to cold temperatures and freezing damage if snow melts early (Inouye 2008; Wipf et al. 2009). Differences in the onset and duration of plant growth can also vary due to differences in plant community composition; for example, deciduous shrub-dominated communities in Arctic tundra were shown to have longer peak growing seasons and greater carbon uptake than evergreen/graminoid communities in the same region (Sweet et al. 2015).

In the Arctic, climate is changing at a faster rate than in other regions, a trend that is expected to continue (Christensen et al. 2013). Rapidly increasing air temperature (~1 °C decade^{-1}) (Christensen et al. 2013), earlier snowmelt (3–5 days decade^{-1}) and later snowfall are changing the seasonality of this ecosystem (Serreze et al. 2000; Ernakovich et al. 2014). Landscape-scale observations via remote sensing suggest that vegetation phenology in the Arctic is indeed advancing and plant production is increasing (Myneni et al. 1997; Jia et al. 2003, 2009; Fraser et al. 2011; Zeng et al. 2011). Earlier snowmelt, especially in combination with warmer temperatures in early spring, should benefit plant growth, since it is the time of year with the greatest light and nutrient availability (Weintraub and Schimel 2005; Edwards and Jefferies 2013). However, experiments are needed to determine how shifts in seasonality will affect phenology of Arctic species, and how changes in phenology affect plant productivity and future community composition.

In Arctic tussock tundra, we established a 3-year study in which we altered seasonality through the independent and combined manipulation of air temperature and timing of snowmelt. We examined the response of spring phenology and plant production for key tundra species and hypothesized that:

(1) The timing of snowmelt and temperature are cues for initiating plant growth. We predicted that leaf appearance and expansion would advance due to early snowmelt and air warming for all species.

(2) The timing of snowmelt and temperature affect plant production. We predicted that early snowmelt and warmer temperatures would increase production of deciduous shrub, graminoid and forb species, but would not change production of evergreen shrubs.

(3) The timing of plant growth affects plant production. We predicted that earlier leaf expansion would lead to greater aboveground biomass at peak season.

Methods

Site description

The experiment was conducted near Imnavait Creek on the North Slope of Alaska, close to the Arctic Long-Term Ecological Research (LTER) site at Toolik Field Station. The plant community at Imnavait is moist acidic tussock tundra, characterized by the tussock forming sedge *Eriophorum vaginatum* and a high moss cover, including *Hylocomium* spp., *Aulacomnuim* spp. and *Dicranum* spp. Associated species include another sedge, *Carex bigelowii*, the deciduous shrubs *Betula nana* and *Salix pulchra*, the evergreen shrubs *Ledum palustre*, *Vaccinium vitis-idaea* and *Cassiope tetragonum*, and a variety of forbs [see Supporting Information—Table S1]. The old (~120 000–600 000 years; Whittinghill and Hobbie 2011), acidic soil (mean pH of 4.5) at this site is underlain by continuous permafrost, with an uneven surface layer of organic material 0–20 cm thick (Walker et al. 1994) and variable soil moisture.

Altered seasonality

For 3 years (2010–12), snowmelt was accelerated in five 8 × 12 m plots using radiation-absorbing black 50 % shade cloth that was placed over the snowpack in late April–early May. The dark fabric accelerated melt and allowed for minimal disturbance of the snowpack. The fabric was removed when plots became 80 % snow-free (determined by daily visual estimates). In 2012, we achieved a 10-day acceleration in the timing of snowmelt with early snowmelt plots becoming snow-free on May 16 and control plots snow-free on May 26. Snow was melted 4 and 15 days earlier in 2010 and 2011, respectively. As plots became snow-free, (OTCs were deployed on subplots within the accelerated snowmelt and control areas. The OTCs are hexagonal chambers with sloping sides, constructed of Plexiglas material that allows transmittance of wavelengths of light in the visible spectrum, enabling passive warming primarily through trapping solar radiation (Marion et al. 1997). Open-top chambers warmed air temperatures by an average of 1.4 °C in 2012. Further details of treatment effects on air temperature, soil temperature and soil moisture are available in **Supporting Information—Table S2**. The approximate area of both control and warming subplots was 1 m². Treatments were replicated five times in a full factorial, randomized split-plot design. Treatment abbreviations are as follows throughout the article: control (C), warming (W), early snowmelt (ES) and combined (W × ES).

Phenology

Five individuals of seven species were marked in each subplot and phenology events were monitored every 2–3 days from snowmelt through mid-August. Observations of 'leaf appearance' and 'leaf expansion' were recorded for each individual. Although definitions of events varied between functional groups, we generally considered leaf appearance to be the first observation of new green leaves and leaf expansion to be when an individual had a leaf that was fully expanded or had reached a previously determined size. For deciduous shrubs (*B. nana* and *S. pulchra*), leaf appearance was recorded at the first observed leaf bud burst, and leaf expansion when an individual had at least one fully unfurled leaf anywhere on the plant. Similarly, evergreen shrub (*L. palustre* and *V. vitis-idaea*) leaf appearance was recorded when the first leaf bud was visible, and full leaf expansion occurred when at least one leaf bud was fully open and leaves unfurled. *Eriophorum vaginatum* retains green leaf material over winter and often begins growth of new leaves and re-greening of old leaves before snow is completely melted (Chapin et al. 1979). Therefore, we recorded leaf appearance (new leaves >1 cm length) for *E. vaginatum* on the day of snowmelt, but did not consider this as a treatment effect. Rather than continuously measuring leaf length to record full leaf expansion, we determined leaf expansion for *E. vaginatum* to have occurred when a new leaf reached >4 cm length. We only considered growth of new leaves, which were identified as those with no senescent material at the leaf tip. We followed similar protocol for *C. bigelowii* leaf appearance (new leaf >1 cm length) and leaf expansion (new leaf >4 cm length), but leaf appearance was considered a treatment effect. First leaf appearance for the forb *P. bistorta* was marked when leaves were visible (generally >1 cm length) and leaf expansion when leaves were fully unrolled and >5 cm length.

Plant production

A destructive harvest to measure plant production, as characterized by growth of individuals in the current year, was carried out on the same species for which phenology was observed. The seven species chosen represented four functional groups and comprised the majority of vascular plant cover at our site [see Supporting Information—Table S1]. The harvest took place in the third year of treatments at peak growing season, which was determined by phenology observations and analysis of daily normalized difference vegetation index measurements showing that peak greenness (i.e. full canopy development) had occurred in each treatment (C. Livensperger and H. Steltzer, unpublished data). Randomly selected individuals were clipped in the field, and then taken back to the laboratory where old and new growth was separated and biomass measured. Eight individuals each of *B. nana*, *S. pulchra* and *L. palustre*, and 16 individuals each of *V. vitis-idea*, *E. vaginatum*, *C. bigelowii* and *P. bistorta* were collected from each subplot and pooled by species. Plant material was separated by tissue type, dried at 60 °C for 48 h and weighed.

Mean individual production for each species was calculated as the sum of current years' biomass divided by the number of individuals collected. Current years' biomass included leaves, new stems and secondary growth for shrub species. For graminoids and a forb, all live aboveground plant tissue was used, which may have included some growth from previous years for *E. vaginatum* and *C. bigelowii*. We calculated current annual secondary stem growth for *B. nana*, *S. pulchra* and *L. palustre* as a proportion of standing stem biomass, using previously determined annual growth rates of woody stems from the nearby Toolik Lake LTER site (Bret-Harte *et al.* 2002). For these species, leaves contributed more to total biomass than the calculated secondary growth. Secondary growth for the remaining shrub species, *V. vitis-idaea*, is negligible and, therefore, was left out of production

calculation for this species (Shaver 1986). Standing stem biomass, excluding current seasonal growth, for individual shrub stems varied among plots and likely is a result of variation prior to when the experiment was established. To control for this variation and better detect treatment effects, individual production data are presented in relation to standing stem biomass excluding current annual growth (i.e. g new production/g standing stem biomass).

Statistical analyses

For all analyses, the experiment was treated as a blocked split-plot design, where a large early snowmelt plot paired with an equally sized control plot comprise a single block. Plant responses and environmental variables were analysed using a mixed-model analysis of variance (ANOVA; SAS v 9.2, SAS Institute, Inc., Cary, NC, USA), with early snowmelt (ES) as the main plot factor and warming (W) as the within plot factor. A random effect of block was included to control for inherent variation between the five replicates. All data were checked for normality and were found to meet the assumptions of ANOVA. Linear regression was used to analyse the relationship between phenology and plant growth.

Results

Plant phenology

Early snowmelt was a strong driver of change in both the timing and rate of leaf appearance and expansion. These events advanced due to early snowmelt alone for all species except the forb, *P. bistorta* (Fig. 1), and the amount of change in timing varied between events, species and functional groups. The largest change in timing was a 10-day advance in leaf expansion for *E. vaginatum* (Fig. 1), corresponding to the 10-day advance in snowmelt through our snow manipulation. Leaf appearance and expansion of evergreen and deciduous shrubs were significantly earlier due to early snowmelt alone, advancing by 1–8 days for *B. nana*, *S. pulchra*, *L. palustre* and *V. vitis-idaea* (Fig. 1, Table 1). The advancement of leaf appearance versus leaf expansion differed in magnitude for *S. pulchra*, *V. vitis-idaea* and *C. bigelowii*, by increasing the number of days between leaf appearance and leaf expansion by 2–5 days. For example, in the early snowmelt treatment, leaf appearance for *S. pulchra* occurred 8 days earlier than the control, while leaf expansion advanced by only 3 days. For deciduous shrubs, evergreen shrubs and the forb, the shift in phenology was less than the 10-day advance in snowmelt, increasing the number of days after snowmelt to when canopy development (i.e. leaf expansion) began; this effectively slowed the

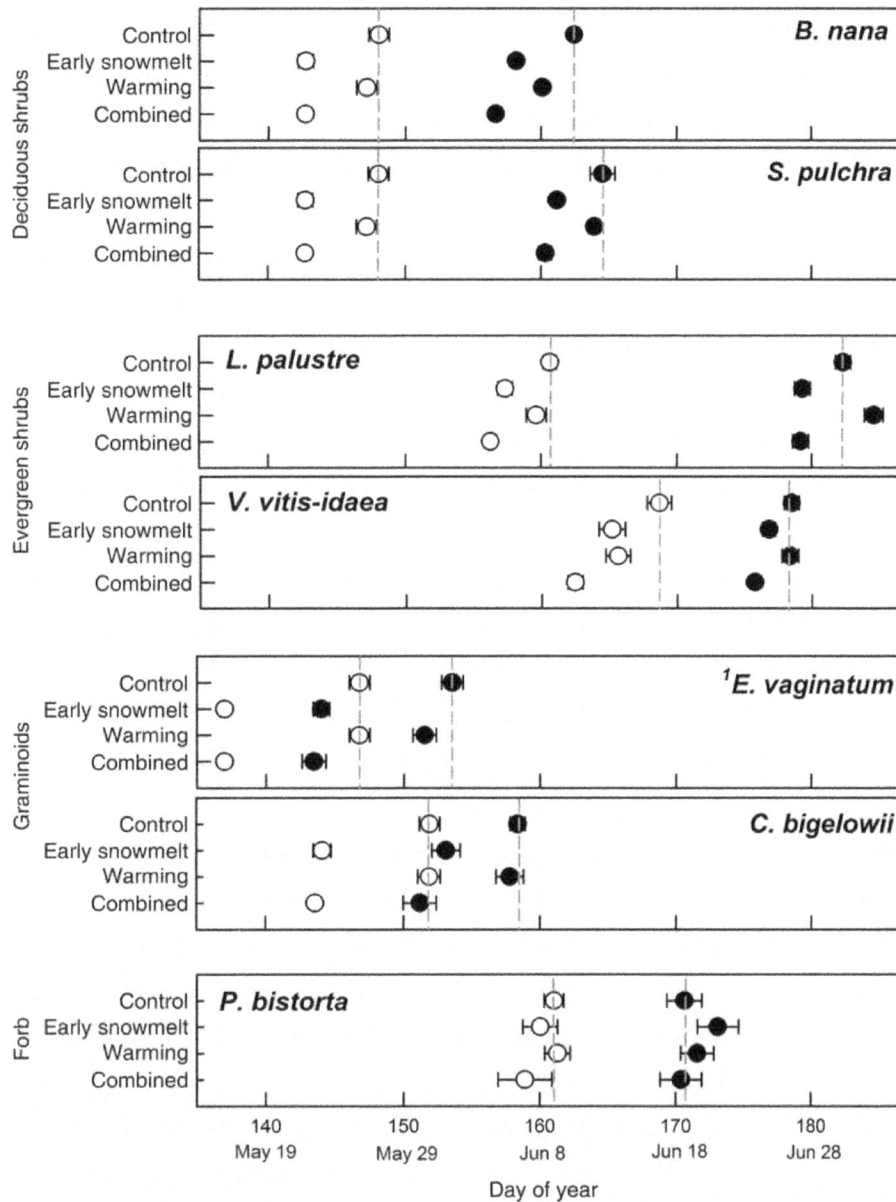

Figure 1. Dates of early-season phenology events, where open circles represent leaf appearance and filled circles represent leaf expansion. Points are the mean date of event ± 1 SEM. Vertical dashed lines denote mean event date for control plots. [1]*E. vaginatum* initiates growth underneath the snowpack so we did not consider leaf appearance to be a treatment effect; however, the event date is shown to signify the presence of new leaves at the time of snowmelt.

rate of plant production (Fig. 2, Table 2). The sedges, *E. vaginatum* and *C. bigelowii*, did not follow this pattern, with no evidence of a change in the number of days between leaf appearance and expansion (Fig. 2).

Warming also advanced the timing of leaf appearance and expansion for most species, but to a lesser extent than early snowmelt (Fig. 1, Table 1). All of the deciduous shrub and graminoid species advanced leaf phenology with warming alone, but only by 1 or 2 days (Fig. 1, Table 1). Evergreen shrubs showed contrasting responses to warming: leaf appearance for *V. vitis-idaea* advanced

by 3 days, while *L. palustre* leaf expansion was delayed for 2 days (Fig. 1, Table 1). Warming generally did not alter phenology in relation to the timing of snowmelt (Fig. 2, Table 2). One exception is that warming led to significantly faster leaf expansion following snowmelt for *B. nana*, effectively speeding plant production.

Phenological responses to the combination of early snowmelt and warming were generally comparable with the response to early snowmelt alone (Figs 1 and 2), and the interactive effect of warming × early snowmelt on phenology was never significant (Table 1). For

Table 1. Results of mixed-model ANOVA on timing of early-season phenology events. Leaf appearance for *E. vaginatum* was not considered a treatment effect and was excluded from the analysis. Bold values indicate a significant main effect of the treatment at $P \leq 0.05$.

	Warming			Early snowmelt			Warming × Early snowmelt		
	df	F	P	df	F	P	df	F	P
Leaf appearance									
B. nana	1, 85	0.88	0.3503	1, 4	9.46	**0.0373**	1, 85	1.49	0.2253
S. pulchra	1, 71	0.21	0.6501	1, 8	20.24	**0.0019**	1, 71	0.15	0.7045
L. palustre	1, 86	5.75	**0.0186**	1, 4	21.53	**0.0099**	1, 86	0.00	0.9775
V. vitis-idaea	1, 83	15.08	**0.0002**	1, 4	8.92	**0.0405**	1, 83	0.01	0.9241
C. bigelowii	1, 83	0.16	0.6910	1, 4	65.58	**0.0012**	1, 83	0.20	0.6533
P. bistorta	1, 65	0.13	0.7212	1, 7	1.34	0.2829	1, 65	0.25	0.6184
Leaf expansion									
B. nana	1, 86	55.71	**<0.0001**	1, 4	79.28	**0.0008**	1, 86	2.33	0.1308
S. pulchra	1, 77	1.86	0.1766	1, 77	42.81	**<0.0001**	1, 77	0.04	0.8329
L. palustre	1, 86	3.26	0.0743	1, 8	30.97	**0.0006**	1, 86	3.95	0.0501
V. vitis-idaea	1, 88	1.59	0.2108	1, 88	19.73	**<0.0001**	1, 88	1.04	0.3100
E. vaginatum	1, 86	3.36	0.0702	1, 4	37.14	**0.0033**	1, 86	1.38	0.2436
C. bigelowii	1, 84	1.64	0.2037	1, 4	31.25	**0.0053**	1, 84	0.55	0.4599
P. bistorta	1, 70	0.13	0.7182	1, 70	0.30	0.5882	1, 70	1.33	0.2525

evergreen shrubs, leaf appearance occurred earliest with the combined treatment, which was 1–3 days earlier than in snowmelt and warming alone (Fig. 1).

Plant production

Although phenological events often occurred earlier in the year due to earlier snowmelt and warming, an increase in individual production was rarely observed. Rather, responses to early snowmelt and warming varied within and among functional groups. Differences were rarely significant (Table 3), in part due to the challenge of quantifying plant production in an ecosystem with high spatial variation.

However, the magnitude of change often represented a high proportion of production in this low productivity system. Deciduous shrub species differed in their response, with *S. pulchra* decreasing individual production by 6–11 % and *B. nana* showing little change across the three treatments (Table 3). Evergreen shrub species increased individual production by 28 and 8 % for *L. palustre* and *V. vitis-idaea*, respectively, due to early snowmelt (Table 3). Individual production of *P. bistorta*, the forb, was highly variable within treatments; for example, control plants ranged from 36 to 297 mg biomass. The most evident response for this species was a large, but non-significant, decrease (36 %) in production due to early snowmelt (Table 3).

The effect of warming on production was statistically significant for two species and led to the largest proportional changes (Fig. 3, Table 3). Both graminoid species responded positively to warming. Mean individual production for *E. vaginatum* increased by 36 %, which was the greatest proportional increase of any species (Fig. 3, Table 3). When early snowmelt and warming were combined, *E. vaginatum* increased individual production by 27 % relative to the control (Fig. 3, Table 3). The other graminoid, *C. bigelowii*, increased individual production by 17 % with warming and 24 % with warming and early snowmelt, although these increases were not significant (Fig. 3, Table 3). An evergreen shrub, *V. vitis-idaea*, had relatively large decreases relative to the control with warming (21 %) and the combined treatment (42 %), and the main effect of warming was significant (Table 3).

Phenology and production relationship

Across all species and for all treatments, earlier leaf expansion was associated with increased production (Fig. 4, $y = 135.52 - 0.81x$, $R^2 = 0.09$, $P = 0.0021$). This relationship reflects differences in the timing of leaf expansion among growth forms and the response of individual plant production to early snowmelt and warming. Species varied in the timing of leaf expansion by 40 days, a range that was expanded by 14 days due to altered seasonality. Early

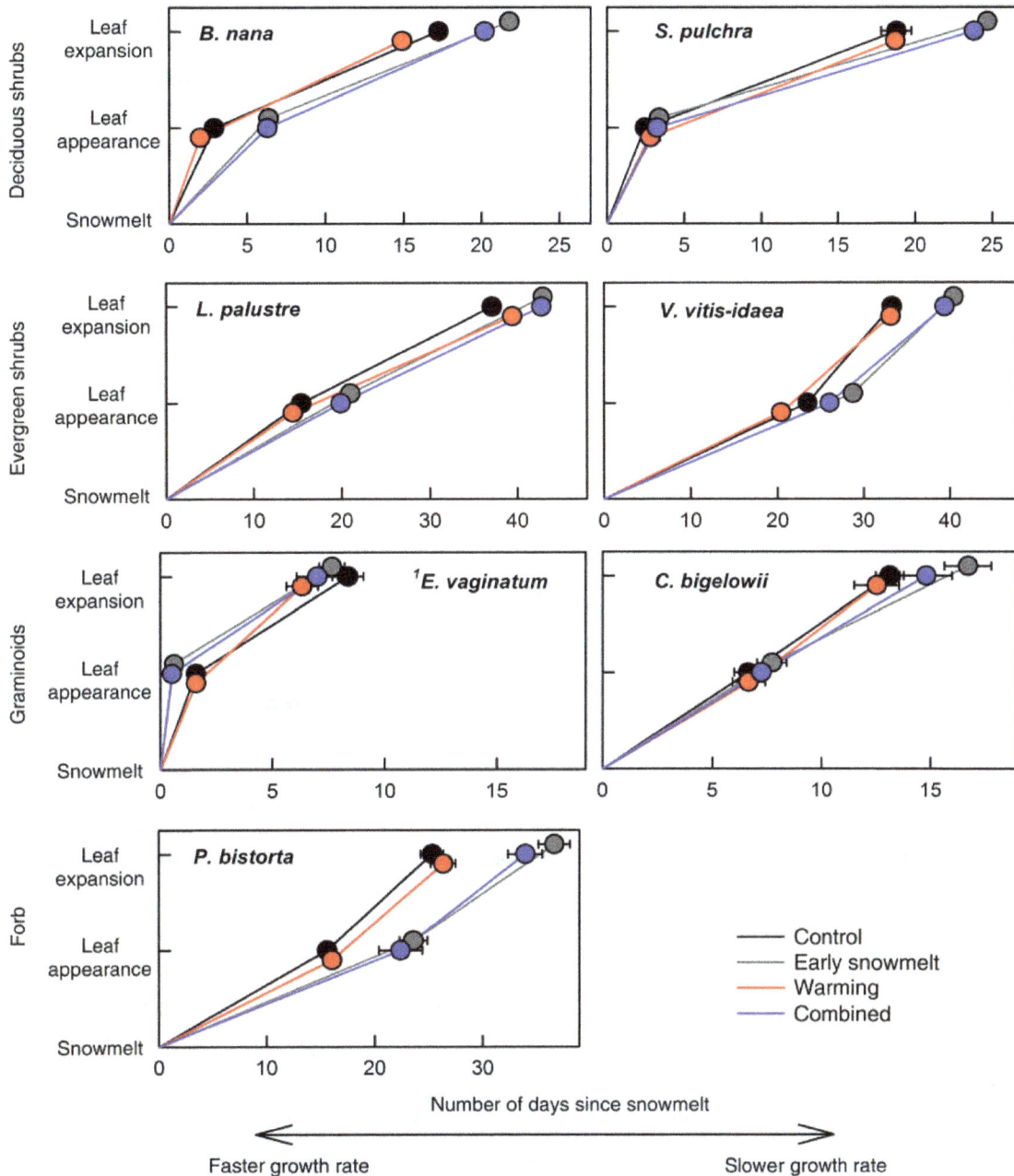

Figure 2. Number of days since snowmelt for early-season phenology events. Points are average number of days since snowmelt ± 1 SEM. A greater number of days until full leaf expansion are interpreted as a slower growth rate. Note different scales on x-axes. [1]As noted in Fig. 1, leaf appearance of *E. vaginatum* is not considered a treatment effect but is shown for clarity.

expanding species (*E. vaginatum* and *C. bigelowii*) had increases in production, while later expanding species (*L. palustre* and *V. vitis-idaea*) had some increases and also large decreases in production as a result of warming. Across functional groups, warming drove the relationship between timing of leaf expansion and individual production, as shown by significantly negative regression slopes within the warming and combined treatments (Fig. 5; C: $y = -62.7 + 0.41x$, $R^2 = 0.01$,

$P = 0.53$; ES: $y = 43.16 - 0.24x$, $R^2 = 0.01$, $P = 0.547$; W: $y = 188.66 - 1.11x$, $R^2 = 0.13$, $P = 0.03$; W × ES: $y = 200.26 - 1.23x$, $R^2 = 0.24$, $P = 0.005$). Within functional groups, there was no relationship between the timing of leaf expansion and individual production, despite earlier leaf expansion due to early snowmelt and warming (Fig. 6; deciduous shrubs: $y = 216.43 - 1.37x$, $R^2 = 0.06$, $P = 0.202$; evergreen shrubs: $y = -606.13 + 3.37x$, $R^2 = 0.08$, $P = 0.129$; graminoids: $y = 50.74 - 0.18x$,

Table 2. Results of mixed-model ANOVA on duration of time since snowmelt for early-season phenology events. Leaf appearance for *E. vaginatum* was not considered a treatment effect and was excluded from the analysis. Bold values indicate a significant main effect of the treatment at $P \leq 0.05$.

	Warming			Early snowmelt			Warming × Early snowmelt		
	df	F	P	df	F	P	df	F	P
Leaf appearance									
B. nana	1, 86	1.00	0.3210	1, 8	22.56	**0.0015**	1, 86	1.35	0.2479
S. pulchra	1, 74	0.14	0.7070	1, 7	1.25	0.2975	1, 74	0.26	0.6112
L. palustre	1, 86	5.73	**0.0188**	1, 4	33.32	**0.0047**	1, 86	0.00	0.9889
V. vitis-idaea	1, 83	15.03	**0.0002**	1, 4	30.73	**0.0057**	1, 83	0.01	0.9179
C. bigelowii	1, 87	0.16	0.6933	1, 87	1.81	0.1823	1, 87	0.19	0.6639
P. bistorta	1, 65	0.10	0.7576	1, 65	36.75	**<0.0001**	1, 65	0.49	0.4859
Leaf expansion									
B. nana	1, 85	55.48	**<0.0001**	1, 4	40.62	**0.0030**	1, 85	2.36	0.1281
S. pulchra	1, 75	0.89	0.3475	1, 4	45.21	**0.0028**	1, 75	0.35	0.5537
L. palustre	1, 90	3.44	0.0669	1, 90	59.34	**<0.0001**	1, 90	4.16	**0.0444**
V. vitis-idaea	1, 85	1.50	0.2234	1, 4	173.94	**0.0002**	1, 85	0.95	0.3323
E. vaginatum	1, 87	3.51	0.0645	1, 4	0.00	0.9497	1, 87	1.33	0.2526
C. bigelowii	1, 84	1.71	0.1951	1, 4	4.42	0.1025	1, 84	0.54	0.4635
P. bistorta	1, 70	0.26	0.6115	1, 70	51.99	**<0.0001**	1, 73	1.66	0.2013

Table 3. Results of mixed-model ANOVA on individual biomass. Bold values indicate a significant main effect of the treatment at $P \leq 0.05$.

	Warming			Early snowmelt			Warming × Early snowmelt		
	df	F	P	df	F	P	df	F	P
Individual biomass									
B. nana	1, 16	0	0.9859	1, 16	0.22	0.6466	1, 16	1.02	0.3287
S. pulchra	1, 12	0.17	0.6900	1, 12	0.18	0.6783	1, 12	1.18	0.2990
L. palustre	1, 8	0.73	0.4185	1, 8	1.35	0.2782	1, 8	0.43	0.5283
V. vitis-idaea	1, 16	**5.39**	**0.0338**	1, 16	0.19	0.6686	1, 16	0.95	0.3454
E. vaginatum	1, 8	**7.48**	**0.0256**	1, 4	0.20	0.6757	1, 8	2.35	0.1635
C. bigelowii	1, 12	1.51	0.2425	1, 12	1.04	0.3277	1, 12	0.16	0.6926
P. bistorta	1, 16	0.85	0.3831	1, 16	0.08	0.7808	1, 16	0.28	0.6106

$R^2 = 0.002$, $P = 0.7$; forb: $y = 276.82 - 1.72x$, $R^2 = 0.03$, $P = 0.582$).

Discussion

Our results showed consistent advancement of leaf appearance and expansion, indicating that spring phenology of moist acidic tundra species is sensitive to early snowmelt and warming, which is consistent with our first hypothesis. Warmer temperatures have been shown to advance spring phenology in other systems, particularly for deciduous shrubs where budburst is well predicted by growing degree-days (Polgar and Primack 2011). However, we found that species were more responsive to early snowmelt, by advancing timing of events to a greater extent than with warming alone. Timing of snowmelt has been shown to be a cue for spring phenology in Arctic and alpine ecosystems (Arft et al.

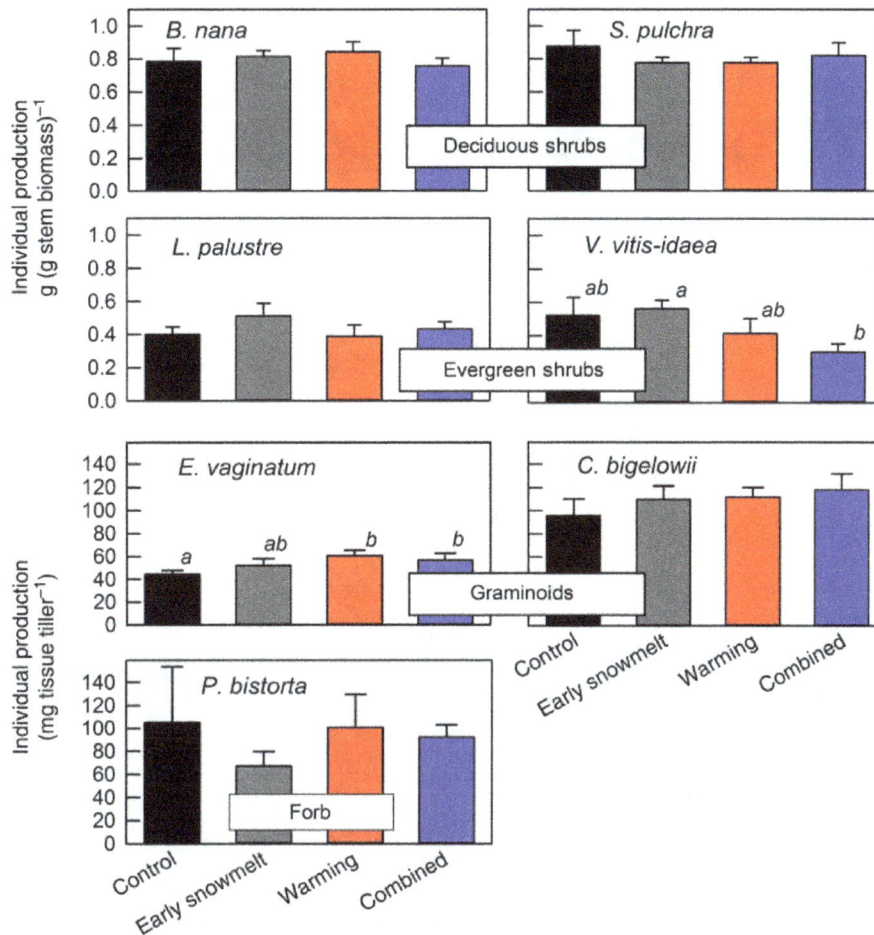

Figure 3. Biomass of individual species harvested in the third year of altered seasonality. For deciduous and evergreen shrubs, bars represent means of proportion of current annual growth to standing stem biomass of the individual, ± 1 SEM. For graminoids and a forb, bars represent means of aboveground biomass, ± 1 SEM. Letters (a and b) represent groupings based on least squares means of the ANOVA mixed model, where bars with the same letter are not statistically different at $P \leq 0.05$.

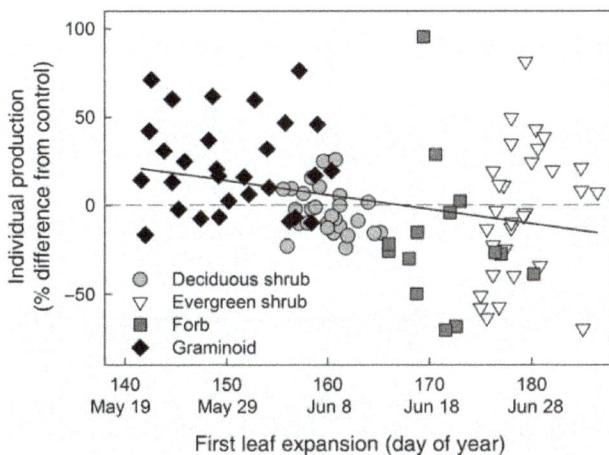

Figure 4. Relationship between phenology and production, with production (y-axis) represented as the per cent difference from the control mean ANPP for each species ($100 \times$ treatment − control mean/control). Each point represents one species, treatment and plot. The solid line is the linear regression (see Results for details).

1999; Steltzer *et al.* 2009), but experiments often confound the effects of warming and timing of snowmelt. Arctic species generally have a wide range of physiological tolerance, allowing spring growth to occur despite temperatures at or near freezing (Billings and Mooney 1968), and our experiment shows that early snowmelt can advance phenology independent of warming.

Along with clear advances in spring phenology, we also observed slower rates of leaf expansion for many species in response to early snowmelt, supporting the conservative growth strategy demonstrated by many Arctic and alpine species to compensate for interannual variation in snowmelt timing (Billings and Mooney 1968). Surprisingly, only one species, *B. nana*, expanded its leaves at a faster rate with warming. It may be that soil temperature, which warmed to a lesser extent than air temperature **[see Supporting Information—Table S2]**, is an additional cue for rate of leaf expansion for most species. Plants that expand leaves early may be susceptible to frost damage if

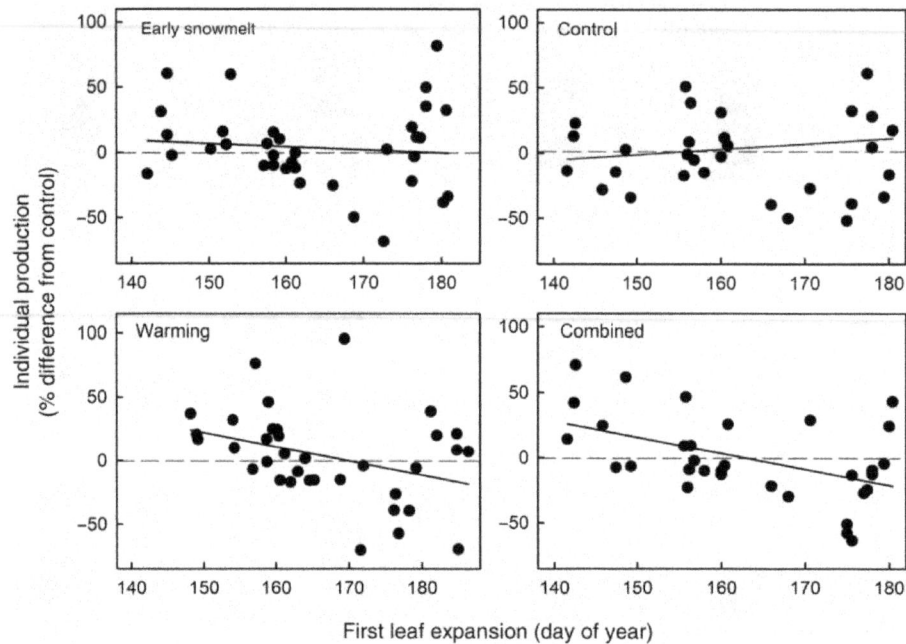

Figure 5. Relationship between phenology and production by treatment type. Data follow Fig. 4. Solid lines are linear regressions for each functional group (see Results for details).

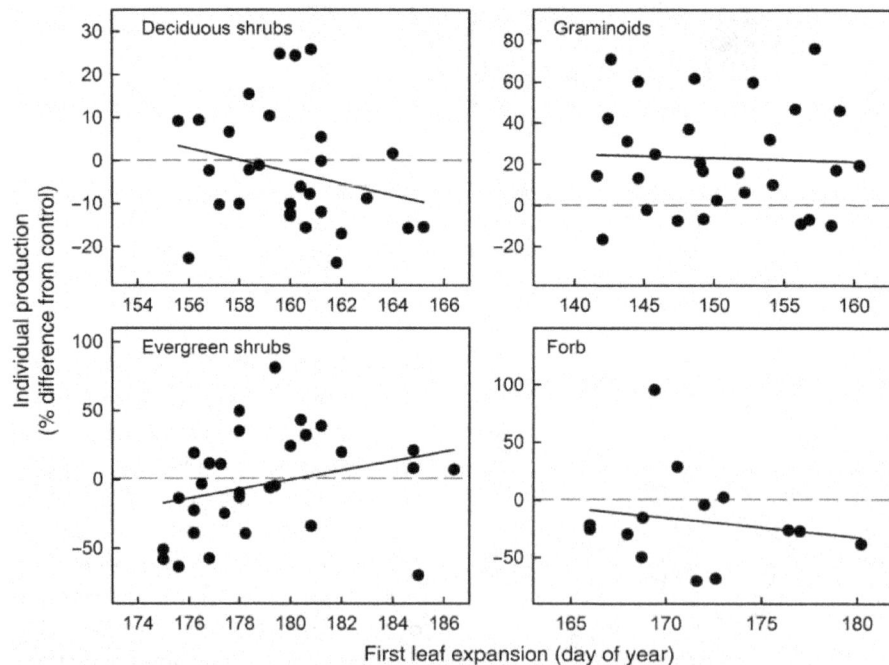

Figure 6. Relationship between phenology and production by functional group. Data follow Fig. 4. Solid lines are linear regressions for each functional group (see Results for details).

temperatures remain cold or freezing events occur (Inouye 2008; Wipf *et al.* 2009).

Production responses to warming and early snowmelt were dependent on growth form and individual species. One functional group (graminoids) matched our predicted direction of response, while others did not (forbs

and deciduous and evergreen shrubs). Previous warming experiments have also shown interspecific variation within tundra communities, with graminoids and deciduous shrubs showing rapid change relative to evergreen shrubs and forbs (Chapin and Shaver 1985; Chapin *et al.* 1995; Hollister *et al.* 2005*b*). The response of graminoids in our study was consistent with these experiments, with both *E. vaginatum* and *C. bigelowii* increasing production in response to warming and early snowmelt. Although we measured biomass of individual tillers, new tiller recruitment is another likely mechanism by which either graminoid species could have increased biomass (Chapin and Shaver 1985). Graminoids were the only functional group that maintained their growth rates when snow was melted early, which may confer an advantage in accessing early-season nutrient pulses, and consequently increasing production in the same year (Shaver *et al.* 1986). Our results are generally consistent with past work on *E. vaginatum*, which showed that early-season air warming leads to accelerated leaf growth and earlier arrival at peak biomass (Sullivan and Welker 2005). Our observation that graminoids were also able to advance timing of early-season phenology to a greater extent than the other functional groups may be due to their ability to initiate growth underneath the snowpack and therefore have new leaves present at snowmelt in addition to green leaves that have overwintered (Chapin *et al.* 1979).

Warming resulted in a large decrease in production for the evergreen shrub, *V. vitis-idaea*, a species that has shown much variability in response to warming in previous experiments (Chapin *et al.* 1995; Arft *et al.* 1999; Zamin *et al.* 2014). A meta-analysis of warming experiments across the Arctic suggests that evergreen shrub response to warming depends on soil moisture regime, with plants in moist soils more often decreasing in abundance (Elmendorf *et al.* 2012a). Regardless, the large change in production that we observed was unexpected because evergreen shrubs have a conservative growth strategy, demonstrated by slower growth rates, lower specific leaf area and lower photosynthetic capacity than other species in the tundra community (Chapin and Shaver 1996; Starr *et al.* 2008). A decrease in new leaf biomass by *V. vitis-idaea* could be related to conditions in previous years, because evergreen shrub growth relies in part on nutrients stored in old leaves (Billings and Mooney 1968). Alternatively, *V. vitis-idaea* may be a poorer competitor than deciduous species (e.g. *B. nana*) for increased nutrients under warmed conditions (Shaver *et al.* 2001). Evergreen shrubs have the ability to access early-season nutrient pulses (McKane *et al.* 2002; Larsen *et al.* 2012) and photosynthesize under the snowpack (Starr and Oberbauer 2003), which may explain why both species increased production in response to early snowmelt, similar to graminoids. However, this does not explain why *V. vitis-idaea* would show the opposite response when early snowmelt was combined with warming.

Production of deciduous shrubs and a forb did not show clear responses to warming or early snowmelt. It may be that the 3-year duration of our study did not allow enough time for *B. nana* or *S. pulchra* to show significant changes in production. Short- and long-term responses to warming in the moist acidic tundra have been shown to vary, in part because of slow recruitment and establishment of new individuals (Hollister *et al.* 2005b). For example, observations from the ITEX experiments showed that community changes in deciduous shrubs did not become significant until after 4 years of warming (Walker *et al.* 2006). However, since we measured growth at an individual (rather than community) level in order to detect within-season changes of biomass accumulation, the response of deciduous shrubs may be more likely attributed to nutrient availability in that year. If evergreen shrubs were able to access nutrient pulses early in the season before deciduous shrubs, it may help explain why the latter showed little response, specifically when snow was melted early. The one forb tested in this experiment, *P. bistorta*, had highly variable results which may have obscured any treatment effects.

While the magnitude of temporal shifts is often a focus of phenological studies, our results suggest that evolved strategies within the plant community also play an important role in determining responses to altered seasonality. We predicted that earlier leaf expansion would lead to greater production, and we found that this was true for early expanding species but not later expanding species. This demonstrates that temporal niche partitioning influences species' responses to environmental change. A previous study (Cleland *et al.* 2012) examined plant responses to warming and found that phenologically flexible species (able to 'track' climate change) had positive performance responses (e.g. increased abundance and production). Our results are only partially consistent with this result. In our study, changes in phenology alone did not always result in a change in production. Rather, community patterns of leaf expansion, along with warming-driven increases and decreases in ANPP (Fig. 5), contributed to a negative relationship between spring phenology and production (Fig. 4). If this relationship was representative of differences in functional groups alone, we would expect the relationship to hold among control plots, which was not the observed result (Fig. 5). Differences in the ability of species to shift the timing and rate of leaf expansion may affect competitive interactions and subsequently influence future plant community composition (Richardson *et al.* 2010; Cleland *et al.* 2012). Specifically, *E. vaginatum*, which was able to green rapidly and maintain its growth rate, may have a competitive advantage. Further, we predict that species that occupy early-season temporal niches across diverse ecosystems may increase in abundance under altered climate conditions.

Conclusions

Changes in vegetative phenology, regardless of changes in production, have important implications for functioning of Arctic ecosystems. Phenological shifts can affect competition among species, and differential responses of individual species may determine future plant community structure. Changes in Arctic plant communities have the potential to affect multiple aspects of ecosystem function, including (i) carbon cycling, by altering the balance between ecosystem-scale photosynthesis and respiration (Shaver et al. 1992; Hobbie et al. 2000); (ii) surface energy balance and feedbacks to the climate system, through change in albedo and seasonal changes in leaf area (Peñuelas et al. 2009; Richardson et al. 2013); and (iii) trophic relationships that may become decoupled if plant phenology responds to a changing climate differently than vertebrate and invertebrate herbivores (Post and Forchhammer 2008; Høye et al. 2013). Our study suggests that an earlier spring as indicated by satellite data may be driven by early greening species such as E. vaginatum and C. bigelowii. These species have the advantage of being able to respond rapidly and positively to changes in seasonality, and may increase in abundance in tundra ecosystems as earlier snowmelt and warmer springs continue.

Sources of Funding

Funding for the Snowmelt Project was provided by the National Science Foundation Office of Polar Programs Grants #PLR-1007672, 0902096 and 0902184. Additional funding for C.L. was provided by a National Science Foundation Graduate Research Fellowship.

Contributions by the Authors

M.N.W., M.W., P.F.S., A.D.-N. and H.S. designed and implemented the experiment. C.L., A.D.-N. and H.S. collected data. C.L. and H.S. analysed the data and wrote the manuscript, and all authors contributed to revisions.

Acknowledgements

We thank two anonymous reviewers for comments that improved the quality of this manuscript.

Table S1. Species composition at Imnavait Creek. Per cent cover estimates are averaged over subplots for the entire experimental site.

Table S2. Microclimate variables in all 3 years of the experiment (2010–12). Air temperature, soil temperature and soil moisture were measured with automated sensors at each subplot throughout spring and summer, and are presented here as means over the observation period ± 1 SEM.

Literature Cited

Arft AM, Walker MD, Gurevitch J, Alatalo JM, Bret-Harte MS, Dale M, Diemer M, Gugerli F, Henry GHR, Jones MH, Hollister RD, Jónsdóttir IS, Laine K, Lévesque E, Marion GM, Molau U, Mølgaard P, Nordenhäll U, Raszhivin V, Robinson CH, Starr G, Stenström A, Stenström M, Totland Ø, Turner PL, Walker LJ, Webber PJ, Welker JM, Wookey PA. 1999. Responses of tundra plants to experimental warming: meta-analysis of the international tundra experiment. Ecological Monographs 69:491–511.

Billings WD, Mooney HA. 1968. The ecology of Arctic and alpine plants. Biological Reviews 43:481–529.

Borner AP, Kielland K, Walker MD. 2008. Effects of simulated climate change on plant phenology and nitrogen mineralization in Alaskan Arctic tundra. Arctic, Antarctic, and Alpine Research 40:27–38.

Bret-Harte MS, Shaver GR, Chapin FS III. 2002. Primary and secondary stem growth in arctic shrubs: implications for community response to environmental change. Journal of Ecology 90:251–267.

Campioli M, Leblans N, Michelsen A. 2012. Twenty-two years of warming, fertilisation and shading of subarctic heath shrubs promote secondary growth and plasticity but not primary growth. PLoS ONE 7:e34842.

Chapin FS III, Shaver GR. 1985. Individualistic growth response of tundra plant species to environmental manipulations in the field. Ecology 66:564–576.

Chapin FS III, Shaver GR. 1996. Physiological and growth responses of Arctic plants to a field experiment simulating climatic change. Ecology 77:822–840.

Chapin FS III, Van Cleve K, Chapin MC. 1979. Soil temperature and nutrient cycling in the tussock growth form of Eriophorum vaginatum. Journal of Ecology 67:169–189.

Chapin FS III, Shaver GR, Giblin AE, Nadelhoffer KJ, Laundre JA. 1995. Responses of Arctic tundra to experimental and observed changes in climate. Ecology 76:694–711.

Christensen JH, Kumar KK, Aldrian E, An S-I, Cavalcanti IFA, De Castro M, Dong W, Goswami P, Hall A, Kanyanga JK, Kitoh A, Kossin J, Lau N-C, Renwick J, Stephenson DB, Xie S-P, Zhou T. 2013. Climate phenomena and their relevance for future regional climate change. In: Stocker TF, Qin D, Plattner G-K, Tignor M, Allen SK, Boschung J, Nauels A, Xia Y, Bex V, Midgley PM, eds. Climate Change 2013: The Physical Science Basis. Contribution of Working Group I to the Fifth Assessment Report of the Intergovernmental Panel on Climate Change. Cambridge, UK: Cambridge University Press, 62.

Cleland EE, Chiariello NR, Loarie SR, Mooney HA, Field CB. 2006. Diverse responses of phenology to global changes in a grassland ecosystem. Proceedings of the National Academy of Sciences of the USA 103:13740–13744.

Cleland EE, Chuine I, Menzel A, Mooney HA, Schwartz MD. 2007. Shifting plant phenology in response to global change. Trends in Ecology and Evolution 22:357–365.

Cleland EE, Allen JM, Crimmins TM, Dunne JA, Pau S, Travers SE, Zavaleta ES, Wolkovich EM. 2012. Phenological tracking enables positive species responses to climate change. Ecology 93:1765–1771.

Dawes MA, Hagedorn F, Zumbrunn T, Handa IT, Hättenschwiler S, Wipf S, Rixen C. 2011. Growth and community responses of alpine dwarf shrubs to in situ CO₂ enrichment and soil warming. New Phytologist 191:806–818.

Dukes JS, Chiariello NR, Cleland EE, Moore LA, Shaw MR, Thayer S, Tobeck T, Mooney HA, Field CB. 2005. Responses of grassland production to single and multiple global environmental changes. PLoS Biology 3:e319.

Edwards KA, Jefferies RL. 2013. Inter-annual and seasonal dynamics of soil microbial biomass and nutrients in wet and dry low-Arctic sedge meadows. *Soil Biology and Biochemistry* **57**:83–90.

Elmendorf SC, Henry GHR, Hollister RD, Björk RG, Bjorkman AD, Callaghan TV, Collier LS, Cooper EJ, Cornelissen JHC, Day TA, Fosaa AM, Gould WA, Grétarsdóttir J, Harte J, Hermanutz L, Hik DS, Hofgaard A, Jarrad F, Jónsdóttir IS, Keuper F, Klanderud K, Klein JA, Koh S, Kudo G, Lang SI, Loewen V, May JL, Mercado J, Michelsen A, Molau U, Myers-Smith IH, Oberbauer SF, Pieper S, Post ES, Rixen C, Robinson CH, Schmidt NM, Shaver GR, Stenstrom A, Tolvanen A, Totland Ø, Troxler T, Wahren C-H, Webber PJ, Welker JM, Wookey PA. 2012a. Global assessment of experimental climate warming on tundra vegetation: heterogeneity over space and time. *Ecology Letters* **15**:164–175.

Elmendorf SC, Henry GHR, Hollister RD, Björk RG, Boulanger-Lapointe N, Cooper EJ, Cornelissen JHC, Day TA, Dorrepaal E, Elumeeva TG, Gill M, Gould WA, Harte J, Hik DS, Hofgaard A, Johnson DR, Johnstone JF, Jónsdóttir IS, Jorgenson JC, Klanderud K, Klein JA, Koh S, Kudo G, Lara M, Lévesque E, Magnússon B, May JL, Mercado-Díaz JA, Michelsen A, Molau U, Myers-Smith IH, Oberbauer SF, Onipchenko VG, Rixen C, Martin Schmidt N, Shaver GR, Spasojevic MJ, Þórhallsdóttir ÞE, Tolvanen A, Troxler T, Tweedie CE, Villareal S, Wahren C-H, Walker X, Webber PJ, Welker JM, Wipf S. 2012b. Plot-scale evidence of tundra vegetation change and links to recent summer warming. *Nature Climate Change* **2**:453–457.

Ernakovich JG, Hopping KA, Berdanier AB, Simpson RT, Kachergis EJ, Steltzer H, Wallenstein MD. 2014. Predicted responses of Arctic and alpine ecosystems to altered seasonality under climate change. *Global Change Biology* **20**:3256–3269.

Fay PA, Carlisle JD, Knapp AK, Blair JM, Collins SL. 2003. Productivity responses to altered rainfall patterns in a C4-dominated grassland. *Oecologia* **137**:245–251.

Fitter AH, Fitter RSR. 2002. Rapid changes in flowering time in British plants. *Science (New York, N.Y.)* **296**:1689–1691.

Forbes BC, Fauria MM, Zetterberg P. 2010. Russian Arctic warming and 'greening' are closely tracked by tundra shrub willows. *Global Change Biology* **16**:1542–1554.

Fraser RH, Olthof I, Carrière M, Deschamps A, Pouliot D. 2011. Detecting long-term changes to vegetation in northern Canada using the Landsat satellite image archive. *Environmental Research Letters* **6**:045502.

Galen C, Stanton ML. 1995. Responses of snowbed plant species to changes in growing-season length. *Ecology* **76**:1546–1557.

Gottfried M, Pauli H, Futschik A, Akhalkatsi M, Barančok P, Alonso JLB, Coldea G, Dick J, Erschbamer B, Calzado MRF, Kazakis G, Krajči J, Larsson P, Mallaun M, Michelsen O, Moiseev D, Moiseev P, Molau U, Merzouki A, Nagy L, Nakhutsrishvili G, Pedersen B, Pelino G, Puscas M, Rossi G, Stanisci A, Theurillat J-P, Tomaselli M, Villar L, Vittoz P, Vogiatzakis I, Grabherr G. 2012. Continent-wide response of mountain vegetation to climate change. *Nature Climate Change* **2**:111–115.

Harte J, Shaw R. 1995. Shifting dominance within a montane vegetation community: results of a climate-warming experiment. *Science (New York, N.Y.)* **267**:876–880.

Hayhoe K, Wake CP, Huntington TG, Luo L, Schwartz MD, Sheffield J, Wood E, Anderson B, Bradbury J, Degaetano A, Troy TJ, Wolfe D. 2007. Past and future changes in climate and hydrological indicators in the US Northeast. *Climate Dynamics* **28**:381–407.

Hobbie SE, Schimel JP, Trumbore SE, Randerson JR. 2000. Controls over carbon storage and turnover in high-latitude soils. *Global Change Biology* **6**:196–210.

Hollister RD, Webber PJ, Bay C. 2005a. Plant response to temperature in Northern Alaska: implications for predicting vegetation change. *Ecology* **86**:1562–1570.

Hollister RD, Webber PJ, Tweedie CE. 2005b. The response of Alaskan arctic tundra to experimental warming: differences between short- and long-term responses. *Global Change Biology* **11**:525–536.

Høye TT, Post E, Meltofte H, Schmidt NM, Forchhammer MC. 2007. Rapid advancement of spring in the High Arctic. *Current Biology* **17**:R449–R451.

Høye TT, Post E, Schmidt NM, Trøjelsgaard K, Forchhammer MC. 2013. Shorter flowering seasons and declining abundance of flower visitors in a warmer Arctic. *Nature Climate Change* **3**:1–5.

Inouye DW. 2008. Effects of climate change on phenology, frost damage, and floral abundance of montane wildflowers. *Ecology* **89**:353–362.

Jia GJ, Epstein HE, Walker DA. 2003. Greening of arctic Alaska, 1981–2001. *Geophysical Research Letters* **30**:3–6.

Jia GJ, Epstein HE, Walker DA. 2009. Vegetation greening in the Canadian Arctic related to decadal warming. *Journal of Environmental Monitoring* **11**:2231–2238.

Knapp AK, Smith MD. 2001. Variation among biomes in temporal dynamics of aboveground primary production. *Science* **291**:481–484.

Körner C, Basler D. 2010. Phenology under global warming. *Science* **327**:1461–1462.

Larsen KS, Michelsen A, Jonasson S, Beier C, Grogan P. 2012. Nitrogen uptake during fall, winter and spring differs among plant functional groups in a Subarctic heath ecosystem. *Ecosystems* **15**:927–939.

Marchin RM, Salk CF, Hoffmann WA, Dunn RR. 2015. Temperature alone does not explain phenological variation of diverse temperate plants under experimental warming. *Global Change Biology* **21**:3138–3151.

Marion GM, Henry GHR, Freckman DW, Johnstone J, Jones G, Jones MH, Lévesque E, Molau U, Mølgaard P, Parsons AN, Svoboda J, Virginia RA. 1997. Open-top designs for manipulating field temperature in high-latitude ecosystems. *Global Change Biology* **3**:20–32.

Mckane RB, Johnson LC, Shaver GR, Nadelhoffer KJ, Rastetter EB, Fry B, Giblin AE, Kielland K, Kwiatkowski BL, Laundre JA, Murray G. 2002. Resource-based niches provide a basis for plant species diversity and dominance in arctic tundra. *Nature* **415**:68–71.

Menzel A, Sparks TH, Estrella N, Koch E, Aasa A, Ahas R, Alm-Kübler K, Bissolli P, Braslavská O, Briede A, Chmielewski FM, Crepinsek Z, Curnel Y, Dahl Å, Defila C, Donnelly A, Filella Y, Jatczak K, Måge F, Mestre A, Nordli Ø, Peñuelas J, Pirinen P, Remišová V, Scheifinger H, Striz M, Susnik A, Van Vliet AJH, Wielgolaski F-E, Zach S, Zust A. 2006. European phenological response to climate change matches the warming pattern. *Global Change Biology* **12**:1969–1976.

Muldavin EH, Moore DI, Collins SL, Wetherill KR, Lightfoot DC. 2008. Aboveground net primary production dynamics in a northern Chihuahuan Desert ecosystem. *Oecologia* **155**:123–132.

Myneni RB, Keeling CD, Tucker CJ, Asrar G, Nemani RR. 1997. Increased plant growth in the northern high latitudes from 1981 to 1991. *Nature* **386**:698–702.

Pau S, Wolkovich EM, Cook BI, Davies TJ, Kraft NJB, Bolmgren K, Betancourt JL, Cleland EE. 2011. Predicting phenology by integrating ecology, evolution and climate science. *Global Change Biology* **17**:3633–3643.

Peñuelas J, Rutishauser T, Filella I. 2009. Phenology feedbacks on climate change. *Science* **324**:887–888.

Polgar CA, Primack RB. 2011. Leaf-out phenology of temperate woody plants: from trees to ecosystems. *The New Phytologist* **191**:926–941.

Post E, Forchhammer MC. 2008. Climate change reduces reproductive success of an Arctic herbivore through trophic mismatch. *Philosophical Transactions of the Royal Society of London. Series B, Biological Sciences* **363**:2369–2375.

Reyes-Fox M, Steltzer H, Trlica MJ, Mcmaster GS, Andales AA, Lecain DR, Morgan JA. 2014. Elevated CO_2 further lengthens growing season under warming conditions. *Nature* **510**:259–262.

Richardson AD, Black TA, Ciais P, Delbart N, Friedl MA, Gobron N, Hollinger DY, Kutsch WL, Longdoz B, Luyssaert S, Migliavacca M, Montagnani L, Munger JW, Moors E, Piao S, Rebmann C, Reichstein M, Saigusa N, Tomelleri E, Vargas R, Varlagin A. 2010. Influence of spring and autumn phenological transitions on forest ecosystem productivity. *Philosophical Transactions of the Royal Society of London. Series B, Biological Sciences* **365**:3227–3246.

Richardson AD, Anderson RS, Arain MA, Barr AG, Bohrer G, Chen G, Chen JM, Ciais P, Davis KJ, Desai AR, Dietze MC, Dragoni D, Garrity SR, Gough CM, Grant R, Hollinger DY, Margolis HA, Mccaughey H, Migliavacca M, Monson RK, Munger JW, Poulter B, Raczka BM, Ricciuto DM, Sahoo AK, Schaefer K, Tian H, Vargas R, Verbeeck H, Xiao J, Xue Y. 2012. Terrestrial biosphere models need better representation of vegetation phenology: results from the North American Carbon Program Site Synthesis. *Global Change Biology* **18**:566–584.

Richardson AD, Keenan TF, Migliavacca M, Ryu Y, Sonnentag O, Toomey M. 2013. Climate change, phenology, and phenological control of vegetation feedbacks to the climate system. *Agricultural and Forest Meteorology* **169**:156–173.

Schwartz MD. 1998. Green-wave phenology. *Nature* **394**:839–840.

Schwartz MD, Ahas R, Aasa A. 2006. Onset of spring starting earlier across the Northern Hemisphere. *Global Change Biology* **12**:343–351.

Serreze MC, Walsh JE, Chapin FS III, Osterkamp T, Dyurgerov M, Romanovsky V, Oechel WC, Morison J, Zhang T, Barry RG. 2000. Observational evidence of recent change in the northern high-latitude environment. *Climatic Change* **46**:159–207.

Shaver GR. 1986. Woody stem production in Alaskan tundra shrubs. *Ecology* **67**:660–669.

Shaver GR, Chapin FS III, Gartner BL. 1986. Factors limiting seasonal growth and peak biomass accumulation in *Eriophorum vaginatum* in Alaskan tussock tundra. *Journal of Ecology* **74**:257–278.

Shaver GR, Billings WD, Chapin FS III, Giblin AE, Nadelhoffer KJ, Oechel WC, Rastetter EB. 1992. Global change and the carbon balance of Arctic ecosystems. *BioScience* **42**:433–441.

Shaver GR, Bret-Harte MS, Jones MH, Johnstone J, Gough L, Laundre J, Chapin FS III. 2001. Species composition interacts with fertilizer to control long-term change in tundra productivity. *Ecology* **82**:3163–3181.

Sherry RA, Zhou X, Gu S, Arnone JA III, Schimel DS, Verburg PS, Wallace LL, Luo Y. 2007. Divergence of reproductive phenology under climate warming. *Proceedings of the National Academy of Sciences of the USA* **104**:198–202.

Sistla SA, Moore JC, Simpson RT, Gough L, Shaver GR, Schimel JP. 2013. Long-term warming restructures Arctic tundra without changing net soil carbon storage. *Nature* **497**:615–618.

Smith MD, La Pierre KJ, Collins SL, Knapp AK, Gross KL, Barrett JE, Frey SD, Gough L, Miller RJ, Morris JT, Rustad LE, Yarie J. 2015. Global environmental change and the nature of aboveground net primary productivity responses: insights from long-term experiments. *Oecologia* 935–947.

Starr G, Oberbauer SF. 2003. Photosynthesis of arctic evergreens under snow: implications for tundra ecosystem carbon balance. *Ecology* **84**:1415–1420.

Starr G, Oberbauer SF, Ahlquist LE. 2008. The photosynthetic response of Alaskan tundra plants to increased season length and soil warming. *Arctic, Antarctic, and Alpine Research* **40**:181–191.

Steltzer H, Landry C, Painter TH, Anderson J, Ayres E. 2009. Biological consequences of earlier snowmelt from desert dust deposition in alpine landscapes. *Proceedings of the National Academy of Sciences of the USA* **106**:11629–11634.

Sullivan PF, Welker JM. 2005. Warming chambers stimulate early season growth of an arctic sedge: results of a minirhizotron field study. *Oecologia* **142**:616–626.

Sweet SK, Gough L, Griffin KL, Boelman NT. 2014. Tall deciduous shrubs offset delayed start of growing season through rapid leaf development in the Alaskan Arctic tundra. *Arctic, Antarctic, and Alpine Research* **46**:682–697.

Sweet SK, Griffin KL, Steltzer H, Gough L, Boelman NT. 2015. Greater deciduous shrub abundance extends tundra peak season and increases modeled net CO_2 uptake. *Global Change Biology* **21**:2394–2409.

Tape K, Sturm M, Racine C. 2006. The evidence for shrub expansion in Northern Alaska and the Pan-Arctic. *Global Change Biology* **12**:686–702.

Thompson R, Clark RM. 2008. Is spring starting earlier? *The Holocene* **18**:95–104.

Wahren C-HA, Walker MD, Bret-Harte MS. 2005. Vegetation responses in Alaskan arctic tundra after 8 years of a summer warming and winter snow manipulation experiment. *Global Change Biology* **11**:537–552.

Walker MD, Walker DA, Auerbach N. 1994. Plant communities of a tussock tundra landscape in the Brooks Range Foothills, Alaska. *Journal of Vegetation Science* **5**:843–866.

Walker MD, Wahren CH, Hollister RD, Henry GHR, Ahlquist LE, Alatalo JM, Bret-Harte MS, Calef MP, Callaghan TV, Carroll AB, Epstein HE, Jónsdóttir IS, Klein JA, Magnússon B, Molau U, Oberbauer SF, Rewa SP, Robinson CH, Shaver GR, Suding KN, Thompson CC, Tolvanen A, Totland Ø, Turner PL, Tweedie CE, Webber PJ, Wookey PA. 2006. Plant community responses to experimental warming across the tundra biome. *Proceedings of the National Academy of Sciences of the USA* **103**:1342–1346.

Walther G-R, Beißner S, Burga CA. 2005. Trends in the upward shift of alpine plants. *Journal of Vegetation Science* **16**:541–548.

Wang S, Duan J, Xu G, Wang Y, Zhang Z, Rui Y, Luo C, Xu B, Zhu X, Chang X, Cui X, Niu H, Zhao X, Wang W. 2012. Effects of warming and grazing on soil N availability, species composition, and ANPP in an alpine meadow. *Ecology* **93**:2365–2376.

Weintraub MN, Schimel JP. 2005. The seasonal dynamics of amino acids and other nutrients in Alaskan Arctic tundra soils. *Biogeochemistry* **73**:359–380.

Whittinghill KA, Hobbie SE. 2011. Effects of landscape age on soil organic matter processing in Northern Alaska. *Soil Science Society of America Journal* **75**:907–917.

Willis CG, Ruhfel B, Primack RB, Miller-Rushing AJ, Davis CC. 2008. Phylogenetic patterns of species loss in Thoreau's woods are driven by climate change. *Proceedings of the National Academy of Sciences of the USA* **105**:17029–17033.

Wipf S, Rixen C. 2010. A review of snow manipulation experiments in Arctic and alpine tundra ecosystems. *Polar Research* **29**:95–109.

Wipf S, Stoeckli V, Bebi P. 2009. Winter climate change in alpine tundra: plant responses to changes in snow depth and snowmelt timing. *Climatic Change* **94**:105–121.

Wolkovich EM, Cook BI, Allen JM, Crimmins TM, Betancourt JL, Travers SE, Pau S, Regetz J, Davies TJ, Kraft NJB, Ault TR, Bolmgren K, Mazer SJ, Mccabe GJ, Mcgill BJ, Parmesan C, Salamin N, Schwartz MD, Cleland EE. 2012. Warming experiments underpredict plant phenological responses to climate change. *Nature* **485**:494–497.

Zamin TJ, Bret-Harte MS, Grogan P. 2014. Evergreen shrubs dominate responses to experimental summer warming and fertilization in Canadian mesic low arctic tundra. *Journal of Ecology.* doi:10.1111/1365-2745.12237

Zavaleta ES, Shaw MR, Chiariello NR, Thomas BD, Cleland EE, Field CB, Mooney HA. 2003. Grassland responses to three years of elevated temperature, CO_2, precipitation, and N deposition. *Ecological Monographs* **73**:585–604.

Zeng H, Jia G, Epstein H. 2011. Recent changes in phenology over the northern high latitudes detected from multi-satellite data. *Environmental Research Letters* **6**:45508.

16

Effects of topoclimatic complexity on the composition of woody plant communities

Meagan F. Oldfather*[1], Matthew N. Britton[2], Prahlad D. Papper[1], Michael J. Koontz[3], Michelle M. Halbur[4], Celeste Dodge[4], Alan L. Flint[5], Lorriane E. Flint[5] and David D. Ackerly[1,6]

[1] Department of Integrative Biology, University of California, Berkeley, CA 94720, USA
[2] Department of Biological Sciences and Bolus Herbarium, University of Cape Town, Private Bag, Rondebosch 7700, South Africa
[3] Department of Plant Sciences, University of California Davis, Davis, CA 95618, USA
[4] Pepperwood Preserve, 2130 Pepperwood Preserve Road Santa Rosa, CA 95404, USA
[5] Water Resources Discipline, U.S. Geological Survey, Placer Hall, 6000 J Street, Sacramento, CA 95819, USA
[6] Jepson Herbarium, University of California, Berkeley, CA 94720, USA

Associate Editor: J. Hall Cushman

Abstract. Topography can create substantial environmental variation at fine spatial scales. Shaped by slope, aspect, hill-position and elevation, topoclimate heterogeneity may increase ecological diversity, and act as a spatial buffer for vegetation responding to climate change. Strong links have been observed between climate heterogeneity and species diversity at broader scales, but the importance of topoclimate for woody vegetation across small spatial extents merits closer examination. We established woody vegetation monitoring plots in mixed evergreen-deciduous woodlands that spanned topoclimate gradients of a topographically heterogeneous landscape in northern California. We investigated the association between the structure of adult and regenerating size classes of woody vegetation and multidimensional topoclimate at a fine scale. We found a significant effect of topoclimate on both single-species distributions and community composition. Effects of topoclimate were evident in the regenerating size class for all dominant species (four *Quercus spp.*, *Umbellularia californica* and *Pseudotsuga menziesii*) but only in two dominant species (*Quercus agrifolia* and *Quercus garryana*) for the adult size class. Adult abundance was correlated with water balance parameters (e.g. climatic water deficit) and recruit abundance was correlated with an interaction between the topoclimate parameters and conspecific adult abundance (likely reflecting local seed dispersal). However, in all cases, the topoclimate signal was weak. The magnitude of environmental variation across our study site may be small relative to the tolerance of long-lived woody species. Dispersal limitations, management practices and patchy disturbance regimes also may interact with topoclimate, weakening its influence on woody vegetation distributions. Our study supports the biological relevance of multidimensional topoclimate for mixed woodland communities, but highlights that this relationship might be mediated by interacting factors at local scales.

Keywords: California; climatic water deficit; community analyses; oak woodlands; topoclimate; woody vegetation.

* Corresponding author's e-mail address: meagan_oldfather@berkeley.edu

Introduction

Woody, canopy-dominant species are crucial, long-lived members of many ecosystems. A wide range of ecological processes determine the landscape patterns of woody vegetation including climate limitations, biotic interactions, priority effects, dispersal and disturbance (Woodward et al. 2004; Bond and Keeley 2005). Changes in elevation, slope and aspect create a complex topoclimatic landscape (Ashcroft and Gollan 2013), and these heterogeneous landscapes have been linked to higher ecological diversity at global scales (Kreft and Jetz 2007). Heterogeneous topoclimates can create a patchwork of diverse woody vegetation over short distances and may shape how species respond to changing climate conditions (Whittaker 1967; Dobrowski 2011; Ackerly et al. 2015). Thus, the influence of topoclimate on local species distributions is of fundamental importance in both basic (MacArthur 1972) and applied ecology (Lawler et al. 2015). Quantification of topoclimate and species diversity at matching scales is a critical first step to understanding the relationship between topoclimate heterogeneity and woody community composition over small spatial extents.

Combinations of topographic features create climatic gradients on the scale of 10s–100s of meters (Geiger et al. 2009). Topoclimate is distinguished here from microclimate, which refers to spatial variations in environmental conditions due to vegetation cover or surface features smaller than 10 m (De Frenne et al. 2013). Across large changes in elevation, lower elevation sites have warmer overall temperatures, as well as higher variation in daily temperature and lower variation in seasonal temperatures (Korner 2003). However, at the topoclimate scale, this pattern can reverse. Lower elevation sites often have cooler minimum temperatures due to cold-air pooling in valleys. Cold-air pooling in steep-sided valleys and basins can greatly lower night-time temperatures, especially in still air and clear sky conditions (Lundquist et al. 2008; Daly et al. 2009).

Slope and aspect influence solar radiation exposure, soil properties and disturbance regimes. Equatorial-facing slopes have increased exposure to solar radiation, which increases light availability and maximum daily temperatures relative to polar-facing slopes (hereafter referred to as south- and north-facing, respectively, as this study was conducted in the Northern Hemisphere) (Holland and Steyn 1975). Southwest-facing slopes generally have higher effective heat loading than southeast-facing slopes, despite similar radiation loads, due to higher afternoon temperatures (McCune and Keon 2002). Steeply sloped areas also have reduced soil depth and greater rates of disturbance-induced erosion

(Heyerdahl et al. 2001; Roering and Gerber 2005). These individual features—elevation, hillslope position, slope and aspect—interact with each other and the regional climate to create complex topoclimate gradients within local landscapes. For instance, increasing slope magnifies the effect of aspect, and increases hill-shading in nearby areas (Flint and Childs 1987).

Topography can also shape topoclimate through local hydrologic processes (Anderson and Kneale 1982). Water runs downhill, evaporates more readily at higher temperature and is less available in the thin soils of steep slopes (Tani 1997; Flint et al. 2013). Measurements of a site's topoclimate should, therefore, incorporate the intensity of solar radiation and availability of soil moisture, as well as their interaction (Stephenson et al. 1990). Water balance variables capture the relationships between these components, including their seasonal availability (Stephenson et al. 1990). Advances in modeling allow estimation of the following water balance variables at the topoclimatic scale: potential evapotranspiration (PET, mm), actual evapotranspiration (AET, mm) and climatic water deficit (CWD, mm) (Flint et al. 2013). CWD is an integrative measure of the cumulative excess of PET relative to AET during the dry season (i.e. when energy availability exceeds water supply), such that $CWD = PET - AET$.

Topoclimate components, considered separately, have well-documented correlations with species distributions. For instance, belts of vegetation occur along elevation contours (Whittaker and Niering 1965), and aspect has variable effects on species diversity and community composition (Armesto and Martinez 1978; Weiss et al. 1988; Sage and Sage 2002; Bennie et al. 2006; Harrison et al. 2010). Furthermore, integrated measures of both temperature and soil moisture in heterogeneous landscapes are strong drivers of vegetation distributions (Stephenson 2015). CWD is a particularly good predictor of woody vegetation distributions, as well as temporal woody vegetation change, in semi-arid landscapes (Lutz et al. 2010; Crimmins et al. 2011; Das et al. 2013; Stephenson 2015). Increasing CWD has been linked to changes in tree recruitment, growth and mortality, as well as community composition (Salzer et al. 2009; Millar et al. 2015). Reductions in large tree densities and shifts toward more oak-dominated landscapes in the last century in California have been strongly correlated with increasing CWD (McIntyre et al. 2015). However, it remains unclear the extent to which variation in multiple topoclimate dimensions, considered in concert, can explain woody vegetation diversity at the local scale (but see Van de Ven et al. 2007).

Landscapes with heterogeneous topoclimates have been championed as valuable conservation units for protecting both current and future biodiversity (Ackerly et al. 2010; Lenoir et al. 2013; Lawler et al. 2015). In the context of rapid climate change, landscape heterogeneity reduces the rate at which a species must move to track its climate niche and increases the availability of cooler, wetter refugia (Loarie et al. 2009; Dobrowski 2011). Heterogeneous landscapes harboring high levels of biodiversity may also provide thermophilic propagules for community reassembly (Ackerly et al. 2010). Most protected areas in North America have a small spatial extent, and land management and acquisition decisions take place at this scale (Chape et al. 2005; Heller and Zavaleta 2009). Thus further research on vegetation–climate relationships at a local scale is a conversation priority, especially in the face of 21st century climate change.

We quantified woody community diversity and topoclimate complexity at a matching local scale in mixed evergreen-deciduous woodlands of Northern California. We established woody vegetation survey plots that span a wide range of topoclimate variability across a local landscape. Woody vegetation may exhibit size-dependent sensitivity to topoclimate (Máliš et al. 2016) and regenerating individuals may require a different suite of climatic conditions to establish than adults require to persist (Grubb 1977; Jackson et al. 2009; Mclaughlin and Zavaleta 2012; Millar et al. 2015). Thus, both adult and regenerating size classes were surveyed and their relationships to topoclimate were analyzed separately.

We asked the following questions: (1) Which components of topoclimate influence local species distributions? (2) Across the species, do the adult and regeneration size classes exhibit different responses to topoclimate gradients? (3) To what extent does topoclimate heterogeneity explain variation in community composition and is this relationship similar for both size classes?

Methods

Study site and plot establishment

This study was conducted across the 1263 ha Pepperwood Preserve in northern California (Sonoma Co., 38.57°N, −122.68°W). The preserve is topographically heterogeneous and features vegetation representative of California Coast Ranges, including chaparral, grasslands, Douglas-fir forest, oak woodland and mixed hardwood forest (Halbur et al. 2013). Pepperwood is in a transition zone between southern and central California woodlands, dominated by Quercus agrifolia along the coast and Q. douglasii inland, and northern woodlands with high abundance of Pseudotsuga menziesii and Q. garryana (a close relative of Q. douglasii). There is an extensive land-use history at this preserve, including logging, charcoal making and livestock grazing from the 1800s to the present (Evett et al. 2013). There were two large fires on the preserve in 1964 and 1965, and no large fires have occurred since that time (Halbur et al. 2013).

Fifty 20 × 20 m woody vegetation-monitoring plots (2 ha in total) were established across Pepperwood Preserve (Ackerly et al. 2013). The plot locations were selected based on two criteria: (1) stratification across the topographic gradients of the preserve, and (2) a balanced spread across deciduous and evergreen woodlands, based on a recently completed vegetation map of the preserve (Halbur et al. 2013). The following topographic variables were used to stratify the plot locations: elevation, slope, aspect, modeled March radiation, topographic position index (TPI), percent lower pixels (PLP) and topographic water index (TOPOID) (see definitions below).

Topographical variables were obtained with GIS analyses of a 10-m digital elevation model for Pepperwood Preserve (Fig. 1). Slope and aspect were calculated using the terrain function in the raster package in R (Hijmans et al. 2015). Average incident solar radiation (kW h/m^2) was calculated for each month, based on slope, aspect and local topographic shading, in the Solar Analyst function of the Spatial Analyst library in ArcGIS (Fu and Rich 2000). March radiation was used for plot selection and for analyses reported here because it represents radiation during the spring growing season. TPI and PLP offered alternative metrics of local topographic relief. TPI (m) is the elevation of a pixel minus the mean elevation of the landscape in a defined radius (positive values indicate upper slope and hilltop positions, negative values are lower slope or valley bottoms) (Weiss 2001). PLP (%) is calculated as the percent of pixels within a specified radius that are lower than the focal pixel (ranges from 0 to 100, with higher values for upper slope positions). Both these topographic relief metrics were calculated with a neighborhood radius of 100 m. The metrics were similar with a neighborhood radius of 500 m, so we only used calculations derived from the 100 m radius. TOPOID was calculated using a hydrologic flow accumulation algorithm that incorporates the amount of 'upstream' area from which water would flow towards a focal pixel and the slope of that area (flatter locations with more upstream area will have a greater TOPOID value).

Across the plots, elevation ranged from 122 m to 462 m, average March radiation ranged from 430 kW h/m^2 to 809 kW h/m^2, TPI ranged from −10 m to 14 m, PLP

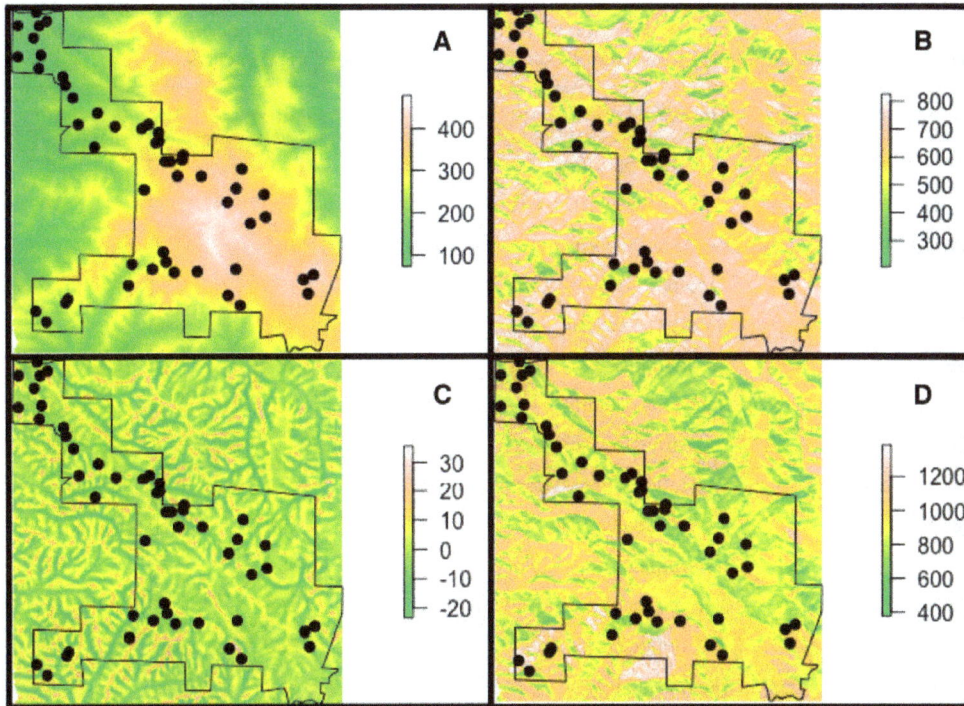

Figure 1. Maps of plots (black circles) across Pepperwood Preserve (black outline) with the following base layers: (A) DEM (m), (B) March Radiation (kWh/m^2), (C) TPI (m) and (D) CWD (mm).

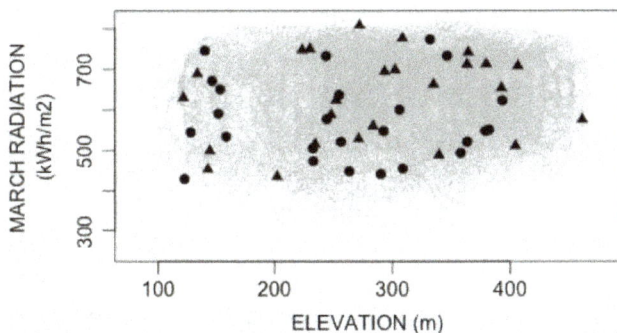

Figure 2. The plot locations span a large amount of the climate space of the preserve. The gray dots represent all combinations of elevation and average March radiation across Pepperwood Preserve. The shapes represent different vegetation types of the plots, with triangles = evergreen woodland, and circles = deciduous woodland. These two climate variables are representative of the two main principal components, illustrating that this figure is representative of many of the combinations of the climate parameters measured across the preserve.

ranged from 14 % to 99 % and TOPOID ranged from 3.5 to 9.6. As plots were installed, we reexamined their distribution across topographic gradients and vegetation types, adding new locations to fill gaps until we achieved a well-stratified distribution across the entire preserve (Fig. 2). Each plot was placed on a homogeneous slope and aspect so that orientation could be clearly characterized, resulting in a bias away from

sampling on ridge tops, valley bottoms or strongly curved hill slopes.

Vegetation monitoring

For this study, we included all species with a woody growth form represented by at least one individual with a greater than 1 cm diameter at breast height (DBH) in at least one plot across the study (Table 1). DBH was measured at 1.4 m. Poison oak (*Toxicodendron diversilobum*) was abundant but not sampled, as it is generally below the threshold size and presents a health hazard. Each 20 × 20 m plot was subdivided into sixteen 5 × 5 m quadrats for vegetation sampling, but all data are reported at the plot level. All individuals of the study species were sampled in the 50 plots and categorized into one of the five size classes: seedling, juvenile, sapling, tree and stump sprout. Seedlings were defined as individuals <10 cm height. Juveniles were defined as individuals with a height of ≥ 10 and < 50 cm. Saplings were defined as individuals with a height ≥ 50 cm and DBH of < 1 cm. Trees were defined as individuals with a DBH of ≥ 1 cm. Stump sprouts were defined as individuals with the same specification as a sapling, but observed to be growing from a larger tree or stump. All individuals of the five size classes were identified to species. Saplings, trees and stump sprouts were tagged with uniquely numbered metal tags, and locations recorded

Table 1 The woody species range in overall abundance for the different size classes. The following metrics are cumulative across all plots: basal area = tree basal area (cm^2), TR = number of individuals in the tree size class, SA = number of individuals in the sapling size class, JU = number of individuals in the juvenile size class, SE = number of individuals in the seedling size class.

Code	Common name	Botanical Latin	Basal area	TR	SA	JU	SE
ACEMAC	Bigleaf Maple	*Acer macrophyllum*	0	0	0	10	6
ADEFAS	Chamise	*Adenostoma fasciculatum*	16	2	3	0	0
AESCAL	Buckeye	*Aesculus californica*	3084	9	5	53	0
AMOCAL	Napa false indigo	*Amorpha californica* var. *napensis*	255	31	242	13	0
ARBMEN	Madrone	*Arbutus menziesii*	31 931	66	101	246	230
ARCMAN	Manzanita	*Arctostaphylos* sp.	4360	15	7	4	0
BACPIL	Coyote Bush	*Baccharis pilularis*	21	1	11	2	0
CEACUN	Wedgeleaf Ceanothus	*Ceanothus cuneatus*	0	0	23	0	0
CERBET	Mountain Mahogany	*Cercocarpus betuloides*	0	0	5	20	1
FRACAL	Coffee berry	*Frangula californica*	46	10	103	183	69
HETARB	Toyon	*Heteromeles arbutifolia*	3393	159	156	99	11
NOTDEN	Tan oak	*Notholithocarpus densiflorus*	564	5	7	12	1
PSEMEN	Douglas-fir	*Pseudotsuga menziesii*	173 320	441	437	800	977
QUEAGR	Coast live oak	*Quercus agrifolia*	208 574	244	273	762	191
QUEBER	Scrub oak	*Quercus berberidifolia*	22	6	31	33	6
QUEDOU	Blue oak	*Quercus douglasii*	55 385	68	0	1561	630
QUEGAR	Oregon oak	*Quercus garryana*	170 152	301	5	2247	606
QUEKEL	Black oak	*Quercus kelloggii*	47 375	58	1	184	48
QUELOB	Valley oak	*Quercus lobata*	1320	1	0	2	0
QUEWIS	Interior live oak	*Quercus wislizenii*	31	1	0	0	0
TORCAL	California nutmeg	*Torreya californica*	0	0	3	3	0
UMBCAL	California bay laurel	*Umbellularia californica*	19 546	168	889	2060	163

to the nearest cm, relative to the corner of each quadrat. For saplings and stump sprouts, we measured the height of the individual, the basal diameter of the main (largest basal area) stem at 10 cm off the ground, and the number of stems that split below 10 cm. Basal area was calculated for trees based on DBH. Seedlings and juveniles of each species were censused in each plot. These methods allow comparison with other standardized woody vegetation monitoring protocols (Condit 1998; Gilbert *et al.* 2010). Plot establishment and baseline data collection for tagged individuals occurred in spring 2013. Abundance of seedlings and juveniles were resurveyed in spring 2015 to confirm identifications, and we used the 2015 seedling and juvenile data in this analysis.

Environmental measurements

Beginning in 2013, temperature and relative humidity were monitored in all plots at 30-minute intervals. A HOBO datalogger (Hobo model U23, Onset Corp., Bourne, MA) nested inside a solar radiation shield was placed at 1.2 m above the ground, and 5 m outside the plot edge in a location of similar light availability and species composition. Annual winter minimum temperature was calculated as the average daily minimum temperature for the months of November and December in 2014, and ranged from 6.4 °C to 10.5 °C across the plots. Annual summer maximum temperature was calculated as the average daily maximum temperature for the months of July and August in 2014, and ranged from 26.0 °C to 32.6 °C across the plots. Soil moisture measurements were taken as volumetric water content in the center of each quadrat in every plot at a depth of 12 cm (Campbell Hydrosense, Model CS659). The mean of all measurements across the 16 quadrats was calculated to determine the average soil moisture of each plot. These readings were taken for the 2013–2015 field seasons in all plots in the first week of May within a 5-day window without precipitation. Thus, measurements were taken prior to the onset of summer, when soils may become uniformly dry. The average soil moisture ranged from 2.6% to 14.5% across the plots.

Water balance parameters for the plots were obtained from a 10 m resolution downscaled analysis of the Basin Characterization Model (Flint et. al. 2013) (Fig. 1D). The Basin Characterization Model is a water balance model that takes into account soils, precipitation, hydrology and temperature to model spatial patterns of AET, PET and CWD. Gridded data were obtained from PRISM and downscaled using the Gradient-Inverse-Distance-Squared algorithm based on the 10 m digital elevation model of the local landscape. Monthly modeled values of AET, PET and CWD were summed to obtain water year totals, and a 30-year (1981–2010) average was used for the analyses in this paper. The 30-year average of each of the modeled water balance parameters were averaged across the four 10 m grid cells within each plot to obtain a single plot value. Across the plots, AET ranged from 128 mm to 455 mm, PET ranged from 878 mm to 1516 mm and CWD ranged from 654 mm to 1314 mm.

Statistical analyses

A principal component analysis (PCA) was performed with *princomp* on 11 of the environmental variables quantified across the plots to reduce the dimensionality, but to still include the contributions of all variables to overall patterns of topographic heterogeneity (Oksanen et al. 2008). The PCA included modeled parameters (AET, PET, CWD, elevation, March radiation, TOPOID, TPI and PLP) and field-measured parameters (2014 winter minimum temperature, 2014 summer maximum temperature and 2014 soil moisture). All climate variables were scaled and centered prior to the PCA. The first two principal components, PC1 and PC2, of this analysis were used as independent variables for subsequent analyses of vegetation distributions. The correlations between these two principal components and the topoclimate variables of interest were quantified with the *dimdesc* function in R (Husson et al. 2015).

Using the first two topoclimate principal component scores, we asked which topoclimate variables affected single-species distributions (Question 1) and whether this relationship depended on size class (Question 2). We considered two size classes: all trees were considered as the "adult" size class and all seedlings and juveniles were considered together as the "regeneration" size class. For these first two questions, we focused on the six most dominant woody species across the

plots: *Q. douglasii, Q. garryana, Q. agrifolia, Q. kellogii, P. menziesii* and *Umbellularia californica*. These six species each have greater than 50% basal area in at least one plot, and together account for 71% of all basal area across the study. For each species, the adult size class and the regeneration size class were analyzed separately and then compared. Linear regressions were used to assess the relationship between adult abundance (total basal area per plot), and the two topoclimate principal components (PC1 and PC2). Binomial regressions were used to assess the probability of presence of individuals in the adult or regeneration size classes across PC1 and PC2. When analyzing the regeneration size class, we also included an additional covariate of conspecific adult basal area as a proxy for local seed rain. Poisson regression was used to examine the sensitivity of regeneration abundance to PC1, PC2 and conspecific basal area. Conspecific basal area was included in the regeneration models as a main effect and as an interaction term with PC1 and PC2. All covariates were scaled and centered, and the model with the best fit, based on AIC, was selected for each species.

We separately assessed the effect of topoclimate on community composition for the adult and regeneration size classes (Question 3) using conditional, constrained redundancy analyses (CCRA). The constrained redundancy analysis is a form of multivariate regression in which the response variable is the community dissimilarity across plots (Anderson et al. 2011). The conditional version of the constrained redundancy analysis removed the effect of evergreen versus deciduous physiognomy on community dissimilarity (Legendre 2007). This conditioning was necessary because a portion of the community dissimilarity arose from our non-random plot placement, which intentionally represented both evergreen and deciduous vegetation types across the topoclimate gradients (Fig. 2). We used the square-root of the Bray–Curtis dissimilarity metric as the response variable in the CCRA with PC1, PC2 and spatial distance between plots as covariates (Anderson et al. 2011). Mantel tests showed spatial autocorrelation in PC1, PC2 and the community matrices which motivated us to include spatial distance between plots as a covariate (Mantel and Valand 1970; Legendre 1993) (**see Supporting Information – Table S2**). The spatial distance between plots was represented in the CCRA by the first principal component of the spatial distance matrix. From the R *vegan* package, we used the *capscale* function to perform the CCRA, and the *adonis* function to determine the variation explained by each significant model parameter (McArdle and Anderson 2001; Oksanen et al. 2008). Lastly, we used PC1 and PC2 to predict adult and regeneration species richness in a multiple regression framework.

Table 2 For most species, there is significant three-way interaction between the topoclimate axes and conspecific adult basal area for models of the regeneration abundance. PC1, first axes of the topoclimate PCA; PC2, second axes of the topoclimate PCA; TR, conspecific adult basal area. ':' indicates an interaction between the parameters. '+' indicates a significant positive effect of that parameter, and '−' indicates a significant negative effect. G^2, the difference in deviance explained by the null model (null deviance) and full model (residual deviance). Wald tests between the null and full models are all highly significant for all species. DF, degrees of freedom.

Species	TR	PC1	PC2	PC1:PC2	PC1: TR	PC2: TR	PC1: PC2: TR	Null deviance; DF = 49	Residual deviance; DF = 42	G^2
Quercus agrifolia	+				−	+	−	2154	637	1517
Quercus garryana	+	+	+		+	+	−	8076	4643	3424
Quercus douglasii	+	+	+	−	+	−		7787	2368	5149
Quercus kellogii	+		+	−	+		+	865	340	525
Pseudotsuga menziesii	+	+	−	+	−	+	−	6937	4091	2846
Umbellularia californica	+	−	−	+	+		+	3880	2207	1673

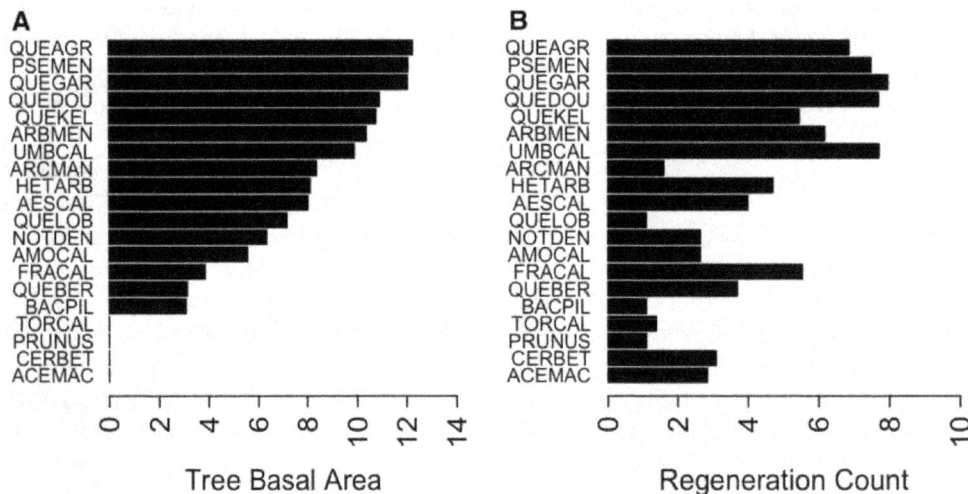

Figure 3. (A) Tree abundance in descending order, based on basal area (cm^2). (B) Regeneration abundance, based on counts. For species names, see Table 1. Both abundance metrics are log-transformed.

Results

Across the 50 plots, a total of 3,900 individuals were tagged and mapped (2312 saplings and 1588 trees), and 11 235 additional individuals ≤50 cm tall were recorded and enumerated by quadrat, but not tagged (2939 seedlings and 8296 juveniles). Twenty-five species were identified and species richness within a plot ranged from 3 to 13 species (Table 1). Tree densities ranged from 3 to 208 individuals per plot. Total basal area (the sum of trunk cross-sectional areas) for trees ranged from 10.8 to 94.8 m^2 ha^{-1}. Across all plots, the number of seedlings and juveniles per plot ranged from 14 to 868. The most abundant tree species based on the number of individuals was *P. menziesii*, followed by *Q. garryana*, *Q. agrifolia* and

U. californica. However, the most abundant tree species based on the basal area was *Q. agrifolia*, followed by *P. menziesii* and *Q. garryana* (Fig. 3). The most common species of the regeneration size class (seedling + juvenile counts) in declining order were *Q. garryana*, *U. californica*, *Q. douglasii* and *P. menziesii* (Fig. 3).

Topoclimate principal components

PC1 and PC2 explained 59% (33% and 26%, respectively) of the environmental variation observed across plots (Fig. 4). PC1 was significantly positively correlated with CWD (R^2=80%), AET (R^2=85%), PET (R^2=95%) and March radiation (R^2=95%). PC2 was significantly positively correlated with maximum summer

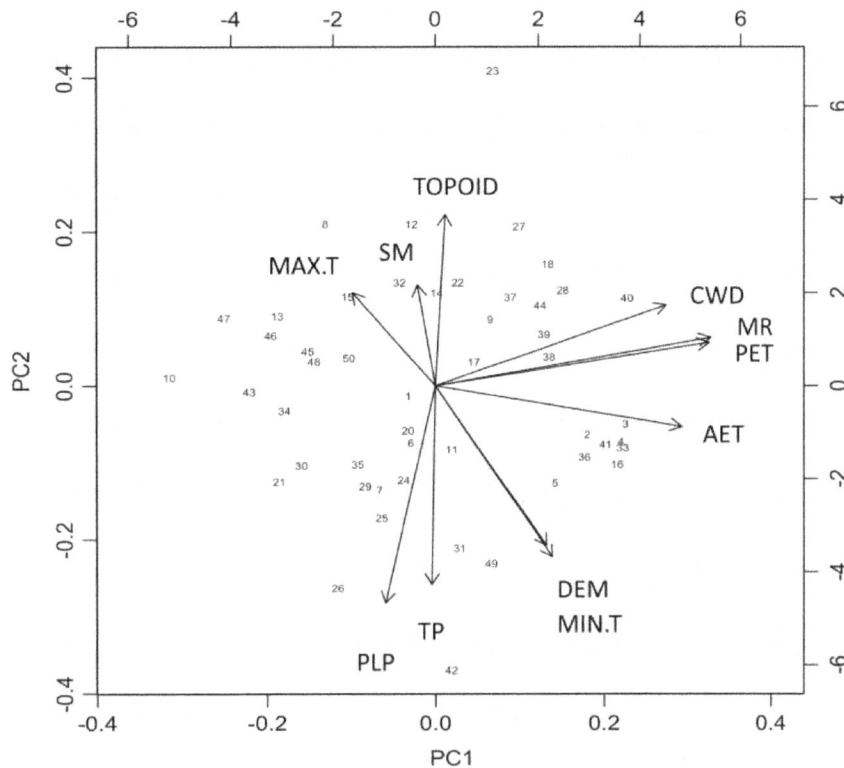

Figure 4. Biplot of the principal components analysis for the 11 environmental variables quantified across the plots. CWD, climatic water deficit (mm); AET, actual evapotranspiration (mm); PET, potential evapotranspiration (mm); MR, March radiation (kWh/m^2); DEM, elevation (m); MIN.T, 2014 annual minimum winter temperature (°C); MAX.T, 2014 annual maximum summer temperature (°C); SM, 2014 soil moisture measurements (%); TPI, topographic position index (m); PLP, percent lower pixels (%); TOPOID, topographic water index.

temperature (R^2=35%), soil moisture (R^2=38%) and TOPOID (R^2=65%), and negatively correlated with elevation (R^2=61%), minimum winter temperature (R^2=65%), TPI (R^2=76%) and PLP (R^2=81%).

Adult responses to topoclimate

Only three of the six dominant woody species distributions showed significant relationships with the preserve's topoclimate for the adult size class. Both *Q. garryana* and *Q. agrifolia* adult abundance were correlated with PC1. *Q. garryana* had a negative relationship with PC1, with 8% of the variation in its abundance explained (P = 0.016). *Q. agrifolia* had a positive relationship with PC1, with 15% of the variation in its abundance explained (P = 0.002). The effect of PC1 and PC2 on the probability of the presence of these two species showed a similar pattern. *Umbellularia californica* abundance was not significantly explained by either PC1 or PC2, however, *U. californica* presence had a weakly significant negative relationship with PC1 (P = 0.048). For the adult size class, none of the dominant species showed any relationship with PC2 for either abundance or presence.

Regeneration responses to topoclimate

The best model fits for regeneration presence/absence included only conspecific adult abundances for all six dominant species. Thus, the effect sizes of PC1 and PC2 on the probability of species presence were not further analyzed. The best model fits for regeneration abundance included PC1, PC2, conspecific adult abundance and their interactions. Conspecific adult abundance had a positive main effect on regeneration abundance for all dominant species (Table 2). The effect of PC1 on regeneration abundance was significant for most of the dominant species, except *Q. kellogii* and *Q.agrifolia*, and the sign of the effect was predominantly positive (higher abundance on south-facing slopes) (Table 2). For most of the *Quercus* spp., PC2 had a significant positive main effect on regeneration abundance (Table 2). For *U. californica* and *P. menziesii*, PC2 had a significant negative effect of regeneration abundance (higher abundance on upper hill slopes) (Table 2). The abundance of conspecific adults also influenced the effect of PC1 and PC2 on regeneration abundance via their interactions (Fig. 5). Many two and three-way interactions between PC1, PC2 and adult conspecific basal area were significant (Table 2 and

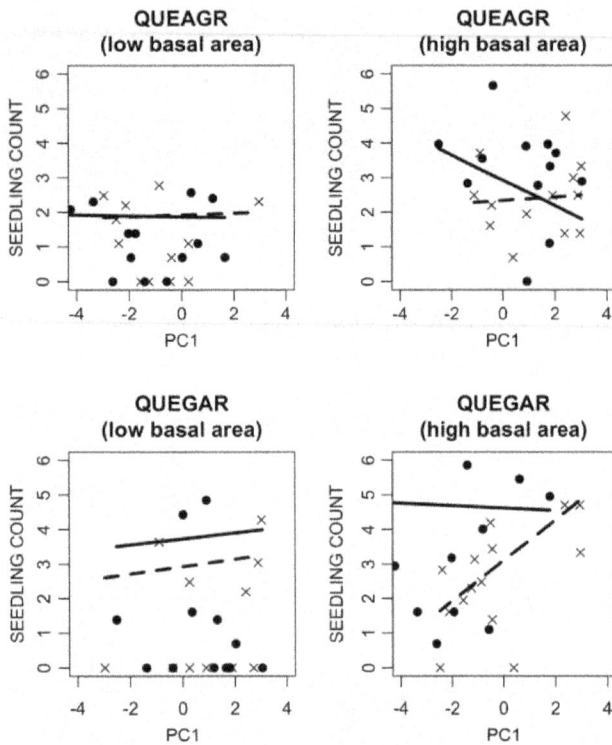

Figure 5. Model predictions show interactions among PC1, PC2, and adult abundance (basal area) for natural log-transformed regeneration abundance of *Q. agrifolia* (QUEAGR) and *Q. garryana* (QUEGAR). Dashed lines are model prediction from the lower 50th percentile PC2 values, and solid lines are from the upper 50th percentile PC1 values. The *x*'s represent data from plots in the lower 50th percentile for PC2 values and points represent data from plots in the upper 50th percentile for PC2 values. The figures on the left are predictions for the lower 50th percentile of conspecific adult basal area and the figures on the right are predictions for the upper 50th percentile of conspecific adult basal area.

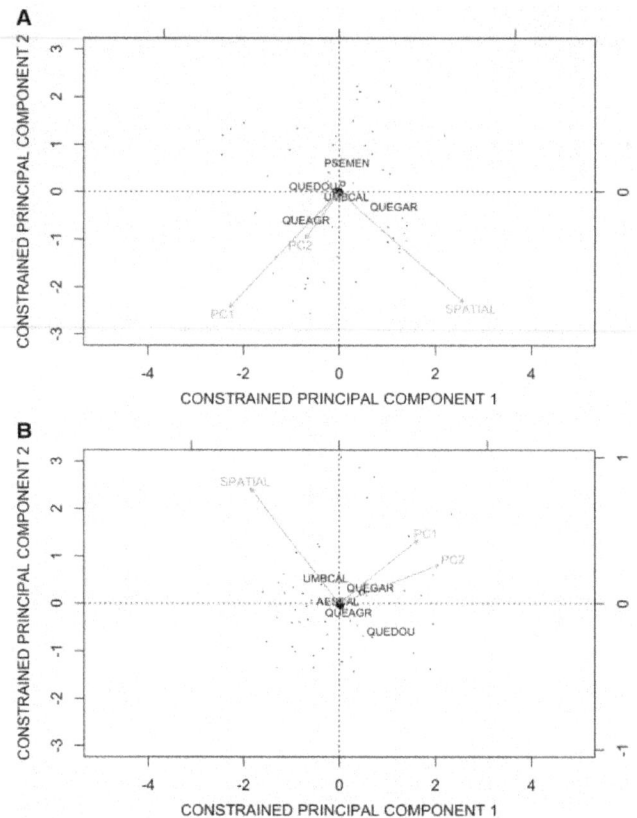

Figure 6. The first two principal components of the constrained redundancy analyses on the adult community (A) and the regeneration community (B). The points represent the 50 sites. Each arrow represents the direction and magnitude, indicated by length, of the effect of the topoclimate axes (PC1 and PC2), and the distance between plots (SPATIAL). The species code locations represent how the species are structured in the constrained ordination space. For species names see Table 1.

Fig. 5). However, no obvious patterns of these effects emerged across species.

Community responses to topoclimate

A small amount of variation in both adult and regeneration community composition was explained by PC1, PC2 and the spatial distance between plots. The adult community structure across the preserve was significantly correlated with PC1 (10 % variation explained, $P = 0.007$) and spatial distance between plots (9 % variation explained, $P = 0.001$) (Fig. 6A). The regeneration community structure was significantly correlated with PC2 (5 % variation explained, $P = 0.013$) spatial distance (4 % variation explained, $P = 0.009$) and PC1 (3 % variation explained, $P = 0.041$) (Fig. 6B). Lastly, adult species richness was negatively correlated with PC1 (10 % variation explained, $P = 0.009$), but had no relationship with PC2. Species richness of the regeneration size class was not correlated with either PC1 or PC2.

Discussion

By examining the distribution of adult and regeneration size classes of woody vegetation across 50 plots that span the climate space of a single preserve, we assessed whether topoclimate heterogeneity was ecologically relevant for woody plant species distributions and community composition at this scale. Overall, we found support for species distributions and community composition being, in part, influenced by the topoclimate variation on the landscape. Although statistically significant, the effects of topoclimate on single-species distributions, community composition and species richness were small. Here, we discuss the observed vegetation patterns related to topoclimate and suggest biological mechanisms that may contribute to the relatively small size of the effects in this study.

Our study adds to a relative paucity of woody vegetation studies in Mediterranean climates with mapped stems including small individuals in regeneration size classes (Gilbert *et al.* 2010; Anderson-Teixeira *et al.* 2015). These vegetation plots were dominated by *Quercus* spp. with high oak basal area, species diversity, and a mix of both evergreen and deciduous oaks. Relative to the UCSC Forest Ecology Research Plot, a single large woody vegetation plot in a comparable climate, our most dominant species contributed to much less of the total basal area (Gilbert *et al.* 2010). This difference perhaps reflects that our sampling strategy encompassed multiple vegetation types. By sampling 2 ha of forest across 50 plots, we were able to capture variation in size structure, abundance and composition in the woody vegetation across the 1263 ha preserve.

The topoclimate principal components (PC1 and PC2) represented combinations of topographic features observed across the preserve. At this local scale, all the PC1 parameters (March radiation, PET, AET and CWD) were highly positively correlated, likely due to strong effects of slope and aspect on solar radiation load. Positive PC1 values were associated with south-facing slopes where the radiation load is higher. Low elevation valleys were associated with positive PC2 values, with wetter soils and cooler temperature minimums due to night-time cold-air pooling. Negative PC2 values were associated with drier, high-elevation ridges and upper hill slopes.

Adults showed species-specific responses to the environmental parameters of PC1. This suggests that the role of water availability and temperature for woody vegetation abundance at the topoclimate scale is primarily reflected in the interaction and seasonality of these climate variables (Stephenson *et al.* 1990). The two dominant oak species, *Q. garryana* and *Q. agrifolia*, showed opposite responses in abundance to PC1. The deciduous *Q. garryana* was more abundant in sites with less solar radiation and lower CWD. The evergreen *Q. agrifolia* was more abundant on sites with higher AET, and occupied sites with the highest CWD. These opposing patterns may reflect differences in these species' ranges; Pepperwood Preserve is near the southern range limit of *Q. garryana*, and the northern range limit of *Q. agrifolia*. These species may be found in topoclimates that exhibit environmental conditions more similar to those found in their range centers (Holland and Steyn 1975). However, geographic range limits may not necessarily be at the edges of the climatic niche of the species (Chardon *et al.* 2015). Further research is needed to assess the degree to which range-wide climate characteristics influence local-scale distributions for this system.

None of the dominant woody species were correlated with the topoclimate principal component associated with cooler, wetter valley bottoms (PC2) for the adult size class. It is possible that the range of climate variation present across the PC2 topoclimate gradient of Pepperwood Preserve is narrow relative to the climate tolerances of these woody species. Although we observed topoclimate sensitivity in the presence and abundance of the adults of some species, the greatest amount of variation explained was only 15 % (*Q. agrifolia*) and three of the six dominant species showed no sensitivity to either topoclimate gradients. The magnitude of topography necessary to cause a response is most likely species specific.

As opposed to the adults, the regeneration size class showed sensitivity to both the topoclimate principal components and the interactions between them. This result aligns with other studies that show that seedlings of woody vegetation may be sensitive to environmental variation at smaller scales than adults (Gray and Spies 1997; Máliš *et al.* 2016). Despite some general trends, there is strong evidence for highly species-specific responses in the two-way and three-way interactions between the topoclimate principal components and conspecific adult basal area. Abundance of seedlings and juveniles generally increased with conspecific adult basal area for all species. In some cases, high seed input by abundant conspecific adults can overwhelm the effects of variable conditions and effectively suppress the topoclimate sensitivity of the regeneration size class (Clark *et al.* 1998; Warren *et al.* 2012). However, we were able to detect a signal of both the conspecific adult abundance and environmental parameters on regeneration abundance. Our results support that local (within 20 m) seed source is the main driver of the probability that recruitment is observed at a site, but the overall abundance of the regeneration class is mediated by the topoclimate.

There may be ecological constraints on the capacity of a heterogeneous landscape to buffer vegetation response to climate change. Replacement of adults by new recruits can be slowed if dispersal is limiting (Aitken *et al.* 2008). We found that conspecific adult abundance, a proxy for local seed input, greatly impacts recruitment, potentially demonstrating dispersal limitation effects on woody species composition at a small spatial extent. Species ability to track their climate niche in response to a changing climate may be delayed due to lack of seed input even at the topoclimate scale. Heterogeneous topoclimates may more likely enable species persistence in favorable sites (i.e. refugia) as the overall species range contracts (Ackerly *et al.* 2010; Dobrowski 2011). This result also warrants further investigation into the microclimate effect of canopy cover on the understory environment occupied by seedlings and juveniles (De Frenne *et al.* 2013; Dobrowski *et al.* 2015).

For both size classes, topoclimate explained some variation in community composition. In concordance with the individual species responses, only the principal component associated with the water balance parameters (PC1) was correlated with variation in the adult community composition. Also the regeneration community composition was correlated with both topoclimate principal components. Previous research has shown a sizable effect of topoclimate on the abundance of short-lived plants (herbs) across a landscape (Harrison *et al.* 2010). For long-lived woody vegetation, we found that in total, only 11 % percent of adult community composition and 7 % of the regeneration community composition were explained by the topoclimate across the preserve. It is possible that the low amount of woody community variation explained by topoclimate is due to the limited range of topoclimate variation at Pepperwood, relative to the environmental tolerances of the studied woody species. We captured a wide range of topoclimate across a small spatial extent, but many of the studied species have broad ranges. Long-lived species, such as woody plants, may also be in disequilibrium with the present climate, impeding our understanding of their climate niche (Svenning and Sandel 2013).

Pepperwood Preserve's history of fire and land-use may also be limiting the role of topoclimate in shaping woody community distributions. Previous disturbances (natural and anthropogenic) interact with climate to shape vegetation distributions, especially at local scales (Delcourt and Delcourt 1988). Pepperwood Preserve had two major fires in the last 50 years (1964 Hanley Fire and 1965 Calistoga Fire) that burned in approximately two-thirds of our plot locations. Fires will have different effects on species, with some species resprouting after fire (e.g. *Quercus* spp. and *U. californica*) while others needing to re-invade a burned area (e.g. *P. menziesii*, in which all but the largest trees are killed by fire) (Keeley *et al.* 2005). Non-pristine habitats, such as managed farm-lands, also may have protracted historical effects on species local distributions. Vegetation in these managed areas may not show sensitivity to topoclimates due to the long-term inertia of vegetation following disturbances (Bodin *et al.* 2013). Evett *et al.* (2013) found increasing tree density, woody encroachment into grasslands and changing community composition (e.g. increased *U. californica*) at Pepperwood Preserve since the early 1900s, and attributed these shifts to changes in land-use and management.

At small spatial extents, ecological processes other than climate limitations may play a prominent role in shaping vegetation communities. Local dispersal limitation may prevent a species from establishing in a suitable topoclimate (Verheyen *et al.* 2003). The stochastic nature of colonization and historical contingencies may also moderate the community dynamics of a site (Duarte *et al.* 2006; Walker and Wardle 2014). Spatial distance between plots, even at this small spatial extent, explains some community variation for both size classes. This pattern may be driven by spatially aggregated land use or disturbances (e.g. fire history), limited seed dispersal at a scale smaller than the preserve or climate patterns at the preserve scale not directly associated with topography (e.g. fog input, as Pepperwood sits at the edge of the Pacific fog belt) (Torregrosa *et al.* 2016).

Future work on the role of topoclimate for woody vegetation distributions and dynamics in this system will focus on range-wide climatic characteristics and functional traits of the species, rather than the species identity per se. This method may reduce the confounding effect of historical contingencies and identify patterns of functional redundancy, in which a topoclimate is equally suitable for different species. Previous work with this trait-based approach has been beneficial in understanding vegetation–climate relationships at the landscape scale (Lenoir *et al.* 2013; De Frenne *et al.* 2013). Comparing results of analyses at the species versus functional level will demonstrate the contribution of topoclimates to the maintenance of both functional and species diversity.

Conclusion

Downscaled climate variables paired with fine scale vegetation data represent a unique opportunity to resolve fundamental ecological questions regarding the maintenance of species distributions and community types across local topography. There has been a resurgence of interest in this question due to the potential role of topography in how species respond to climate change (Rapacciuolo *et al.* 2014), and the potential importance of small-scale topography in future conservation strategies (Lawler *et al.* 2015). To protect our forests, we need to have a better understanding of topography–vegetation relationships in local landscapes with past disturbances (Millar and Stephenson 2015). We show that disturbed (both naturally and through management) lands can capture community diversity with topographic complexity. However, wide climate tolerances of species and historical contingencies may weaken the relationship between topoclimate and woody vegetation. Our study not only addresses impacts of topography on woody vegetation on small spatial extents but also serves as a baseline for long-term studies of vegetation dynamics in response to climate change in heterogeneous landscapes.

Sources of Funding

Our work was funded by the Gordon and Betty Moore Foundation (California, USA) Grant nos. 4430 and 2861. Additional support was provided by the US National Science Foundation Graduate Research Fellowship Grant DGE-1106400 (to M.F.O.) and Graduate Research Fellowship Grant DGE-1321845 Amend. 3 (to M.J.K.).

Contributions by the Authors

D.D.A., M.F.O., M.N.B. and M.M.H. conceived and designed the research. M.F.O., M.N.B., P.D.P., M.J.K., M.M.H., C.D., A.L.F. and L.E.F. performed the research. M.F.O. analyzed the data. M.F.O., M.J.K. and D.D.A. wrote the paper.

Acknowledgements

The authors thank all the volunteers and stewards of Pepperwood Preserve that aided in the establishment and monitoring of the plots, as well as Pepperwood Preserve President, Lisa Micheli and manager, Michael Gillogly. The authors also thank the Ackerly Lab members for many great discussions. Finally, the authors thank our anonymous reviewers for their suggestions, which greatly improved this manuscript. This paper is a contribution of the Terrestrial Biodiversity and Climate Change Collaborative (www.tbc3.org).

Supporting Information

The following additional information is available in the online version of this article —

File 1. Table. Lists the coordinates, field-measured environmental variables, vegetation statistics, and modeled topographic variables for all 50 plots. UTM.N = Northing coordinate, UTM.E = Easting coordinate, MIN.T = minimum winter temperatures (°C), MAX.T = maximum summer temperatures (°C), SM = 2014 average volumetric water content (%), DEC = percentage of deciduous individuals, DIV = species richness, ELE = elevation (m), MR = March Radiation (kWh/m^2), TOPOID = topographical water index, TPI = topographic position index (m), PLP = percent lower pixels (%), AET = actual evapotranspiration (mm), PET = potential evapotranspiration (mm), CWD = climatic water deficit (mm); the last three are derived from analysis of the Basin Characterization Model on a 10 m digital elevation model (Flint *et al.* 2013).

File 2. Figure. Regeneration model predictions for the additional dominant woody species (*Q. douglasii, Q. kellogii, P. menziesii,* and *U. californica*). All counts of regeneration abundance are log-transformed. Dashed lines are model prediction from the lower 50th percentile PC2 values, and solid lines are from the upper 50th percentile PC1 values. The x's represent data from plots in the lower 50th percentile for PC2 values and points represent data from plots in the upper 50th percentile for PC2 values. The figures on the left are predictions for the lower 50th percentile of conspecific adult basal area and the figures on the right are predictions for the upper 50th percentile of conspecific adult basal area.

File 3. Table. Mantel test results based on Pearson's product-moment correlation with 1e + 05 permutations. These tests measure the correlation between the spatial distance between plots (SPATIAL) and the dissimilarity of the topoclimate principal components (PC1, PC2), and the correlation between SPATIAL and the vegetation community Bray-Curtis dissimilarity for both adult basal area (ADULT) and regeneration counts (REGEN).

Literature Cited

Ackerly DD, Cornwell WK, Weiss SB, Flint LE, Flint AL. 2015. A geographic mosaic of climate change impacts on terrestrial vegetation: which areas are most at risk? *PloS One* **10**:e0130629.

Ackerly DD, Loarie SR, Cornwell WK, Weiss SB, Hamilton H, Branciforte R, Kraft NJB. 2010. The geography of climate change: implications for conservation biogeography. *Diversity and Distributions* **16**:476–487.

Ackerly D, Oldfather M, Britton M, Halbur M, Micheli L. 2013. Establishment of Woodland Vegetation Research Plots at Pepperwood Preserve. Techinical Report for Moore Foundation.

Aitken SN, Yeaman S, Holliday JA, Wang T, Curtis-McLane S. 2008. Adaptation, migration or extirpation: climate change outcomes for tree populations. *Evolutionary Applications* **1**:95–111.

Anderson MG, Kneale PE. 1982. The influence of low-angled topography on hillslope soil–water convergence and stream discharge. *Journal of Hydrology* **57**:65–80.

Anderson MJ, Crist TO, Chase JM, Vellend M, Inouye BD, Freestone AL, Sanders NJ, Cornell HV, Comita LS, Davies KF, Harrison SP, Kraft NJB, Stegen JC, Swenson NG. 2011. Navigating the multiple meanings of β diversity: a roadmap for the practicing ecologist. *Ecology Letters* **14**:19–28.

Anderson-Teixeira KJ, Davies SJ, Bennett AC, Gonzalez-Akre EB, Muller-Landau HC, Joseph Wright S, Abu Salim K, Almeyda Zambrano AM, Alonso A, Baltzer JL, Basset Y, Bourg NA, Broadbent EN, Brockelman WY, Bunyavejchewin S, Burslem DFRP, Butt N, Cao M, Cardenas D, Chuyong GB, Clay K, Cordell S, Dattaraja HS, Deng X, Detto M, Du X, Duque A, Erikson DL, Ewango CEN, Fischer GA, Fletcher C, Foster RB, Giardina CP, Gilbert GS, Gunatilleke N, Gunatilleke S, Hao Z, Hargrove WW, Hart TB, Hau BCH, He F, Hoffman FM, Howe RW, Hubbell SP, Inman-Narahari FM, Jansen PA, Jiang M, Johnson DJ, Kanzaki M, Kassim AR, Kenfack D, Kibet S, Kinnaird MF, Korte L, Kral K, Kumar J, Larson AJ, Li Y, Li X, Liu S, Lum SKY, Lutz JA, Ma K, Maddalena DM, Makana J-R, Malhi Y, Marthews T, Mat Serudin R, McMahon SM, McShea WJ, Memiaghe HR, Mi X, Mizuno T, Morecroft M, Myers JA, Novotny V, de Oliveira AA, Ong PS, Orwig DA, Ostertag R, den Ouden J, Parker GG, Phillips RP, Sack L, Sainge MN, Sang W, Sri-ngernyuang K, Sukumar R, Sun I-F,

Sungpalee W, Suresh HS, Tan S, Thomas SC, Thomas DW, Thompson J, Turner BL, Uriarte M, Valencia R, Vallejo MI, Vicentini A, Vrška T, Wang X, Wang X, Weiblen G, Wolf A, Xu H, Yap S, Zimmerman J. 2015. CTFS-ForestGEO: a worldwide network monitoring forests in an era of global change. *Global Change Biology* **21**:528–549.

Armesto JJ, Martinez JA. 1978. Relations between vegetation structure and slope aspect in the Mediterranean Region in Chile. *Journal of Ecology* **66**:881–889.

Ashcroft MB, Gollan JR. 2013. Moisture, thermal inertia, and the spatial distributions of near-surface soil and air temperatures: understanding factors that promote microrefugia. *Agricultural and Forest Meteorology* **176**:77–89.

Bennie J, Hill MO, Baxter R, Huntley B. 2006. Influence of slope and aspect on long-term vegetation change in British chalk grasslands. *Journal of Ecology* **94**:355–368.

Bodin J, Badeau V, Bruno E, Cluzeau C, Moisselin JM, Walther GR, Dupouey JL. 2013. Shifts of forest species along an elevational gradient in Southeast France: climate change or stand maturation? *Journal of Vegetation Science* **24**:269–283.

Bond WJ, Keeley JE. 2005. Fire as a global "herbivore": the ecology and evolution of flammable ecosystems. *Trends in Ecology & Evolution* **20**:387–394.

Chape S, Harrison J, Spalding M, Lysenko I. 2005. Measuring the extent and effectiveness of protected areas as an indicator for meeting global biodiversity targets. *Philosophical Transactions of the Royal Society B: Biological Sciences* **360**: 443–455.

Chardon NI, Cornwell WK, Flint LE, Flint AL, Ackerly DD. 2015. Topographic, latitudinal and climatic distribution of *Pinus coulteri*: geographic range limits are not at the edge of the climate envelope. *Ecography* **38**:590–601.

Clark JS, Macklin E, Wood L. 1998. Stages and spatial scales of recruitment limitation in southern Apalachian forests. *Ecological Monographs* **68**:213–235.

Condit R. 1998. *Tropical forest census plots*. Berlin, Heidelberg: Springer.

Crimmins SM, Dobrowski SZ, Greenberg JA, Abatzoglou JT, Mynsberge AR. 2011. Changes in climatic water balance drive downhill shifts in plant species' optimum elevations. *Science* **331**:324–327.

Daly C, Conklin DR, Unsworth MH. 2009. Local atmospheric decoupling in complex topography alters climate change impacts. International Journal of Climatology **3**:1857–1864.

Das AJ, Stephenson NL, Flint A, Das T, van Mantgem PJ. 2013. Climatic correlates of tree mortality in water- and energy-limited forests. *PloS One* **8**:1–11.

De Frenne P, Rodríguez-Sánchez F, Coomes DA, Baeten L, Verstraeten G, Vellend M, Bernhardt-Römermann M, Brown CD, Brunet J, Cornelis J, Decocq GM, Dierschke H, Eriksson O, Gilliam FS, Hédl R, Heinken T, Hermy M, Hommel P, Jenkins MA, Kelly DL, Kirby KJ, Mitchell FJG, Naaf T, Newman M, Peterken G, Petrík P, Schultz J, Sonnier G, Van Calster H, Waller DM, Walther GR, White PS, Woods KD, Wulf M, Graae BJ, Verheyen K. 2013. Microclimate moderates plant responses to macroclimate warming. *Proceedings of the National Academy of Sciences of the United States of America* **110**:18561–18565.

Delcourt HR, Delcourt PA. 1988. Quaternary landscape ecology: relevant scales in space and time. *Landscape Ecology* **2**:23–44.

Dobrowski SZ. 2011. A climatic basis for microrefugia: the influence of terrain on climate. *Global Change Biology* **17**: 1022–1035.

Dobrowski SZ, Swanson AK, Abatzoglou JT, Holden ZA, Safford HD, Schwartz MK, Gavin DG. 2015. Forest structure and species traits mediate projected recruitment declines in western US tree species. *Global Ecology and Biogeography* **24**:917–927.

Duarte LDS, Machado RE, Hartz SM, Pillar VD. 2006. What saplings can tell us about forest expansion over natural grasslands. *Journal of Vegetation Science* **17**:799–808.

Evett RR, Dawson A, Bartolome JW. 2013. Estimating vegetation reference conditions by combining historical source analysis and soil phytolith analysis at pepperwood preserve, Northern California Coast Ranges, USA. *Restoration Ecology* **21**: 464–473.

Flint AL, Childs SW. 1987. Calculation of solar radiation in mountainous terrain. *Agricultural and Forest Meteorology* **40**: 233–249.

Flint LE, Flint AL, Thorne JH, Boynton R. 2013. Fine-scale hydrologic modeling for regional landscape applications: the California Basin Characterization Model development and performance. *Ecologial Processes* **2**:1–21.

Fu P, Rich PM. 2000. The Solar Analyst 1.0 user manual. Helios Environmental Modeling Institute http://www.hemisoft.com (26 June 2016).

Geiger R, Aron RH, Todhunter P. 2009. *The climate near the ground*, 7th edn. Lanham, MD: Rowman & Littlefield.

Gilbert GS, Howard E, Ayala-Orozco B, Bonilla-Moheno M, Cummings J, Langridge S, Parker IM, Pasari J, Schweizer D, Swope S. 2010. Beyond the tropics: forest structure in a temperate forest mapped plot. *Journal of Vegetation Science* **21**:388–405.

Gray AN, Spies TA. 1997. Microsite controls on tree seedling establishment in conifer forest canopy gaps. *Ecology* **78**: 2458–2473.

Grubb PJ. 1977. The maintenance of species-richness in plant communities: *the importance of the regeneration niche. *Biological Reviews* **52**:107–145.

Halbur M, Kennedy M, Ackerly D, Micheli L, Thorne J. 2013. Creating a Detailed Vegetation Map for Pepperwood Preserve Creating a Detailed Vegetation Map for Pepperwood Preserve. Moore Foundation Technical Report.

Harrison S, Damschen EI, Grace JB. 2010. Ecological contingency in the effects of climatic warming on forest herb communities.

Heller NE, Zavaleta ES. 2009. Biodiversity management in the face of climate change: a review of 22 years of recommendations. *Biological Conservation* **142**:14–32.

Heyerdahl EK, Brubaker LB, Agee JK. 2001. Spatial controls of historical fire regimes: a multiscale example from the interior West, USA. *Ecology* **82**:660–678.

Hijmans R, van Etten J, Mattiuzzi M. 2015. R Package "raster."

Holland PG, Steyn DG. 1975. To latitudinal responses vegetational in slope angle and aspect variations. *Journal of Biogeography* **2**: 179–183.

Husson AF, Josse J, Le S, Mazet J, Husson MF. 2015. Package "FactoMineR."

Jackson ST, Betancourt JL, Booth RK, Gray ST. 2009. Ecology and the ratchet of events: climate variability, niche dimensions, and species distributions. *Proceedings of the National Academy of Sciences* **106**:19685–19692.

Keeley JE, Fotheringham CJ, Baer-Keeley M. 2005. Factors affecting plant diversity during post-fire recovery and succession of Mediterranean-climate shrublands in California, USA. *Diversity and Distributions* **11**:525–537.

Korner C. 2003. *Alpine plant life: functional plant ecology of high mountain ecosystems*. Heidelberg, Germany: Springer.

Kreft H, Jetz W. 2007. Global patterns and determinants of vascular plant diversity. *Proceedings of the National Academy of Sciences of the United States of America* **104**: 5925–5930.

Lawler JJ, Ackerly DD, Albano CM, Anderson MG, Dobrowski SZ, Gill JL, Heller NE, Pressey RL, Sanderson EW, Weiss SB. 2015. The theory behind, and the challenges of, conserving nature's stage in a time of rapid change. *Conservation Biology : The Journal of the Society for Conservation Biology* **29**:618–629.

Legendre P. 1993. Spatial autocorrelation: trouble or new paradigm? *Ecology* **74**:1659–1673.

Legendre P. 2007. Studying beta diversity: ecological variation partitioning by multiple regression and canonical analysis. *Journal of Plant Ecology* **1**:3–8.

Lenoir J, Graae BJ, Aarrestad PA, Alsos IG, Armbruster WS, Austrheim G, Bergendorff C, Birks HJB, Bråthen KA, Brunet J, Bruun HH, Dahlberg CJ, Decocq G, Diekmann M, Dynesius M, Ejrnaes R, Grytnes JA, Hylander K, Klanderud K, Luoto M, Milbau A, Moora M, Nygaard B, Odland A, Ravolainen VT, Reinhardt S, Sandvik SM, Schei FH, Speed JDM, Tveraabak LU, Vandvik V, Velle LG, Virtanen R, Zobel M, Svenning JC. 2013. Local temperatures inferred from plant communities suggest strong spatial buffering of climate warming across Northern Europe. *Global Change Biology* **19**:1470–1481.

Loarie SR, Duffy PB, Hamilton H, Asner GP, Field CB, Ackerly DD. 2009. The velocity of climate change. *Nature* **462**:1052–1055.

Lundquist JD, Pepin N, Rochford C. 2008. Automated algorithm for mapping regions of cold-air pooling in complex terrain. *Journal of Geophysical Research* **113**:1:15.

Lutz JA, van Wagtendonk JW, Franklin JF. 2010. Climatic water deficit, tree species ranges, and climate change in Yosemite National Park. *Journal of Biogeography* **37**:936–950.

MacArthur R. 1972. *Geographical ecology*. New York: Harper and Row.

Máliš F, Kopecký M, Petřík P, Vladovič J, Merganič J, Vida T. 2016. Life-stage, not climate change, explains observed tree range shifts. Global Change Biology.

Mantel N, Valand RS. 1970. A technique of nonparametric multivariate analysis. *Biometrics* **26**:547–558.

McArdle B, Anderson MJ. 2001. Fitting multivariate model to community data: a comment on distance-based redundancy analyses. *Ecology* **82**:290–297.

McCune B, Keon D. 2002. Equations for potential annual direct incident radiation and heat load. *Journal of Vegetation Science* **13**: 603–606.

McIntyre PJ, Thorne JH, Dolanc CR, Flint AL, Flint LE, Kelly M, Ackerly DD. 2015. Twentieth-century shifts in forest structure in California: denser forests, smaller trees, and increased dominance of oaks. *Proceedings of the National Academy of Sciences* **112**:1458–1463.

Mclaughlin BC, Zavaleta ES. 2012. Predicting species responses to climate change: demography and climate microrefugia in California valley oak (Quercus lobata). *Global Change Biology* **18**: 2301–2312.

Millar CI, Stephenson NL. 2015. Temperate forest health in an era of emerging megadisturbance. *Science* **349**:823–826.

Millar CI, Westfall RD, Delany DL, Flint AL, Flint LE. 2015. Recruitment patterns and growth of high-elevation pines in response to climatic variability (1883–2013), western Great Basin, USA. *Canadian Journal of Forest Research* **45**:1–51.

Oksanen J, Kindt R, Legendre P, O'Hara RB, Simpson GL, Solymos P, Stevens HH, Wahner H. 2008. The vegan Package.

Rapacciuolo G, Maher SP, Schneider AC, Hammond TT, Jabis MD, Walsh RE, Iknayan KJ, Walden GK, Oldfather MF, Ackerly DD, Beissinger SR. 2014. Beyond a warming fingerprint: individualistic biogeographic responses to heterogeneous climate change in California. *Global Change Biology* **20**:2841–2855.

Roering JJ, Gerber M. 2005. Fire and the evolution of steep, soil-mantled landscapes. *Geology* **33**:349–352.

Sage T, Sage R. 2002. Microsite characteristics of Muhlenbergia richardsonis (Trin.) Rydb., an alpine C 4 grass from the White Mountains, California. *Oecologia* **132**:501–508.

Salzer MW, Hughes MK, Bunn AG, Kipfmueller KF. 2009. Recent unprecedented tree-ring growth in bristlecone pine at the highest elevations and possible causes. *Proceedings of the National Academy of Sciences* **106**:20348–20353.

Stephenson NL. 2015. Climatic control of vegetation distribution: the role of the water balance. *The American Naturalist* **135**: 649–670.

Stephenson NL, The S, Naturalist A, May N. 1990. Climatic control of vegetation distribution: the role of the water balance. *The American Naturalist* **135**:649–670.

Svenning JC, Sandel B. 2013. Disequilibrium vegetation dynamics under future climate change. *American Journal of Botany* **100**: 21–21.

Tani M. 1997. Runoff generation processes estimated from hydrological observations on a steep forested hillslope with a thin soil layer. *Journal of Hydrology* **200**:84–109.

Torregrosa A, Combs C, Peters J. 2016. GOES-derived fog and low cloud indices for coastal north and central ecological analyses. *Earth and Space Science* **3**:46–67.

Van de Ven CM, Weiss SB, Ernst WG. 2007. Plant species distributions under present conditions and forecasted for warmer climates in an arid mountain range. *Earth Interactions* **11**:1–33.

Verheyen K, Guntenspergen GR, Beisbrouck B, Hermy M. 2003. An integrated analysis of the effects of past land use on forest herb colonization at the landscape scale. *Journal of Ecology* **91**: 731–742.

Walker LR, Wardle DA. 2014. Plant succession as an integrator of contrasting ecological time scales. *Trends in Ecology and Evolution* **29**:504–510.

Warren RJ, Bahn V, Bradford MA. 2012. The interaction between propagule pressure, habitat suitability and density-dependent reproduction in species invasion. *Oikos* **121**:874–881.

Weiss A. 2001. Topographic position and landforms analysis. Poster presentation, ESRI User Conference, San Diego, CA.

Weiss SB, Murphy DD, White RR. 1988. Sun, slope, and butterflies: topographic determinants of habitat quality for Euphydryas Editha. *Ecology* **69**:1486–1496.

Whittaker RH. 1967. Gradient analysis of vegetation. *Biological Reviews* **42**:207–264.

Whittaker RH, Niering WA. 1965. Vegetation of the Santa Catalina

17

The defensive role of foliar endophytic fungi for a South American tree

Marcia González-Teuber*

Departamento de Biología, Universidad de La Serena, Casilla 554, La Serena, Chile

Associate Editor: James F. Cahill

Abstract. Fungal endophytes colonize living internal plant tissues without causing any visible symptoms of disease. Endophytic fungi associated with healthy leaves may play an important role in the protection of hosts against herbivores and pathogens. In this study, the diversity of foliar endophytic fungi (FEF) of the southern temperate tree *Embothrium coccineum* (Proteaceae), as well as their role in plant protection in nature was determined. Fungal endophytes were isolated from 40 asymptomatic leaves by the culture method for molecular identification of the 18S rRNA gene. A relationship between FEF frequency and plant protection was evaluated in juveniles of *E. coccineum*. Fungal endophyte frequency was estimated using real-time PCR analyses to determine endophyte DNA content per plant. A total of 178 fungal isolates were identified, with sequence data revealing 34 different operational taxonomic units (OTUs). A few common taxa dominated the fungal endophyte community, whereas most taxa qualified as rare. A significant positive correlation between plant protection (evaluated in terms of percentage of leaf damage) and FEF frequency was found. Furthermore, in vitro confrontation assays indicated that FEF were able to inhibit the growth of fungal pathogens. The data showed a relatively high diversity of fungal endophytes associated with leaves of E. coccineum, and suggest a positive relationship between fungal endophyte frequencies in leaves and host protection in nature.

Keywords: *Embothrium coccineum*; endophyte diversity; fungal endophytes; inhibitory effects; pathogens; plant protection.

Introduction

Plants interact with a variety of microbes in their roots, stems and leaves (Partida-Martinez and Heil 2011). Fungal endophytes frequently occur in aerial plant structures, living inter-cellularly in leaf and stem tissue (Clay 1990) for at least part of their life cycle without causing any apparent sign of disease (Wilson 1995). These fungal associations are common in angiosperms, but have been particularly found and described for grasses (Poaceae family) (Saikkonen *et al.* 2004). The associations between grasses and 'type I' endophytic fungi of the Clavicipitaceae family have been well documented, as the latter colonize the host systemically, and are vertically transmitted in a classic example of mutualism. Non-clavicipitaceous 'type II' endophytes are highly diverse, and in contrast to type I, are non-systemic and mostly

* Corresponding author's e-mail address: mfgonzalez@userena.cl

horizontally transmitted (Rodriguez *et al.* 2009). Despite the fact that type II endophytes are ubiquitous, having been found in all plant species to date (Arnold *et al.* 2000; Arnold and Herre 2003; Albrectsen *et al.* 2010; Gazis and Chaverri 2010), they are far less well studied and their ecological roles are not yet fully understood.

Increased resistance to pathogens and/or herbivores may be a consequence of plant colonization by fungal endophytes (Partida-Martinez and Heil 2011). Host protection from natural enemies provided by clavicipitaceous endophytes, as well as their mechanisms of action, has been extensively studied (Clay 1988; Wilkinson *et al.* 2000; Cheplick and Faeth 2009). Much less is known about the role of non-clavicipitaceous endophytes in this regard, although there is evidence that type II also contribute to plant protection from pathogens (Arnold *et al.* 2003; Adame-Álvarez *et al.* 2014) and herbivores (Van Bael *et al.* 2009; Bittleston *et al.* 2011; Estrada *et al.* 2015). Several studies have reported high infection rates by type II FEF in plants (Arnold *et al.* 2000; Arnold and Herre 2003; Gazis and Chaverri 2010). Since infection intensity (frequency of FEF in host tissues) by type II endophytes is not systematic, infection intensity may be an important factor determining plant resistance to enemies. Accordingly, an earlier study showed that a higher infection frequency of the dominant fungal endophyte in oak trees was negatively correlated with the performance of a leaf mining herbivore (Preszler *et al.* 1996).

Embothrium coccineum (Proteaceae) is a small tree endemic to South American temperate forests commonly occurring in open sites (Díaz and Armesto 2007). Embothrium *coccineum* suffers relatively high levels of damage by pathogens and/or herbivores under natural conditions (Baldini and Pancel 2002; Salgado-Luarte 2010; approximately 30% of leaf damage, personal observations). Interestingly, marked variations in the amount of leaf damage in juvenile plants of this species are evident; whereas leaves of some juveniles show significant damage, no signs of damage are evident in others (Fig. 1, **Supporting Information—Fig. s1**). This raises the question of whether higher FEF frequencies may relate to improved plant protection in nature. In light of the above, this study assessed the following questions: (1) How diverse is the community of FEF associated with asymptomatic leaves of *E. coccineum*? (2) Is there a relationship between FEF frequency and host plant resistance in nature? (3) Do foliar endophytic fungi inhibit the growth of fungal pathogens *in vitro*?

Methods

Study site

The study site (Anticura) was situated in mature temperate rainforest at Puyehue National Park (−40.65S, −72.18W; 350–400 m a.s.l.), in the western foothills of the Andes in southern Chile. Mean annual precipitation at the site is 2800 mm, with a mean temperature of 9.8 °C (Dorsch 2003). Old-growth lowland forest in this region is composed of broad-leaved evergreen trees

Figure 1. Differences in leaf damage among juveniles of *Embothrium coccineum* in the field. Four leaves from three individuals (I: individuals 1, 5 and 8) are shown.

(Lusk and Del Pozo 2002). *Embothrium coccineum* (Chilean firetree, Proteaceae) is one of the dominant species in the rainforest, and is susceptible to damage by herbivores and pathogens at the juvenile stage (estimated 30% under natural conditions). *Embothrium coccineum* shows fast growth in comparison to other native species (Escobar *et al.*, 2006).

Isolation of fungal endophytes

Ten *E. coccineum* individuals were randomly selected for leaf collection, undertaken in March 2014. All plants were juveniles between 1.5 and 2.0 m in height, at similar developmental stage (plant diameter and number of leaves). Four healthy (0% damage) mature leaves per individual were collected and immediately transported to the lab for further endophyte isolation and molecular characterization. Leaves were disinfected under laboratory conditions with different washes of ethanol (70%), sodium hypochlorite (1%) and sterilized water (according to Arnold et al. 2001, 2003). The success of the surface sterilization method was confirmed by the absence of any microbial growth on PDA (potato-dextrose-agar) (Phyto Technology Laboratories) agar plates from the plating of last washing water. Small sections of sterilized leaves (2–3 mm) were subsequently cultivated on PDA petri dishes plates. Plates were then incubated at room temperature for 2–3 weeks. After that time, emerging colonies were subcultured to obtain pure isolates. Pure isolates were grown on PDA plates (Phyto Technology Laboratories) at room temperature for one month for further DNA extraction and molecular identification.

Molecular characterization of endophytes

Genomic DNA was extracted from the mycelial mat using a modified method described by Nicholson *et al.* (2001). Fresh mycelium was ground on Mini-BeadBeater-16 (BioSpec, USA). Ground micelyum was suspended in extraction buffer (10 mM Tris buffer pH 8.0, 10 mM EDTA, 0.5% SDS, NaCl 250 mM). To this aqueous solution, phenol:chloroform:isoamyl alcohol (25:24:1) was added and mixed slowly for 3 min. The phases were separated by centrifugation at 13.000 rpm for 10 min at room temperature. Traces of phenol were removed by treating the aqueous layer with chloroform:isoamyl alcohol (24:1) (this step was repeated twice). Phases were separated as before. DNA was precipitated from the aqueous phase with 2.0 volumes of isopropanol. The DNA was recovered by centrifugation at 10 000 rpm for 15 min at 4 °C. The pellet was then washed with 70% ethanol and resuspended in molecular biology grade water (Mo Bio Laboratories, Inc). Species identification of endophytic fungi was performed using the primers ITS1-F

(CTTGGTCATTTAGAGGAAGTAA) (Gardes and Bruns 1993) and ITS4 (TCCTCCGCTTATTGATATGC) (White *et al.* 1990). Amplification of the ITS (internal transcribed spacers) region (around 680 kbp) was conducted using 50 µL of PCR reaction mixtures, each containing 7 µL of total fungal genomic DNA, 1 µL of each primer (at a concentration of 10 µM for each primer), 27.5 µL of SapphireAmp Fast PCR Master Mix (Takara) and 13.5 µL of sterilized water. PCR was performed in a Techne TC-5000 Thermal Cycler (Fisher Scientific) with the following program: 94 °C for 3 min, followed by 35 cycles of denaturation at 94 °C for 1 min, annealing at 54 °C for 30 s and primer extension at 72 °C for 1 min, completed with a final extension at 72 °C for 7 min. PCR products were sent to Macrogen (South Korea) for purification and sequencing. Sequences were assembled using SeqTrace software. Consensus sequences were used for BLAST searches at the NCBI (http://www.ncbi.nlm.nih.gov).

Relationship between endophytic frequency and leaf damage under natural conditions

Twenty *E. coccineum* juveniles were selected in the field in March 2014, and three leaves from each were randomly collected for determination of fungal endophyte DNA content (infection frequencies) and leaf damage (percentage of leaf damage was considered as a proxy for plant protection in nature). To calculate the percentage of leaf damage, leaves from each plant were scanned and stored as digital image files. Total leaf area (TA) and damage area (DA) were determined for each sample using the program ImageJ (http://imagej.nih.gov/ij/). Average TA and DA for the three leaves were used to derive per plant values. Given the difficulty of differentiating pathogen from herbivore damage, they are referred hereafter collectively as damage by natural enemies.

The FEF DNA content was only evaluated for those endophyte species that exhibited the highest frequencies in *E. coccineum*, and accounted for 70% of the fungal community (according to Fig. 2): *Mycosphaerella sp.*, *Diaporthe sp.*, *Xylaria sp.* and *Penicillum sp.* Total DNA (2 g per individual plant, according to Lambertini *et al.* 2008) was extracted from disinfected leaves (three leaves were collected and pooled per plant) of the same 20 juveniles mentioned above and subjected to real-time PCR (q-PCR) assays using primers and probes, which were specifically designed to recognize the four FEF genera (*Mycosphaerella, Diaporthe, Penicillium* and *Xylaria*) (**See Supporting Information—Table S1**). Some of the leaves showed a high degree of damage, potentially by pathogenic fungi, and the use of universal fungal primers for q-PCR analyses was, therefore, inappropriate, since this

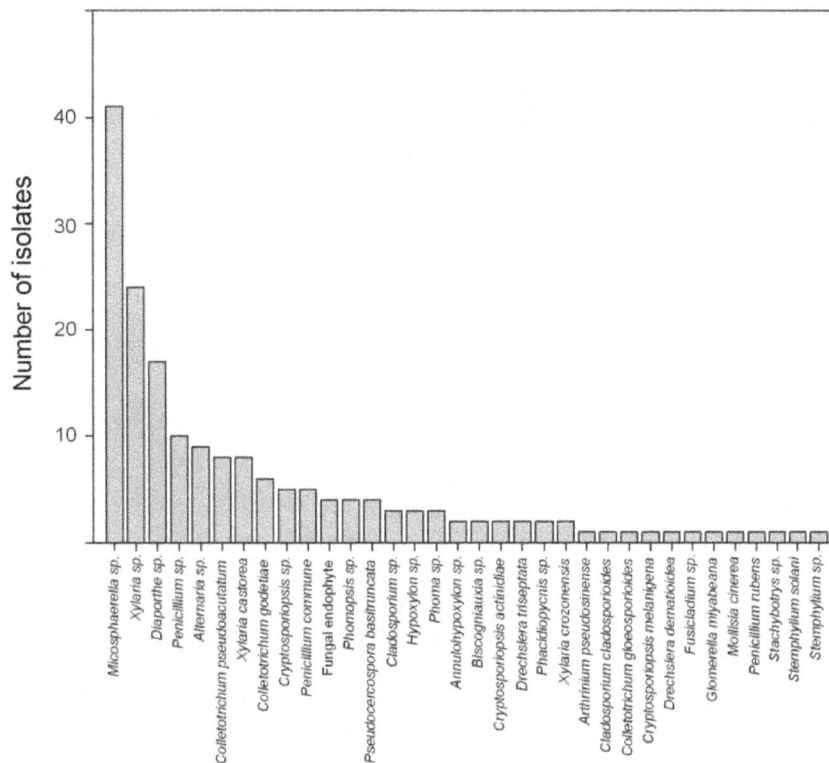

Figure 2. Frequency of endophyte taxa isolated from asymptomatic leaves of *Embothrium coccineum*.

would also result in the amplification of fungal pathogens. Primers and probes were designed using Primer3 (Koressaar and Remm 2007; Untergasser *et al.* 2012). Primers were commercially synthesized in Macrogen (South Korea). q-PCR analyses were performed in a Stratagene Mx3000P (Thermo Fisher Scientific), using the kit Brillant II QPCR Master mix (Agilent Technologies) and the primers and probes TaqMan with fluorophores FAM-TAMRA. A standard curve based on threshold cycle (Ct) was constructed for each fungal endophyte. Samples (100 ng DNA per plant) were then run in triplicate in each plate. PCR cycling parameters were 95 °C for 10 min, 40 cycles at 95 °C for 15 s, 55 °C for 30 s and 72 °C for 1 min. The cycle at which a sample's signal exceeds the threshold, the CT value, was used to calculate total amplified DNA, using a formula obtained from the slope of the regression line from the standard curve. Potential differences in ribosomal RNA gene copies among endophytes might relate to an overestimation of total amplified DNA observed for each fungus.

Correlations between the percentage of leaf damage and FEF frequency (fungal endophyte DNA content per plant in ng) were tested with the non-parametric Spearman's rho test. Correlations were done for each endophytic fungus (*Mycosphaerella*, *Diaporthe*, *Penicillium* and *Xylaria*), and for the four endophytic fungi taken as a whole. The latter was calculated by adding the DNA

contents of the four dominant endophytes. All analyses were performed in Statistica 7.0 software (StatSoft, Inc.).

Confrontation assays

For the confrontation assays, the four most abundant endophytes were confronted against three common fungal pathogens of endemic woody plants in Chile (*Botrytis cinerea*, *Fusarium oxysporum* and *Ceratocystis pilifera*) (Baldini and Pancel 2002). Fungal pathogens were kindly provided by the Laboratorio de Química de Productos Naturales, Universidad de Concepción, Chile. Endophytes and pathogenic fungi were co-cultured on 90 mm petri dishes containing PDA (potato dextrose agar) at 21 °C over a period of 11 days. Sections (1 cm²) of each fungal mycelium were placed 4 cm apart on a fresh potato dextrose agar plate. Pathogens cultivated in isolation served as controls. Microbial growth was determined by measuring the diameter of the colonies every day over the 11-day period, and comparing colony sizes in the confrontation situation to that of the controls. The percentage of growth inhibition (GI) was calculated with the following formula: GI = ((A−B)/A) * 100, where *A* is the radial diameter of the control pathogen; *B* is the radial diameter of test pathogen.

In order to evaluate potential inhibitory or stimulatory effects on growth among endophytes themselves, the four endophytes (*Mycosphaerella sp.*, *Diaporthe sp.*, *Penicillium sp.* and *Xylaria sp.*, named confronted endophytes) were confronted against themselves (named tested endophytes). Assays were carried out as described above, again with endophytes cultivated in isolation used as controls. The percentage of growth was evaluated over a 5-day period using the same process described earlier. Whereas positive values would indicate inhibitive effects on growth of confronted endophytes on tested endophytes, negative values would by contrast suggest a stimulative effect.

Results

Diversity and composition of fungal endophytes

Of the 10 *E. coccineum* juveniles screened in this study, a total of 178 fungal isolates were purified into individual cultures. All plants (incidence = 100%), but not all leaves (incidence = 82.5%) contained fungal endophytes. A total of 34 OTUs were identified (Table 1). The distribution of fungal isolates revealed a few common taxa, and many rare taxa (Fig. 2). The endophytic fungal community was dominated by the genera *Mycosphaerella sp.* (23%), *Xylaria sp.* (18%) and *Diaporthe sp.* (10%). *Penicillium* and *Colletotrichum* genera occurred at frequencies lower than 10%, whereas other genera were found in rare instances, with frequencies between 1% and 5%. Total species richness was 34, with an evenness of 0.80. The Simpson index suggested a relatively high diversity of fungal endophytes associated with *E. coccineum*. Simpson diversity was 0.90 (with a range of 0–1, with 1 representing higher sample diversity).

Protection of fungal endophytes in nature and in vitro

A significant negative correlation between FEF DNA content and percentage of leaf damage was found when frequencies of the four endophytic fungi were summed ($r = -0.53$, $P = 0.0156$) (Fig. 3). Nevertheless, no significant relationship between the DNA amount of any of the four fungal endophytes and the % of leaf damage was detected (*Mycosphaerella sp.*: $r = -0.34$, $P = 0.135$; *Diaporthe sp.*: $r = -0.36$, $P = 0.113$; *Xylaria sp.*: $r = -0.35$, $P = 0.119$; *Penicillium* sp.; $r = -0.27$, $P = 0.232$).

All three pathogen species demonstrated significant reductions in growth relative to experimental controls when exposed to endophytes; *Penicillium* showed the strongest inhibitory effects against the three pathogens, whereas the FEF *Diaporthe sp.*, *Mycosphaerella sp.* and *Xylaria sp.* showed weaker inhibitory effects (Table 2). Of the three pathogens, inhibitory effects by all endophytes were most pronounced in *Fusarium sp.* and *Ceratoystis*

Table 1. Best BLAST matches for isolated fungal endophyte OTUs collected from asymptomatic leaves of *Embothrium coccineum* in the Valdivian rainforest.

Endophytic fungi	Accession number	Identity (%)
Alternaria sp.	KP985749.1	99
Annulohypoxylon sp.	JQ327866.1	99
Arthrinium sp.	NR_121559.1	93
Biscogniauxia sp.	JN225898.1	97
Cladosporium cladosporioides	AF455442.1	98
Cladosporium sp.	KF367501.1	99
Colletotrichum gloeosporioides	AJ301972.1	100
Colletotrichum godetiae	KC860043.1	97
Colletotrichum pseudoacutatum	NR_111756.1	99
Cryptosporiopsis actinidiae	KF727420.1	99
Cryptosporiopsis melanigena	AF141196.1	98
Cryptosporiopsis sp.	JF288555.1	99
Diaporthe sp.	JN225920.1	98
Drechslera dematioidea	JN712465.1	99
Drechslera triseptata	AF163059.1	99
Fungal endophyte	EU686153.1	96
Fusicladium sp.	GU446639.1	98
Glomerella miyabeana	JN943455.1	99
Hypoxylon sp.	KJ406985.1	89
Mollisia cinerea	DQ491498.1	98
Mycosphaerella sp.	JN225927.1	100
Penicillium commune	KR012904.1	99
Penicillium rubens	LC015689.1	99
Penicillium sp.	KT264644.1	99
Phacidiopycnis sp.	JN944643.1	99
Phoma sp.	JN225888.1	100
Phomopsis sp.	AY518680.1	98
Pseudocercospora basitruncata	DQ267600.1	99
Stachybotrys sp.	KR081400.1	100
Stemphylium solani	AF203448.1	98
Stemphylium sp.	JX164072.1	99
Xylaria castorea	JN225908.1	99
Xylaria crozonensis	GU324748.1	99
Xylaria sp.	JN225909.1	100

pilifera, with *Botrytis cinerea* demonstrating relatively lower levels of inhibition. Tests for effects on growth among endophytes themselves demonstrated an overall positive effect between the most dominant endophytes

(Table 3); *Mycosphaerella* was the only endophyte that demonstrated inhibitory effects, although negligible.

Discussion

The main aims of this study were to investigate the diversity of foliar endophytic fungi associated with *E. coccineum*, and to determine whether correlations between endophyte infection frequencies and plant resistance

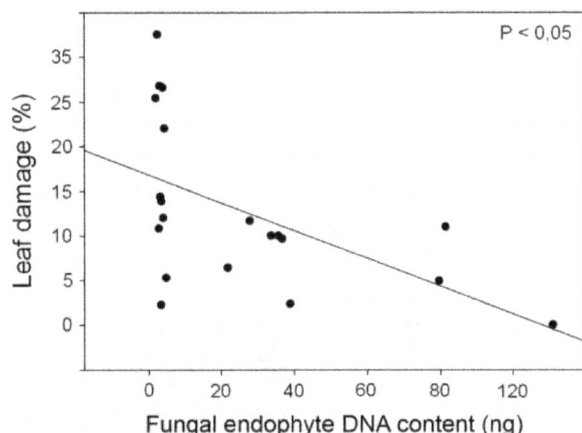

Figure 3. Correlation between fungal endophyte DNA content (per 100 ng of plant DNA) and the percentage of leaf damage in twenty individuals of *Embothrium coccineum*.

occur. The present study provides evidence that infection by the dominant endophyte community confer host protection from natural enemies in the field.

The endophytic fungal community associated with *E. coccineum* was represented primarily by four genera, *Mycosphaerella*, *Xylaria*, *Diaporthe* and *Penicillium*, which collectively accounted for 70% of the culturable community. Culture-based methods may underestimate diversity and misrepresent taxonomic composition of endophyte communities (Arnold 2007; Hyde and Soytong 2008) in comparison with culture-independent approaches, such as DNA cloning (Guo *et al.* 2001) or PCR product pyrosequencing (Nilsson *et al.* 2009). Despite the fact that culture-independent approaches demonstrate certain advantages over the culture-based techniques, including the capacity to detect unculturable species or species with low abundances (Hyde and Soytong 2008), comparative studies of both methods indicate that many of the proportionally dominant microbial taxa identified by culture-independent approaches are accurately represented by the culture method (see Pitkäranta *et al.*, 2011; Jackson *et al.*, 2013; Bodenhausen *et al.*, 2013).

Data here support earlier findings that a relatively small number of endophyte species dominate the fungal community, and that most taxa, with few isolations, qualify as rare (Arnold *et al.* 2001; Gazis and Chaverri

Table 2 Inhibition effects (%) of the four most abundant fungal endophytes isolated from *E. coccineum* against the fungal pathogens *Fusarium oxysporum*, *Ceratocystis pirifera* and *Botrytis cinerea* in confrontation assays in Petri dishes. Values indicate the mean ± SE (standard error) ($N = 3$).

	Fusarium oxysporum	*Ceratocystis pilifera*	*Botrytis cinerea*
Xylaria sp.	17.7 ± 1.96	7.6 ± 0.49	4.05 ± 0.65
Penicillium sp.	28.7 ± 2.08	31.8 ± 3.58	17.8 ± 3.52
Diaporthe sp.	30.4 ± 1.70	33.8 ± 6.69	27.7 ± 1.62
Mycosphaerella sp.	20.0 ± 0.04	22.5 ± 2.06	8.1 ± 2.04

Table 3. Growth (%) results from confrontation assays of fungal endophytes themselves. Positive values in brackets indicate growth inhibition effects of confronted endophytes on tested endophytes. Negative values in brackets indicate a positive growth effect of confronted endophytes on tested endophytes. Values indicate the mean ± SE (standard error) ($N = 3$).

Confronted endophytes	Tested endophytes			
	Xylaria sp.	*Penicillium sp.*	*Diaporthe sp.*	*Mycosphaerella sp.*
Xylaria sp.		(-) 7.8 ± 5.3	(-) 10.9 ± 4.0	(-) 9.1 ± 3.6
Penicillium sp.	(-) 3.9 ± 3.4		(-) 7.1 ± 3.5	(-) 3.1 ± 1.4
Diaporthe sp.	(-) 9.1 ± 2.6	(+) 0.2 ± 0.7		(+) 3.0 ± 3.4
Mycosphaerella sp.	(+) 2.5 ± 1.9	0.0 ± 2.9	(+) 3.6 ± 4.1	

2010). Endophyte genera *Xylaria, Mycosphaerella* and *Penicillium* are commonly isolated from leaf tissues of different plant species at a range of latitudes (Arnold and Lutzoni 2007; Gazis and Chaverri 2010; Douanla-Meli *et al.* 2013), which indicates their ability to colonize a broad host range. Despite the fact that FEF diversity associated with *E. coccineum* was probably underestimated, OTU richness found here appears similar to that reported for other woody plants. For example, 58 OTUs were isolated from leaves and stems of the Peruvian tropical tree *Hevea brasiliensis* (Gazis and Chaverri 2010); 242 and 259 FEF morphospecies were recovered from the neotropical trees *Heisteria concinna* and *Ouratea lucens*, respectively (Arnold *et al.* 2001), and 89 OTUs were obtained from photosynthetic tissue of four woody desert plants (Massimo *et al.* 2015).

Host plant benefits conferred by FEF are predominantly reported in studies investigating endophyte effects on plant resistance based on experimental applications of endophytic fungi. These studies demonstrate that FEF are able to effectively protect hosts from pathogen and herbivore attack (Arnold *et al.* 2003; Van Bael *et al.* 2009; Bittleston *et al.* 2011). For example, in the tropical tree *Theobroma cacao* (Malvaceae), a combination of the seven most dominant endophytes (previously isolated from the host plant) effectively prevented host pathogen damage (Arnold *et al.* 2003). Furthermore, Van Bael *et al.* (2009) demonstrated that high endophyte frequencies in leaves of the tropical vine *Merremia umbellate* had a negative effect on herbivore fecundity relative to endophyte-free plants. Using q-PCR analyses, the present study tested the relationship between endophyte infection frequencies and host plant protection in nature. Results showed that higher infection rates of the dominant endophyte genera correlate with lower levels of leaf damage in *E. coccineum*. q-PCR analyses were unable to detect dominant endophyte DNA in several collected samples. The endophyte incidence in leaves of *E. ccocineum* was of 82.5%, which might help to explain this result. An incidence of foliar endophytic fungi of less than 100% appears to be not atypical (see Gazis and Chaverri 2010).

A significant negative correlation between FEF frequency and plant protection was found for the four endophyte genera as a whole, although no correlation between individual endophyte genera and leaf damage was evident. The latter result strongly supports the 'multiple defender effect' hypothesis (McKeon *et al.* 2012), which states that the presence of multiple mutualists act to produce protective benefits synergistically or additively. Synergistic protective effects by fungal endophytes have been described (Gazis and Chaverri 2015); for example, the abundance, as opposed to just the

presence, of competitive endophyte strains and species has been shown to confer protective benefits in wild *Hevea* trees (Gazis and Chaverri 2015). Furthermore, combined effects on growth of arbuscular mycorrhizal fungi and fungal endophytes were found to be additive in the grass *Elymus hystrix* (Larimer et al. 2012). Results here provide further evidence that cooperation among endophytes leads to enhanced protective effects for the host plant.

Endophytes are known to produce a large number of specific toxins (Schulz *et al.* 2002; Strobel and Daisy 2003; Mousa and Raizada 2013) that directly affect the growth and performance of herbivores and pathogens. Furthermore, by changing leaf chemical characteristics, it has been shown that FEF negatively influence herbivore preferences for *Cucumis sativus* (Estrada *et al.* 2013). Confrontation assays in Petri dishes showed that FEF investigated in this study were able to effectively inhibit the growth of several woody plant pathogens, which is likely a result of active inhibition, due to the presence of an inhibition zone. *Xylaria, Diaporthe* and *Penicillium* genera are known for their chemical diversity and ability to produce bioactive compounds (Schneider et al. 1996; Oliveira *et al.* 2009; Specian *et al.* 2012; Baraban *et al.* 2013; Lai *et al.* 2013), which demonstrate antipathogen properties *in vitro* (Arnold *et al.* 2003; Thirumalesh *et al.* 2014). Tests for inter-endophyte interactions demonstrated overall positive growth effects in endophytic partners. This suggests that positive interactions between endophytic mutualists within the host plant might occur, which would help to explain their enhanced benefit to the host plant (Afkhami et al. 2014).

Since the endophytic fungi *Micosphaerella, Xylaria, Diaporthe* and *Penicillium* dominate the endophyte community in *E. coccineum*, and considering their ability to reduce the growth of pathogens *in vitro*, results here provide convincing evidence to suggest that these genera play a role in the protection of *E. coccineum* under natural conditions. Further studies should consider inoculation experiments *in planta* in order to reliably detect positive host plant endophyte effects (Adame-Álvarez *et al.* 2014).

Conclusions

This study represents the first attempt to link FEF infection intensity and host protection under natural conditions using molecular approaches. Data here showed a relatively high diversity of fungal endophytes associated with leaves of *E. coccineum*. The fungal endophyte community was dominated by just four endophyte genera, which were able to reduce the growth of common pathogens in *vitro*. In addition, colonization by these genera resulted in lower levels of predation of the host plant in natural conditions. These results provide further evidence that colonization by multiple foliar endophytic fungi confers important benefits to host plants in terms of resistance to natural enemies.

Accession Numbers

Sequence data were submitted to NCBI (accession numbers: KU743942-KU743975).

Sources of Funding

This study was funded by FONDECYT (Fondo Nacional de Desarrollo Científico y Tecnológico, Chile) Grant no. 11130039.

Contributions by the Authors

I am the only author of the manuscript.

Acknowledgements

The author thank Mónica Cisternas and Andrea Morales for their valuable help in the lab. I am also very grateful to Craig Weideman for comments on an earlier version of this manuscript. This study was funded by Fondecyt Project no. 11130039.

Supporting Information

The following additional information is available in the online version of this article —

Fig. S1. Distribution of the percentage of leaf damage in 35 individuals (in chronological order) of *Embothrium coocineum* in the Valdivian rainforest.

Table S1. Specific primers and probes designed for the four most abundant endophytic fungi (*Xylaria, Diaporthe, Penicillium* and *Micosphaerella*) isolated from leaves of *Embothrium coccineum*.

Literature Cited

Adame-Álvarez RM, Mendiola-Soto J, Heil M. 2014. Order of arrival shifts endophyte–pathogen interactions in bean from resistance induction to disease facilitation. *FEMS Microbiology Letters* **355**: 100–107.

Afkhami ME, Rudgers JA, Stachowicz JJ. 2014. Multiple mutualist effects: conflict and synergy in multispecies mutualisms. *Ecology* **95**:833–844.

Albrectsen BR, Bjorken L, Varad A, Hagner A, Wedin M, Karlsson J, Jansson S. 2010. Endophytic fungi in European aspen (*Populus tremula*) leaves-diversity, detection, and a sug-

gested correlation with herbivory resistance. *Fungal Diversity* **41**: 17–28.

Arnold AE, Maynard Z, Gilbert GS, Coley PD, Kursar TA. 2000. Are tropical fungal endophytes hyperdiverse? *Ecology Letters* **3**:267–274.

Arnold AE, Maynard Z, Gilbert GS. 2001. Fungal endophytes in dicotyledonous neotropical trees: patterns of abundance and diversity. *Mycological Research* **105**:1502–1507.

Arnold AE. 2007. Understansing the diversity of foliar endophytic fungi: progress, challenges, and frontiers. *Fungal Biology Reviews* **21**:51–66.

Arnold AE, Herre EA. 2003. Canopy cover and leaf age affect colonization by tropical fungal endophytes: ecological pattern and process in *Theobroma cacao* (Malvaceae). *Mycologia* **95**: 388–398.

Arnold AE, Mejia LC, Kyllo D, Rojas EI, Maynard Z, Robbins N, Herre EA. 2003. Fungal endophytes limit pathogen damage in a tropical tree. *Proceedings of the National Academy of Sciences of the United States of America* **100**:15649–15654.

Arnold AE, Lutzoni F. 2007. Diversity and host range of foliar fungal endophytes: are tropical leaves biodiversity hotspots? *Ecology* **88**:541–549.

Baldini A, Pancel L. 2002. *Agentes de daño en el bosque nativo*. Santiago, Chile: Editorial Universitaria.

Baraban EG, Morin JB, Phillips GM, Phillips AJ, Strobel SA, Handelsman J. 2013. Xyolide, a bioactive nonenolide from an Amazonian endophytic fungus, *Xylaria feejeensis*. *Tetrahedron Letters* **54**:4058–4060.

Bittleston LS, Brockmann F, Wcislo W, Van Bael SA. 2011. Endophytic fungi reduce leaf-cutting ant damage to seedlings. *Biology Letters* **7**:30–32.

Bodenhausen N, Horton MW, Bergelson J. 2013. Bacterial communities associated with the leaves and roots of *Arabidopsis thaliana*. *PLoS One* **8**:e56329.

Clay K. 1988. Fungal endophytes of grasses – a defensive mutualism between plants and fungi. *Ecology* **69**:10–16.

Clay K. 1990. Fungal endophytes of grasses. *Annual Review of Ecology and Systematics* **21**:275–297.

Cheplick GP, Faeth SH. 2009. *Ecology and evolution of the grass-endophyte symbiosis*. Oxford: Oxford University Press.

Díaz MF, Armesto JJ. 2007. Physical and biotic constraints on tree regeneration in secondary shrublands of Chiloe Island, Chile. *Revista Chilena De Historia Natural* **80**:13–26.

Dorsch K. 2003. *Hydrogeologische Untersuchungen der Geothermalfelder von Puyehue und Cordón Caulle*, Chile. Dissertation, Ludwig-Maximilians-Universität, Germany.

Douanla-Meli C, Langer E, Mouafo FT. 2013. Fungal endophyte diversity and community patterns in healthy and yellowing leaves of *Citrus limon*. *Fungal Ecology* **6**:212–222.

Escobar E, Donoso C, Souto C, Alberdi M, Zúñiga A. 2006. *Embothrium coccineum* JR. et. G. Foster. In: Donoso C, ed. *Las especies arbóreas de los bosques templados de Chile y Argentina*. Valdivia: Autoecología Marisa Cúneo Ediciones, 233–245.

Estrada C, Wcislo WT, Van Bael SA. 2013. Symbiotic fungi alter plant chemistry that discourages leaf-cutting ants. *New Phytologist* **198**:241–251.

Estrada C, Degner EC, Rojas EI, Wcislo WT, Van Bael SA. 2015. The

Gardes M, Bruns TD. 1993. ITS primers with enhanced specificity for basidiomycetes – application to the identification of mycorrhizae and rusts. *Molecular Ecology* **2**:113–118.

Gazis R, Chaverri P. 2010. Diversity of fungal endophytes in leaves and stems of wild rubber trees (*Hevea brasiliensis*) in Peru. *Fungal Ecology* **3**:240–254.

Gazis R, Chaverri P. 2015. Wild trees in the Amazon basin harbor a great diversity of beneficial endosymbiotic fungi: is this evidence of protective mutualism? *Fungal Ecology* **17**:18–29.

Guo LD, Hyde KD, Liew EC. 2001. Detection and taxonomic placement of endophytic fungi within frond tissues of *Livistona chinensis* based on rDNA sequences. *Molecular Phylogenetics and Evolution* **20**:1–13.

Hyde KD, Soytong K. 2008. The fungal endophyte dilemma. *Fungal Diversity* **33**:163–173.

Jackson CR, Randolph KC, Osborn SL, Tyler HL. 2013. Culture dependent and independent analysis of bacterial communities associated with commercial salad leaf vegetables. *BMC Microbiology* **13**:274.

Koressaar T, Remm M. 2007. Enhancements and modifications of primer design program Primer3. *Bioinformatics* **23**:1289–1291.

Lai DW, Brotz-Oesterhelt H, Müller WEG, Wray V, Proksch P. 2013. Bioactive polyketides and alkaloids from *Penicillium citrinum*, a fungal endophyte isolated from *Ocimum tenuiflorum*. *Fitoterapia* **91**:100–106.

Lambertini C, Frydenberg J, Gustafsson MHG, Brix H. 2008. Herbarium specimens as a source of DNA for AFLP fingerprinting of *Phragmites* (Poaceae): possibilities and limitations. *Plant Systematics and Evolution* **272**:223–231.

Larimer AL, Bever JD, Clay K. 2012. Consequences of simultaneous interactions of fungal endophytes and arbuscular mycorrhizal fungi with a shared host grass. *Oikos* **121**:2090–2096.

Lusk CH, Del Pozo A. 2002. Survival and growth of seedlings of 12 Chilean rainforest trees in two light environments: gas exchange and biomass distribution correlates. *Austral Ecology* **27**:173–182.

Massimo NC, Nandi Devan MM, Arendt KR, Wilch MH, Riddle JM, Furr SH, Steen C, U'Ren JM, Sandberg DC, Arnold AE. 2015. Fungal endophytes in aboveground tissues of desert plants: infrequent in culture, but highly diverse and distinctive symbionts. *Microbial Ecology* **70**:61–76.

McKeon CS, Stier AC, McIlroy SE, Bolker BM. 2012. Multiple defender effects: synergistic coral defense by mutualist crustaceans. *Oecologia* **169**:1095–1103.

Mousa WK, Raizada MN. 2013. The diversity of anti-microbial secondary metabolites produced by fungal endophytes: an interdisciplinary perspective. *Frontiers in Microbiology* **4**:65.

Nicholson TP, Rudd BAM, Dawson MJ, Lazarus CM, Simpson TJ, Cox RJ. 2001. Design and utility of oligonucleotide gene probes for fungal polyketide synthases. *Chemistry and Biology* **8**:157–178.

Nilsson RH, Ryberg M, Abarenkov K, Sjökvist E, Kristiansson E. 2009. The ITS region as a target for characterization of fungal communities using emerging sequencing technologies. *FEMS Microbiology Letters* **296**:97–101.

Oliveira CM, Silva GH, Regasini LO, Zanardi LM, Evangelista AH, Young MCM, Bolzani VS, Araujo AR. 2009. Bioactive metabolites produced by *Penicillium* sp.1 and sp.2, two endophytes associated with *Alibertia macrophylla* (Rubiaceae). *Zeitschrift Für Naturforschung C* **64**:824–830.

Partida-Martinez LP, Heil M. 2011. The microbe-free plant: fact or artifact? *Frontiers in Plant Science* **2**:100.

Pitkäranta M, Meklin T, Hyvärinen A, Nevalainen A, Paulin L, Auvinen P, Lignell U, Rintala H. 2011. Molecular profiling of fungal communities in moisture damaged buildings before and after remediation – a comparison of culture-dependent and culture-independent methods. *BMC Microbiology* **11**:235.

Preszler RW, Gaylord ES, Boecklen WJ. 1996. Reduced parasitism of a leaf-mining moth on trees with high infection frequencies of an endophytic fungus. *Oecologia* **108**:159–166.

Rodriguez RJ, White JF, Arnold AE, Redman RS. 2009. Fungal endophytes: diversity and functional roles. *New Phytologist* **182**:314–330.

Saikkonen K, Wali P, Helander M, Faeth SH. 2004. Evolution of endophyte-plant symbioses. *Trends in Plant Science* **9**:275–280.

Salgado-Luarte C. 2010. *Los herbívoros como agentes selectivos sobre atributos de valor funcional en ambientes de sombra.* PhD Thesis, Universidad de Concepción, Chile.

Schneider G, Anke H, Sterner O. 1996. Xylaramide, a new antifungal compound, and other secondary metabolites from *Xylaria longipes*. *Zeitschrift Für Naturforschung C* **51**:802–806.

Schulz B, Boyle C, Draeger S, Rommert AK, Krohn K. 2002. Endophytic fungi: a source of novel biologically active secondary metabolites. *Mycological Research* **106**:996–1004.

Specian V, Sarragiotto MH, Pamphile JA, Clemente E. 2012. Chemical characterization of bioactive compounds from the endophytic fungus *Diaporthe helianthi* isolated from *Luehea divaricata*. *Brazilian Journal of Microbiology* **43**:1174–1182.

Strobel G, Daisy B. 2003. Bioprospecting for microbial endophytes and their natural products. *Microbiology and Molecular Biology Reviews* **67**:491–502.

Thirumalesh BV, Thippeswamy B, Krishnappa M. 2014. Antibacterial activity of *Xylaria* species in vitro against *Xanthomonas campestris* pv. *mangiferaeindicae* isolated from bacterial black spot of mango fruit. *International Journal of Life Sciences Biotechnology and Pharma Research* **3**:125–130.

Untergasser A, Cutcutache I, Koressaar T, Ye J, Faircloth BC, Remm M, Rozen SG. 2012. Primer3-new capabilities and interfaces. *Nucleic Acids Research* **40**:e115.

Van Bael SA, Valencia MC, Rojas EI, Gomez N, Windsor DM, Herre EA. 2009. Effects of foliar endophytic fungi on the preference and performance of the leaf beetle *Chelymorpha alternans* in Panama. *Biotropica* **41**:221–225.

White TJ, Bruns T, Lee S, Taylor J. 1990. Amplification and direct sequencing of fungal ribosomal RNA genes for phylogenetics. In: Innis MA, Gelfand DH, Shinsky JJ, White TJ, eds. *PCR protocols: a guide to methods and applications.* San Diego: Academic Press, 315–322.

Wilkinson HH, Siegel MR, Blankenship JD, Mallory AC, Bush LP, Schardl CL. 2000. Contribution of fungal loline alkaloids to protection from aphids in a grass–endophyte mutualism. *Molecular Plant-Microbe Interactions* **13**:1027–1033.

Wilson D. 1995. Endophyte – the evolution of a term, and clarification of its use and definition. *Oikos* **73**:274–276.

18

Reproductive traits affect the rescue of valuable and endangered multipurpose tropical trees

Viviane Sinébou[1,2], Muriel Quinet[1], Bonaventure C. Ahohuendo[2] and Anne-Laure Jacquemart*[1]

[1] Research Group Genetics, Reproduction, Populations, Earth and Life Institute – Agronomy, Université Catholique de Louvain, Croix du Sud 2 Box L7.05.14, B-1348 Louvain-la-Neuve, Belgium
[2] Département de Productions Végétales, Faculté des Sciences Agronomiques, Université Abomey-Calavi, Cotonou 01 BP 526, Benin

Associate Editor: Dennis F. Whigham

Abstract. Conservation strategies are urgently needed in Tropical areas for widely used tree species. Increasing numbers of species are threatened by overexploitation and their recovery might be poor due to low reproductive success and poor regeneration rates. One of the first steps in developing any conservation policy should be an assessment of the reproductive biology of species that are threatened by overexploitation. This work aimed to study the flowering biology, pollination and breeding system of *V. doniana*, a multipurpose threatened African tree, as one step in assessing the development of successful conservation strategies. To this end, we studied (1) traits directly involved in pollinator attraction like flowering phenology, flower numbers and morphology, and floral rewards; (2) abundance, diversity and efficiency of flower visitors; (3) breeding system, through controlled hand-pollination experiments involving exclusion of pollinators and pollen from different sources; and (4) optimal conditions for seed germination. The flowering phenology was asynchronous among inflorescences, trees and sites. The flowers produced a large quantity of pollen and nectar with high sugar content. Flowers attracted diverse and abundant visitors, counting both insects and birds, and efficient pollinators included several Hymenoptera species. We detected no spontaneous self-pollination, indicating a total dependence on pollen vectors. *Vitex doniana* is self-compatible and no inbreeding depression occurred in the first developmental stages. After extraction of the seed from the fruit, seed germination did not require any particular conditions or pre-treatments and the seeds showed high germination rates. These pollination and breeding characteristics as well as germination potential offer the required conditions to develop successful conservation strategies. Protection, cultivation and integration in agroforestry systems are required to improve the regeneration of the tree.

Keywords: Bees; breeding system; conservation biology; mating system; pollinator efficiency; seed germination.

Introduction

Global biodiversity is decreasing at an unprecedented rate as a complex response to several human-induced changes (Sala *et al.* 2000). Land use change is the driver that has the largest global impact on biodiversity, mostly due to habitat destruction and fragmentation. Land use change is particularly the most important driver in tropical forests (Sala *et al.* 2000). The African forests are

* Corresponding author's e-mail address: anne-laure.Jacquemart@uclouvain.be

subject of haphazard modification following anthropogenic pressures including tree cutting and clearing for agriculture. In West Africa, logging and seasonal fires set by farmers and hunters have increased deforestation, destroying an average of 870 000 ha/year between 2000 and 2010 (FAO 2011). Consequently, several food and medicinal tree species have been declared locally endangered and are priorities for conservation (Eyog-Matig et al. 2002; FAO 2011). Forest harvesting directly decreases the rates of survival, growth and reproduction of individual trees, affecting the structure and dynamics of harvested populations (Oumorou et al. 2010). In harvested stands, only large, old trees and a few seedlings and saplings generally survive (Oumorou et al. 2010). Maintaining these local remnant populations will require informed management and conservation practices.

Vitex doniana, commonly called black plum, is one of the most important wild-harvested, multipurpose trees in tropical Africa (Akoègninou et al. 2006; N'Danikou et al. 2015). The tree is widely used for food, medicinal purposes and as firewood (Achigan-Dako et al. 2011). The species also has a great socio-cultural and mythological importance for local people (Dadjo et al. 2012; N'Danikou et al. 2015).

Despite its high value for local populations, there is no evidence of any conservation initiatives from harvesters (N'Danikou et al. 2011). People usually harvest the desired resources without considering regeneration and management or attempting silvicultural practices (e.g. planting and sowing) (Dadjo et al. 2012; N'Danikou et al. 2015). Mainly due to overexploitation, *V. doniana* is one of the most-threatened food tree species with high priority for conservation in Benin, Kenya, Niger and Burkina Faso (Eyog-Matig et al. 2002; Oumorou et al. 2010; N'Danikou et al. 2015).

The threats to the tree affect multiple reproductive processes. Even, if the species is partially spared during land clearing, because of its many uses and high market value (Dadjo et al. 2012; N'Danikou et al. 2015), fruit harvesting reduces available seeds for recruitment, and overexploitation of the leaves, branches, bark and wood increases the pressures on remnant populations (Oumorou et al. 2010). Also, seed germination creates a bottleneck due to dormancy, which hampers the species' natural regeneration (N'Danikou et al. 2011). *Vitex doniana* has a hard seed coat, reducing germination rates, and seedlings are scarce in natural habitats due to fires and grazing pressure. Moreover, leaf harvest reduces reproductive success by delaying flowering and decreasing seed set (Dadjo et al. 2012; N'Danikou et al. 2015). Therefore, a sustainable conservation strategy is urgently needed to preserve this valuable tree.

Successful conservation depends on sexual reproduction and seedling recruitment (Vodouhè et al. 2011). The failure of reproductive processes such as pollination often causes species loss (Faegri and van der Pijl 1979; Moza and Bhatnagar 2007). The failure of pollination has diverse causes, including pollinator rarity or infidelity, poor quantity or quality of pollen deposited on the stigmas, delayed stigma receptivity and self-incompatibility (Faegri and van der Pijl 1979). Other key factors of plant reproductive biology include floral phenology, the periodicity in the production of flowers, which follows a species-specific schedule, ranging from complete synchrony within the population to full asynchrony, thus shaping mating possibilities among individuals (Ewédjè et al. 2015). The breeding system also shapes exchanges among individuals, mainly by self-incompatibility (Faegri and van der Pijl 1979; Kalinganire et al. 2000).

Despite its importance for conservation, the reproductive biology of *V. doniana* remains unexplored. Thus this work aimed to study the flower biology, pollination, and breeding system of *V. doniana* and to assess reproductive outcomes to develop successful conservation strategies. To this end, we studied traits that directly influence reproductive success including (1) flowering phenology, flower numbers and morphology, and floral rewards; (2) abundance, diversity and efficiency of flower visitors; (3) breeding system; and (4) seed germination.

Methods

Study species and studied sites

The genus *Vitex* (Lamiaceae) includes over 270 species, predominantly trees and shrubs, and is mainly present in tropical and subtropical regions (Eyog-Matig et al. 2002). *Vitex doniana* is the most widespread species in the genus (Ky 2008). This medium-sized (8–20 m tall) deciduous tree grows in coastal woodlands, riverine and lowland forests, and savannahs of sub-Saharan Africa, east from Senegal to Somalia and south to South Africa (Kapooria and Aime 2005; Ky 2008; Dansi et al. 2008; Maundu et al. 2009; Dadjo et al. 2012; N'Danikou et al. 2015; Fig. 1). The tree is widely used for food, medicinal purposes and as firewood (Achigan-Dako et al. 2011). Processing the fleshy fruits provides juice, jam, syrup and alcohol (Agbede and Ibitoye 2007; Mapongmetsem et al. 2005, 2008; Oumorou et al. 2010; Ajenifujah-Solebo and Aina 2011; Vunchi et al. 2011; Dadjo et al. 2012). Also, new shoots and leaves are traditionally consumed as leafy vegetables (Codjia et al. 2003; Dansi et al. 2008; Achigan-Dako et al. 2011), and almost all parts of the plant have medicinal properties (anti-bacterial, anti-inflammatory, analgesic, anti-fungal, hepato-protective,

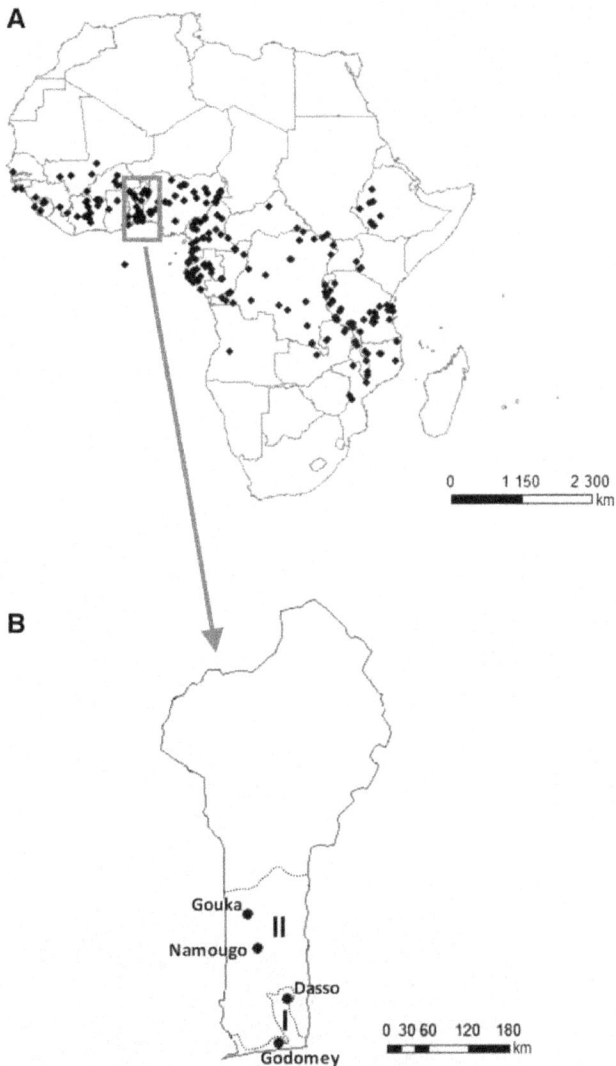

Figure 1. Spatial distribution of *Vitex doniana*. (A) Africa: map based on geographical coordinates from herbarium specimens from the Botanical Gardens of Meise, Université Libre de Bruxelles, Wageningen and Leiden (October 2015); (B) The four studied populations in Benin. Agroecological zones are adapted from Ky (2008): (I) Guineo-Congolian zone and (II) Guineo-Sudanian transition zone.

anti-oxidant and hypotensive; Ladedji and Okoye 1996; Ladedji *et al.* 1996; Ganapaty and Vidyadhar 2005; Iwueke *et al.* 2006; Kilani 2006; Padmalatha *et al.* 2009; Lagnika *et al.* 2012; Adetoro *et al.* 2013; Hamzah *et al.* 2013). Overall, 25 diseases are known to be treated with Black plum (Oumorou *et al.* 2010; Dadjo *et al.* 2012). The wood is used for firewood, charcoal, and furniture.

Our observations and experiments were conducted on four sites situated in Southern and Central Benin, West Africa (Fig. 1 and Table 1). The Godomey and Dasso sites, in the South of Benin, were located in the Guineo-Congolian climatic zone, which is characterized by two rainy seasons. The Guineo-Congolian zone is located

between latitudes 6°25′N and 7°30′N. Total mean annual rainfall averaged 1200 mm. These sites have relative humidity of 69–97 % and mean daily temperatures of 25–29 °C. The Namougo and Gouka sites were in the Guineo-Sudanian climate regime which is characterized by one rainy season (from mid-April to October), approximately 7 months of dry season, annual rainfall of 900–1200 mm and mean daily temperatures of 19–36 °C (Yédomonhan *et al.* 2012).

All trees were located in agricultural landscapes, in close proximity to farms and traditional fields of maize, cassava, and peanut, with palm trees (in the Guineo-Congolian zone) or yams (in the Guineo-Sudanian zone).

The trees in Namougo and Gouka sites were protected, as leaf collection has been forbidden since 2013. Leaf collection occurred before and during the study at the Godomey and Dasso sites.

Phenology and floral synchrony

We randomly sampled 10 adult trees per site during the flowering and fruiting periods over two successive years (2014 and 2015). Flowering was observed weekly, from January to May, and fruit development was observed every 2 weeks on the same individuals, from February to April. Four stages were defined for flowering: young bud (fl1), well-developed bud (fl2), flower at anthesis (fl3) and flower withering (fl4, Fig. 2A). Two stages were defined for fruiting: fruit initiation (persistent expanded ovary, fr1) and young green fruits (fr2).

Floral phenology was followed daily on 20 flowers between stage fl2 and fl4 per tree on each site for 2 weeks to assess the flower life span and dichogamy (protogyny or protandry).

The quantity of flowers and fruits in a given stage per tree was quantified visually by using binoculars to inspect all branches. Data were collected in ordinal classes: 0 = absence of flowers or fruits in a given stage; 1 = 1–25 % of branches bearing organs in a given stage; 2 = 26–50 %; 3 = 51–75 %; 4 = 76–100 %. The dates of first flowering and the duration of each stage were recorded to construct the temporal development of phenophases at each site.

Flower number and morphology

Four inflorescences per branch (on 4–11 branches per tree depending on the tree size) from five trees per site were examined to determine the mean number of flowers per inflorescence. The number of flowers per tree was extrapolated for each tree by multiplying numbers of flowers per inflorescence by the numbers of inflorescences per branch and the numbers of branches per tree.

Table 1. *Vitex doniana* studied site location: geographical coordinates, agroecological zones, main crops on the sites, type of sites, leaf collection intensity and distance among trees per site.

Sites	Geographical coordinates	Agroecological zones	Main crops	Type of site	Exploitation intensity	Distance among trees (m)
Godomey	6°25'15.9" N 2°20'48.4"E	Guineo-Congolian	Palm tree, maize, cassava, cowpea	Farm	Young leaves poorly collected	50–2000
Dasso	7°00'42.8"N 2°27'59.5"E	Guineo-Congolian	Palm tree, maize, cassava, peanut	Farm	Branches hardly cut, young leaves regularly collected (~70 %)	10–200
Namougo	7°41'38.2"N 2°04'32.8"E	Guineo-Sudanian	Maize, cassava, yam, peanut	Woodland	Branches cut but foliage in recovery from overexploitation one year before	10–1500
Gouka	8°09'23.7"N 1°56'09.3"E	Guineo-Sudanian	Maize, cassava, yam, peanut	Degraded forest gallery	Branches cut but foliage in recovery from overexploitation one year before	10–200

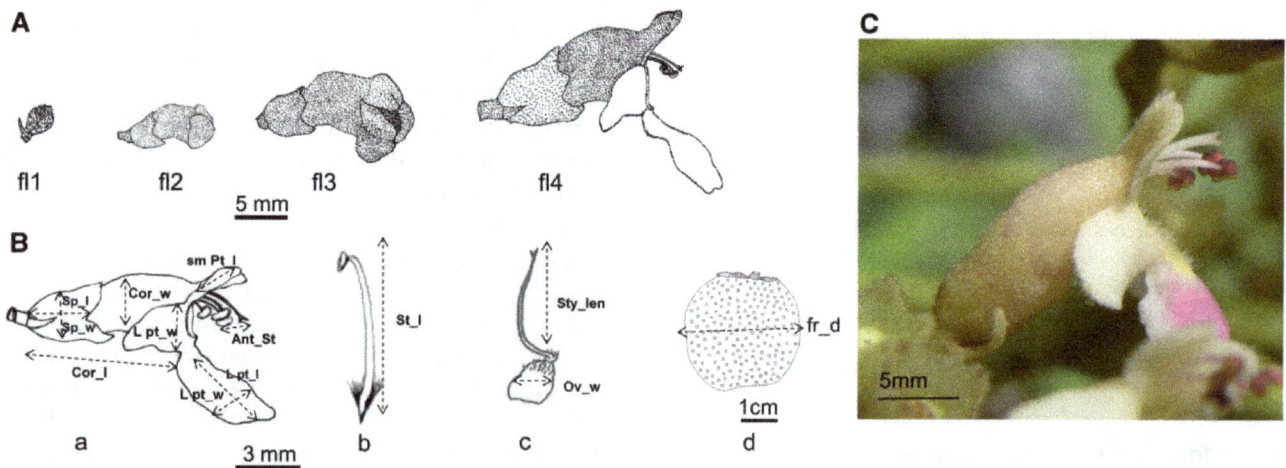

Figure 2. Flower morphology of *Vitex doniana*. (A) Phenological stages: fl1 – young bud, fl2 – well-developed bud, fl3 – flower at anthesis, fl4 – flower 1 day after anthesis. (B) Measured floral and fruit traits: (a) flower traits, (b) androceum, (c) gynoecium, (d) fruit. Cor_l: corolla length, Sp_w: Sepal width, Sp_l: Sepal length, sm pt_l: small fused petal length, L pt_w: lateral fused petal width, L pt_w: lower petal width, L pt_l: lower petal length, Ant_St: anther stigma distance, St_l: stamen length, Sty_len: style length, Ov_w: ovary width, Fr_d: fruit diameter. (C) Photograph of the flower showing the relative position of the stigma and the anthers.

Eighty inflorescences, four inflorescences per tree and per site, were dissected to determine whether all flowers were hermaphroditic, to describe the structure of the androecium and the gynoecium, and to document the number of stamens and ovules and the number of nectaries per flower. The flower colour was also observed.

Two hundred flowers (randomly selected from two different inflorescences from 20 trees, five trees per site) were dissected to describe their structure. A caliper was used to measure the length and width of the sepals, corolla tube, lateral fused petals, inferior petal and small fused superior petals. We also measured the lengths of the exterior and inner cycles of stamens, the diameter and length of the ovary, and the distances between the stamens and the stigma (Fig. 2B).

For quantification of pollen grains, 20 anthers were collected from ten flowers buds (fl2) per tree and per site. Pollen grains were counted under a light microscope (AO Spencer). Anthers were individually crushed in a microcentrifuge tube containing 100 μL of Alexander's stain (Kearns and Inouye 1993), mixed and sonicated to homogenize the pollen grains into the solution. For each anther, the number of pollen grains in 10 μl was counted in triplicate under a light microscope (AO Spencer). To assess pollen viability, one anther per flower stage was collected from each of 10 flower buds (fl2) or flowers (fl3) the morning and afternoon of the day of anthesis and (fl4) the day after anthesis from 10 different trees in each site (200 anthers in total). Pollen grains from one anther were then dispersed by squeezing on a glass slide in a drop of Alexander's solution. A minimum of 200 pollen grains were observed per anther in triplicate and counted according to their colour. Viable grains had a red protoplasm, whereas aborted grains were empty exhibiting a green colouration in their wall. Percentage of pollen viability was calculated based on the number of viable and total grains.

Stigma receptivity was estimated by fruit initiation (persistent expanded ovary) following hand pollinations on a total of 120 flower buds or flowers from three developmental stages (fl2, fl3 and fl4) on the five study trees in each site ($N = 2$ per stage and per tree).

For the estimation of nectar production, 100 flowers from five different trees per site were covered with exclusion bags (Delnet pollination bags, USA) 24 h before sampling to prevent any visit. Nectar production was assessed twice a day (morning and afternoon) on the same flowers. Nectar was collected with 5 μL glass capillary tubes (Hirschmann Laborgerate, Eberstadt, Germany), and nectar volume was estimated by measuring the length of the nectar column in the capillary tube. Sugar concentration was measured with a low-volume hand refractometer (Eclipse Handheld refractometer, Bellingham and Stanley Ltd, Tunbridge Wells, UK) and expressed as the percentage of sucrose (w/w). To determine the total sugar content of nectar per flower, sugar concentration (%) was converted to mg/μL according to the following formula: $y = 0.00226 + (0.00937x) + (0.0000585x^2)$ where y is the sugar concentration (mg/μL) and x is the sugar concentration (%) (Dafni et al. 2005). The total sugar content of nectar per flower (mg) was then calculated as volume of nectar (μL) × sugar concentration (mg/μL).

The sugar composition of the nectar was determined by high-performance liquid chromatography with a Shimadzu HPLC system coupled to a RID10A refractometer (Shimadzu, 's-Hertogenbosch, Netherlands) using a Hypersil gold amino (150 × 4.6 mm) column (Thermo Scientific, Aalst, Belgium) at 26 °C. The mobile phase consisted of 83 % acetonitrile in water and the flow was 1.0 mL min^{-1}. Analyses for nectar composition were performed in the Groupe de Recherche en Physiologie Végétale (Université catholique de Louvain, Louvain-la-Neuve, Belgium).

Floral visitors and their pollination potential

On two trees per site, four inflorescences per tree were marked for observations of insect and bird visitors. Insects and birds visiting flowers were observed from January to May 2014 for five consecutive days per site. The observations were conducted at 0700, 0900, 1100, 1400 and 1700 h, for 5 min per inflorescence for each time period. A total of 17 h of observations was performed at each site.

Visitor abundance and behaviour (nectar and/or pollen collection, and contacts with reproductive organs) were assessed. Insects that were not identified in the field were collected by hand net, killed with ethyl acetate, identified and stored separately in small vials. Precise identifications were performed at the Centre régional de Biodiversité des insectes from the Institut International d'Agriculture Tropicale (Abomey-Calavi, Benin) and at the Institut Royal des Sciences Naturelles de Belgique (Brussels, Belgium).

The potential of the different insect visitors to be effective pollinators was estimated by determining their relative abundance, fidelity and capacity to carry pollen. Their fidelity was indirectly estimated by the proportion of pollen from V. doniana versus other plant species in their corbicular pollen loads. Their capacity for carrying pollen was directly assessed by counting the pollen grains on the different parts of the insect body. For insect fidelity, in each site, insects captured with pollen loads were immobilized in a bee-marking cage and one pollen load was removed by toothpick per individual. For the four sites, a total of 60 pollen loads was collected and acetolyzed (Erdtman 1960, modified). From each sample, approximately 400 randomly chosen pollen grains were identified by light microscopy (Leitz Wetzlar). Pollen identification was based on a reference slide of V. doniana pollen. A total of 13 insect species per site (20 individuals per species) were examined for their carrying capacity. For each visitor type, pollen identity and quantity on insect bodies were assessed. The pollen grains were removed from the different insect body parts (head, abdomen and legs) using a small cube of gelatin passed over each part of the insect body. The gelatin was melted and observed under a light microscope (Ernst Leitz Wetzlar).

Breeding system experiments

To assess the breeding system, 1600 flowers were marked and allocated to one of the four pollination treatments. Each treatment was performed on five flowers per inflorescence, four inflorescences per tree and five trees in each of the four sites. All hand-pollinated flowers were emasculated at bud stage.

One treatment was carried out in the presence of pollinators (T1; free exposure for control or open pollination). Two treatments involved hand-pollinations: (T2) self-pollination with pollen from another flower of the same tree, to test for self-compatibility and (T3), cross-pollination with pollen from another tree in the same site, to assess inbreeding depression when comparing with T2. To quantify spontaneous self-pollination, flowers in the fourth treatment (T4) were bagged and left unmanipulated.

Except for the control treatments, all flowers were bagged before flower anthesis to exclude visitors (Delnet pollination bags, USA). All bags were removed at the start of fruit maturation. Pollination was carried out by brushing anthers of the donor flower on the stigma of the recipient flower. Hand pollinations were performed daily from 0700 to 1100 h.

Pollen germination and pollen tube growth

Following hand-pollination treatments, 10 pistils were removed at different times after pollination (15 and 30 min, 1, 2, 4, 6, 24, and 48 h). Pistils were fixed in FAA (ethanol 70 %: formaldehyde 35 %: acetic acid; 8:1:1) and stored at room temperature. Before observation, the pistils were rinsed with distilled water, softened and clarified in 4M NaOH for 4 h at room temperature. The pistils were rinsed again and stained for 2 h in 0.1 % aniline blue solution in 0.1M KH_2PO_4 (Kearns and Inouye 1993). Pollen germination on stigmas and pollen tube growth in the styles were examined under a fluorescence microscope (Nikon Optiphot-2/LH-M100C-1) with a 420-nm to 440-nm excitation filter and a 480-nm emission filter, according to Kearns and Inouye (1993). The numbers of pollen grains per stigma, of pollen grains initiating a pollen tube and of pollen tubes reaching the end of the style were counted.

Fructification and germination

Mature fruits were harvested 4 months after pollination. Fruit set (percentage of flowers giving mature fruits), fruit fresh weight, numbers of viable seeds, weight of 100 viable seeds and germination rate were measured.

In September 2013, a preliminary *ex situ* germination test was performed under controlled conditions on 240 ripe fresh fruits. Fruit samples were divided into batches of 10 fruits for germination treatments. Besides control batches (T0), three pre-treatments were tested: soaking in distilled water for 24 h (T1), soaking in sulfuric acid 95 % for 1 h and rinsed with demineralized water (T2), and scarification with a scalpel (T3). Six replicates of 10 fruits per treatment were put on filter paper in jars (300 mL volume), watered with demineralized water and overwrapped with a plastic film. They were then placed in two germination chambers (Snijders Scientific, Netherlands). Two temperature regimes were then applied to the seeds (30/25 °C or 35/30 °C with a photoperiod cycle 12 h dark: 12 h light). Temperature conditions were chosen since they correspond to the soil temperature in the natural environment of *V. doniana* (N'Danikou et al. 2014). Germination was recorded once a week during 6 months. No germination was observed and the test was stopped due to fungal infection.

In September 2014, germination tests were performed *in situ* following hand pollination treatments on 10 samples of five seeds for hand self- and cross-pollinated fruits. Extracted dried seeds were sown in 30 black LDPE (low-density polyethylene) nursery bags (29.6 cm^3) filled with ground soil. Bags were left under ambient conditions, and watered every 2 d. The total germination rate was assessed 45 d after sowing.

In September 2015, seeds were extracted from fruits and sun-dried for 10 h before being stored in paper bags until use. Seed samples were divided into four batches of 100 seeds for germination treatments under controlled conditions. Seeds were or were not submitted to a pre-treatment: soaking (S) or not (NS) in demineralized water for 2 h (Ahoton et al. 2011; N'Danikou et al. 2014). Two temperature regimes were then applied to the seeds (30/25 °C or 35/30 °C with a photoperiod cycle 12 h dark: 12 h light). Five replicates of 20 seeds per treatment were put on filter paper in petri dishes (90 mm diameter) and watered with demineralized water. They were then placed in germination chambers (Snijders Scientific, Netherlands). Seeds were considered germinated when the radicle protruded. Germination was recorded every 2 days. Final germination rate was estimated after 21 days, after several days without any new seedling emergence.

Data analysis

Results were compared by analyses of variance (ANOVA, one- and two-way). Normality of the data was estimated with Shapiro–Wilk tests and homoscedasticity was verified using the Levene test. Data were transformed when required to ensure normal distribution and Welch's correction was applied when homoscedasticity was not met. Differences among sites or conditions were assessed using Tukey's HSD post-hoc tests.

Pearson correlations were performed among flower and fruit measurements. Significant correlations were detected among several floral traits: length and width of lateral fused petals ($R = 0.24$, $P = 0.01$), length and width of small fused superior petals ($R = 0.81$, $P < 0.0001$), lengths of the exterior and the inner cycles of stamens ($R = 0.69$, $P < 0.0001$), and diameter and length of ovary ($R = 0.27$, $P = 0.0001$). Thus, 10 independent measures were analyzed: width and length of sepals, width of the lateral fused petals, length and width of the inferior petal, corolla tube length, length of the long stamens, diameter of the ovary, length of the style, and the distance between the long stamens and the stigma. In the same way, fruit diameter and length were significant correlated ($R = 073$, $P < 0.0001$), and only diameter was presented.

Chi-square tests were used to analyze the proportion of visiting insects among sites. Differences in the composition of pollen loads and pollen carried by different visitor species were visually assessed with heatmaps ('heatmap.2' command, 'gplots' package).

The self-compatibility index (SCI) and self-fertility index (SFI) were calculated according to Lloyd and Schoen (1992). SCI determines the capacity of a plant to produce zygotes following self-pollination relative to that following outcrossing. It is calculated as the ratio between the fruit set (or seed set) produced after hand self-pollination and the fruit set (or seed set) produced after hand cross-pollination (Lloyd and Schoen 1992). SFI gives an estimation of the capacity of a plant to produce fruits and seeds without any pollen vector. It is calculated as the fruit set (or seed set) of autonomous self-pollination relative to that of hand cross-pollination (Kouonon et al. 2009). SCI values above 0.2 indicate self-compatibility and SFI values above 0.2 indicate autonomous selfing (Lloyd and Schoen 1992).

Levels of inbreeding depression at different developmental stages (fruit set, seed set and seed germination) were determined as the ratio between relative performance of selfed progeny (ws) and outcrossed progeny (wc) ($\delta = 1 - (ws/wc)$; Charlesworth and Charlesworth 1987). Values around 0 indicate no inbreeding depression.

All analyses were performed under R version 2.2.1 (R Development Core Team 2013). Data are presented as means ± standard deviation.

Results

Phenology and floral biology

Floral synchrony: Vitex doniana flowered during the dry season, from December to April, and flowering was not synchronous among sites. Flowering began earlier in the Guineo-Congolian than in the Guineo-Sudanian sites: Godomey trees started and ended flowering first, while Namougo and Gouka trees initiated flowering 20–30 days later (Fig. 3A). Flowering was not synchronous within each site: flowering differences among trees were approximately 25 days, with differences especially pronounced in Dasso (41 days). Moreover, flowering was asynchronous within a tree, among inflorescences.

Flowers first appeared on new twigs, after leaves developed. After initiation, buds (fl1) reached stage fl2 (perianth developed, petals fused, buds elongated to 4-6 mm long x 2.6-2.8 mm wide) in 13–20 days. Stage fl3 with well-developed flower buds (central inferior petal deviated from the others, flower size of 7–9 mm long and 3–4 mm wide), occurred 22–28 days after bud initiation. At the fl3 stage, the flower opened as soon as any petal was touched, especially by insects or birds. The stage fl4 (all petals expanded, flowers 12–14 mm long and 3–4 mm wide) occurred 26–35 days after fl1. Flowers started to open in the morning between 0800 and 1200 h, with a high frequency around 1000 h. During this first step of anthesis, the anthers presented numerous white pollen grains along the longitudinal dehiscence splits. Stigmas reached receptivity with flower opening and remained receptive for 72 h. Flower life span (from opening to corolla drop) extended for 3–4 days.

Within an inflorescence, flowers displayed asynchronous anthesis: the terminal flower was the first that bloomed and it took 4–6 days for all flowers to open.

Fruit maturation: Fruit maturation was observed from the end of January to May, while fruits began to fall from June to September and fruit harvesting extended from July to September. The first signs of fruit development (fr1) were registered 5–7 days after corolla drop. At stage fr2 (21–34 days after fr1), young green fruits measured

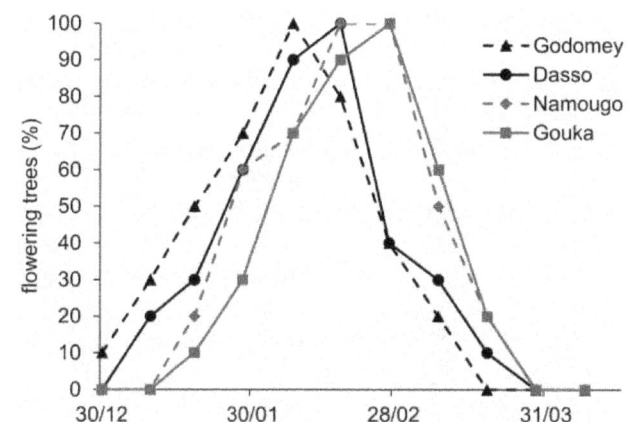

Figure 3. Percentage of Vitex doniana flowering trees in the four studied sites.

Table 2. Morphological characteristics of inflorescences, flowers, and fruits of *vitex doniana* in the four studied sites ($n = 200$, means ± SD).

Parameter	Godomey	Dasso	Namougo	Gouka	ANOVA/Kruskal- Wallis
Floral characteristics					
Flowers per tree	52938 ± 9786^a	16322 ± 6738^b	35635 ± 8396^{ab}	43019 ± 4174^{ab}	$F = 5.23, P = 0.0104$
Sepal length (mm)	4.01 ± 0.32^b	3.98 ± 0.40^b	4.19 ± 0.38^a	4.10 ± 0.33^{ab}	$F = 3.31, P = 0.02$
Sepal width (mm)	2.86 ± 0.34^a	2.64 ± 0.24^b	2.67 ± 0.25^b	2.58 ± 0.15^b	$F = 11.39, P < 0.0001$
Lateral fused petal width (mm)	3.35 ± 0.23^a	3.65 ± 0.51^a	4.10 ± 0.51^a	3.58 ± 0.50^a	$F = 1.7, P = 0.19$
Lower lip length (mm)	4.21 ± 0.48^a	4.15 ± 0.47^a	4.88 ± 0.53^a	4.52 ± 0.84^a	$F = 1.8, P = 0.18$
Lower lip width (mm)	7.52 ± 0.71^a	7.14 ± 0.20^a	7.02 ± 0.26^a	6.90 ± 0.48^a	$F = 1.63, P = 0.2$
Corolla tube length (mm)	7.34 ± 0.40^a	7.29 ± 0.55^a	6.73 ± 0.90^b	6.65 ± 0.75^b	$F = 14.36, P < 0.0001$
Ovary diameter (mm)	1.64 ± 0.20^a	1.56 ± 0.24^b	1.47 ± 0.17^c	1.53 ± 0.16^{bc}	$F = 6.14, P = 0.0005$
Stamen length (mm)	6.59 ± 0.50^a	6.66 ± 0.54^a	5.86 ± 0.54^c	6.25 ± 0.66^b	$F = 20.69, P < 0.0001$
Style length (mm)	7.49 ± 0.59^{ab}	7.72 ± 0.70^a	7.41 ± 0.64^b	7.66 ± 0.41^a	$F = 2.97, P = 0.03$
Anther-stigma distance (mm)	0.65 ± 0.15^a	0.69 ± 0.12^a	0.67 ± 0.17^a	0.66 ± 0.11^a	$F = 0.84, P = 0.47$
Pollen grains/anther	3903 ± 385^a	3502 ± 391^a	3952 ± 389^a	3978 ± 364^a	$F = 0.64, P = 0.42$
Fruit characteristics					
Fruit width (mm)	3.03 ± 0.37^a	2.09 ± 0.25^b	1.87 ± 0.24^c	1.85 ± 0.23^c	$F = 382.94, P < 0.0001$
Fruit weight (g)	16.33 ± 4.49^a	6.01 ± 1.73^b	4.92 ± 1.57^c	4.78 ± 1.40^c	$F = 448.92, P < 0.0001$

Means followed by different letters within a line are significantly different at $P = 0.05$.

10–15 mm wide. Fruits remained green until full maturity. The ripe fruits started to fall 102–141 days after fr2 (March to July). The colour only became black 5–8 days after lying on the ground. Fruit size and weight were significantly higher in Godomey than in other sites (Table 2).

Inflorescence and floral biology: Over all sites, inflorescences (dense and axillary cymes) averaged 79 ± 28 flowers. Based on the flower number per inflorescence and per tree, we calculated that a single tree produced between 16 300 (Dasso) and 52 900 (Godomey) flowers (Table 2). A significant proportion of flowers were white (30 %) at Godomey, but flowers were invariably purple at the other sites. Excepting anther-stigma distance and petal sizes, flower sizes differed significantly among sites (Table 2). Flowers from the Guineo-Sudanian sites (Namougo and Gouka) were significantly smaller than those from the Guineo-Congolian zone (Table 2). The number of pollen grains per flower did not differ among sites and averaged approximately 15 500 (Table 2). All flowers contained four ovules. At the flower morphology level, the two lobes of the stigma were divergent and faced the upper fused petals, whereas the dehiscent side of the anthers faced the lower petal, which could preclude spontaneous selfing (Fig. 2C).

Pollen viability: No pollen viability was detected at bud stages (fl1 and fl2). During the first day of anthesis, the viability of the pollen grains reached 88–95 %. This proportion decreased at the end of the first day (1700 h., 66–72 %) and was lower the second day (45–48 %, Fig. 4). The percentage of pollen viability varied among sites ($F_{3,108} = 2.80$, $P = 0.0435$), whereas its decrease with time was similar ($F_{6,108} = 0.61$, $P = 0.1257$). The highest pollen viability was observed in Godomey and the lowest in Namougo (95 ± 3 % vs. 88 ± 7 % on the first day of anthesis).

Stigma receptivity: Stigma receptivity did not differ among sites ($F_{3\ 48} = 0.33$, $P = 0.8013$) nor among floral stages ($F_{2,48} = 0.54$, $P = 0.5853$). The stigmas were receptive just before (fl2) and during flower anthesis (Fig. 4). Pollen germination reached 68–74 % for well-developed buds and 74–80 % at flower anthesis.

Nectar production: The volume of nectar produced per flower per day varied among sites and was higher in Godomey (1.2 µL) compared with Dasso and Gouka (< 1 µl; $F_{3,192} = 8.80$, $P < 0.0001$, Fig. 5A). Similarly, mean sugar concentration in nectar differed among sites ($F_{3,192} = 102.93$, $P < 0.0001$). It was approximately 39 ± 17 % in Godomey, 41 ± 17 % in Dasso, 51 ± 11 % in Namougo and 45 ± 16 % in Gouka. Thus, the total sugar content in nectar was higher in Gouka (0.5 mg/flower) and lower in Dasso (< 0.4 mg/flower; $F_{3,192} = 4.78$, $P = 0.0031$, Fig. 5B). Regardless of site, the volume of nectar produced was higher in the morning than in the afternoon ($F_{1,192} = 179.46$, $P < 0.0001$, Fig. 5A), but the morning nectar had lower concentrations of sugars

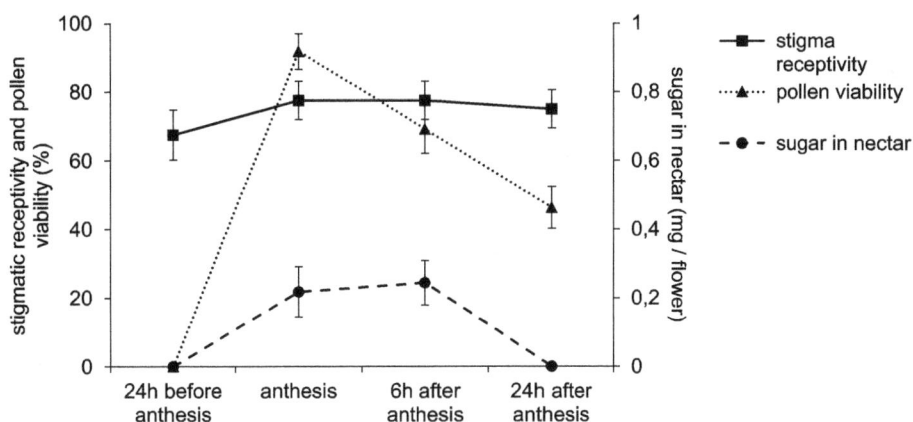

Figure 4. Stigma receptivity, pollen viability and sugar amount in the nectar of *Vitex doniana* flowers for all sites pooled according to flower age. Stigma receptivity (full line) is given as the percentage of pollinated flowers that set fruit; pollen viability (dotted line); sugar amount in nectar per flower (dashed line).

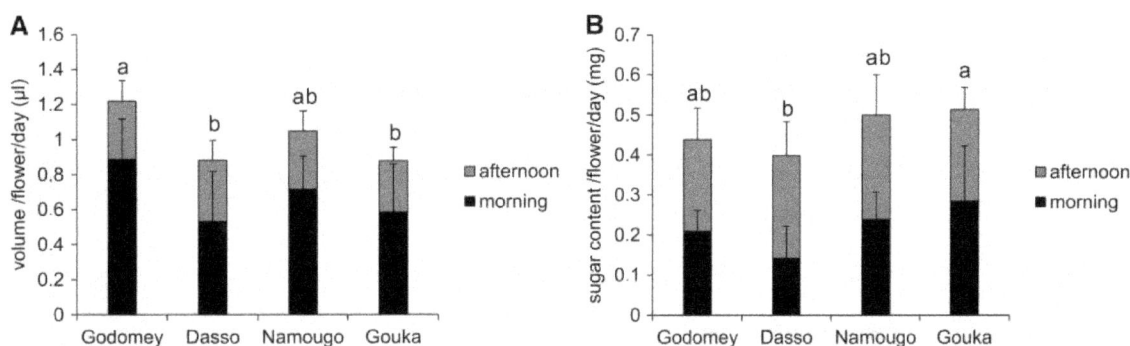

Figure 5. Nectar production of *Vitex doniana* flowers on the four studied sites, according to time of day. (A) Volume of nectar produced per flower per day; (B) total sugar content in nectar per flower per day. Different letters indicate significant difference at $P < 0.05$ according to Tukey's HSD post-hoc analysis.

$(29 \pm 8$ % vs. 58 ± 4 %; $F_{1,192} = 2875.74$, $P < 0.0001$). Nectar contained sucrose (39–58 %), glucose (23–35 %) and fructose (19–25 %).

Visitor diversity, efficiency and fidelity

Abundance and diversity of visitors: During a total of 68 h of observation, we recorded 7311 insect visitors, belonging to 27 species (Table 3). We also recorded 207 bird visitors belonging to the Nectariniideae family (one species, the Splendid sunbird, *Cinnyris coccinigastrus*, Table 3). The numbers of visitors (over 17 h of observation at each site) was significantly lower in Dasso (total = 900) compared with the other sites (Fig. 6 and Table 3). The diversity of visitor fauna differed among sites ($X^2 = 374.801$, $df = 15$, $P = 0$) and included several orders and families: Hymenoptera (78–89 %), Lepidoptera (2–6 %), Heteroptera (1–3 %), Coleoptera (0–4 %), Diptera (2–8 %) and Passeriformes (1–7 %, Fig. 6). Hymenoptera were mainly represented by the families Apidae and Megachilidae (Table 3). Within Apidae, the native honeybees (*Apis mellifera adansonii*) were the main visitors,

representing 26–40 % of the observed visitors over all sites, with a higher abundance in Godomey. *Xylocopa olivacea* individuals were also abundant (24–32 %), especially in Gouka. Individuals from *Megachile cincta* (44–68 %) and *Megachile rufipes* (19–35 %) were the main visitors from the Megachilidae family. Within the Syrphidae family, *Eumerus vestitus* (40–64 %) and *Allobaccha* sp. (29–36 %) were particularly abundant. The Dasso site was characterized by the highest abundance of the sunbird, *Cinnyris coccinigastrus* ($X^2 = 52.652$, $df = 3$, $P = 0$, Table 3).

Fidelity and pollen loads: The proportion of *V. doniana* pollen in corbicular loads differed significantly among visitor species ($F_{3,33} = 10.40$, $P < 0.0001$) and among sites ($F_{3,33} = 3.04$, $P = 0.0426$), with a marginally significant interaction among sites and species ($F_{3,33} = 2.20$, $P = 0.0270$). The pollen loads from *Dactylurina staudingeri* contained little pollen from *V. doniana* (25 %, Fig. 7A). By contrast, 17 % of *Apis mellifera*, 40 % of *Megachile torrida* and 60 % of *Ceratina (Simioceratina)* individuals carried monospecific pollen loads of *V. doniana* (Fig. 7A).

Pollen carrying capacity: Insect visitors carried 42–87 % *V. doniana* pollen grains, depending on the insect

Table 3. Insect and bird visitors on *Vitex doniana* in the four studied sites. Most abundant visitors for the different sites are indicated in grey.

	Family	Species	Total number recorded	Godomey	Dasso	Namougo	Gouka
Hymenoptera	Apidae	*Xylocopa olivacea* Fabricius	955	263	119	282	291
		Xylocopa nigrita Fabricius	127	22	11	43	51
		Apis mellifera adansonii Latreille	1189	407	110	383	289
		Ceratina (Pithitis) viridis Guérin Méneville	544	173	30	160	181
		Ceratina (Simioceratina) sp. Latreille	151	29	12	48	62
		Dactylurina staudingeri Gribodo	359	111	34	101	113
		Braunsapis ghanae Michener	368	85	71	92	120
		Amegilla albocaudata Dours	148	34	18	41	55
	Megachilidae	*Megachile cincta* Fabricius	916	223	87	236	370
		Megachile rufipes Fabricius	529	132	94	166	137
		Megachile torrida Smith	452	131	64	116	141
		Euaspis abdominalis Fabricius	153	37	16	47	53
		Chalicodoma maxillosa Guérin Méneville	98	27	8	16	47
		Anthidium sp. Fabricus	85	45	0	27	13
	Halictidae	*Lipotriches sp* Gerstaecker	125	67	5	34	19
Heteroptera	Lygaeidae	*Graptostethus servus* Fabricius	131	33	3	41	54
Lepidoptera	Arctiidae	*Euchromia lethe* Fabricius	53	9	13	13	18
	Lycaenidae	*Axiocerses harpax* Fabricius	68	30	4	15	19
	Hesperiidae	*Borbo perobscura* Druce	104	34	24	15	31
	Sphingidae	*Cephonodes hylas virescens* Wallengren	55	12	3	19	21
Coleoptera	Tenebrionidae	*Alogista serricorne* Kolbe	93	28	15	50	0
	Chrysomelidae	*Podagrixena decolorata* Duvivier	35	5	8	22	0
Diptera	Tephritidae	*Dacus ciliatus* Loew	136	74	21	16	25
	Syrphidae	*Allobaccha sp* Curran	51	12	8	11	20
		Eumerus sp Fabricus	34	13	5	0	16
		Eumerus vestitus Bezzi	76	33	8	14	21
	Bombyliidae	*Eurycarenus sp* Loew	69	0	16	22	31
Passeriformes	Nectariniidae	*Cinnyris coccinigastrus* Latham	207	54	93	24	36

species ($F_{14,206} = 5.11$; $P < 0.0001$) but not on the site ($F_{3,206} = 1.62$, $P = 0.1849$). *Vitex doniana* pollen grains constituted the majority of the carried pollen grains on 63 % of insects caught (Fig. 7B). The pollen grain numbers found on the insects differed significantly among body parts ($F_{2,618} = 28.48$, $P < 0.0001$), among insect species ($F_{14,618} = 87.91$, $P < 0.0001$), and among sites, with the insects caught in the Guineo-Sudanian sites carrying more pollen than those caught in the Guineo-Congolian sites ($F_{3,618} = 20.49$, $P < 0.0001$, Fig. 7C).

The large bees like *Xylocopa* spp. and the medium-sized bees *Megachile* spp., *Apis mellifera* and *Amegilla albocaudata* used the lower petal as a 'landing platform' and during foraging, their heads contacted the anthers and stigma. They, therefore, received more pollen grains on their heads (39–65 % of total pollen load) than on their abdomens and legs (Fig. 7C). Small bees like *Braunsapis ghanae* or *Ceratina* spp., which directly plunged into the corolla tube, received more pollen grains (36–60 % of total pollen load) on their abdomens.

Pollination trials

Pollen germination and pollen tube growth: Pollen germination was observed 30 and 60 min after pollination and

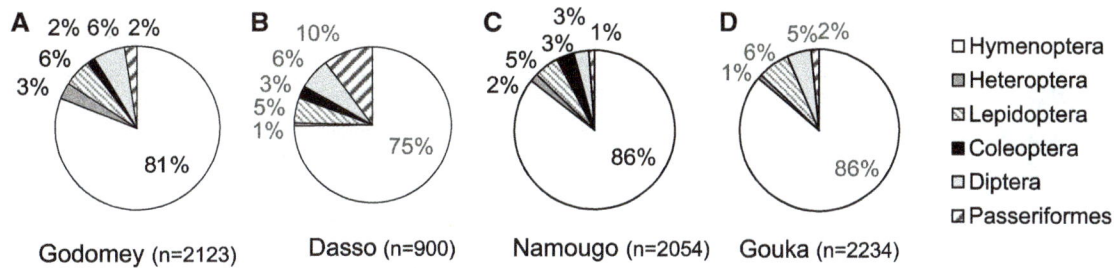

Figure 6. Visitor proportions on *Vitex doniana* recorded per site, on the four studied sites in (A) Godomey, (B) Dasso, (C) Namougo and (D) Gouka (*n* = total number of different visitors in 17 h of observation per site).

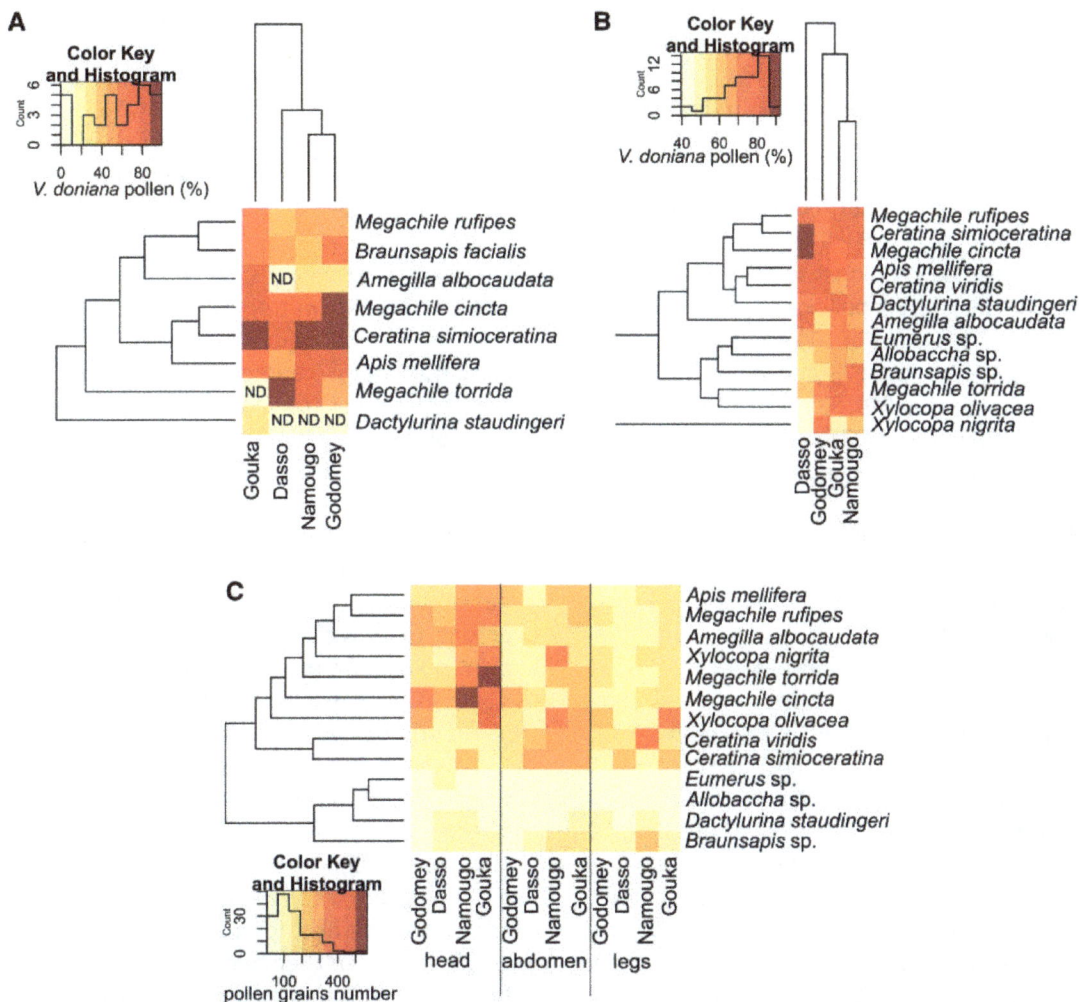

Figure 7. Insect visitor efficiency and fidelity on *Vitex doniana* recorded on the four studied sites. (A) Heatmap of the percentage of *V. doniana* pollen observed in the insect pollen loads according to insect species; (B) heatmap of the percentage of *V. doniana* pollen observed on the insect body according to insect species; (C) heatmap of the number of pollen grains on the different insect body parts according to insect species. ND, no data.

pollen tubes reached the extremity of the style after 4–8 h (Fig. 8). Signs of fertilization in the ovary were observed 1–2 days after pollination. No differences were observed between self- and cross-pollinated styles for the number of pollen tubes reaching the ovary ($X^2 = 0.349$, d$f = 2$, $P = 0.8397$).

Fruit and seed production: Since no differences were detected in fruit set ($F_{3,308} = 2.33$, $P = 0.0744$), seed set ($F_{3,308} = 0.08$, $P = 0.9728$) or fruit weight ($F_{3,308} = 1.85$, $P = 0.1388$) among sites, we pooled the results. No fruits were produced after spontaneous self-pollination ($SFI = 0$) at any site, indicating that pollen vectors are

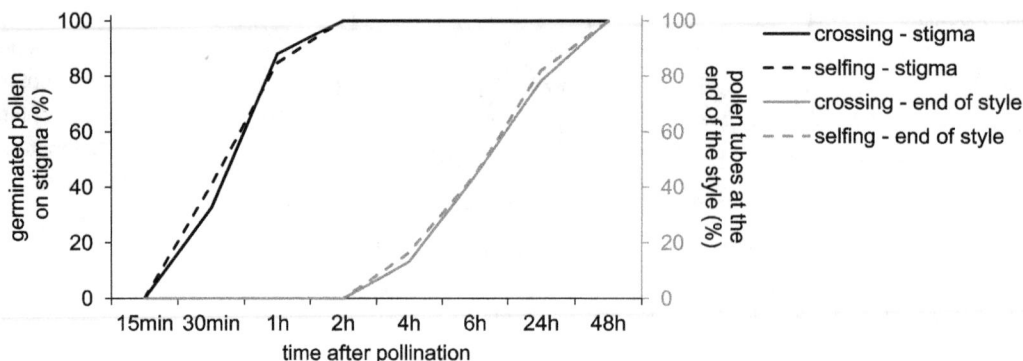

Figure 8. Cumulative percentage of germinated pollen on stigma (in black) and of pollen tubes reaching the end of the style (in grey) after hand cross-pollination (solid lines) or hand self-pollination (dashed lines) in *Vitex doniana*, according to time after pollination.

essential for pollination. Open-pollination treatment produced significantly more fruits (37 %, $F_{2,308} = 39.54$, $P < 0.0001$, Fig. 9A) than hand-pollination, but fruit weight was lower after cross-pollination compared with open- and self-pollination ($F_{2,308} = 4.63, P = 0.0104$, Fig. 9B).

Seed set was significantly higher after open- and self-pollination than after cross-pollination ($F_{2,308} = 3.45$, $P = 0.0329$, Fig. 9C). The weight of 100 seeds was lower after crossing compared with self- and open- pollination ($F_{2,36} = 4.65, P = 0.0158$, Fig. 9D). Self-compatibility indices were high ($SCI_f = 0.80 \pm 1.26$, $SCI_s = 1.22 \pm 0.84$). Germination of the resulting seeds did not differ between treatments ($F_{2,36} = 0.82, P = 0.4482$) and reached approximately 50–62 % (Fig. 9E). No inbreeding depression was detected during the first developmental stages for seed set (-0.2 ± 0.16) or germination rate (-0.04 ± 0.1); indeed, selfed progeny performed better than outcrossed ones. Cumulative inbreeding depression reached $\delta = -0.04$.

Seed viability and germination: Under controlled conditions, no germination occurred with pre-treated fruits but seeds that were extracted from the seed coat germinated within 2 days, regardless of pre-treatment (soaking or not) or temperature conditions (30/25 °C, 35/30 °C; Fig. 10A). After 21 days, germination rate reached 70 % and did not differ by pre-treatment ($F_{1,16} = 0.03$, $P = 0.8715$, Fig. 10B) or temperature conditions ($F_{1,16} = 0.00, P = 0.10$, Fig. 10B). Germination tests *in situ* resulted in a 36–52 % germination rate and did not differ among sites ($F_{1,377} = 2.05, P = 0.1142$, Fig. 10C).

Discussion

What is the floral phenology and morphology of *Vitex doniana*?

Vitex doniana flowering occurred once a year in the dry season (December–April) as generally found in other woody species from Sudanian and Sahelian zones (for example *Parkia biglobosa, Ziziphus mauritiana* and *Vitex*

fischeri) while fruit maturation takes place during the rainy season which might favor seed dispersion, seed germination and sapling development (Ahenda 1999). Flowering patterns showed a shift according to sites and latitude (in December in the Guineo-Congolian sites, two weeks later in Guineo-Sudanian sites). These patterns resulted from the wetter conditions in Guineo-Congolian zone than in the Guineo-Sudanian one. This is congruent with the general assumption that the main determinants of phenology of tropical trees are temperature, insolation, rainfall and relative humidity (Ewédjè et al. 2015).

Flowering asynchrony was high at all levels, among sites, among trees and within trees. This low flowering overlap increases the attraction of pollinators, as floral rewards are offered during an extended period, and facilitates pollen exchange among flowers within a tree and among trees within a site. Thus with the observed efficient pollination rate, a low level of genetic differentiation could be expected within populations of this tree species (Ewédjèet al. 2015).

Several morphological traits of *V. doniana* flowers predispose them to selfing, including hermaphroditism, and a short anther–stigma distance (0.65–0.69 mm). Moreover, pollen viability is high on the day of anthesis and coincides with the stigma receptivity; this homogamy trait was also found on *Vitex negundo, V. fischeri* and *V. kenyensis* (Reddy and Reddi 1994; Ahenda 1999). Nevertheless, the divergent positions of the lobes of the stigma and the anthers could preclude spontaneous selfing. Such a floral morphology, combined with flowering asynchrony could suggest a mixed mating system, combining both geitonogamy and outcrossing.

Is *Vitex doniana* efficiently pollinated by visitors?

The flower morphology and the production of floral rewards indicate that *V. doniana* could be attractive to visitors. The zygomorphic and tubular flowers present the typical gullet blossom of the genus *Vitex*, with a large

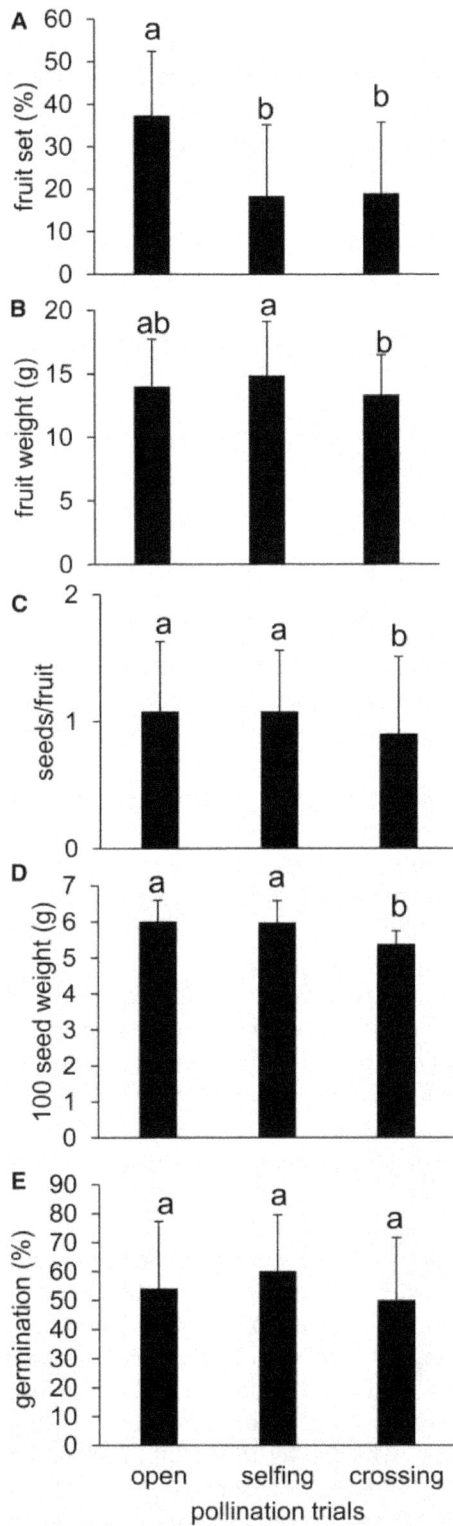

Figure 9. Fruit and seed parameters derived from pollination trials. (A) Fruit set, (B) fruit weight, (C) seed number per fruit, (D) weight of 100 seeds and (E) seed germination rate after open pollination (open), hand self-pollination (selfing) and hand cross-pollination (crossing) in *Vitex doniana*. Data are shown as means ± SD. Pollination trials followed by different letters are significantly different at $P < 0.05$ according to Tukey's HSD post-hoc analysis.

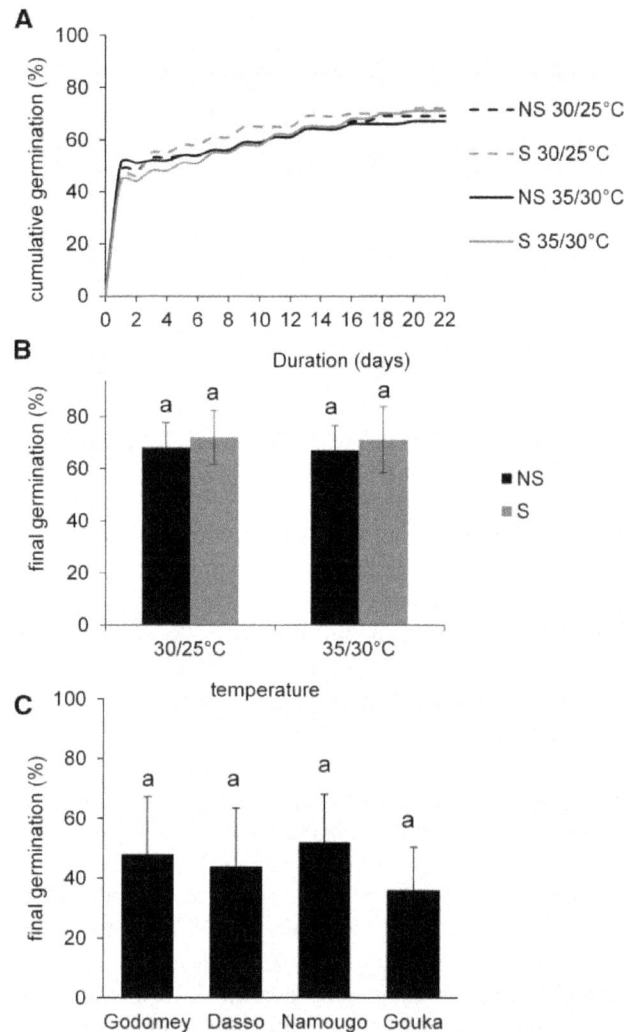

Figure 10. *Vitex doniana* seed germination. (A) Cumulative germination rate as a function of time; (B) final germination rate of seeds subjected to different temperatures (30/25 °C, 35/30 °C) and soaking treatments (NS: not soaking seeds, S: soaking seeds); (C) final germination rate of seeds sown in the four studied sites in Benin. Data are shown as means ± SD. Sites or treatments followed by different letters are significantly different at $P < 0.05$ according to Tukey's HSD post-hoc analysis.

inferior petal where insects can land. The morphology also suggests that most pollen-bearing insects likely contact the stigmas during a visit and could thus efficiently contribute to pollination.

Most flowers of *V. doniana* opened as soon as an insect or bird visitor touched any petal. This stage of flower (fl3) coincides with the peak of nectar production (0.75–0.89 µL), which might be an important reward. As with many Lamiaceae species, nectar is offered at the base of the corolla tube (Faegri and van der Pijl 1979; Jorge *et al.* 2015). The sugar concentration of nectar (39–51 % sugars), negatively correlated with the volume, remained high in all sites and during the entire flower life span.

Such a high sugar concentration mainly attracts nectar-feeding insects. Birds usually prefer more dilute but more abundant nectar (Kulloti et al. 2011). In addition to nectar, V. doniana produced copious pollen (15 000–16 000 pollen grains per flower), which is attractive for pollen-collecting insects. Vitex doniana flowers, therefore, appear extremely attractive to insect visitors. Indeed, the species is considered a valuable melliferous plant and honey resource, and trees are sometimes chosen for hanging bark beehives (Djonwangwe et al. 2011; Yédomonhan et al. 2012; Dukku 2013).

We found that V. doniana flowers were extensively visited by insects and birds from early in the morning to late afternoon. Vitex doniana was visited by 27 insect species belonging to 13 families from five orders, and by one species of sunbird from the family Nectariniidae. It thus has a generalist pollination system. Other Lamiaceae species (Salvia spp., Leonurus spp., etc.) are also pollinated by both birds and insects (Proctor et al. 1996; Wester and Claßen-Bockhoff 2006; Kulloti et al. 2011). However, birds remained marginal visitors for Vitex doniana since they accounted only for 3 % of the observed visitors.

The Hymenoptera were the most abundant visitors at the four sites. Apideae species (45–56 % of visitors), mainly represented by Apis mellifera and Xylocopa spp., were the main visitors. Similar insect guilds were observed on Vitex negundo (Reddy and Reddi 1994), V. fischeri and V. kenyensis (Ahenda 1999). Pollen load analyses showed that most bee individuals recorded were foraging on V. doniana even if other plant pollen grains were found in their corbicular loads. Fidelity to V. doniana pollen differed significantly among species and among sites. Individuals of the species Apis mellifera, Ceratina sp., Megachile cincta and Megachile torrida collected V. doniana pollen in higher proportion than did other species. Pollen grains carried by the most abundant V. doniana visitors were predominantly those of this plant species whatever the site. The high proportions of V. doniana pollen grains on insect bodies indicated that these four species might be more efficient pollinators. Hymenoptera are prevalent pollen-bearing insect visitors, which typically carry large quantities of pollen on their bodies (and in their pollen loads).

What is the breeding system of Vitex doniana?

Most fruits contained only one seed for four ovules. As with many Lamiaceae species, including others from the genus Vitex, three out of the four ovules usually abort in response to post-zygotic competition within the developing fruit (Ahenda 1999; Eyog-Matig et al. 2002; Ky 2008).

In controlled hand-pollinations, V. doniana produced similar or even higher numbers of fruit and seed after self- than after cross-pollinations, indicating that the species is self-compatible, as observed in other species of the genus such as Vitex negundo (Reddy and Reddi 1994). Being a self-compatible plant with many flowers simultaneously open, the species might experience selfing both by autogamy and by geitonogamy, mediated by pollinators. However, seed set and seed viability did not differ among self- and cross-pollinations, indicating that the species does not suffer from inbreeding depression.

Despite the observed self-compatibility, V. doniana did not produce any fruits and seeds after spontaneous selfing (SFI = 0). The species, therefore, requires pollen vectors for its pollination.

What are the best germination conditions for Vitex doniana?

This species has a hard seed coat, which causes very low germination rates (Ahoton et al. 2011). Indeed no germination was observed without extracting seeds from the fruits. Moreover, germination generally occurs over a long period (Ahoton et al. 2011; N'Danikou et al. 2014). The hard seed coat prevents the entry of water and oxygen, which could break seed dormancy (Mapongmetsem 2006; Ahoton et al. 2011). Germination is most successful with fresh seeds. Many seed treatments have been suggested to break dormancy and improve germination, without any conclusive recommendations. In the literature, a higher germination rate (50–58 %) was obtained with the dormancy-breaking treatment of alternation of 8 h sun-drying and 1 h soaking in tap water for 3 days (N'Danikou et al. 2014), or with physical shock (Ahoton et al. 2011). Our results in the field (50 % germination) were similar and also probably due to the alternation to drying and soaking, with suitable water uptake by water-impermeable layers due to palisade dislocation. Furthermore, alternately drying and soaking seeds would have induced a temperature fluctuation within the seed coat that is reported to promote germination. The germination rates obtained under controlled conditions in our study were significantly higher (70 %) and faster (< 21 days) than all previous studies. Time to the first germination (2 days) was shorter than that obtained by Ahoton et al. (2011, 11 days) or N'Danikou et al. (2014, 18 days). Germination rates were also higher than those obtained for scarified seeds of other related Vitex species, as a previous study found 25–50 % germination for V. keniensis and V. fischeri (Ahenda 1999) or 6 % for V. madiensis (Mapongmetsem 2006). Our high germination rates most probably resulted from the prior destruction of the almost impermeable and hard seed coat, which delays and inhibits seed germination. Soaked and not-soaked seeds displayed no significant differences under the incubation conditions of 30/25 °C and 35/30 °C, probably due to the relatively high water content of seeds freshly extracted from fruits. These incubation temperatures corresponded to the soil temperatures that the dispersed seeds might experience in their natural environment (N'Danikou et al. 2014; Ewédjè et al. 2015).

Implications for conservation strategies

While not measured directly during this study, the reproductive traits of the species are most likely to be impacted by extensive leaf harvesting, which delays or prevents flowering of trees, which reduce fruit and seed availability. Even if the flowering pattern of this species is influenced by climatic variables, as found for many other plant species, the regularly delayed or cancelled flowering of trees from the site Dasso suggests that leaf collection puts significant biological stress on the reproductive performance of V. doniana (Oumorou et al. 2010). Leaf harvesting induces low flower production and extends the flowering shift among trees. To make matters worse, when collecting new leaves, harvesters cut the entire twigs with floral buds. Moreover, to facilitate leaf collection, branches and even trees are cut down (N'Danikou et al. 2015).

Nevertheless, fruit and seed sets in exploited sites, such as Dasso, remained similar to those in protected sites (Namougo and Gouka), showing that pollination remained sufficient to ensure suitable reproductive success. The major threat seems thus not a decrease of reproductive success at the flower level but the decrease of flower production. A decrease in flower production induced a lower number of fruits. Moreover, as fruits are harvested, there are no seeds left for natural recruitment. Further studies on V. doniana population biology would help to evaluate the real impact of overexploitation on local population survival. Moreover, studies are ongoing to investigate conservation strategies in Benin.

The low level of natural seed germination and poor recruitment (Ahoton et al. 2011, N'Danikou et al. 2015) can be compensated by the high germination rates obtainable by destruction of the hard seed coat. Apart from vegetative propagation (Oumorou et al. 2010; Mapongmetsem et al. 2012; Sanoussi et al. 2012; Achigan-Dako et al. 2014), the enrichment of residual populations of V. doniana could be fostered using imbibed seeds extracted from fruits. Survival rates in experimental plantations are generally good, reaching 80–90 % after 3 years (N'Danikou et al. 2015). As multiplication by seeds is a tractable conservation method, a strategic action might be to organize both ex situ and in situ

conservation approaches, studying the possibilities for cultivation and domestication in orchards or in managed agroforestry systems (Oumorou et al. 2010; Vodouhè et al. 2011; N'Danikou et al. 2015). Producing V. doniana in horticultural or in peri-urban settings would enlarge the resource base and keep pressure off natural stands. Based on the current demand for V. doniana products and the intensity of the harvest, cultivation would be the best option to sustain the species (N'Danikou et al. 2015). Authorities could encourage local people to plant, protect and promote natural regeneration in their fields or in their home gardens (Oumorou et al. 2010). The protection, cultivation or domestication of this species and its integration into diverse agroforestry systems are important components of a strategy for the improvement of land use in Africa (Vodouhè et al. 2011).

Conclusion

Vitex doniana is a hermaphroditic, homogamous and self-compatible species, presenting a predominant selfing through geitonogamy. Moreover, V. doniana trees are abundantly visited by a wide array of insect and bird species. These reproductive traits assure sexual reproduction even for isolated and distant trees in small, remnant populations. These pollination and breeding characteristics offer the required conditions to develop successful conservation strategies. This study proved that reproductive biology is primordial before any conservation strategy planning for poorly studied species.

Sources of Funding

This work was founded by the Université catholique de Louvain (ADRI) with a Ph.D. grant to V. Sinébou.

Contributions by the Authors

V.S. performed all the experiments both in the field (Benin) and in the lab (Belgium) as a part of her PhD dissertation. M.Q. performed the statistical analyses, drawn the figures and co-wrote the manuscript. B.C.A. supervised the field experiments in Benin. A-L J. designed the experiments, analysed the results and wrote the manuscript. All authors approved the final version of the manuscript.

Acknowledgements

The authors are indebted to the local people for their help in the field during the experiment. All our thanks to E.E. Ewédjè for the drawings, to C. Buyens for the follow-up of the preliminary tests of germination at Louvain-la-Neuve, to A. Pauly, G. Goergen (Institut Royal des

Sciences Naturelles de Belgique, Brussels, Belgium) and O. T. Lougbégnon (Institut International d'Agriculture Tropicale, Abomey-Calavi, Benin) for their help in insect identification, M. Tossou (Laboratoire d'écologie et de biologie végétale, University of Abomey-Calavi, Benin) for granting access to her lab, H. Dailly (Centre Apicole de Recherche et d'Information, Louvain-la-Neuve, Belgium) for the analyses of sugar composition in nectar, P.V. Baret, C. Descamps, L. Moquet and P. Ouvrard for discussion about the first drafts of this manuscript, J. Mach and K. Sherrard for language improvement, and two anonymous reviewers for their constructive comments on the manuscript.

Literature Cited

Achigan-Dako EG, N'Danikou S, Assogba-Komlan F, Bianca A, Ahanchede A, Margaret WP. 2011. Diversity, geographical, and consumption patterns of traditional vegetables in sociolinguistic communities in Benin: implications for domestication and utilization. *Economic Botany* **65**:129–145.

Achigan-Dako EG, N'Danikou S, Tchokponhoue DA, Assogba-Komlan F, Larwanou M, Vodouhe RS, Ahanchede A. 2014. Sustainable use and conservation of *Vitex doniana* Sweet: *unlocking the propagation ability using stem cuttings. *Journal of Agriculture and Environment for International Development* **108**: 43–62.

Adetoro KO, Bolanle DG, Balarebe SA, Ahmed OA. 2013. *In vivo* antioxidant effect of aqueous root bark, stem bark and leaf extracts of *Vitex doniana* in CCl4 induced liver damage rats. *Asian Pacific Journal of Tropical Biomedicine* **3**:395–400.

Agbede JO, Ibitoye AA. 2007. Chemical composition of black plum (*Vitex doniana*): an underutilized fruit. *Journal of Food Agriculture and Environment* **5**:95.

Ahenda JO. 1999. Taxonomy and genetic structure of Meru oak populations, Vitex keniensis Turrill and Vitex fischeri Gürke, in East Africa. Wageningen University, Netherlands. http://library.wur.nl/WebQuery/clc/967678 (19 July 2016).

Ahoton LE, Adjakpa JB, Gouda M, Daïnou O, Akpo EL. 2011. Effet des prétraitements de semences du prunier des savanes (*Vitex doniana* Sweet) sur la germination et la croissance des plantules. *Annales Des Sciences Agronomiques* **15**: http://www.ajol.info/index.php/asab/article/view/67366 (19 July 2016).

Ajenifujah-Solebo SO, Aina JO. 2011. Physico-chemical properties and sensory evaluation of jam made from black-plum fruit (*Vitex doniana*). *African Journal of Food, Agriculture, Nutrition and Development* **11**:4784.

Akoègninou A, Van der Burg VJ, Van der Maesen LJG. 2006. *Flore analytique du Bénin*. Wageningen: Backhuys.

Charlesworth D, Charlesworth B. 1987. Inbreeding depression and its evolutionary consequences. *Annual Review of Systematics and Ecology* **18**:237–268.

Codjia JTC, Assogbadjo AE, Mensah Ekué MR. 2003. Diversité et valorisation au niveau local des ressources végétales forestières alimentaires du Bénin. *Cahiers Agricultures* **12**:321–331.

Dadjo C, Assogbadjo AE, Fandohan B, Kakaï RG, Chakeredza S, Houehanou T, Van Damme P, Sinsin B. 2012. Uses and management of Black plum (*Vitex doniana Sweet*) in Southern Benin. *Fruits* **67**:239–248.

Dafni A, Kevan PG, Husband BC. 2005. *Practical pollination biology*. Cambridge, Ontario: Enviroquest.

Dansi A, Adjatin A, Adoukonou-Sagbadja H, Faladé V, Yedomonhan H, Odou D, Dossou B. 2008. Traditional leafy vegetables and their use in the Benin Republic. *Genetic Resources and Crop Evolution* **55**:1239–1256.

Djonwangwe D, Tchuenguem Fohouo FN, Messi J, Brückner D. 2011. Foraging and pollination activities of Apis mellifera adansonii Latreille (Apidae) on Syzygium guineense var. guineense (Myrtaceae) flowers at Ngaoundéré (Cameroon). *Journal of Animal and Plant Sciences* **10**:1325–1333.

Dukku UH. 2013. Identification of plants visited by the honeybee, *Apis mellifera* L. in the Sudan Savanna zone of northeastern Nigeria. *African Journal of Plant Science* **7**:273–284.

Erdtman G. 1960. The acetolysis method. A revised description. *Svensk Botanisk Tidskrift* **54**:561–564.

Ewédjè EEBK, Ahanchédé A, Hardy OJ, Ley AC. 2015. Reproductive biology of *Pentadesma butyracea* (Clusiaceae), source of a valuable non timber forest product in Benin. *Plant Ecology and Evolution* **148**:213–228.

Eyog-Matig O, Gaoué OG, Dossou B. 2002. Espèces ligneuses alimentaires, Compte rendu de la première réunion du réseau tenue 11–13 Décembre 2000 au CNSF Ouagadougou, Burkina Faso. Institut International des Ressources Phytogénétiques, Nairobi, Kenya.

Faegri K, van der Pijl L. 1979. *The principles of pollination ecology*, 3rd edn. Oxford: Pergamon Press.

FAO. 2011. *Situation des forêts du monde 2011*. Roma: FAO.

Ganapaty S, Vidyadhar KN. 2005. Phytoconstituents and biological activities of Vitex. A review. *Journal of Natural Remedies* **5**:75–95.

Hamzah RU, Egwim EC, Kabiru AY, Muazu MB. 2013. Phytochemical and *in vitro* antioxidant properties of the methanolic extract of fruits of *Blighia sapida*, *Vitellaria paradoxa* and *Vitex doniana*. *Oxidants and Antioxidants in Medical Science* **2**:217–223.

Iwueke AV, Nwodo OFC, Okoli CO. 2006. Evaluation of the anti-inflammatory and analgesic activities of *Vitex doniana* leaves. *African Journal of Biotechnology* **5**:1929–1935.

Jorge A, Loureiro J, Castro S. 2015. Flower biology and breeding system of *Salvia sclareoides* Brot. (Lamiaceae). *Plant Systematics and Evolution* **301**:1485–1497.

Kalinganire A, Harwood CE, Slee MU, Simons AJ. 2000. Floral structure, stigma receptivity and pollen viability in relation to protandry and self-incompatibility in silky oak (*Grevillea robusta* A. Cunn.). *Annals of Botany* **86**:133–148.

Kapooria RG, Aime MC. 2005. First report of the rust fungus *Olivea scitula* on *Vitex doniana* in Zambia. *Plant Disease* **89**:431–431.

Kearns CA, Inouye DW. 1993. *Techniques for pollination biologists*. Niwot: University Press of Colorado.

Kilani AM. 2006. Antibacterial assessment of whole stem bark of *Vitex doniana* against some Enterobacteriaceae. *African Journal of Biotechnology* **5**:958–959.

Kouonon L, Jacquemart AL, Zoro Bi A, Bertin P, Baudoin JP, Djé Y. 2009. Reproductive biology of the andromonoecious *Cucumis melo* subsp. *agrestis* (Cucurbitaceae). *Annals of Botany* **104**: 1129–1139.

Kulloti SK, Chandore AR, Aitawade MM. 2011. Nectar dynamics and pollination studies in three species of Lamiaceae. *Current Science* **100**:509–516.

Ky KJM. 2008. *Vitex doniana* Sweet. In: Louppe D, Oteng-Amoako AA, Brink M, eds. *PROTA plant resources of tropical Africa*. Wageningen,

Netherlands. http://database.prota.org/recherche.htm

Ladedji O, Okoye Z. 1996. Anti-hepatotoxic properties of *Vitex doniana* bark extract. *International Journal of Pharmacognosy* **34**: 355–358.

Ladedji O, Udoh FV, Okoye Z. 1996. Effects of *Vitex doniana* stem bark extract on blood pressure. *Phytotherapy Research* **10**: 245–247.

Lagnika L, Amoussa M, Adjovi Y, Sanni A. 2012. Antifungal, antibacterial and antioxidant properties of *Adansonia digitata* and *Vitex doniana* from Benin pharmacopeia. *Journal of Pharmacognosy and Phytotherapy* **4**:44–52.

Lloyd DG, Schoen DJ. 1992. Self-and cross-fertilization in plants. I. Functional dimensions. *International Journal of Plant Sciences* **153**:358–369.

Mapongmetsem PM, Loura Benguellah B, Nkongmeneck BA, Marin Ngassoum B, Gübbük H, Baye–Niwah C, Ongmou J. 2005. Litterfall, decomposition and nutrients release in *Vitex doniana* Sweet and *Vitex madiensis* Oliv. in the Sudano–Guinean Savannah. *Akdeniz Üniversitesi Ziraat Fakültesi Dergisi* **18**:75.6.

Mapongmetsem PM. 2006. Domestication of *Vitex madiensis* in the Adamawa highlands of Cameroon: phenology and propagation. *Akdeniz Universitesi Ziraat Fakultesi Dergisi* **19**:269–278.

Mapongmetsem PM, Djoumessi MC, Yemele TM, Fawa G, Doumara GD, Noubissie TJB, Avana TML, Bellefontaine R. 2012. Domestication de *Vitex doniana* Sweet. (Lamiaceae): influence du type de substrat, de la stimulation hormonale, de la surface foliaire et de la position du nœud sur l'enracinement des boutures uninodales. *Journal of Agriculture and Environment for International Development* **106**:23–45.

Maundu P, Achigan-Dako E, Yasuyuki M. 2009. Biodiversity of African vegetables. In: Shackleton CM, Pasquini MW, Drescher AW, eds. *African indigenous vegetables in urban agriculture*, Vol. **6**. London: Earthscan, 104.

Moza MK, Bhatnagar AK. 2007. Plant reproductive biology studies crucial for conservation. *Current Science* **92**:1207.

N'Danikou S, Achigan-Dako GE, Wong JLG. 2011. Eliciting local values of wild edible plants in southern Benin to identify priority species for conservation. *Economic Botany* **65**:381–395.

N'Danikou S, Achigan-Dako EG, Tchokponhoué DA, Assogba Komlan F, Gebauer J, Vodouhè RS, Ahanchédé A. 2014. Enhancing germination and seedling growth in *Vitex doniana* Sweet for horticultural prospects and conservation of genetic resources. *Fruits* **69**:279–291.

N'Danikou S, Achigan-Dako EG, Tchokponhoué DA, Agossou CO, Houdegbe CA, Vodouhè RS, Ahanchédé A. 2015. Modelling socioeconomic determinants for cultivation and in-situ conservation of *Vitex doniana* Sweet (Black plum), a wild harvested economic plant in Benin. *Journal of Ethnobiology and Ethnomedicine* **1**:43.2.

Oumorou M, Sinadouwirou T, Kiki M, Glele Kakaï R, Mensah GA, Sinsin B. 2010. Disturbance and population structure of *Vitex doniana* Sweet in northern Benin, West Africa. *International Journal of Biological and Chemical Sciences* **4**:624–632.

Padmalatha K, Jayaram K, Raju NL, Prasad MNV, Rajesh A. 2009. Ethnopharmacological and biotechnological significance of *Vitex*. *Bioremediation, Biodiversity and Bioavailability* **3**:6–14.

Proctor M, Yeo P, Lack A. 1996. *The natural history of pollination*. London: Harper Collins.

R Development Core Team. 2013. *R: a language and environment for statistical computing*. USA: R Foundation for Statistical Computing.

Reddy TB, Reddi CS. 1994. Pollination ecology of *Vitex negundo* (Lamiaceae). *Proceedings of the Indian Natural Science Academy, Section B, Biological Sciences* **60**:66.

Sala OE, Chapin FS, Armesto JJ, Berlow E, Bloomfield J, Dirzo R, Huber-Sanwald E, Huenneke LF, Jackson RB, Kinzig A, Leemans R, Lodge DM, Mooney HA, Oesterheld M, Poff NL, Sykes MT, Walker BH, Walker M, Wall DH. 2000. Global biodiversity scenarios for the year 2010. *Science* **287**:1770–1774.

Sanoussi A, Ahoton LE, Odjo T. 2012. Propagation of Black Plum (*Vitex donania* Sweet) using stem and root cuttings in the ecological conditions of South Benin. *Tropicultura* **30**: 107–112.

Vodouhè R, Dansi A, Avohou HT, Kpéki B, Azihou F. 2011. Plant domestication and its contributions to in situ conservation of genetic resources in Benin. *International Journal of Biodiversity and Conservation* **3**:40–56.

Vunchi MA, Umar AN, King MA, Liman AA, Jeremiah G, Aigbe CO. 2011. Proximate, vitamins and mineral composition of *Vitex doniana* (black plum) fruit pulp. *Nigerian Journal of Basic and Applied Sciences* **19**:97–101.

Wester P, Claβen-Bockhoff R. 2006. Bird pollination in South African Salvia species. *Flora* **201**:396–406.

Yédomonhan H, Houenon GJ, Akoègninou A, Adomou AC, Tossou GM, van der Maesen LJG. 2012. The woody flora and its importance for honey production in the Sudano-Guinean zone in Benin. *International Journal of Science and Advanced Technology* **2**:64–74.

Permissions

All chapters in this book were first published in AOB PLANTS, by Oxford University Press; hereby published with permission under the Creative Commons Attribution License or equivalent. Every chapter published in this book has been scrutinized by our experts. Their significance has been extensively debated. The topics covered herein carry significant findings which will fuel the growth of the discipline. They may even be implemented as practical applications or may be referred to as a beginning point for another development.

The contributors of this book come from diverse backgrounds, making this book a truly international effort. This book will bring forth new frontiers with its revolutionizing research information and detailed analysis of the nascent developments around the world.

We would like to thank all the contributing authors for lending their expertise to make the book truly unique. They have played a crucial role in the development of this book. Without their invaluable contributions this book wouldn't have been possible. They have made vital efforts to compile up to date information on the varied aspects of this subject to make this book a valuable addition to the collection of many professionals and students.

This book was conceptualized with the vision of imparting up-to-date information and advanced data in this field. To ensure the same, a matchless editorial board was set up. Every individual on the board went through rigorous rounds of assessment to prove their worth. After which they invested a large part of their time researching and compiling the most relevant data for our readers.

The editorial board has been involved in producing this book since its inception. They have spent rigorous hours researching and exploring the diverse topics which have resulted in the successful publishing of this book. They have passed on their knowledge of decades through this book. To expedite this challenging task, the publisher supported the team at every step. A small team of assistant editors was also appointed to further simplify the editing procedure and attain best results for the readers.

Apart from the editorial board, the designing team has also invested a significant amount of their time in understanding the subject and creating the most relevant covers. They scrutinized every image to scout for the most suitable representation of the subject and create an appropriate cover for the book.

The publishing team has been an ardent support to the editorial, designing and production team. Their endless efforts to recruit the best for this project, has resulted in the accomplishment of this book. They are a veteran in the field of academics and their pool of knowledge is as vast as their experience in printing. Their expertise and guidance has proved useful at every step. Their uncompromising quality standards have made this book an exceptional effort. Their encouragement from time to time has been an inspiration for everyone.

The publisher and the editorial board hope that this book will prove to be a valuable piece of knowledge for researchers, students, practitioners and scholars across the globe.

List of Contributors

Selena Ahmed
Sustainable Food and Bioenergy Systems Program, Department of Health and Human Development, Montana State University, Bozeman, MT 59715, USA Department of Biology, Tufts University, Medford, MA 02155, USA

Colin M. Orians and Anne Elise Stratton
Department of Biology, Tufts University, Medford, MA 02155, USA

Timothy S. Griffin and Sean Cash
Friedman School of Nutrition Science and Policy, Tufts University, Boston, MA 02111, USA

Sarabeth Buckley
Department of Earth Sciences, Boston University, Boston, MA 02215, USA

Uchenna Unachukwu
Department of Biochemistry, The Graduate Center of the City University of New York, New York, NY 10016, USA

John Richard Stepp
Department of Anthropology, University of Gainesville, Gainesville, FL 32611, USA

Albert Robbat Jr.
Department of Chemistry, Tufts University, Medford, MA 02155, USA

Edward J. Kennelly
Department of Biochemistry, The Graduate Center of the City University of New York, New York, NY 10016, USA
Department of Biological Sciences, Lehman College, Bronx, NY 10468, USA

Denise Johnstone and Gregory Moore
Department of Resource Management and Geography, University of Melbourne, Burnley Campus, Richmond 3012, Australia

Michael Tausz
Department of Forest and Ecosystem Science, University of Melbourne, Creswick Campus, Creswick 3363, Australia

Marc Nicolas
Department of Agriculture and Food Systems, University of Melbourne, Parkville Campus, Parkville 3010, Australia

Barbara Düll, Christiane Reinbothe and Erwin Beck
Department of Plant Physiology, University of Bayreuth, Universitätsstrasse 30, 95440 Bayreuth, Germany

Ulrike Schaz
Department of Plant Physiology, University of Bayreuth, Universitätsstrasse 30, 95440 Bayreuth, Germany
Department of Anatomy and Cell Biology, University of Ulm, Albert-Einstein-Allee 11, D-89081 Ulm, Germany

Catherine A. Offord and Patricia F. Meagher
The Royal Botanic Gardens and Domain Trust, The Australian Botanic Garden, Mount Annan, NSW 2567, Australia

Heidi C. Zimmer
Department of Forest and Ecosystem Science, University of Melbourne, Richmond, VIC 3121, Australia

Ryan P. Walsh, Paige M. Arnold and Helen J. Michaels
Department of Biological Sciences, Bowling Green State University, Bowling Green, OH 43402, USA

Helena J. R. Einzmann and Joachim Beyschlag
Department of Biology and Environmental Sciences, Carl von Ossietzky University of Oldenburg, Carl-von-Ossietzky-Straße 9-11, D-26111 Oldenburg, Germany

Florian Hofhansl and Wolfgang Wanek
Department of Microbiology and Ecosystem Science, University of Vienna, Althanstrasse 14, A-1090 Vienna, Austria

Gerhard Zotz
Department of Biology and Environmental Sciences, Carl von Ossietzky University of Oldenburg, Carl-von-Ossietzky-Straße 9-11, D-26111 Oldenburg, Germany Smithsonian Tropical Research Institute, Apartado 0843-03092, Balboa, Ancon, Panamá, República de Panamá

M. Affan Baig, M. Irfan Qureshi, Humayra Bashir and Rita Bagheri
Proteomics and Bioinformatics Lab, Department of Biotechnology, Jamia Millia Islamia, New Delhi 110025, India

Mohamed M. Ibrahim, Javed Ahmad and Ibrahim A. Arif
Department of Botany and Microbiology, Science College, King Saud University, Riyadh, Saudi Arabia
Department of Botany and Microbiology, Faculty of Science, Alexandria University, Alexandria, Egypt

Helene Fotouo-M., Elsa S. du Toit and Petrus J. Robbertse
Department of Plant Production and Soil Science, University of Pretoria, Pretoria 0002, South Africa

Junmin Li, Beifen Yang and Qiaodi Yan
Zhejiang Provincial Key Laboratory of Plant Evolutionary Ecology and Conservation, Taizhou 318000, China
Institute of Ecology, Taizhou University, Taizhou 318000, China

Jing Zhang
Institute of Ecology, Taizhou University, Taizhou 318000, China
School of Life Science, Shanxi Normal University, Linfen 041004, China

Min Yan
School of Life Science, Shanxi Normal University, Linfen 041004, China

Maihe Li
Ecophysiology Group, Forest Dynamics, Swiss Federal Research Institute WSL, 8903 Birmensdorf, Switzerland

Lloyd L. Nackley, Corey Barnes and Lorence R. Oki
Department of Plant Sciences, University of California, Davis, CA 95616, USA

Hana Skálová and Lenka Moravcová
Department of Invasion Ecology, Institute of Botany, The Czech Academy of Sciences, CZ-252 43, Průhonice, Czech Republic

Anthony F. G. Dixon
School of Biological Sciences, University of East Anglia, Norwich NR4 7TJ, Norfolk, UK
Department of Biodiversity Research, Global Change Research Centre, The Czech Academy of Sciences, Bělidla 4a, CZ-602 00, Brno, Czech Republic

P. Kindlmann
Department of Biodiversity Research, Global Change Research Centre, The Czech Academy of Sciences, Bělidla 4a, CZ-602 00, Brno, Czech Republic
Institute for Environmental Studies, Charles University in Prague, Benátská2, CZ-128 01, Prague, Czech Republic

Petr Pyšek
Department of Invasion Ecology, Institute of Botany, The Czech Academy of Sciences, CZ-252 43, Průhonice, Czech Republic
Department of Ecology, Faculty of Science, Charles University in Prague, Viničná7, CZ-128 44, Prague, Czech Republic

M. Paniw, R. Casimiro-Soriguer and F. Ojeda
Departamento de Biología and IVAGRO, Universidad de Cádiz, Campus Río San Pedro, E-11510 Puerto Real, Spain

A. Salces-Castellano
Departamento de Biología and IVAGRO, Universidad de Cádiz, Campus Río San Pedro, E-11510 Puerto Real, Spain
IPNA-CSIC, C/Astrofísico Francisco Sánchez 3, 38206-La Laguna, Tenerife, Canary Islands, Spain

Alejandra Guisande-Collazo, Luís González and Pablo Souza-Alonso
Department of Plant Biology and Soil Science, University of Vigo, 36310 Vigo, Spain

Tomasz E. Koralewski
Department of Ecosystem Science and Management, Texas A&M University, 2138 TAMU, College Station, TX 77843-2138, USA

Mariana Mateos
Department of Wildlife and Fisheries Sciences, Texas A&M University, 2258 TAMU, College Station, TX 77843-2258, USA

Konstantin V. Krutovsky
Department of Ecosystem Science and Management, Texas A&M University, 2138 TAMU, College Station, TX 77843-2138, USA
Department of Forest Genetics and Forest Tree Breeding, Büsgen-Institute, Georg-August University of Göttingen, Büsgenweg 2, D-37077 Göttingen, Germany
N.I. Vavilov Institute of General Genetics, Russian Academy of Sciences, Moscow 119333, Russia
Genome Research and Education Center, Siberian Federal University, 50a/2 Akademgorodok, Krasnoyarsk 660036, Russia

Carolyn Livensperger and Matthew Wallenstein
Department of Ecosystem Science and Sustainability, Colorado State University, Fort Collins, CO 80523, USA

Heidi Steltzer
Biology Department, Fort Lewis College, Durango, CO 81301, USA

Anthony Darrouzet-Nardi
Department of Biological Sciences, University of Texas at El Paso, El Paso, TX 79968, USA

Patrick F. Sullivan
Environment and Natural Resources Institute, University of Alaska Anchorage, Anchorage, AK 99508, USA

Michael N. Weintraub
Department of Environmental Sciences, University of Toledo, Toledo, OH 43606, USA

Meagan F. Oldfather and Prahlad D. Papper
Department of Integrative Biology, University of California, Berkeley, CA 94720, USA

Matthew N. Britton
Department of Biological Sciences and Bolus Herbarium, University of Cape Town, Private Bag, Rondebosch 7700, South Africa

Michael J. Koontz
Department of Plant Sciences, University of California Davis, Davis, CA 95618, USA

Michelle M. Halbur and Celeste Dodge
Pepperwood Preserve, 2130 Pepperwood Preserve Road Santa Rosa, CA 95404, USA

Alan L. Flint and Lorriane E. Flint
Water Resources Discipline, U.S. Geological Survey, Placer Hall, 6000 J Street, Sacramento, CA 95819, USA

David D. Ackerly
Department of Integrative Biology, University of California, Berkeley, CA 94720, USA
Jepson Herbarium, University of California, Berkeley, CA 94720, USA

Marcia González-Teuber
Departamento de Biología, Universidad de La Serena, Casilla 554, La Serena, Chile

Muriel Quinet and Anne-Laure Jacquemart
Research Group Genetics, Reproduction, Populations, Earth and Life Institute – Agronomy, Université Catholique de Louvain, Croix du Sud 2 Box L7.05.14, B-1348 Louvain-la-Neuve, Belgium

Viviane Sinébou
Research Group Genetics, Reproduction, Populations, Earth and Life Institute – Agronomy, Université Catholique de Louvain, Croix du Sud 2 Box L7.05.14, B-1348 Louvain-la-Neuve, Belgium
Département de Productions Végétales, Faculté des Sciences Agronomiques, Université Abomey-Calavi, Cotonou 01 BP 526, Benin

Bonaventure C. Ahohuendo
Département de Productions Végétales, Faculté des Sciences Agronomiques, Université Abomey-Calavi, Cotonou 01 BP 526, Benin

Index